Case Studies in Bayesian Statistical Modelling and Analysis

Case Studies in Bayesian Statistical Modelling and Analysis

Edited by

Clair L. Alston, Kerrie L. Mengersen and Anthony N. Pettitt

Queensland University of Technology, Brisbane, Australia

WILEY

A John Wiley & Sons, Ltd., Publication

This edition first published 2013

Registered office
John Wiley & Sons Ltd, The Atrium, Southern Gate, Chichester, West Sussex, PO19 8SQ, United Kingdom

For details of our global editorial offices, for customer services and for information about how to apply for permission to reuse the copyright material in this book please see our website at www.wiley.com.

Library of Congress Cataloging-in-Publication Data

Case studies in Bayesian statistical modelling and analysis / edited by Clair Alston, Kerrie Mengersen, and Anthony Pettitt.
pages cm
Includes bibliographical references and index.
ISBN 978-1-119-94182-8 (cloth)
1. Bayesian statistical decision theory. I. Alston, Clair. II. Mengersen, Kerrie L. III. Pettitt, Anthony (Anthony N.)
QA279.5.C367 2013
519.5'42–dc23

2012024683

A catalogue record for this book is available from the British Library.

ISBN: 978-1-119-94182-8

Typeset in 10/12pt Times Roman by Thomson Digital, Noida, India
Printed and bound in Singapore by Markono Print Media Pte Ltd

Contents

Preface

Bayesian statistics is now an established statistical methodology in almost all research disciplines and is being applied to a very wide range of problems. These approaches are endemic in areas of health, the environment, genetics, information science, medicine, biology, industry, remote sensing, and so on. Despite this, most statisticians, researchers and practitioners will not have encountered Bayesian statistics as part of their formal training and often find it difficult to start understanding and employing these methods. As a result of the growing popularity of Bayesian statistics and the concomitant demand for learning about these methods, there is an emerging body of literature on Bayesian theory, methodology, computation and application. Some of this is generic and some is specific to particular fields. While some of this material is introductory, much is at a level that is too complex to be replicated or extrapolated to other problems by an informed Bayesian beginner.

As a result, there is still a need for books that show how to do Bayesian analysis, using real-world problems, at an accessible level.

This book aims to meet this need. Each chapter of this text focuses on a real-world problem that has been addressed by members of our research group, and describes the way in which the problem may be analysed using Bayesian methods. The chapters generally comprise a description of the problem, the corresponding model, the computational method, results and inferences, as well as the issues arising in the implementation of these approaches. In order to meet the objective of making the approaches accessible to the informed Bayesian beginner, the material presented in these chapters is sometimes a simplification of that used in the full projects. However, references are typically given to published literature that provides further details about the projects and/or methods.

This book is targeted at those statisticians, researchers and practitioners who have some expertise in statistical modelling and analysis, and some understanding of the basics of Bayesian statistics, but little experience in its application. As a result, we provide only a brief introduction to the basics of Bayesian statistics and an overview of existing texts and major published reviews of the subject in Chapter 2, along with references for further reading. Moreover, this basic background in statistics and Bayesian concepts is assumed in the chapters themselves.

Of course, there are many ways to analyse a problem. In these chapters, we describe how we approached these problems, and acknowledge that there may be alternatives or improvements. Moreover, there are very many models and a vast number of applications that are not addressed in this book. However, we hope that the material presented here provides a foundation for the informed Bayesian beginner to engage with Bayesian modelling and analysis. At the least, we hope that beginners will become better acquainted with Bayesian concepts, models and computation, Bayesian ways of thinking about a problem, and Bayesian inferences. We hope that this will provide them with confidence in reading Bayesian material in their own discipline or for their own project. At the most, we hope that they will be better equipped to extend this learning to do Bayesian statistics. As we all learn about, implement and extend Bayesian statistics, we all contribute to ongoing improvement in the philosophy, methodology and inferential capability of this powerful approach.

This book includes an accompanying website. Please visit `www.wiley.com/go/statistical_modelling`

Clair L. Alston
Kerrie L. Mengersen
Anthony N. Pettitt

List of contributors

Clair L. Alston
School of Mathematical Sciences
Queensland University of Technology
Brisbane, Australia

Hassan Assareh
School of Mathematical Sciences
Queensland University of Technology
Brisbane, Australia

Carla Chen
School of Mathematical Sciences
Queensland University of Technology
Brisbane, Australia

Samuel Clifford
School of Mathematical Sciences
Queensland University of Technology
Brisbane, Australia

David A. Cook
Princess Alexandra Hospital
Brisbane, Australia

Susanna M. Cramb
School of Mathematical Sciences
Queensland University of Technology
Brisbane, Australia
and
Viertel Centre for Research in
Cancer Control
Cancer Council Queensland
Australia

Robert J. Denham
Department of Environment and
Resource Management
Brisbane, Australia

Margaret Donald
School of Mathematical Sciences
Queensland University of Technology
Brisbane, Australia

Christopher C. Drovandi
School of Mathematical Sciences
Queensland University of Technology
Brisbane, Australia

Arul Earnest
Tan Tock Seng Hospital, Singapore &
Duke–NUS Graduate Medical School
Singapore

Graham E. Gardner
School of Veterinary and Biomedical
Sciences
Murdoch University
Perth, Australia

Philip Gharghori
Department of Accounting and Finance
Monash University
Melbourne, Australia

Petra L. Graham
Department of Statistics
Macquarie University
North Ryde, Australia

Candice M. Hincksman
School of Mathematical Sciences
Queensland University of Technology
Brisbane, Australia

Wenbiao Hu
School of Population Health and
Institute of Health and Biomedical
Innovation
University of Queensland
Brisbane, Australia

Katja Ickstadt
Faculty of Statistics
TU Dortmund University
Germany

Helen Johnson
School of Mathematical Sciences
Queensland University of Technology
Brisbane, Australia

Sandra Johnson
School of Mathematical Sciences
Queensland University of Technology
Brisbane, Australia

Jonathan M. Keith
School of Mathematical Sciences
Queensland University of Technology
Brisbane, Australia
and
Monash University
Melbourne, Australia

Jeong E. Lee
School of Computing and
Mathematical Sciences
Auckland University of Technology
New Zealand

Samantha Low Choy
Cooperative Research Centre for
National Plant Biosecurity, Australia
and
School of Mathematical Sciences
Queensland University of Technology
Brisbane, Australia

James M. McGree
School of Mathematical Sciences
Queensland University of Technology
Brisbane, Australia

Clare A. McGrory
School of Mathematical Sciences
Queensland University of Technology
Brisbane, Australia
and
School of Mathematics
University of Queensland
St. Lucia, Australia

Kerrie L. Mengersen
School of Mathematical Sciences
Queensland University of Technology
Brisbane, Australia

Rebecca A. O'Leary
Department of Agriculture and Food
Western Australia, Australia

Anthony N. Pettitt
School of Mathematical Sciences
Queensland University of Technology
Brisbane, Australia

Jegar O. Pitchforth
School of Mathematical Sciences
Queensland University of Technology
Brisbane, Australia

Christian P. Robert
Université Paris-Dauphine
Paris, France
and
Centre de Recherche
en Économie et Statistique
(CREST), Paris, France

Margaret Rolfe
School of Mathematical Sciences
Queensland University of Technology
Brisbane, Australia

Judith Rousseau
Université Paris-Dauphine
Paris, France
and
Centre de Recherche
en Économie et Statistique
(CREST), Paris, France

Peter Silburn
St. Andrew's War Memorial
Hospital and Medical Institute
Brisbane, Australia

Ian Smith
St. Andrew's War Memorial
Hospital and Medical Institute
Brisbane, Australia

Christopher M. Strickland
School of Mathematical Sciences
Queensland University of Technology
Brisbane, Australia

Sri Astuti Thamrin
School of Mathematical Sciences
Queensland University of Technology
Brisbane, Australia
and
Hasanuddin University, Indonesia

Cathal D. Walsh
Department of Statistics
Trinity College Dublin
Ireland

Mary Waterhouse
School of Mathematical Sciences
Queensland University of Technology
Brisbane, Australia
and
Wesley Research Institute
Brisbane, Australia

Nicole M. White
School of Mathematical Sciences
Queensland University of Technology
Brisbane, Australia
and
CRC for Spatial Information, Australia

Rick Young
Tamworth Agricultural Institute
Department of Primary Industries
Tamworth, Australia

1

Introduction

Clair L. Alston, Margaret Donald, Kerrie L. Mengersen and Anthony N. Pettitt

Queensland University of Technology, Brisbane, Australia

1.1 Introduction

This book aims to present an introduction to Bayesian modelling and computation, by considering real case studies drawn from diverse fields spanning ecology, health, genetics and finance. As discussed in the Preface, the chapters are intended to be introductory and it is openly acknowledged that there may be many other ways to address the case studies presented here. However, the intention is to provide the Bayesian beginner with a practical and accessible foundation on which to build their own Bayesian solutions to problems encountered in research and practice.

In the following, we first provide an overview of the chapters in the book and then present a list of texts for further reading. This book does not seek to teach the novice about Bayesian statistics per se, nor does it seek to cover the whole field. However, there is now a substantial literature on Bayesian theory, methodology, computation and application that can be used as support and extension. While we cannot hope to cover all of the relevant publications, we provide a selected review of texts now available on Bayesian statistics, in the hope that this will guide the reader to other reference material of interest.

1.2 Overview

In this section we give an overview of the chapters in this book. Given that the models are developed and described in the context of the particular case studies, the first

Case Studies in Bayesian Statistical Modelling and Analysis, First Edition. Edited by Clair L. Alston, Kerrie L. Mengersen and Anthony N. Pettitt.

two chapters focus on the other two primary cornerstones of Bayesian modelling: computational methods and prior distributions. Building on this foundation, Chapters 4–9 describe canonical examples of Bayesian normal linear and hierarchical models. The following five chapters then focus on extensions to the regression models for the analysis of survival, change points, nonlinearity (via splines) and spatial data. The wide class of latent variables models is then illustrated in Chapters 15–19 by considering multivariate linear state space models, mixtures, latent class analysis, hidden Markov models and structural equation models. Chapters 20 and 21 then describe other model structures, namely Bayesian classification and regression trees, and Bayesian networks. The next four chapters of the book focus on different computational methods for solving diverse problems, including approximate Bayesian computation for modelling the transmission of infection, variational Bayes methods for the analysis of remotely sensed data and sequential Monte Carlo to facilitate experimental design. Finally, the last chapter describes a software package, PyMCMC, that has been developed by researchers in our group to provide accessible, efficient Markov chain Monte Carlo algorithms for solving some of the problems addressed in the book.

The chapters are now described in more detail.

Modern Bayesian computation has been hailed as a 'model-liberating' revolution in Bayesian modelling, since it facilitates the analysis of a very wide range of models, diverse and complex data sets, and practically relevant estimation and inference. One of the fundamental computational algorithms used in Bayesian analysis is the Markov chain Monte Carlo (MCMC) algorithm. In order to set the stage for the computational approaches described in subsequent chapters, Chapter 2 provides an overview of the Gibbs and Metropolis–Hastings algorithms, followed by extensions such as adaptive MCMC, approximate Bayesian computation (ABC) and reversible jump MCMC (RJMCMC).

One of the distinguishing features of Bayesian methodology is the use of prior distributions. In Chapter 3 the range of methodology for constructing priors for a Bayesian analysis is described. The approach can broadly be categorized as one of the following two: (i) priors are based on mathematical criteria, such as conjugacy; or (ii) priors model the existing information about the unknown quantity. The chapter shows that in practice a balance must be struck between these two categories. This is illustrated by case studies from the author's experience. The case studies employ methodology for formulating prior models for different types of likelihood models: binomial, logistic regression, normal and a finite mixture of multivariate normal distributions. The case studies involve the following: time to submit research dissertations; surveillance for exotic plant pests; species distribution models; and delineating ecoregions. There is a review of practical issues. One aim of this chapter is to alert the reader to the important and multi-faceted role of priors in Bayesian inference. The author argues that, in practice, the prior often assumes a silent presence in many Bayesian analyses. Many practitioners or researchers often passively select an 'inoffensive prior'. This chapter provides practical approaches towards more active selection and evaluation of priors.

Chapter 4 presents the ubiquitous and important normal linear regression model, firstly under the usual assumption of independent, homoscedastic, normal residuals, and secondly for the situation in which the error covariance matrix is not necessarily diagonal and has unknown parameters. For the latter case, a first-order serial correlation model is considered in detail. In line with the introductory nature of this chapter, two well-known case studies are considered, one involving house prices from a cross-sectional study and the other a time series of monthly vehicle production data from Australia. The theory is extended to the situation where the error covariance matrix is not necessarily diagonal and has unknown parameters, and a first-order serial correlation model is considered in detail. The problem of covariate selection is considered from two perspectives: the stochastic search variable selection approach and a Bayesian lasso. MCMC algorithms are given for the various models. Results are obtained for the two case studies for the fixed model and the variable selection methods.

The application of Bayesian linear regression with informed priors is described in Chapter 5 in the context of modelling patient risk. Risk stratification models are typically constructed via 'gold-standard' logistic regressions of health outcomes of interest, often based on a population that has different characteristics to the patient group to which the model is applied. A Bayesian model can augment the local data with priors based on the gold-standard models, resulting in a locally calibrated model that better reflects the target patient group.

A further illustration of linear regression and variable selection is presented in Chapter 6. This concerns a case study involving a genome-wide association (GWA) study. This involves regressing the trait or disease status of interest (a continuous or binary variable) against all the single nucleotide polymorphisms (SNPs) available in order to find the significant SNPs or effects and identify important genes. The case studies involve investigations of genes associated with Type 1 diabetes and breast cancer. Typical SNP studies involve a large number of SNPs and the diabetes study has over 26 000 SNPs while the number of cases is relatively small. A main effects model and an interaction model are described. Bayesian stochastic search algorithms can be used to find the significant effects and the search algorithm to find the important SNPs is described, which uses Gibbs sampling and MCMC. There is an extensive discussion of the results from both case studies, relating the findings to those of other studies of the genetics of these diseases.

The ease with which hierarchical models are constructed in a Bayesian framework is illustrated in Chapter 7 by considering the problem of Bayesian meta-analysis. Meta-analysis involves a systematic review of the relevant literature on the topic of interest and quantitative synthesis of available estimates of the associated effect. For one of the case studies in the chapter this is the association between red meat consumption and the incidence of breast cancer. Formal studies of the association have reported conflicting results, from no association between any level of red meat consumption to a significantly raised relative risk of breast cancer. The second case study is illustrative of a range of problems requiring the synthesis of results from time series or repeated measures studies and involves the growth rate and size of fish. A multivariate analysis is used to capture the dependence between parameters of interest. The chapter illustrates the use of the WinBUGS software to carry out the computations.

Mixed models are a popular statistical model and are used in a range of disciplines to model complex data structures. Chapter 8 presents an exposition of the theory and computation of Bayesian mixed models.

Considering the various models presented to date, Chapter 9 reflects on the need to carefully consider the way in which a Bayesian hierarchical model is constructed. Two different hierarchical models are fitted to data concerning the reduction in bone mineral density (BMD) seen in a sample of patients attending a hospital. In the sample, one of three distinct methods of measuring BMD is used with a patient and patients can be in one of two study groups, either outpatient or inpatient. Hence there are six combinations of data, the three BMD measurement methods and in- or outpatient. The data can be represented by covariates in a linear model, as described in Chapter 2, or can be represented by a nested structure. For the latter, there is a choice of two structures, either method measurement within study group or vice versa, both of which provide estimates of the overall population mean BMD level. The resulting posterior distributions, obtained using WinBUGS, are shown to depend substantially on the model construction.

Returning to regression models, Chapter 10 focuses on a Bayesian formulation of a Weibull model for the analysis of survival data. The problem is motivated by the current interest in using genetic data to inform the probability of patient survival. Issues of model fit, variable selection and sensitivity to specification of the priors are considered.

Chapter 11 considers a regression model tailored to detect change points. The standard model in the Bayesian context provides inferences for a change point and is relatively straightforward to implement in MCMC. The motivation of this study arose from a monitoring programme of mortality of patients admitted to an intensive care unit (ICU) in a hospital in Brisbane, Australia. A scoring system is used to quantify patient mortality based on a logistic regression and the score is assumed to be correct before the change point and changed after by a fixed amount on the odds ratio scale. The problem is set within the context of the application of process control to health care. Calculations were again carried out using WinBUGS software.

The parametric regression models considered so far are extended in Chapter 12 to smoothing splines. Thin-plate splines are discussed in a regression context and a Bayesian hierarchical model is described along with an MCMC algorithm to estimate the parameters. B-splines are described along with an MCMC algorithm and extensions to generalized additive models. The ideas are illustrated with an adaptation to data on the circle (averaged 24 hour temperatures) and other data sets. MATLAB code is provided on the book's website.

Extending the regression model to the analysis of spatial data, Chapter 13 concerns disease mapping which generally involves modelling the observed and expected counts of morbidity or mortality and expressing each as a ratio, a standardized mortality/morbidity rate (SMR), for an area in a given region. Crude SMRs can have large variances for sparsely populated areas or rare diseases. Models that have spatial correlation are used to smooth area estimates of disease risk and the chapter shows how appropriate Bayesian hierarchical models can be formulated. One case study involves the incidence of birth defects in New South Wales, Australia. A conditional autoregressive (CAR) model is used for modelling the observed number of defects in an area and various neighbour weightings considered and compared. WinBUGS is used for computation. A second case study involves survival from breast cancer in Queensland and excess mortality, a count, is modelled using a CAR model. Various priors are used and sensitivity analyses carried out. Again WinBUGS is used to estimate the relative excess risk. The approach is particularly useful when there are sparsely populated areas, as is the situation in the two case studies.

The focus on spatial data is continued in Chapter 14 with a description of the analysis carried out to investigate the effects of different cropping systems on the moisture of soil at varying depths up to 300 cm below the surface at 108 different sites, set out in a row by column design. The experiment involved collecting daily data on about 60 occasions over 5 years but here only one day's data are analysed. The approach uses a Gaussian Markov random field model defined using the CAR formulation to model the spatial dependence for each horizontal level and linear splines to model the smooth change in moisture with depth. The analysis was carried out using the WinBUGS software and the code on the book's website is described.

Complex data structures can be readily modelled in a Bayesian framework by extending the models considered to data to include latent structures. This concept is illustrated in Chapter 15 by describing a Bayesian analysis for multivariate linear state space modelling. The theory is developed for the Fama–French model of excess return for asset portfolios. For each portfolio the excess return is explained by a linear model with time-varying regression coefficients described by a linear state space model. Three different models are described which allow for different degrees of dependence between the portfolios and across time. A Gibbs algorithm is described for the unknown parameters while an efficient algorithm for simulating from the smoothing distribution for the system parameters is provided. Discrimination between the three possible models is carried out using a likelihood criterion. Efficient computation of the likelihood is also considered. Some results for the regression models for different contrasting types of portfolios are given which confirm the characteristics of these portfolios.

The interest in latent structure models is continued in Chapter 16 with an exposition of mixture distributions, in particular finite normal mixture models. Mixture models can be used as non-parametric density estimates, for cluster analysis and for identifying specific components in a data set. The latent structure in this model indicates mixture components and component membership. A Gibbs algorithm is described for obtaining samples from the posterior distribution. A case study describes the application of mixtures to image analysis for computer tomography (CT) for scans taken from a sheep's carcase in order to determine the quantities of bone, muscle and fat. The basic model is extended so that the spatial smoothness of the image can be taken into account and a Potts model is used to spatially cluster the different components. A brief description of how the method can be extended to estimate the volume of bone, muscle and fat in a carcase is given. Some practical hints on how to set up the models are also given.

Chapter 17 again involves latent structures, this time through latent class models for clustering subgroups of patients or subjects, leading to identification of meaningful clinical phenotypes. Between-subject variability can be large and these differences can be modelled by an unobservable, or latent, process. The first case study involves the identification of subgroups for patients suffering from Parkinson's disease using symptom information. The second case study involves breast cancer patients and their cognitive impairment possibly as a result of therapy. The latent class models involving finite mixture models and trajectory mixture models are reviewed, and various aspects of MCMC implementation discussed. The finite mixture model is used to analyse the Parkinson's disease data using binary and multinomial models in the mixture. The trajectory mixture model is used with regression models to analyse the cognitive impairment of breast cancer patients. The methods indicate two or three latent classes in the case studies. Some WinBUGS code is provided for the trajectory mixture model on the book's website.

A related form of latent structure representation, described in Chapter 18, is hidden Markov models (HMMs) which have been extensively developed and used for the analysis of speech data and DNA sequences. Here a case study involves electrophysiology and the application of HMMs to the identification and sorting of action potentials in extracellular recordings involving firing neurons in the brain. Data have been collected during deep brain stimulation, a popular treatment for advanced Parkinson's disease. The HMM is described in general and in the context of a single neuron firing. An extension to a factorial HMM is considered to model several neurons firing, essentially each neuron having its own HMM. A Gibbs algorithm for posterior simulation is described and applied to simulated data as well as the deep brain stimulation data.

Bayesian models can extend to other constructs to describe complex data structures. Chapter 19 concerns classification and regression trees (CARTs) and, in particular, the Bayesian version, BCART. The BCART model has been found to be highly rated in terms of interpretability. Classification and regression trees give sets of binary rules, repeatedly splitting the predictor variables, to finally end at the predicted value. The case studies here are from epidemiology, concerning a parasite living in the human gut (cryptosporidium), and from medical science, concerning disease of the spine (kyphosis), and extensive analyses of the data sets are given. The CART approach is described and then the BCART is detailed. The BCART approach employs a stochastic search over possible regression trees with different structures and parameters. The original BART employed reversible jump MCMC and is compared with a recent implementation. MATLAB code is available on the book's website and a discussion on implementation is provided. The kyphosis data set involves a binary indicator for disease for subjects after surgery and a small number of predictor variables. The cryptosporidiosis case study involves predicting incidence rates of the disease. The results of the BCART analyses are described and details of implementation provided.

As another example of alternative model constructs, the idea of a Bayesian network (BN) for modelling the relationship between variables is introduced in Chapter 20. A BN can also be considered as a directed graphical model. Some details about software for fitting BNs are given. A case study concerns MRSA transmission in hospitals (see also Chapter 19). The mechanisms behind MRSA transmission and containment have many confounding factors and control strategies may only be effective when used in combination. The BN is developed to investigate the possible role of high bed occupancy on transmission of MRSA while simultaneously taking into account other risk factors. The case study illustrates the use of the iterative BN development cycle approach and then can be used to identify the most influential factors on MRSA transmission and to investigate different scenarios.

In Chapter 21 the ideas of design from a Bayesian perspective are considered in particular in the context of adaptively designing phase I clinical trials which are aimed at determining a maximum tolerated dose (MTD) of a drug. There are only two possible outcomes after the administration of a drug dosage: that is, whether or not a toxic event (or adverse reaction) was observed for the subject and that each response is available before the next subject is treated. The chapter describes how sequential designs which choose the next dose level can be found using SMC (Sequential Monte Carlo). Details of models and priors are given along with the SMC procedure. Results of simulation studies are given. The design criteria considered are based on the posterior distribution of the MTD, and also ways of formally taking into account the safety of subjects in the design are discussed. This chapter initiates the consideration of other computational algorithms that is the focus of the remaining chapters of the book.

Chapter 22 concerns the area of inference known as approximate Bayesian computation (ABC) or likelihood-free inference. Bayesian statistics is reliant on the availability of the likelihood function and the ABC approach is available when the likelihood function is not computationally tractable but simulation of data from it is relatively easy. The case study involves the application of infectious disease models to estimate the transmission rates of nosocomial pathogens within a hospital ward and in particular the case of Methicillin-resistant *Staphylococcus aureus* (MRSA). A Markov process is used to model the data and simulations from the model are straightforward, but computation of the likelihood is computationally intensive. The ABC inference methods are briefly reviewed and an adaptive SMC algorithm is described and used. Results are given showing the accuracy of the ABC approach.

Chapter 23 describes a computational method, variational Bayes (VB), for Bayesian inference which provides a deterministic solution to finding the posterior instead of one based on simulation, such as MCMC. In certain circumstances VB provides an alternative to simulation which is relatively fast. The chapter gives an overview of some of the properties of VB and application to a case study involving levels of chlorophyll in the waters of the Great Barrier Reef. The data are analysed using a VB approximation for the finite normal mixture models described in Chapter 14 and details of the iterative process are given. The data set is relatively large with over 16 000 observations but results are obtained for fitting the mixture model in a few minutes. Some advice on implementing the VB approach for mixtures, such as initiating the algorithm, is given.

The final investigation into computational Bayesian algorithms is presented in Chapter 24. The focus of this chapter is on ways of developing different MCMC algorithms which combine various features in order to improve performance. The approaches include a delayed rejection algorithm (DRA), a Metropolis adjusted Langevin algorithm (MALA), a repulsive proposal incorporated into a Metropolis–Hastings algorithm, and particle Monte Carlo (PMC). In the regular Metropolis–Hastings algorithm (MHA) a single proposal is made and either accepted or rejected, whereas in this algorithm the possibility of a second proposal is considered if the first proposal is rejected. The MALA uses the derivative of the log posterior to direct proposals in the MHA. In PMC there are parallel chains and the iteration values are known as particles. The particles usually interact in some way. The repulsive proposal (RP) modifies the target distribution to have holes around the particles and so induces a repulsion away from other values. The PMC avoids degeneracy of the particles by using an importance distribution which incorporates repulsion. So here two features are combined to give a hybrid algorithm. Other hybrids include DRA in MALA, MHA with RP. The various hybrid algorithms are compared in terms of statistical efficiency, computation and applicability. The algorithms are compared on a simulated data set and a data set concerning aerosol particle size. Some advantages are given and some caution provided.

The book closes with a chapter that describes PyMCMC, a new software package for Bayesian computation. The package aims to provide a suite of efficient MCMC algorithms, thus alleviating some of the programming load on Bayesian analysts while still providing flexibility of choice and application. PyMCMC is written in Python and takes advantage of Python libraries Numpy, Scipy. It is straightforward to optimize, extensible to C or Fortran, and parallelizable. PyMCMC also provides wrappers for a range of common models, including linear models (with stochastic search), linear and generalized linear mixed models, logit and probit models, independent and spatial mixtures, and a time series suite. As a result, it can be used to address many of the problems considered throughout the book.

1.3 Further reading

We divide this discussion into parts, dealing with books that focus on theory and methodology, those focused on computation, those providing an exposition of Bayesian methods through a software package, and those written for particular disciplines.

1.3.1 Bayesian theory and methodology

Foundations

There are many books that can be considered as foundations of Bayesian thinking. While we focus almost exclusively on reviews of books in this chapter, we acknowledge that there are excellent articles that provide a review of Bayesian statistics. For example, Fienberg (2006) in an article 'When did Bayesian inference become "Bayesian?"' charts the history of how the proposition published posthumously in the *Transactions of the Royal Society of London* (Bayes 1763) became so important for statistics, so that now it has become perhaps the dominant paradigm for doing statistics.

Foundational authors who have influenced modern Bayesian thinking include De Finetti (1974, 1975), who developed the ideas of subjective probability, exchangeability and predictive inference; Lindley (1965, 1980, 1972) and Jeffreys and Zellner (1980), who set the foundations of Bayesian inference; and Jaynes (2003), who developed the field of objective priors. Modern Bayesian foundational texts that have eloquently and clearly embedded Bayesian theory in a decision theory framework include those by Bernardo and Smith (1994, 2000), Berger (2010) and Robert (1994, 2001) which all provide a wide coverage of Bayesian theory, methods and models.

Other texts that may appeal to the reader are the very readable account of Bayesian epistemology provided by Bovens and Hartmann (2003) and the seminal discussion of the theory and practice of probability and statistics from both classical and Bayesian perspectives by DeGroot *et al.* (1986).

Introductory texts

The number of introductory books on Bayesian statistics is increasing exponentially. Early texts include those by Schmitt (1969), who gives an introduction to the field through the focal lens of uncertainty analysis, and by Martin (1967), who addresses Bayesian decision problems and Markov chains.

Box and Tiao (1973, 1992) give an early exposition of the use of Bayes' theorem, showing how it relates to more classical statistics with a concern to see in what way the assumed prior distributions may be influencing the conclusions. A more modern exposition of Bayesian statistics is given by Gelman *et al.* (1995, 2004). This book is currently used as an Honours text for our students in Mathematical Sciences.

Other texts that provide an overview of Bayesian statistical inference, models and applications include those by Meyer (1970), Iversen (1984), Press (1989, 2002) and Leonard and Hsu (1999). The last of these explicitly focuses on interdisciplinary research. The books by Lee (2004b) and Bolstad (2004) also provide informative introductions to this field, particularly for the less mathematically trained.

Two texts by Congdon (2006, 2010) provide a comprehensive coverage of modern Bayesian statistics, and include chapters on such topics as hierarchical models, latent trait models, structural equation models, mixture models and nonlinear regression models. The books also discuss applications in the health and social sciences. The chapters typically form a brief introduction to the salient theory, together with the many references for further reading. In both these books a very short appendix is provided about software ('Using WinBUGS and BayesX').

Compilations

The maturity of the field of Bayesian statistics is reflected by the emergence of texts that comprise reviews and compilations. One of the most well-known series of such texts is the Proceedings of the Valencia Conferences, held every 4 years in Spain. Edited by Bernardo and co-authors (Bernardo *et al.* 2003, 2007, 2011, 1992, 1996, 1999, 1980, 1985, 1988), these books showcase frontier methodology and application over the course of the past 30 years.

Edited volumes addressing general Bayesian statistics include *The Oxford Handbook of Applied Bayesian Data Analysis* by O'Hagan (2010). Edited volumes within specialist areas of statistics are also available. For example, Gelfand *et al.* (2010)'s *Handbook of Spatial Statistics* is a collection of chapters from prominent researchers in the field of spatial statistics, and forms a coherent whole while at the same time pointing to the latest research in each contributor's field. Mengersen *et al.* (2011) have recently edited a series of contributions on methods and applications of Bayesian mixtures. Edited volumes in specialist discipline areas are discussed below.

1.3.2 Bayesian methodology

Texts on specific areas of Bayesian methodology are also now quite common, as given in Table 1.1.

1.3.3 Bayesian computation

There is a wide literature on Monte Carlo methods in general, from different perspectives of statistics, computer science, physics, and so on. There are also many books that contain sections on Bayesian computation as part of a wider scope, and similarly books that focus on narrow sets of algorithms. Finally, the books and documentation associated with Bayesian software most often contain descriptions of the underlying

Table 1.1 Bayesian methodology books.

Author and Year	Topic
Broemeling (1985)	Bayesian analysis of linear models
Spall (1988)	Bayesian analysis of time series and dynamic models
West and Harrison (1989, 1997)	Bayesian forecasting and dynamic models
Berry and Stangl (1996)	Bayesian biostatistics
Neal (1996)	Bayesian learning for neural networks
Kopparapu and Desai (2001)	Bayesian approach to image interpretation
Denison (2002)	Bayesian methods for nonlinear classification and regression
Ghosh and Ramamoorthi (2003)	Bayesian non-parametrics
Banerjee *et al.* (2004)	Hierarchical modelling and analysis for spatial data
Lee (2004a)	Bayesian non-parametrics via neural networks
Congdon (2005)	Bayesian models for categorical data
O'Hagan *et al.* (2006)	Uncertain judgements: eliciting expert probabilities
Lee *et al.* (2008)	Semi-parametric Bayesian analysis of structural equation models
Broemeling (2009)	Bayesian methods for measures of agreement
Ando (2010)	Bayesian model selection and statistical modelling
Fox (2010)	Bayesian item response modelling (free e-book)
Hjort *et al.* (2010)	Bayesian non-parametrics
Ibrahim (2010)	Bayesian survival analysis

computational approaches. In light of this, here we review a selected set of books targeted at the Bayesian community by Christian Robert, who is a leading authority on modern Bayesian computation and analysis.

Three books by Robert and co-authors provide a comprehensive overview of Monte Carlo methods applicable to Bayesian analysis. The earliest, *D*iscretization and MCMC Convergence Assessment (Robert 1998), describes common MCMC algorithms as well as less well-known ones such as perfect simulation and Langevin Metropolis–Hastings. The text then focuses on convergence diagnostics, largely grouped into those based on graphical plots, stopping rules and confidence bounds. The approaches are illustrated through benchmark examples and case studies.

The second book, by Robert and Casella, *M*onte Carlo Statistical Methods (Robert and Casella 1999, 2004), commences with an introduction (statistical models, likelihood methods, Bayesian methods, deterministic numerical methods, prior distributions and bootstrap methods), then covers random variable generation, Monte Carlo approaches (integration, variance, optimization), Markov chains, popular algorithms (Metropolis–Hastings, slice sampler, two-stage and multi-stage Gibbs, variable selection, reversible jump, perfect sampling, iterated and sequential importance sampling) and convergence.

The more recent text by Robert and Casella, *I*ntroducing Monte Carlo Methods in R (Robert and Casella 2009), presents updated ideas about this topic and comprehensive R code. The code is available as freestanding algorithms as well as via an R package, mcsm. This book covers basic R programs, Monte Carlo integration, Metropolis–Hastings and Gibbs algorithms, and issues such as convergence, optimization, monitoring and adaptation.

1.3.4 Bayesian software

There is now a range of software for Bayesian computation. In the following, we focus on books that describe general purpose software, with accompanying descriptions about Bayesian methods, models and application. These texts can therefore act as introductory (and often sophisticated) texts in their own right. We also acknowledge that there are other texts and papers, both hard copy and online, that describe software built for more specific applications.

WinBUGS at http://www.mrc-bsu.cam.ac.uk/bugs/winbugs/contents.shtml, a free program whose aim is to 'make practical MCMC methods available to applied statisticians', comes with two manuals, one for WinBUGS (Spiegelhalter *et al.* 2003) (under the Help button) and the other for GeoBUGS (Thomas *et al.* 2004) (under the Map button), which together with the examples (also under the Help and Map buttons) explain the software and show how to get started. Ntzoufras (2009) is a useful introductory text which looks at modelling via WinBUGS and includes chapters on generalized linear models and also hierarchical models.

In Albert (2009), a paragraph suffices to introduce us to Bayesian priors, and on the next page we are modelling in R using the LearnBayes R package. This deceptive start disguises an excellent introductory undergraduate text, or 'teach yourself' text, with generally minimal theory and a restricted list of references. It is a book to add

Table 1.2 Applied Bayesian books.

Discipline/Author and year	Title
Economics	
Jeffreys and Zellner (1980)	*Bayesian Analysis in Econometrics and Statistics*
Dorfman (1997, 2007)	*Bayesian Economics through Numerical Methods*
Bauwens *et al.* (1999)	*Bayesian Inference in Dynamic Econometric Models*
Koop (2003)	*Bayesian Econometrics*
Business	
Neapolitan (2003)	*Learning Bayesian Networks*
Rossi *et al.* (2005)	*Bayesian Statistics and Marketing*
Neapolitan and Jiang (2007)	*Probabilistic Methods for Financial & Marketing Informatics*
Health	
Spiegelhalter (2004)	*Bayesian Approaches to Clinical Trials and Health-Care Evaluation*
Berry (2011)	*Bayesian Adaptive Methods for Clinical Trials*
Earth sciences	
Koch (1990)	*Bayesian Inference with Geodetic Applications*
Ecology	
McCarthy (2007)	*Bayesian Methods for Ecology*
King (2009)	*Bayesian Analysis for Population Ecology*
Link and Barker (2009)	*Bayesian Inference with Ecological Applications*
Space	
Hobson *et al.* (2009)	*Bayesian Methods in Cosmology*
Social sciences	
Jackman (2009)	*Bayesian Analysis for the Social Sciences*
Bioinformatics	
Do *et al.* (2006)	*Bayesian Inference for Gene Expression and Proteomics*
Mallick et al. (2009)	*Bayesian Analysis of Gene Expression Data*
Dey (2010)	*Bayesian Modeling in Bioinformatics*
Engineering	
Candy (2009)	*Bayesian Signal Processing*
Yuen (2010)	*Bayesian Methods for Structural Dynamics and Civil Engineering*
Archaeology	
Buck *et al.* (1996)	*The Bayesian Approach to Interpreting Archaeological Data*
Buck and Millard (2004)	*Tools for Constructing Chronologies*

to the shelf if you are unfamiliar with R and even integrates complex integrals using the Laplace approximation for which there is a function in LearnBayes.

1.3.5 Applications

The number of books on Bayesian statistics for particular disciplines has grown enormously in the past 20 years. In this section we do not attempt a serious review of this literature. Instead, in Table 1.2 we have listed a selection of books on a selection of subjects, indicating the focal topic of each book. Note that there is some inevitable overlap with texts described above, where these describe methodology applicable across disciplines, but are strongly adopted in a particular discipline. The aim is thus to illustrate the breadth of fields covered and to give some pointers to literature within these fields.

References

Albert J 2009 *Bayesian Computation with R*. Springer, Dordrecht.

Ando T 2010 *Bayesian Model Selection and Statistical Modeling*. CRC Press, Boca Raton, FL.

Banerjee S, Carlin BP and Gelfand AE 2004 *Hierarchical Modeling and Analysis for Spatial Data*. Monographs on Statistics and Applied Probability. Chapman & Hall, Boca Raton, FL.

Bauwens L, Richard JF and Lubrano M 1999 *Bayesian Inference in Dynamic Econometric Models*. Advanced Texts in Econometrics. Oxford University Press, Oxford.

Bayes T 1763 An essay towards solving a problem in the doctrine of chances. *Philosophical Transactions of the Royal Society of London* **53**, 370–418.

Berger J 2010 *Statistical Decision Theory and Bayesian Analysis*, 2nd edn. Springer Series in Statistics. Springer, New York.

Bernardo JM and Smith AFM 1994 *Bayesian Theory*, Wiley Series in Probability and Mathematical Statistics. John Wiley & Sons, Inc., New York.

Bernardo JM and Smith AFM 2000 *Bayesian Theory*. John Wiley & Sons, Inc., New York.

Bernardo JM, Bayarri MJ, Berger JO, Dawid AP, Heckerman D, Smith AFM and West M (eds) 2003 *Bayesian Statistics 7*. Oxford University Press, Oxford.

Bernardo JM, Bayarri MJ, Berger JO, Dawid AP, Heckerman D, Smith AFM and West M (eds) 2007 *Bayesian Statistics 8*. Oxford University Press, Oxford.

Bernardo JM, Bayarri MJ, Berger JO, Dawid AP, Heckerman D, Smith AFM, and West M (eds) 2011 *Bayesian Statistics 9*. Oxford University Press, Oxford.

Bernardo JM, Berger JO, Dawid AP and Smith AFM (eds) 1992 *Bayesian Statistics 4*. Oxford University Press, Oxford.

Bernardo JM, Berger J, Dawid A and Smith AFM (eds) 1996 *Bayesian Statistics 5*. Oxford University Press, Oxford.

Bernardo JM, Berger JO, Dawid AP and Smith AFM (eds) 1999 textitBayesian Statistics 6. Oxford University Press, Oxford.

Bernardo JM, DeGroot MH, Lindley DV and Smith AFM (eds) 1980 *Bayesian Statistics*. University Press, Valencia.

Bernardo JM, DeGroot MH, Lindley DV and Smith AFM (eds) 1985 *Bayesian Statistics 2*. North-Holland, Amsterdam.

Bernardo JM, DeGroot MH, Lindley DV and Smith AFM) 1988 *Bayesian Statistics 3*. Oxford University Press, Oxford.

Berry D and Stangl D 1996 *Bayesian Biostatistics*. Marcel Dekker, New York.
Berry SM 2011 *Bayesian Adaptive Methods for Clinical Trials*. CRC Press, Boca Raton, FL.
Bolstad W 2004 *Introduction to Bayesian Statistics*. John Wiley & Sons, Inc., New York.
Bovens L and Hartmann S 2003 *Bayesian Epistemology*. Oxford University Press, Oxford.
Box GEP and Tiao GC 1973 *Bayesian Inference in Statistical Analysis*. Wiley Online Library.
Box GEP and Tiao GC 1992 *Bayesian Inference in Statistical Analysis*, Wiley Classics Library edn. John Wiley & Sons, Inc., New York.
Broemeling LD 1985 *Bayesian Analysis of Linear Models*. Marcel Dekker, New York.
Broemeling LD 2009 *Bayesian Methods for Measures of Agreement*. CRC Press, Boca Raton, FL.
Buck C and Millard A 2004 *Tools for Constructing Chronologies: Crossing disciplinary boundaries*. Springer, London.
Buck CE, Cavanagh WG and Litton CD 1996 *The Bayesian Approach to Interpreting Archaeological Data*. John Wiley & Sons, Ltd, Chichester.
Candy JV 2009 *Bayesian Signal Processing: Classical, Modern and Particle Filtering Methods*. John Wiley & Sons, Inc., Hoboken, NJ.
Congdon P 2005 *Bayesian Models for Categorical Data*. John Wiley & Sons, Inc., New York.
Congdon P 2006 *Bayesian Statistical Modelling*, 2nd edn. John Wiley & Sons, Inc., Hoboken, NJ.
Congdon PD 2010 *Applied Bayesian Hierarchical Methods*. CRC Press, Boca Raton, FL.
De Finetti B 1974 *Theory of Probability*, Vol. 1 (trans. A Machi and AFM Smith). John Wiley & Sons, Inc., New York.
De Finetti B 1975 *Theory of Probability*, Vol. 2 (trans. A Machi and AFM Smith). Wiley, New York.
DeGroot M, Schervish M, Fang X, Lu L and Li D 1986 *Probability and Statistics*. Addison-Wesley, Boston, MA.
Denison DGT 2002 *Bayesian Methods for Nonlinear Classification and Regression*. John Wiley & Sons, Ltd, Chichester.
Dey DK 2010 *Bayesian Modeling in Bioinformatics*. Chapman & Hall/CRC, Boca Raton, FL.
Do KA, Mueller P and Vannucci M 2006 *Bayesian Inference for Gene Expression and Proteomics*. Cambridge University Press, Cambridge.
Dorfman JH 1997 *Bayesian Economics through Numerical Methods*. Springer, New York.
Dorfman JH 2007 *Bayesian Economics through Numerical Methods*, 2nd edn. Springer, New York.
Mengersen KL, Robert CP and Titterington DM (eds) 2011 *Mixtures: Estimation and Applications*. John Wiley & Sons, Inc., Hoboken, NJ.
O'Hagan A (ed.) 2010 *The Oxford Handbook of Applied Bayesian Analysis*. Oxford University Press, Oxford.
Fienberg SE 2006 When did Bayesian inference become 'Bayesian'?. *Bayesian Analysis* **1**, 1–40.
Fox JP 2010 *Bayesian Item Response Modeling*. Springer, New York.
Gelfand AE, Diggle PJ, Fuentes M and Guttorp P 2010 *Handbook of Spatial Statistics*, Handbooks of Modern Statistical Methods. Chapman & Hall/CRC, Boca Raton, FL.
Gelman A, Carlin JB, Stern HS and Rubin DB 1995 *Bayesian Data Analysis*, Texts in statistical science. Chapman & Hall, London.
Gelman A, Carlin JB, Stern HS and Rubin DB 2004 *Bayesian Data Analysis*, 2nd edn. Texts in Statistical Science. Chapman & Hall/CRC, Boca Raton, FL.
Ghosh JK and Ramamoorthi RV 2003 *Bayesian Nonparametrics*. Springer, New York.

Hjort NL, Holmes C, Moller P and Walker SG 2010 *Bayesian Nonparametrics*. Cambridge University Press, Cambridge.

Hobson MP, Jaffe AH, Liddle AR, Mukherjee P and Parkinson D 2009 *Bayesian Methods in Cosmology*. Cambridge University Press, Cambridge.

Ibrahim JG 2010 *Bayesian Survival Analysis*. Springer, New York.

Iversen GR 1984 *Bayesian Statistical Inference*. Sage, Newbury Park, CA.

Jackman S 2009 *Bayesian Analysis for the Social Sciences*. John Wiley & Sons, Ltd, Chichester.

Jaynes E 2003 *Probability Theory: The Logic of Science*. Cambridge University Press, Cambridge.

Jeffreys H and Zellner A 1980 *Bayesian Analysis in Econometrics and Statistics: Essays in Honor of Harold Jeffreys*, Vol. 1. Studies in Bayesian Econometrics. North-Holland, Amsterdam.

King R 2009 *Bayesian Analysis for Population Ecology*. Interdisciplinary Statistics, 23. CRC Press, Boca Raton, FL.

Koch KR 1990 *Bayesian Inference with Geodetic Applications*. Lecture Notes in Earth Sciences, 31. Springer, Berlin.

Koop G 2003 *Bayesian Econometrics*. John Wiley & Sons, Inc., Hoboken, NJ.

Kopparapu SK and Desai UB 2001 *Bayesian Approach to Image Interpretation*. Kluwer Academic, Boston, MA.

Lee HKH 2004a *Bayesian Nonparametrics via Neural Networks*. Society for Industrial and Applied Mathematics, Philadelphia, PA.

Lee P 2004b *Bayesian Statistics*. Arnold, London.

Lee SY, Lu B and Song XY 2008 *Semiparametric Bayesian Analysis of Structural Equation Models*. John Wiley & Sons, Inc., Hoboken, NJ.

Leonard T and Hsu JSJ 1999 *Bayesian Methods: An Analysis for Statisticians and Interdisciplinary Researchers*. Cambridge Series in Statistical and Probabilistic Mathematics. Cambridge University Press, Cambridge.

Lindley D 1965 *Introduction to Probability and Statistics from a Bayesian Viewpoint*, 2 vols. Cambridge University Press, Cambridge.

Lindley D 1980 *Introduction to Probability and Statistics from a Bayesian Viewpoint*, 2nd edn, 2 vols. Cambridge University Press, Cambridge.

Lindley DV 1972 *Bayesian Statistics: A Review*. Society for Industrial and Applied Mathematics, Philadelphia, PA.

Link W and Barker R 2009 *Bayesian Inference with Ecological Applications*. Elsevier, Burlington, MA.

Mallick BK, Gold D and Baladandayuthapani V 2009 *Bayesian Analysis of Gene Expression Data*. Statistics in Practice. John Wiley & Sons, Ltd, Chichester.

Martin JJ 1967 *Bayesian Decision Problems and Markov Chains*. Publications in Operations Research, no. 13. John Wiley & Sons, Inc., New York.

McCarthy MA 2007 *Bayesian Methods for Ecology*. Cambridge University Press, Cambridge.

Meyer DL 1970 *Bayesian Statistics*. Peacock, Itasca, IL.

Neal RM 1996 *Bayesian Learning for Neural Networks*. Lecture Notes in Statistics, 118. Springer, New York.

Neapolitan RE 2003 *Learning Bayesian Networks*. Prentice Hall, Englewood Cliffs, NJ.

Neapolitan RE and Jiang X 2007 *Probabilistic Methods for Financial and Marketing Informatics*. Elsevier, Amsterdam.

Ntzoufras I 2009 *Bayesian Modeling Using WinBUGS*. John Wiley & Sons, Inc., Hoboken, NJ.

O'Hagan A, Buck CE, Daneshkhah A, Eiser R, Garthwaite P, Jenkinson DJ, Oakley J and Rakow T 2006 *Uncertain Judgements Eliciting Experts' Probabilities*. John Wiley & Sons, Ltd, Chichester.

Press SJ 1989 *Bayesian Statistics: Principles, Models, and Applications*. John Wiley & Sons, Inc., New York.

Press SJ 2002 *Bayesian Statistics: Principles, Models, and Applications*, 2nd edn. John Wiley & Sons, Inc., New York.

Robert C 1998 *Discretization and MCMC Convergence Assessment*. Lecture Notes in Statistics, 135. Springer, New York.

Robert C and Casella G 2009 *Introducing Monte Carlo Methods in R*. Springer, New York.

Robert CP 1994 *The Bayesian Choice: A Decision-Theoretic Motivation*. Springer Texts in Statistics. Springer, New York.

Robert CP 2001 *The Bayesian Choice: A Decision-Theoretic Motivation*, 2nd edn. Springer Texts in Statistics. Springer, New York.

Robert CP and Casella G 1999 *Monte Carlo Statistical Methods*. Springer Texts in Statistics. Springer, New York.

Robert CP and Casella G 2004 *Monte Carlo Statistical Methods*, 2nd edn. Springer Texts in Statistics. Springer, New York.

Rossi PE, Allenby GM and McCulloch RE 2005 *Bayesian Statistics and Marketing*. John Wiley & Sons, Inc., Hoboken, NJ.

Schmitt SA 1969 *Measuring Uncertainty: An Elementary Introduction to Bayesian Statistics*. Addison-Wesley, Reading, MA.

Spall JC 1988 *Bayesian Analysis of Time Series and Dynamic Models*. Statistics, Textbooks and Monographs, Vol. 94. Marcel Dekker, New York.

Spiegelhalter D, Thomas A, Best N and Lunn D 2003 *WinBUGS User Manual Version 1.4, January 2003*. http://www.mrc-bsu.cam.ac.uk/bugs/winbugs/manual14.pdf (accessed 9 May 2012).

Spiegelhalter DJ 2004 *Bayesian Approaches to Clinical Trials and Health-Care Evaluation*. Statistics in Practice. John Wiley & Sons, Ltd, Chichester.

Thomas A, Best N, Lunn D, Arnold R and Spiegelhalter D 2004 *GeoBUGS User Manual Version 1.2, September 2004*. http://www.mrc-bsu.cam.ac.uk/bugs/winbugs/geobugs12manual.pdf (accessed 9 May 2012).

West M and Harrison J 1989 *Bayesian Forecasting and Dynamic Models*. Springer Series in Statistics. Springer, New York.

West M and Harrison J 1997 *Bayesian Forecasting and Dynamic Models*, 2nd edn. Springer Series in Statistics. Springer, New York.

Yuen KV 2010 *Bayesian Methods for Structural Dynamics and Civil Engineering*. John Wiley & Sons (Asia) Pte Ltd.

2

Introduction to MCMC

Anthony N. Pettitt and Candice M. Hincksman

Queensland University of Technology, Brisbane, Australia

2.1 Introduction

Although Markov chain Monte Carlo (MCMC) techniques have been available since Metropolis and Ulam (1949), which is almost as long as the invention of computational Monte Carlo techniques in the 1940s by the Los Alamos physicists working on the atomic bomb, they have only been popular in mainstream statistics since the pioneering paper of Gelfand and Smith (1990) and the subsequent papers in the early 1990s. Gelfand and Smith (1990) introduced Gibbs sampling to the statistics community. It is no coincidence that the BUGS project started in 1989 in Cambridge, UK, and was led by David Spiegelhalter, who had been a PhD student of Adrian Smith's at Oxford. Both share a passion for Bayesian statistics. Recent accounts of MCMC techniques can be found in the book by Gamerman and Lopes (2006) or in Robert and Casella (2011).

Hastings (1970) generalized the Metropolis algorithm but the idea had remained unused in the statistics literature. It was soon realized that Metropolis–Hastings could be used within Gibbs for those situations where it was difficult to implement so-called pure Gibbs. With a clear connection between the expectation–maximization (EM) algorithm, for obtaining modal values of likelihoods or posteriors where there are missing values or latent values, and Gibbs sampling, MCMC approaches were developed for models where there are latent variables used in the likelihood, such as mixed models or mixture models, and models for stochastic processes such as those involving infectious diseases with various unobserved times. Almost synonymous with MCMC is the notion of a hierarchical model where the probability model,

Case Studies in Bayesian Statistical Modelling and Analysis, First Edition. Edited by Clair L. Alston, Kerrie L. Mengersen and Anthony N. Pettitt.

likelihood times prior, is defined in terms of conditional distributions and the model can be described by a directed acyclic graph (DAG), a key component of generic Gibbs sampling computation such as BUGS. WinBUGS has the facility to define a model through defining an appropriate DAG and the specification of explicit MCMC algorithms is not required from the user. The important ingredients of MCMC are the following. There is a target distribution, π, of several variables $x_1, \ldots, x_k$. The target distribution in Bayesian statistics is defined as the posterior, $p(\theta|y)$, which is proportional to the likelihood, $p(y|\theta)$, times the prior, $p(\theta)$. The unknown variables can include all the parameters, latent variables and missing data values. The constant of proportionality is the term which implies that the posterior integrates or sums to 1 over all the variables and it is generally a high-dimensional calculation.

MCMC algorithms produce a sequence of values of the variables by generating the next set of values from just the current set of values by use of a probability transition kernel. If the variables were discrete then the transition kernel would be the transition probability function or matrix of a discrete Markov chain.

2.2 Gibbs sampling

In a Gibbs algorithm each variable is updated in turn using a value simulated from the full conditional distribution of that variable with values of the other variables given by their most up-to-date values. It is then a remarkable result that simulating from the set of full conditional distributions results in a set of values being distributed from the joint distribution. For a practical Gibbs algorithm, these full conditionals have to be standard or straightforward distributions to sample from.

2.2.1 Example: Bivariate normal

Suppose the target distribution for x_1, x_2 is taken as the bivariate normal distribution with means 0, unit variances and correlation ρ. The two conditional distributions are $x_1|x_2 \sim \mathrm{N}(\rho x_2, (1-\rho^2))$ and $x_2|x_1 \sim \mathrm{N}(\rho x_1, (1-\rho^2))$. The Gibbs sampling can start with an initial value for x_2, $x_2^{(0)}$, then $x_1^{(1)}$ is generated from $\mathrm{N}(\rho x_{(2)}^{0}, (1-\rho^2))$, then $x_2^{(1)}$ is generated from $\mathrm{N}(\rho x_1^{(1)}, (1-\rho^2))$, using the most recently generated value of x_1. Then consequent values are generated as follows for $j = 2, \ldots, N$: The chain is

---

$x_1^{(j)}$ is generated from $\mathrm{N}(\rho x_2^{(j-1)}, (1-\rho^2))$
$x_2^{(j)}$ is generated from $\mathrm{N}(\rho x_1^{(j)}, (1-\rho^2))$

---

run for a burn-in phase so that the chain is deemed to have converged (this is discussed in greater detail below) and it is assumed that pairs (x_1, x_2) are being drawn from the bivariate normal distribution. These dependent pairs of values can be retained. Averages of these retained values can be used to estimate the corresponding population values such as moments, but also probabilities such as $pr(x_1 < 0.5, x_2 < 0.5)$ which are estimated by the corresponding sample proportions.

2.2.2 Example: Change-point model

This involves inference for a change point as given in Carlin *et al.* (1992). The model assumes data $y_1, \ldots, y_n$ have a Poisson distribution but the mean could change at m, with m taking a value in $\{1, \ldots, n\}$. For $i = 1, \ldots, m$ it is assumed $(y_i|\lambda) \sim \text{Poisson}(\lambda)$ while for $i = m+1, \ldots, n$ it is assumed $(y_i|\phi) \sim \text{Poisson}(\phi)$. Independent priors are chosen for λ, ϕ, m with $\lambda \sim \text{Gamma}(a, b)$, $\phi \sim \text{Gamma}(c, d)$ and is discrete uniform over $\{1, \ldots, n\}$. Here a, b, c, d are known constants. The posterior

$$p(\lambda, \phi, m|y_1, \ldots, y_n) \propto p(y_1, \ldots, y_n|\lambda, \phi, m)p(\lambda, \phi, m).$$

With the Poisson likelihood terms and the priors we obtain the posterior

$$p(\lambda, \phi, m|y_1, \ldots, y_n) \propto \lambda^{a+s_m-1}e^{(b+m)\lambda}\phi^{c+s_n-s_m-1}e^{-(d+n-m)\phi}$$

with $s_k = \sum_{i=1}^{k} y_i$. The full conditional distributions for λ and ϕ are recognized as Gamma distributions

$$(\lambda|\phi, m, y_1, \ldots, y_n) \sim \text{Gamma}(a + s_m, b + m)$$
$$(\phi|\lambda, m, y_1, \ldots, y_n) \sim \text{Gamma}(c + s_n - s_m, d + n - m).$$

The full conditional distribution for m is discrete and found up to the normalizing constant as

$$p(m, \phi|\lambda, y_1, \ldots, y_n) \propto \lambda^{s_m}e^{-m\lambda}\phi^{s_n-s_m}e^{(-n+m)\phi}.$$

The probability can be normalized by summing the right hand side over $m = 1, \ldots, n$. Gibbs sampling can proceed by allocating initial values to ϕ and m and then proceeding to sample values of λ, ϕ, m from the full conditionals, which is straightforward to do.

2.3 Metropolis–Hastings algorithms

A key ingredient of Metropolis–Hastings algorithms is that values are first proposed and then randomly accepted or rejected as the next set of values. If rejected, then the next set of values is just taken as the current set. With target $\pi(x)$, current value x, next iterate x' and proposed value x^* generated from $q(x^*|x)$, the probability of acceptance, α, is given by $\alpha = \min(1, A)$ where A is the acceptance ratio given by

$$A = \frac{\pi(x^*)q(x|x^*)}{\pi(x)q(x^*|x)}.$$

So the next iterate x' is set equal to x^* with probability α and remains equal to x with probability $1 - \alpha$. The remarkable property of a Metropolis–Hastings algorithm is that the proposal distribution q is arbitrary. Obviously the choice of q very much affects the performance of the algorithm. A special choice of q so that it is symmetric, $q(x|x^*) = q(x^*|x)$, results in the Metropolis algorithm with α defined by

$$\alpha = \min\left\{1, \frac{\pi(x^*)}{\pi(x)}\right\}.$$

For continuous x, such a choice is that $q(x^*|x)$ is normal with a mean of x and a fixed variance, a so-called random walk proposal. Here we have assumed that all values of the variable are updated together. However, a subset of variables or a single variable could be updated, with either systematic or random choice of the variables to be updated. Such an algorithm is known as Metropolis–Hastings (MH) within Gibbs.

2.3.1 Example: Component-wise MH or MH within Gibbs

A key issue in designing an MH algorithm is the choice of proposal density and there are several ways to do this including random walk and independent proposals. The proposal has to be correctly tuned for the moves it makes in order that the parameter space is covered; proposed changes should not be too small, giving high acceptance probability, and not too large, giving small acceptance probability. Suggestions have been in the range of 20% to 50% and we return to this point below. Another criterion for deciding on better proposals is to consider the resulting sample autocorrelation function of the iteration values or, as a simplification, the expected squared jump distance, $E(x - x')^2$, which incorporates the proposed change in value and the probability of acceptance. There are several practical issues in running an MCMC algorithm to approximate a posterior distribution. These include burn-in and convergence, Monte Carlo accuracy and effective sample size (ESS) and multiple chains. Markov Chain convergence is both a theoretical topic and one for which the practitioner needs statistical methods in order to apply them to the observed values generated by a chain to determine convergence. Values from the target are only obtained when the number of past iterations approaches infinity. In practice it is assumed that convergence to a good enough approximation has occurred after a finite number, say 10 000 iterations, the burn-in phase. But is this number large enough? The burn-in phase values are discarded and the next 20 000, say, iterations are collected (the sample phase) and averages found to estimate the characteristics of the posterior distribution of interest, for example mean, standard deviation, quantiles. Practical diagnostics to determine the burn-in phase consist of a mix of graphical and statistical methods. Plots of the chain iterations can be used to determine whether the later iterations have a similar distribution compared with the early ones. This comparison can be carried out formally using a test statistic such as that proposed by Geweke (1992) which compares the average in the early part of the sequence with that in the latter part and uses time series methods to estimate the standard errors. A Z-statistic is constructed which has a standard normal distribution if convergence has occurred, with large values indicating non-convergence. Small values may not necessarily imply (practical) convergence. There are many different convergence methods and Brooks and Roberts (1998) gives a review. Running multiple independent chains provides a cross-chain comparison which is a useful graphical diagnostic. Chains should converge to a common distribution after which time it is assumed the burn-in is over. Gelman and Rubin (1992) provides a statistic which compares within- and between-chain variability and convergence is evaluated by the closeness of its value to 1. The sample size to collect is determined by the size of the Monte Carlo standard error of estimates of the posterior distribution summaries. This can be determined using time series techniques as the chain values are dependent. It would seem prudent to take a sufficiently large sample size so that the Monte Carlo standard error was equal to, say, 0.01 times the estimated posterior standard deviation of an important parameter. The idea of the ESS is useful to decide on the number of values in the sample. The ESS is defined for MCMC as the number of independent Monte Carlo samples which would be required to obtain the same variance as that obtained from an MCMC sample. The ESS uses the quantity known as the integrated autocorrelation time, which is a large-sample approximation to the variance of a sample mean from a time series (Green and Han 1992). Many of the methods for burn-in, convergence and assessing Monte Carlo precision are implemented in WinBUGS (Lunn *et al.* 2000) and other purpose-built software for Bayesian MCMC computation.

2.3.2 Extensions to basic MCMC

Population-based MCMC ideas are reviewed in Jasra *et al.* (2007a). The general idea here is that the single target distribution π is replaced by a product of K, say, distributions π_j, $j = 1, \ldots, K$, where at least one of the π_j is equal to π. MCMC samples from the product then make a population of samples. Standard single chain MCMC methods can be used to make moves within a chain for a given distribution, π_j. However, the idea is that samples learn from one another as follows.

The sequence of distributions π_j, $j = 1, \ldots, K$, are selected so that they are all related and, in general, easier to simulate than from π. This can provide valuable information for simulating from π.

The usage of a population of samples will allow more global moves (than a single chain MCMC) to be constructed resulting in faster mixing MCMC algorithms.

The standard way to swap information between chains is to use the exchange move. This is an MH move that proposes to swap the values, x_j and x_k, of two chains, j and k, respectively; this move is accepted with probability min(1,A) with

$$A = \frac{\pi_j(x_k)\pi_k(x_j)}{\pi_j(x_j)\pi_k(x_k)}.$$

It is assumed that the two chains are selected with equal probability. There are other moves such as crossover, where, with a vector-valued x, part of x_j is swapped with x_k.

The sequence of distributions π_j, $j = 1, \ldots, K$, can be defined using the idea of tempering, that is powering the original target by a value η in (0, 1] so that with

$0 < \eta_1 \leq \eta_2 \leq \ldots \leq \eta_K = 1$ we take $\pi_j \propto \pi^{\eta_j}$. Population-based MCMC methods are relatively easy to implement and can provide better exploration of the target distribution and improved convergence over a single chain approach but at the cost of running the complete population.

2.3.3 Adaptive MCMC

As mentioned above, one of the challenges of implementing an MH algorithm is the tuning of the proposal distributions, such as the variance of a random walk proposal, so that the chain mixes well. This can be done in an exploratory fashion by running different settings for proposal distributions, but a more satisfactory approach would be for a chain to learn or discover itself what the good values are. Roberts and Rosenthal (2009) describe some ways of making MH algorithms adaptive while still having a chain that converges to the correct target distribution. One way involves using the values generated by the chain so far to estimate the variance of the target distribution and use the estimate in a proposal that would have close to optimal properties, that is a random walk proposal which has variance $(2.38)^2 V/d$, where V is the estimated variance and d is the dimension, mixed with a fixed variance proposal.

In a situation where the MCMC algorithm amounts to one-dimensional MH updates then proposals could be adapted to make the acceptance rate of proposals for a variable as close as possible to 0.44 (which is optimal for one-dimensional proposals in certain settings (Roberts *et al.* 1997). The variance of a random walk proposal is changed after each batch of 50 iterations so that if the acceptance rate is more than 0.44 the variance is increased by a small amount while if the acceptance rate is less, the variance is decreased. These two examples illustrate simple ways in which MH algorithms can be modified so that close to optimal performance is learnt as iterations become available.

2.3.4 Doubly intractable problems

There is a class of distributions for which the likelihood is difficult or intractable to normalize, see for example Møller *et al.* (2006). In practical terms these distributions tend to be multivariate discrete distributions in the exponential family. Distributions used in spatial statistics include the Ising, the autologistic and the Potts distributions. As an example, the Ising distribution for binary data is given by

$$p(y_1, \ldots, y_n|\theta) = \frac{\exp(\sum_{i\sim j} \theta y_i y_j)}{z(\theta)}$$

where the sum is over neighbouring locations denoted by $i \sim j$ and $z(\theta)$ is the normalizing constant for the distribution and $y_i \in \{-1, 1\}$. For a two-dimensional lattice with smaller dimension in excess of 20 the normalizing constant becomes computationally intractable (Reeves and Pettitt 2005). Other examples occur in models for networks where exponential random graph models (ERGMs) are used. Here $y_i = 1$ if there is a connection between two nodes and 0 otherwise. Models involve statistics

$s_1(y), s_2(y), \ldots, s_m(y)$ which count the number of small-order patterns or motifs, such as connections between two nodes, a cycle connecting three nodes, and so on. The distribution is given by a member of the exponential family,

$$p(y_1, \ldots, y_n|\theta) = \frac{\exp(\sum_j s_j(y)\theta_j)}{z(\theta)},$$

where the normalizing constant is found by a summation over all possible values of node links for all possible pairs of nodes, or a sum over $2^{\binom{n}{2}}$ terms which soon becomes intractable for even small values of n. It is possible to use Gibbs sampling to generate distributions from these binary exponential family distributions where the full conditional distribution for y_i is itself binary. For the Ising distribution this becomes

$$p(y_i|y_i, \theta) \propto \exp\left(y_i \sum_{i \sim j} \theta y_j\right)$$

and is normalized by simply evaluating the right hand side for $y_i = -1, 1$. More elaborate and efficient simulation schemes have been developed. An estimate of $z(\theta)/z(\theta')$ can be found by using the identity

$$E_{y|\theta'}\left(\frac{h(y|\theta)}{h(y|\theta')}\right) = \frac{z(\theta)}{z(\theta')}$$

where we have expressed $p(y|\theta)$ as $h(y|\theta)/z(\theta)$. The left hand side can be estimated using importance sampling by drawing samples y from $p(y|\theta')$ using, for example, Gibbs sampling. In order to make inferences for θ with data y the posterior is given by

$$p(\theta|y) \propto p(y|\theta)p(\theta)$$

which involves the computationally intractable $z(\theta)$. A Metropolis algorithm with symmetric proposal density $q(\theta|\theta^*)$ would have acceptance probability min(1,A) with

$$A = \frac{h(y|\theta^*)}{z(\theta^*)} \frac{z(\theta)}{h(y \mid \theta)}.$$

This involves the intractable ratio $z(\theta)/z(\theta^*)$ which could be estimated using the above approach but the resulting Metropolis algorithm would be approximate. Møller *et al.* (2006) provide an exact MH algorithm by introducing a latent variable x which has the same support as data y. The extended target is given by

$$p(x, \theta|y) \propto p(x|y, \theta)p(y|\theta)p(\theta).$$

The proposal distribution, $q(\theta^*, x^*|\theta, x)$, is chosen to remove the need for explicit computation of the intractable ratio $z(\theta)/z(\theta^*)$ from the acceptance ratio A. With

$$q(\theta^*, x^*|\theta, x) = q(x^*|\theta^*)q(\theta^*|\theta, x)$$

then $q(x^*|\theta^*)$ is chosen to be a draw from the same distribution as the data y but with θ taken equal to θ^*. This leads to an MH algorithm with acceptance probability min(1,A) with

$$A = \frac{h(y|\theta^*)}{h(y|\theta)} \frac{h(x|\theta)}{h(x^*|\theta')} \frac{q(\theta|\theta^*)}{q(\theta^*|\theta)} \frac{p(x|y, \theta)}{p(x^*|y, \theta^*)}$$

free of the normalizing constant $z(\theta)$. The auxiliary distribution is arbitrary but taking $p(x|y, \theta) = p(x|y)$ and equal to the sampling distribution of y with θ equal to a pseudolikelihood estimate and then updated to an approximate posterior mean from previous MCMC output works well. Here a perfect simulation is required from the sampling distribution $p(y|\theta)$ which is often approximated by taking the last iteration of an MCMC run. Cucala *et al.* (2009) apply the approach to probabilistic clustering using a discrete exponential family model. The exchange algorithm (Murray *et al.* 2006) also works by introducing an auxiliary variable x but chosen to have the same distribution as data y. Additionally an auxiliary parameter value θ^* is also introduced to give an augmented target distribution as

$$p(\theta, \theta^*, x|y) \propto p(y|\theta)p(\theta)p(\theta^*|\theta)p(x|\theta^*)$$

with $p(\theta^*|\theta)$ being any distribution such as a random walk distribution. An MCMC algorithm is constructed made up of first a Gibbs proposal for (θ^*, x), then an exchange of θ and θ^* with an MH step, as outlined in the algorithm below. It can be seen readily

---

1. Gibbs update of (θ^*, x). Generate θ^* from $p(\theta^*|\theta)$. Generate x from $p(x|\theta^*)$.
2. Propose a deterministic swap of θ and θ^*. This is accepted with probability min(1,A), where

$$A = \frac{p(y|\theta^*)p(\theta^*)p(\theta|\theta^*)p(x|\theta)}{p(y|\theta)p(\theta)p(\theta^*|\theta)p(x|\theta^*)}.$$

---

that the expression for A does not involve the normalizing constant $z(\theta)$ as cancellation takes place. Caimo and Friel (2011) use this algorithm for Bayesian inference for ERGMs.

2.4 Approximate Bayesian computation

An interesting extension of the above ideas is where the likelihood is completely intractable but it is possible to simulate data from the likelihood. To solve problems like this, techniques known as approximate Bayesian computation (ABC) have been developed and were originally developed in the area of biology; see Marjoram *et al.* (2003). They provide approximations to the posterior by simulating data from the likelihood and the data are close to the observed data. Denoting by x the data simulated from the likelihood $p(y|\theta)$, a posterior target distribution including x is given by

$$p(x, \theta|y) \propto p(x|\theta)p(\theta)p(y|x, \theta).$$

Here $p(y|x, \theta)$ is a distribution that acts as a weighting function which links the observed data y to the simulated data x and has high weight where x and y are similar; see Reeves and Pettitt (2005), for example. A simplifying choice is one that uses a summary of the data $s(y)$ and a measure of distance between y and x, $\rho(s(x), s(y))$, so that $p(y|x, \theta) \propto \text{Ind}(\rho(s(x), s(y)) < \epsilon)$ for some appropriately chosen ϵ. An MH algorithm can be developed by using a proposal $q(\theta^*|\theta)$ and for x^* the likelihood with θ^* is chosen. This gives the acceptance probability equal to min(1,A) with

$$A = \frac{\text{Ind}(\rho(s(x^*), s(y)) < \epsilon)}{\text{Ind}(\rho(s(x), s(y)) < \epsilon)} \frac{p(\theta^*)}{p(\theta)} \frac{q(\theta|\theta^*)}{q(\theta^*|\theta)}$$

for the pair (x^*, θ^*); see Marjoram *et al.* (2003). The terms involving the intractable likelihood $p(x|\theta)$, $p(x^*|\theta)$ cancel. The resulting MCMC algorithm can have poor properties with low acceptance in regions of low posterior probability as x^* close to the data y have to be generated and this can have small probability. One way to improve the chain's performance is to consider population MCMC (see above) and use chains with different values of ϵ and swaps between chains. Another way is to replace the uniform weighting $\text{Ind}(\rho(s(x), s(y)) < \epsilon)$ by one based on the normal density; see Sisson and Fan (2010). It must be added that the MCMC approach to ABC appears to be outperformed by a sequential Monte Carlo version developed first by Sisson *et al.* (2007, 2009).

2.5 Reversible jump MCMC

The MH algorithm has been extended to allow for the situation where there are a number of possible different models, such as in regression when the covariates to include in the linear model are not specified. Another example is with mixture modelling and the number of components in the mixture is not specified but unknown. The extension is known as reversible jump MCMC (RJMCMC) and proposed by Green (1995). The likelihood depends on a model, M_m, and corresponding parameter θ_m giving likelihood $p(y|m, \theta_m)$ and prior $p(M_m, \theta_m)$ and hence posterior $p(m, \theta_m|y)$ up to normalization over m and θ_m. Green's algorithm can be considered for just two models, M_1, M_2, and parameters θ_1 and θ_2. In the regression context M_1 could be a model with two covariates and M_2 with four covariates. The challenge is to propose moves from one model of a given dimension to another model of different dimension (but possibly the same). If θ_1 has smaller dimension r_1 with θ_2 having dimension r_2 and $s = r_2 - r_1$ then the proposal or jump from M_1 to M_2 has to be reversible.

This can be achieved by introducing a random quantity u of dimension s in order to define a bijection (one-to-one) function $g(\theta_1, u) = \theta_2$. If the models are nested then one can take $(\theta_1, u) = \theta_2$ assuming the first r_1 components of θ_2 correspond to θ_1. The proposal density for u is $q(u|\theta_1)$. At each iteration a change of model can be proposed or not. Then let c_{ij} be the probability that a move from model M_i to model M_j is proposed. The acceptance probability for the move from M_1 to M_2 is then min(1,A) with

$$A = \frac{p(y|2, \theta_2)p(2, \theta_2)}{p(y|1, \theta_1)p(1, \theta_1)} \frac{c_{21}}{c_{12}q(u|\theta_1)} \left| \frac{\partial g(\theta_1, u)}{\partial(\theta_1, u)} \right|.$$

The first fraction is the usual term in the MH ratio of proposed target over current target. The second fraction gives the ratio of the probability of the proposal from M_2 to M_1, which does not require u, and the probability of the proposal from M_1 to M_2 multiplied by the density for u which is required to propose θ_2 from θ_1. Note that the dimension of the variables in the numerators of the first two terms match those in the denominators, that is they are both r_2. The last term is the Jacobian for the transformation from (θ_1, u) to $g(\theta_1, u)$. If the parameters are nested as above then it is possible to take $g(\theta_1, u) = (\theta_1, u)$ so the third fraction is equal to 1. Similarly, the acceptance probability for the move from M_2 to M_1 is then min(1,1/A).

The ideas above can be extended to consider a number of possible models with proposals typically moving from one model to a neighbouring model. There have been many implementations of RJMCMC in complex problems but, in general, there are challenges in terms of designing proposal distributions to move across models and dimensions. These tend to be specific to a problem but have generic names including split-or-merge or birth-and-death proposals, which, for example, are applied to mixture models with an unknown number of components. Monitoring of MCMC convergence of RJMCMC is more challenging than within-model convergence. An approach to extend the multiple chain approach of Gelman and Rubin (1992) is given by Brooks and Giudici (2000). This considers the overall variation of a suitable function of a parameter decomposed into terms associated with between and within runs and between- and within-model variation. Useful functions are those that retain their definition across models such as those based on fitted or predicted values of observations. A way to improve the mixing of RJMCMC include population RJMCMC (Jasra *et al.* 2007b) where a number of chains are run in parallel and some of the chains are tempered to allow better mixing. The chains are allowed to interact, allowing the swapping of states. A recent review of RJMCMC is given in Green and Hastie (2009).

2.6 MCMC for some further applications

An MCMC kernel has the (obvious) property that if a kernel is applied to a random variable which has a stationary distribution of the chain then the resulting outcome also has the stationary distribution. Such a property is useful in many circumstances other than those described already. For example, in particle systems where the current target distribution is approximated by a weighted sample of values, the particle values can be diversified by application of an MCMC kernel which is stationary for the current target and applied to all the particles in parallel; see, for example, Chopin (2002). MCMC is also used in likelihood inference in various ways. For the doubly intractable distributions mentioned above, an estimate of the normalizing constant ratio $z(\theta)/z(\theta')$ can be found by using MCMC sample values generated from the distribution with parameter θ. Additionally, derivatives of the logarithm of the normalizing constant can be approximated by MCMC draws and these used in an approximate iterative algorithm to find the maximum likelihood estimate of θ; see Geyer and Thompson (1995). Hunter *et al.* (2008) describe how these methods are used in the analysis of social network data. MCMC algorithms which give posterior distributions for missing values in a random sample of values from a multivariate normal distribution with unknown parameters are used in missing value methods, where prediction for missing values is known as imputation. In particular, the method of multiple imputation, for example Rubin (1996), uses sufficiently lagged output from MCMC to give approximately independent realizations of the imputed missing values. The emphasis is on imputing the missing values rather than posterior inference for the unknown parameters. These methods have been implemented on or in several different software platforms. Interestingly, they are used in frequentist regression analyses of data where the imputation method for missing covariate values is seen as a first stage and the frequentist model fitting as a second stage. A fully Bayesian approach would combine the imputation model with the inference model into one overall model. MCMC has a long history of use in optimization. Kirkpatrick *et al.* (1983) introduced simulated annealing for finding function maxima which uses a form of MH algorithm. In the statistical context this would mean finding the posterior mode or maximum likelihood estimate. For the posterior mode, the target distribution $p(y|\theta)p(\theta)$ is incrementally powered up to higher and higher powers and eventually the chain's value converges pointwise to the mode. These ideas have been combined with RJMCMC to search for an optimum over multidimensional spaces as well involving, for example, a model selection or complexity measure such as the Bayesian information criterion (BIC); see Brooks *et al.* (2003).

References

Brooks SP and Giudici P 2000 Markov chain Monte Carlo convergence assessment via two-way analysis of variance. *Journal of Computational Graphical Statistics* **9**, 266–285.

Brooks SP and Roberts G 1998 Assessing convergence of Markov chain Monte Carlo algorithms. *Statistics and Computing* **8**, 319–335.

Brooks SP, Friel N and King R 2003 Classical model selection via simulated annealing. *Journal of the Royal Statistical Society, Series B* **65**(3), 503–520.

Caimo A and Friel N 2011 Bayesian inference for exponential random graph models. *Social Networks* **33**(3), 41–55.

Carlin B, Gelfand A and Smith A 1992 Hierarchical Bayesian analysis of changepoint problems. *Applied Statistics* **41**, 389–405.

Chopin N 2002 A sequential particle filter method for static models. *Biometrika* **89**(3), 539–551.

Cucala J, Marin J, Robert C and Titterington D 2009 Bayesian reassessment of nearest-neighbour classification. *Journal of the American Statistical Association* **104**(3), 263–273.

Gamerman D and Lopes H (eds) 2006 *Markov Chain Monte Carlo: Stochastic Simulation for Bayesian Inference*, 2nd edn. Chapman and Hall/CRC, Boca Raton, FL.

Gelfand A and Smith A 1990 Sampling-based approaches to calculating marginal densities. *Journal of the American Statistical Association* **85**, 398–409.

Gelman A and Rubin D 1992 Inference from iterative simulation using multiple sequences (with discussion and re-joinder). *Statistical Science* **7**(3), 457–511.

Geweke J 1992 Evaluating the accuracy of sampling-based approaches to the calculation of posterior moments (with discussion). In *Bayesian Statistics 4* (ed. JM Bernando, JO Berger, AP Dawid and AFM Smith), pp. 169–193. Oxford University Press, Oxford.

Geyer C and Thompson E 1995 Annealing Markov chain Monte Carlo with applications to ancestral inference. *Journal of the American Statistical Association* **90**, 909–920.

Green P and Han XL 1992 Metropolis methods, Gaussian proposals and antithetic variables. In *Stochastic Models*. Lecture Notes in Statistics, 74 (eds A Barone, A Frigessi and M Piccioni), pp. 142–164. Springer, Berlin.

Green P and Hastie D 2009 Reversible jump MCMC. *Genetics* **155**(3), 1391–1403.

Green PJ 1995 Reversible jump Markov chain Monte Carlo computation and Bayesian model determination. *Biometrika* **82**, 711–732.

Hastings W 1970 Monte Carlo sampling methods using Markov chains and their applications. *Biometrika* **57**, 97–109.

Hunter DR, Goodreau S and Handcock M 2008 Goodness of fit of social network models. *Journal of the American Statistical Association* **103**(3), 248–258.

Jasra A, Stephens D and Holmes C 2007a On population based simulation for statistical inference. *Statistics and Computing* **17**(3), 263–279.

Jasra A, Stephens D and Holmes C 2007b Population-based reversible jump Markov chain Monte Carlo. *Biometrika* **94**(4), 787–807.

Kirkpatrick S, Gelatt C and Vecchi M 1983 Optimization by simulated annealing. *Science* **220**(3), 671–680.

Lunn DJ, Thomas A, Best N and Spiegelhalter D 2000 WinBUGS – A Bayesian modelling framework: concepts, structure, and extensibility. *Statistics and Computing* **10**(4), 325–337.

Marjoram P, Molitor J, Plagnol V and Tavare S 2003 Markov chain Monte Carlo without likelihoods. *Proceedings of the National Academy of Sciences* **100**(3), 15324–15328.

Metropolis N and Ulam S 1949 The Monte Carlo method. *Journal of the American Statistical Association* **44**(3), 335–341.

Møller J, Pettitt A, Reeves R and Berthelsen K 2006 An efficient Markov chain Monte Carlo method for distributions with intractable normalising constants. *Biometrika* **93**, 451–458.

Murray I, Ghahramani Z and MacKay D 2006 MCMC for doubly-intractable distributions. *Proceedings of the 22nd Annual Conference on Uncertainty in Artificial Intelligence (UAI-06)*. AUAI Press, Corvallis, OR.

Reeves RW and Pettitt AN 2005 A theoretical framework for approximate Bayesian computation *Proceedings of the 20th International Workshop for Statistical Modelling, Sydney, Australia*, pp. 393–396.

Robert C and Casella G 2011 A history of Markov chain Monte Carlo-subjective recollections from incomplete data. *Statistical Science* **26**(1), 102–115.

Roberts G and Rosenthal J 2009 Examples of adaptive MCMC. *Journal of Computational and Graphical Statistics* **18**(3), 349–367.

Roberts GO, Gelman A, and Gilks WR 1997 Weak convergence and optimal scaling of random walk Metropolis algorithms. *Annals of Applied Probability* **7**(3), 110–120.

Rubin D 1996 Multiple imputation after 18+ years (with discussion). *Journal of the American Statistical Association* **91**(3), 473–489.

Sisson S and Fan Y 2010 *Likelihood-free Markov chain Monte Carlo*. Chapman & Hall, Boca Raton, FL.

Sisson SA, Fan Y and Tanaka MM 2007 Sequential Monte Carlo without likelihoods. *Proceedings of the National Academy of Sciences* **104**(3), 1760–1765.

Sisson SA, Fan Y and Tanaka MM 2009 Sequential Monte Carlo without likelihoods. Errata. *Proceedings of the National Academy of Sciences* **106**, 168.

3

Priors: Silent or active partners of Bayesian inference?

Samantha Low Choy[1,2]

[1] *Cooperative Research Centre for National Plant Biosecurity, Australia*
[2] *Queensland University of Technology, Brisbane, Australia*

3.1 Priors in the very beginning

In Bayes' celebrated paper, posthumously submitted and added to by Price (1763), there is no explicit statement of Bayes' theorem that highlights the role of the prior. Instead Bayes' argument focused on the need to evaluate credible intervals. Thus began the quiet but important role of priors in Bayesian inference.

In the appendix, co-author Price did argue why assumptions, equivalent to assuming a uniform prior distribution, would be appropriate, while allowing the possibility that priors could be informed by other information (Bayes and Price 1763): '*What has been said seems sufficient to shew us what conclusions to draw from uniform experience. . . . supposing our only data derived from experience, we shall find additional reason for thinking thus if we apply other principles, or have recourse to such considerations as reason, independently of experience, can suggest.*' In contrast de la Place (1774) simply accepted a uniform prior as an '*intuitively obvious axiom*'. These two approaches marked the beginning of '*efforts to discover the statistical holy grail: prior distributions reflecting ignorance*' (Fienberg 2006), and of a continuing debate (Section 3.1.2) on objective and informative priors (Sections 3.2 and 3.3).

Case Studies in Bayesian Statistical Modelling and Analysis, First Edition. Edited by Clair L. Alston, Kerrie L. Mengersen and Anthony N. Pettitt.

From the premise of a uniform prior for a binomial model, de la Place demonstrated that the posterior distribution for an unknown parameter θ given the observed data y is proportional to (what is now known as) the likelihood (Stigler 1986): $p(\theta|y) \propto Lik(y|\theta)$. In this special case, the posterior distribution is proportional to the likelihood only due to an implicit assumption of a uniform prior. More generally, Bayesian inference explicitly permits priors $p(\theta)$ that are not necessarily uniform:

$$p(\theta|y) \propto Lik(y|\theta)p(\theta). \tag{3.1}$$

Indeed, when the premise of a uniform prior is ignored, this can lead to a problem with logical reasoning, so common that it has earned the name of the 'inversion fallacy': people assume that $\Pr(A|B) = \Pr(B|A)$ without acknowledging that this requires that $\Pr(A)$ is uniform across all possible values (Low Choy and Wilson 2009). Using Bayes' theorem, the prior can provide a mechanism for acknowledging these important sample or baseline weights (Section 3.2.1), providing a bridge between classical 'frequentist' and Bayesian statistics.

Today, most Bayesian researchers acknowledge that where possible, if prior information is available, then this information ought to be used to construct an *informative* prior (e.g. Berger 2006; Box and Tiao 1982; Goldstein 2006; O'Hagan *et al.* 2006; Spiegelhalter *et al.* 2004). Even when informative priors are sought, priors that are not so informative can also be assessed, and provide a baseline for evaluating the contribution of prior information to the posterior. The latter approach is embraced within an *objective* approach to Bayesian statistics, whose general aim is to allow the empirical data to dominate the analysis, by minimizing the information contained in the prior compared with the data. Nevertheless Berger (2006) advocates '*both objective Bayesian analysis and subjective Bayesian analysis to be indispensable, and to be complementary parts of the Bayesian vision*'.

Typically, the methodology for constructing prior models follows one of two main goals: (i) priors are based on mathematical criteria (Section 3.2), such as conjugacy or objective criteria; or (ii) priors model the existing information (Section 3.3). In practice, a balance must be struck between these two goals, as illustrated by case studies from our experience (Section 3.4). The case studies employ methodology for formulating prior models for different types of likelihood models: binomial, logistic regression, normal, and a finite mixture of multivariate normals (Section 3.3). Each case study demonstrates the application of these concepts to real-world problems from our experience: time to submit research dissertations; surveillance for exotic plant pests; species distribution models; and delineating ecoregions. The chapter closes in Section 3.5, with a review of practical issues.

One aim of this chapter is to alert the reader to the important and multi-faceted role of priors in Bayesian inference. In practice, they often assume a silent presence in many Bayesian analyses. Many practitioners or researchers often passively select an inoffensive prior. This chapter provides practical approaches towards more active selection and evaluation of priors.

The remainder of this section (3.1.1–3.1.4) outline issues common to selecting priors whether based on mathematical criteria (3.2) or modelling (3.3).

3.1.1 Priors as a basis for learning

The role of the prior is to capture knowledge about θ, denoted D_0, which existed *prior*, in time order, to consideration of new empirical data D_1. This can be explicitly acknowledged by denoting the prior as $p(\theta|D_0)$ instead of $p(\theta)$. Thus the posterior of this study depends on both D_0 and D_1. Suppose that the researcher conducts a further study and obtains new empirical data D_2; then the information about θ prior to this study is encapsulated in the previous posterior $p(\theta|D_0, D_1)$. Hence this defines a sequence of posterior analyses, each informing the next:

$$p(\theta|D_1, D_0) \propto Lik(D_1|\theta)p(\theta|D_0) \tag{3.2}$$
$$p(\theta|D_1, D_2, D_0) \propto Lik(D_2|\theta)p(\theta|D_1, D_0).$$

This embodies the Bayesian cycle of learning, where the researcher's understanding of parameters $p(\theta|\cdot)$ is continually updated, and we explicitly state that inference is predicated on specific sets of data $D_0, D_1, \ldots$. Indeed, one way of looking at Bayes' theorem is as a 'model' for rational thought itself (Goldstein 2011). Some people, however, may tend to give more emphasis to the role of the data or of the prior (El-Gamal and Grether 1995). These two stages may be integrated into a single analysis:

$$p(\theta|D_1, D_2, D_0) \propto Lik(D_2|\theta)Lik(D_1|\theta)p(\theta|D_0). \tag{3.3}$$

Separating these two stages may better reflect and inform learning. It also reflects the essentially hierarchical nature of Bayesian statistical modelling.

Hence the formulation of priors within this Bayesian cycle of learning provides a flexible basis for accumulating learning across several sources of information. One motivation for the objective approach is that new data D_1 often have larger sample size and may also adopt more advanced measurement techniques, leading to greater accuracy compared with prior evidence D_0. In such cases it seems fair that prior information ought to be assigned lower weight in the posterior analysis. This may hold true for scientific investigations where knowledge discovery is the main aim. However it need not apply in other situations where the most 'up-to-date' inferences are required that harness all the knowledge at hand, for example to inform decision making based on incomplete evidence. This has been called 'post-normal' science (Funtowicz and Ravetz 1993), as distinct from Popperian science structured around a logical sequence of hypothesis generation and testing (Dawid 2004). In such cases prior evidence may deserve greater weight in the posterior analysis, for instance when new data have smaller sample size. Examples of this abound in the ecological context where it is difficult to collate information that is representative of the complex diversity of the landscape (e.g. Case studies 3.4.3 and 3.4.4).

3.1.2 Priors and philosophy

The main approaches to formulating priors can be viewed on a spectrum from 'ob jective' to 'subjective':

Objective Bayesian priors ensure that inference is dominated by empirical data, and are defined mathematically via an information criterion (Section 3.2.4).

Empirical–subjective Bayesian priors can be informed by a single pilot study or meta-analysis to provide a synthesis of relevant studies (Hobbs and Hilborn 2006) (Section 3.2.4–3.2.4).

Subjective Bayesian priors apply a philosophy that all probabilities are subjective and represent a degree of belief (Dawid 2004) (Section 3.3).

For the latter two subjective options, conjugate mathematical forms are common but not necessary, and may also adopt the mathematical form of an objective prior. More broadly objective priors have several potential roles in a subjective analysis (Berger 2006):

1. Provide a *baseline* prior, as a kind of 'control', for examining the impacts of subjective priors, and therefore help assess the benefits and credibility of subjective priors.
2. Determine for which *nuisance* parameters in the model objective priors are (likely) sufficient, and hence *target subjective elicitation* to focus on those components of the model requiring more information, where these efforts will have the greatest impact.
3. *Suggest alternative mathematical forms* for the prior, which may not be conjugate but have clear interpretation.

Historically, where computing has been a challenge:

Empirical Bayesian priors can pragmatically be constructed using summary statistics of data that inform the likelihood.

Strictly speaking, however, this violates Bayes' theorem, which requires independence between prior and likelihood models. Estimating summary statistics from an alternative information source can circumvent this problem, for example for setting overall bounds (Fisher *et al.* 2011). In practice other priors are common but not advisable (Berger 2006):

Pseudo-Bayesian priors including constant priors, vague proper priors, and post-hoc tuning of the hyperparameters of priors.

3.1.3 Prior chronology

It is often assumed that the prior is stipulated at the outset of the Bayesian analysis, *before* inspecting the data. This makes sense when the aim of Bayesian analysis is to update the current knowledge (represented by a prior), using new empirical data (Equation 3.2). In this situation, an explicit statement of the current knowledge can be clearly attributed to a source in the literature, to a particular expert or to a group of experts (Goldstein 2011).

An alternative view is that the prior is '*simply intended to summarize reasonable uncertainty given evidence external to the study in question*' and may therefore reflect the stance of a hypothetical expert (Spiegelhalter *et al.* 2004, p. 73). The issue is whether the data are sufficient to change the prior stance of an expert who, for instance, may be sceptical or enthusiastic about the potential efficacy of a new treatment (Kass and Greenhouse 1989). This approach focuses on how the information provided affects the utility of decisions. It evaluates how convincing an argument is, rather than focusing on its logical merits, and hence is a function of the 'audience' being convinced.

In summary, subjective prior information may arise from:

explicit testimonial of current knowledge with specific and named information sources (Goldstein 2011); or

hypothetical current knowledge reflected by a '*community of priors*', for example ranging from sceptical to enthusiastic (Kass and Greenhouse 1989; Spiegelhalter *et al.* 2004) or conservatively adhering to, versus ignoring, the prior (El-Gamal and Grether 1995).

Careful management of elicitation needs to ensure that experts do *not* view the new empirical data prior to their subjective assessments, to avoid dependence between the prior and the likelihood.

3.1.4 Pooling prior information

There are several approaches for compiling information, ranging from objective to subjective:

Psychological pooling is a subjective approach, where the analyst seeks consensus from the experts, although group dynamics need to be managed (O'Hagan *et al.* 2006; Plous 1993), with potential to escalate in electronic media (French 2011).

Objective pooling based on linear pooling (Stone 1961) provides a democratic synthesis (avoiding evaluation) of prior information sources using a mathematical model (Genest and Zidek 1986; O'Hagan *et al.* 2006).

Weighted pooling extends objective pooling to weight each source by its quality (Cooke and Goossens 2008).

The Delphi approach elicits contributions individually in the first round, collates and presents these responses across individuals, then permits participants to modify their responses (Strauss and Zeigler 1975).

Bayesian models provide a natural framework for analysing a sequence of assessments (French 1986), and could require the assessor (termed the Decision Maker or Supra-Bayesian) to evaluate the bias and precision of each expert (Lindley 1983; Roback and Givens 2001).

Hierarchical models have used random effects to capture consensus and diversity among experts (Lipscomb *et al.* 1998) or several empirical sources (Tebaldi *et al.* 2011).

3.2 Methodology I: Priors defined by mathematical criteria

Mathematically, priors may be defined by concerns such as conjugacy (Section 3.2.1), impropriety (Section 3.2.2), and the application of information criteria to define objective priors (Section 3.2.4).

3.2.1 Conjugate priors

Definition of conjugacy

For a particular form of the likelihood model for the data given the parameters, a prior is *conjugate* for a parameter if the posterior distribution of the same parameter follows the same distributional form as the prior. $\diamond$

Utilizing a conjugate prior intuitively makes it easier to make direct comparisons between the prior and posterior, to essentially evaluate what has been learnt from the data. For some cases, a conjugate prior also helps to emphasize the relative contribution of information from the prior and from the data in the likelihood. This is especially clear for location parameters in the normal and binomial likelihoods (Section 3.2.1). More generally, conjugate priors have been tabulated for several different likelihoods (e.g. Gelman *et al.* 2004).

Conjugacy for a normal prior on the mean, in a normal likelihood

Consider a normal likelihood $y \sim \mathrm{N}(\mu, \sigma^2)$, with a normal prior on the mean $\mu \sim \mathrm{N}(\mu_0, \sigma_0^2)$. Assume that the precision of the data $\omega = 1/\sigma^2$ and of the prior estimate $\omega_0 = 1/\sigma_0^2$ are both known. Applying Bayes' theorem, and eliminating terms not involving μ, yields the posterior:

$$\begin{aligned} p(\mu|y) &\propto \frac{1}{\sqrt{2\pi}\sigma} \exp\left\{-\frac{1}{2}\frac{(y-\mu)^2}{\sigma^2}\right\} \frac{1}{\sqrt{2\pi}\sigma_0} \exp\left\{-\frac{1}{2}\frac{(\mu-\mu_0)^2}{\sigma_0^2}\right\} \\ &\propto \exp\left\{-\frac{1}{2\sigma_1^2}(\mu-\mu_1)^2\right\} \equiv \mathrm{N}(\mu_1, \sigma_1^2) \end{aligned}$$

where $\mu_1 = (\omega_0\mu_0 + w_n y)/\omega_1$ and $\omega_1 = \omega_0 + \omega$. Thus the posterior mean is a weighted average of the prior and empirical means, where weights are the prior and empirical precisions. For multiple independent observations $y_1, y_2, \ldots, y_n$, we may replace y by the mean $\bar{y}_n$ and ω by n/σ^2.

Conjugacy for a beta prior on the probability of success, with a binomial likelihood; see Gelman *et al.* (2004)

Consider a binomial likelihood $y \sim \mathrm{Bin}(\mu, n)$, with a beta[1] prior on the mean $\mu \sim \mathrm{Beta}(\alpha, \beta)$. Then after eliminating terms that do not involve μ we find that

[1]In a beta distribution $X \sim \mathbf{B}(a, b)$ the two parameters a, b may be interpreted as the effective number of successes and failures expected in $n = a + b$ trials.

the posterior also has a Beta distribution.

$$\begin{aligned} p(\mu|y) &\propto \mu^y(1-\mu)^{n-y}\mu^{\alpha-1}(1-\mu)^{\beta-1} \\ &\propto \mu^{\alpha_1}(1-\mu)^{\beta_1} \end{aligned}$$

where the posterior can be interpreted as having an effective number of successes $\alpha_1 = \alpha - 1 + y$ and failures $\beta_1 = \beta - 1 + n - y$, accumulated over both the prior and empirical sources of information.

Conjugate prior for normal linear regression; see Gelman *et al.* (2004)

Consider a normal linear regression model, with likelihood $[y|X, \beta, \sigma^2] \sim \text{N}_n(X\beta, \sigma^2)$. Then conjugate (hierarchical) priors for the regression parameters β of dimension $k+1$ and variance σ^2 are

$$[\beta|\sigma^2] \sim \text{N}_{k+1}(\beta_0, \sigma^2 R^{-1}) \quad \text{and} \quad [\sigma^2] \sim \text{Inv-Gamma}(a, b) \tag{3.4}$$

where R is a $(k+1) \times (k+1)$ positive definite symmetric matrix.

Definition of conditional conjugacy

It can be more advantageous to consider conjugacy in a conditional rather than a marginal way. This is useful in a random effects model, where an inverse-gamma prior on the variance for the random effects is conditionally conjugate (see Section 3.2.1).

For a specific likelihood model, a prior is *conditionally conjugate* for a parameter if its conditional posterior distribution follows the same distributional form as its prior. ◇

Conjugate priors may impose unintended constraints on parameters, so in practice it is advisable to gauge their effect by comparing their informativeness relative to baseline priors (e.g. Sections 3.4.3 and 3.4.4). For example, in a (normal) random effects model, an inverse-gamma prior on the variance for the random effects is conditionally conjugate, but may also be highly informative. This prior distribution $\tau = \sigma^{-2} \sim \text{Gamma}\,(a, \tau)$ can be parameterized as $f(\tau) = \mu^a / \Gamma(a)\tau^{a-1}e^{-\mu\tau}$. Setting the two parameters equal to 0.001 can be highly informative that the value is close to 0 (Section 3.2.1).

The conjugacy property does not necessarily lead to a unique definition of the conjugate prior for a specified likelihood (Section 3.2.1).

Conditionally conjugate priors for random effects variances

Gelman (2006) considers priors for σ_α^2 in a random effects model:

$$y_{ij} \sim \text{N}(\mu + \alpha_j, \sigma_y^2), \quad \alpha_j \sim \text{N}(0, \sigma_\alpha^2) \tag{3.5}$$

for $i = 1, \ldots, n;\ j = 1, \ldots, J$. These parameters can be augmented to speed up computation and also provide more flexibility:

$$y_{ij} \sim \mathrm{N}(\mu + \xi\eta_j, \sigma_y^2), \quad \eta_j \sim \mathrm{N}(0, \sigma_\eta^2) \tag{3.6}$$

where $\alpha_j = \xi\eta_j$ and $\sigma_\alpha = |\xi|\sigma_\eta$. A conditionally conjugate prior for σ_η is inverse-gamma, and for ξ is normal, leading to a folded non-central t for σ_α, with three parameters corresponding to the prior mean of η, and the prior scale and degrees of freedom for σ_η. A useful two-parameter version constrains the mean to zero, which yields a half-t:

$$p(\sigma_\alpha) \propto \left[1 + \frac{1}{\nu}\left(\frac{\sigma_\alpha}{A}\right)^2\right]^{-(\nu+1)/2}. \tag{3.7}$$

The corresponding conditional posterior for σ_α is not half-t, but still within the folded non-central t family. The half-Cauchy distribution is a special case with $\nu = 1$:

$$p(\sigma_\alpha) \propto (\sigma_\alpha^2 + s_\alpha^2)^{-1}. \tag{3.8}$$

3.2.2 Impropriety and hierarchical priors

It is important that the posterior distribution in a Bayesian inference is proper, that is

$$\int_\theta p(\theta)Lik(y|\theta)d\theta < \infty \tag{3.9}$$

so that $p(\theta)Lik(y|\theta)$ can be normalized with respect to θ. For proper priors, the posterior will be proper; however, for improper priors the posterior may also be improper, and hence it is necessary to check the condition above. This is a particular issue for mixture models (Marin and Robert 2007).

Hierarchical prior distributions can be useful: for example, avoiding improper priors in mixture models (Section 3.4.3), and avoiding specification of correlation among prior components in regression (Section 3.2.3). In general, the explicit conditioning in the likelihood model $p(D|\theta)$ clearly identifies θ as the parameter for which prior distributions must be defined. When these prior distributions $p(\theta|\phi)$ themselves rely on additional parameters ϕ, then these are called *hyperparameters*, and lead to hierarchical prior distributions. For an *hierarchical* prior, the prior relationship among parameters θ induces a priori dependence among them. In contrast, for the *independent* prior, components of parameter θ are unrelated a priori.

3.2.3 Zellner's g-prior for regression models

For linear regression, Zellner's g-prior applies a normal prior to the regression coefficient, with precision defined in terms of the 'hat' matrix, and a Jeffreys prior on the variance. This avoids the difficult task of specifying a prior correlation structure (Marin and Robert 2007):

$$[\beta|\sigma^2, X] \sim \mathrm{N}_{k+1}(\tilde{\beta}, g\sigma^2 X(X^T X)^{-1} X^T) \quad \text{and} \quad p(\sigma^2|X) \propto \sigma^{-2}. \tag{3.10}$$

This prior depends on the covariates X, as does the likelihood, but does not violate the assumption that the prior and likelihood are informed by independent information sources. The influence of the prior is determined by g, as reflected in the posterior means:

$$\begin{aligned} \mathbf{E}[\beta|y, X] &= \frac{\tilde{\beta} + g\tilde{\beta}}{g+1} \\ \mathbf{E}[\sigma^2|X] &= \frac{s^2 + (g+1)^{-1}(\widehat{\beta} - \tilde{\beta})^T (X^T X)^{-1} (\widehat{\beta} - \tilde{\beta})}{n-2} \end{aligned} \tag{3.11}$$

where $\widehat{\beta}$ is the usual least squares estimate of β.

3.2.4 Objective priors

Non-informative priors evolved to represent a prior state of ignorance about model parameters, to provide a data-driven analysis. However '*the older terminology of noninformative priors is no longer in favor among objective Bayesians because a complete lack of information is hard to define*' (Ghosh *et al.* 2006). More broadly, *objective* priors are constructed to ensure that the posterior depends more on the data than the prior, and are often motivated by a sociological or regulatory imperative to seek (at least the appearance of) objective science (Berger 2006). Because an objective prior is defined in terms of the information contributed to the posterior, it is model specific, so strictly can be said to violate the likelihood principle. The main choices among objective priors are:

Democratic or uniform priors where each value of the parameter (on a suitable scale) has equal probability a priori.

Jeffreys' priors allow the information in the prior to approach zero, or the variance of the prior to approach infinity (Section 3.2.4).

Other non-informative priors include left and right invariant priors and reference priors without entropy maximization, and general forms exist for location-scale families and exponential families of likelihood distributions (Ghosh *et al.* 2006).

Reference priors minimize the information in the prior relative to the posterior using some criterion, such as Bernardo's reference prior based on Kullbach–Liebler distance and its extensions (Section 3.2.4).

Other objective priors include matching priors that have optimal frequentist properties, invariance priors and admissible priors (Bayarri and Berger 2004).

Vaguely informative priors provide an informal mechanism for specifying that the information in the prior is large, with a scale parameter that provides a flexible basis for increasing the information in the prior (e.g. Section 3.3).

These priors are not always distinct. Jeffreys' priors are a special case of Bernardo's reference priors and in many situations are equivalent to the uniform.

Priors for model selection

The use of uniform priors is often motivated by a desire for non-informativeness (Section 3.2.4). In variable selection, uniform priors are common, but care is required in how model space is enumerated. For instance, a uniform prior defined over the combinations of regression coefficients may place subtly different emphasis on variables than a series of uniform priors defined on each variable individually (Congdon 2001). The latter approach provides a convenient shortcut that avoids specification of *all* combinations of variables, whereas implementation of the former approach often specifies only a few combinations. The g parameter of Zellner's g-prior can be tuned between these two extremes (Zellner and Siow 1980). Priors have been constructed separately on the number of important variables and the choice of variables, in the context of classification and regression trees (O'Leary *et al.* 2008).

Non-informative priors, Jeffreys' priors and maximum likelihood

As the weight on the likelihood increases *relative* to the weight on the prior, the posterior parameter estimates become increasingly dominated by the likelihood. For a normal likelihood and a normal prior on the mean (Section 3.2.1), if the prior weight is substantially smaller than the likelihood weight, $w_0 \leq w_n$, then the posterior distribution approximates the likelihood $\mathbf{N}(\bar{y}, \sigma_n^2)$. This limit is only actually attained when $w_0 = 0$. In practice this would mean claiming that the prior information has no weight whatsoever, requiring $\sigma_0^2 \to \infty$, which leads to a distribution that is uniform over the whole real line $(-\infty, \infty)$. Such a distribution cannot integrate to 1, so is improper (Equation 3.9). This exposes a theoretical difference between a likelihood-centred analysis (such as maximum likelihood), which in this case is the limit as the prior weight approaches zero, and a Bayesian analysis, which could assign a very small (but never a zero) prior weight (Box and Tiao 1982).

Non-informative priors apply this asymptotic argument by allowing the information in the prior to approach zero, so that the prior variance approaches infinity. Consider a log likelihood $\ell = \log Lik(y|\theta)$ for data $y = (y_1, y_2, \ldots, y_n)$. Then large-sample theory leads to Jeffreys' rule that a prior for a parameter θ is approximately non-informative when it is proportional to the square root of Fisher's information (Box and Tiao 1982):

$$p(\theta) \propto \sqrt{\mathbf{I}(\theta)}, \text{ where } \mathbf{I}(\theta) = \mathbf{E}_{y|\theta}\left[-\frac{\partial^2 \ell}{\partial \theta^2}\right]. \tag{3.12}$$

Jeffreys' rule was originally derived to fulfil the invariance criterion: the prior distribution is invariant (the same) under one-to-one transformations of the parameter θ.

In the estimation setting, the non-informative prior arises naturally from considering the limit of conjugate inferences. However, in the testing setting, these limits are not so useful since they lead to the Jeffreys–Lindley paradox (Marin and Robert 2007).

Reference priors

Shannon's entropy provides a natural basis for measuring uncertainty (the opposite of information) for discrete distributions:

$$H(p) = -\sum_{i=1}^{n} p_i \log p_i, \quad H(x) = -\mathbf{E}_{\mathbf{p}}[\log p(x)]. \tag{3.13}$$

The discrete uniform arises as the distribution that maximizes entropy, and therefore minimizes information. However Ghosh *et al.* (2006) showed that this approach is problematic for continuous distributions.

Bernardo (1979) suggested the use of the Kullbach–Liebler divergence between prior and posterior:

$$\begin{aligned} J(p(\theta), y) &= \mathbf{E}\left[\log \frac{p(\theta|y)}{p(\theta)}\right] \qquad (3.14)\\ &= \int_{\Theta}\int_{\mathbf{Y}}\int_{\theta} \log \frac{p(\theta^*|y)}{p(\theta^*)} p(\theta^*|y)d\theta^* \; Lik(y|\theta)dy \; p(\theta)d\theta \end{aligned}$$

so that $-J$ measures the information in the prior. When the prior is diffuse, J will be large, but when the prior is highly concentrated, J will be close to zero. Hence maximizing J defines an objective prior. As explained in Ghosh *et al.* (2006) this maximization should be applied asymptotically, rather than for fixed n, which would provide a multimodal prior. These authors note that this same measure of information has been used (Lindley 1956) to guide Bayesian design of experiments, although the aim was to choose the design X to minimize information for fixed $p(\theta)$.

This approach has continued to evolve (Berger and Bernardo 1992), leading to a modern version that maximizes the contribution of each additional data item (Bernardo 2006). Unfortunately, the triple integral makes it difficult to evaluate in many situations.

3.3 Methodology II: Modelling informative priors

3.3.1 Informative modelling approaches

Informative priors are constructed using information elicited from experts, the literature or previous studies. This process of construction, where the prior knowledge is translated into a statistical distribution, is called *encoding*. Informative priors are typically encoded via two main approaches (Low Choy *et al.* 2011):

deterministic encoding where a minimum number of summary statistics are elicited, and equations relating these to the prior hyperparameters are solved (deterministically); or

statistical encoding where prior information is analysed, like 'data', to estimate the hyperparameters in the prior.

Modelling issues come into play when prior information is available for formulation of informative priors. Priors may be used to represent:

sampling weights $w_j = \Pr(\theta_j)$ may reflect prevalence of strata which affect model parameters, where $\Pr(\theta_k|y) = w_k \Pr(y|\theta_k)/\sum_j w_j \Pr(y|\theta_j)$;

results from previous studies which may be updated via the Bayesian cycle of learning (Equation 3.2); or

information elicited from experts typically as a starting point, or a new direction, within a Bayesian cycle of learning.

The latter approach of expert elicitation is particularly useful when data sets are small or limited in terms of extent, coverage or quality. There are two main approaches to eliciting expert knowledge (Low Choy *et al.* 2010):

Direct elicitation simplifies the effort by the analyst, by asking the expert(s) to express their knowledge directly in terms of values for the Bayesian model's hyperparameters.

Indirect elicitation seeks to reduce the effort by the expert(s), by asking them questions about quantities they are familiar with, then using this information as a basis for inferring the model's hyperparameters.

Example 1: Time to submit dissertation Consider elicitation of a subjective prior for the standard deviation σ to encode a normal distribution in X (Section 3.4.1). Asking an expert, even a mathematical expert, to *directly* communicate a point estimate for σ is a challenge. Instead, *indirect* elicitation of quantiles $F^{-1}(q|\mu, \sigma)$ is an easier task (Spetzler and Staël von Holstein 1975).

Example 2a: Ecological prior options for mortality of powerful owls Here three individuals were observed, with one dying after 8 years, and two still being alive after 10 and 17 years. McCarthy (2007) proposed a simple binomial model with 36 trials (years) and probability of mortality being 1 death in 36 years. This assumes that every year of life is equivalent, both within the lifetime of a single bird and between birds across the times when the owls were alive. The different sets of prior information considered (McCarthy 2007) were:

1. A democratic prior about probability of mortality in a given year, expressed as a Unif(0, 1) prior.
2. Given such a large bird, the expected lifespan is between 5 and 50 years, which can be expressed as a Unif(5, 50) prior on lifespan.

3. Inverting lifespan gives annual mortality rate. Inverting the range endpoints provides a Unif(0.02, 2) prior on mortality rate.
4. The predicted average adult mortality (for diurnal and nocturnal birds of prey) is estimated at 0.11 (SE = 0.05) using a logistic regression based on body mass (≈ 1.35 kg). This can be directly encoded as a Beta(4.198, 33.96) prior (O'Hagan *et al.* 2006, Chapter 2).

The data are all censored, so that it makes sense to perform the analysis in terms of annual mortality rates. The informative priors 2 and 3 exemplify indirect (on scale of lifespan) and direct (in terms of mortality) encoding of ranges as uniform distributions, whereas prior 4 illustrates direct encoding of a beta distribution.

3.3.2 Elicitation of distributions

Experts can typically supply at least one fundamental piece of information about a parameter, which can be used as the basis for constructing an informative prior distribution:

Ranges representing absolute (theoretical) bounds for parameters are often feasible to elicit; if exceeded, experts would be extremely surprised. Eliciting such endpoints first can improve accuracy (Hora *et al.* 1992), circumvent anchoring and adjustment bias (Morgan *et al.* 2001), as well as avoid misinterpretation as a confidence interval on the best estimate (Fisher *et al.* 2011).

The best estimate of a parameter can often be provided when experts have a well-developed understanding of its role in the model. Depending on the wording of questions put to the experts (Low Choy and Wilson 2009), best estimates may correspond mathematically to the mean, the median with a 50:50 chance of a value above or below (Denham and Mengersen 2007), or the mode as the most likely value (Low Choy *et al.* 2008).

Scale parameters can be difficult to elicit directly, unless experts have considerable experience with them (e.g. O'Hagan *et al.* 2006).

The elicited range can be combined with a best estimate to directly encode a distribution, such as a triangular or PERT (beta) distribution in risk assessment (Burgman 2005). It can be difficult to elicit variability directly, so indirect approaches are often used. Strategies include eliciting:

Quantiles can be sought for fixed levels of probability (Spetzler and Staël von Holstein 1975, *Q* method). Quartiles are common targets (e.g. Kadane *et al.* 1980).

Tertiles divide the distribution into thirds, so make it possible to elicit a plausible range without obtaining a point estimate (Garthwaite and O'Hagan 2000, *T* method).

Cumulative probabilities can be sought for fixed quantiles (Spetzler and Staël von Holstein 1975, *P* method).

Plausible range of parameter values is a probability interval with fixed probability of a value lower or higher than the lower and upper bounds.

'Drop-off' interval The experts are provided feedback, for instance on the central 50% interval, using parameters encoded from their previous assessments. The feedback is based on a 'drop-off', being a small increase or decrease in a particular parameter, such as the mode; see description in Gill and Walker (2005) based on Chaloner and Duncan (1983).

Using a deterministic encoding method (analytic or numerical) just two Ps or Qs are sufficient to encode most two-parameter distributions, such as the beta (over a dozen options in O'Hagan *et al.* 2006), gamma and lognormal (Low Choy *et al.* 2008). Where more elicitation effort is viable, several strategies are hybrids of P and Q elicitation:

Graphical *P*s and *Q*s Rather than estimating Ps and Qs individually, they can simultaneously be specified using a graphical representation of the distribution: a boxplot (Leal *et al.* 2007), probability or cumulative density curve (e.g. Denham and Mengersen 2007) or all of the above (James *et al.* 2010). This implicitly conditions on the expert's underlying conceptual model (Albert *et al.* In press).

Histogram This extended form of cumulative probability elicitation asks experts to estimate relative frequencies in pre-specified 'bins'. Allowing experts to allocate frequencies by gambling with poker chips, or the Roulette method (Oakley and O'Hagan 2010, *R* method), can make the process more concrete.

Histogram as Dirichlet West (1985) modelled an elicited histogram using a Dirichlet distribution.

Histogram with error Rounding errors in an histogram elicited via the Roulette or *R* method can be accommodated using a non-parametric approach based on a Gaussian process (Oakley and O'Hagan 2007), or a parametric Gaussian measurement error model (Albert *et al.* In press) as discussed in the next section.

Variability of prior parameters is distinct from the expert's confidence in their assessment. Elicitation errors including bias (in assessing their best estimate) or imprecision (in assessing variability) can arise from the expert's numerical inaccuracy in quantifying their knowledge. This expert uncertainty can be targeted via:

Self-assessment of sureness The expert's self-calibration can also be elicited as a kind of coverage probability: 'Of all the questions you are 90% sure that you've provided the correct response, we will find that 90% are correct' (Fisher *et al.* 2011).

Supra-Bayesian assessment designates the responsibility of evaluating experts (in terms of bias, imprecision and correlation) to the Decision Maker or Supra-Bayesian (Lindley 1983).

Confidence intervals have been used to describe the accuracy of the best estimate (Speirs-Bridge *et al.* 2010).

Calibration is an empirical approach, for example using *seed* questions with known 'correct' responses to both train and evaluate experts (Cooke and Goossens 2008).

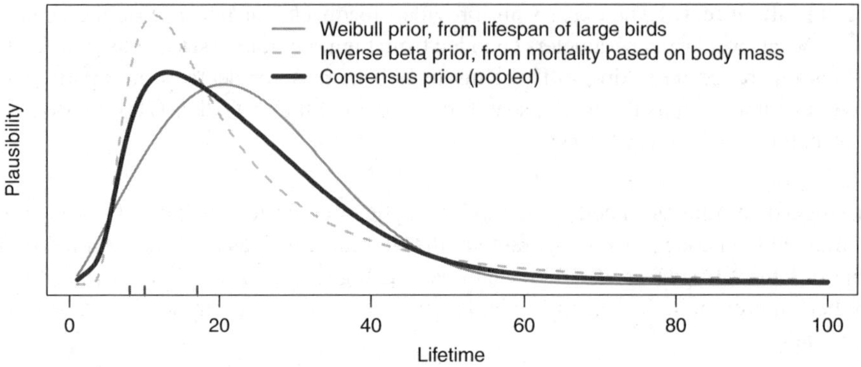

Figure 3.1 Priors on lifespan of powerful owls, encoded and pooled for two sources of ecological information, based on example from McCarthy (2007).

Example 2b: Ecological prior options for lifespan in powerful owls The priors proposed in McCarthy (2007) were formulated to complement a binomial likelihood for annual mortality of powerful owls. Instead we imagine a future situation where lifetime data are not censored, so that a likelihood model relates to survival times rather than mortality rates. The elicited lifespan range (prior 2) may then be reinterpreted as a 95% central probability interval from a Weibull distribution, and the parameters encoded using least squares (Low Choy *et al.* 2008), yielding a Weib(2.2, 27) distribution (Figure 3.1, grey line). The beta prior encoded from the body mass to mortality rate relationship is inverted, to provide a prior distribution skewed closer towards zero, giving higher weight to smaller lifespans (Figure 3.1, dashed line). The consensus prior calculated via linear pooling (Section 3.1.4), as an evenly weighted mixture of these Weibulls, provides a skewed distribution which falls midway between the two prior options. The mode is most similar to the mode under prior 4, based on body mass (Figure 3.1, black line). This example illustrates the difficult trade-off between fidelity to the scale of the prior information and the form of the empirical data available.

3.4 Case studies

We use case studies from our experience, where expert knowledge was considered just as, or even more, valuable than empirical data. These provide an opportunity to illustrate constructing a range of priors, for various likelihood models, both objective and informative, as shown in Table 3.1.

3.4.1 Normal likelihood: Time to submit research dissertations

We asked research fellows for their considered opinion on how long the current and next few cohorts of doctoral students would take to submit their thesis (Albert *et al.* In press). Submission times were analysed in terms of the logarithm of the time beyond

Table 3.1 Overview of case studies.

Application	Likelihood	Baseline/informative priors
1. Time to submit research dissertations	Normal, lognormal	Jeffreys Conjugate priors encoded using hybrid *PQ* measurement error model
2. Surveillance for exotic pests	Binomial, with zero counts	Bayes–Laplace, Haldane, Jeffreys Conjugate beta distribution encoded from the mode and effective sample size
3. Delineatingecoregions	Multivariate normal mixture model	Vaguely informative prior encoded from range of all data Indirect classification, inferred from expert delineation of regions
4. Mapping habitat of rock wallabies	Logistic regression	Conjugate prior, vaguely informative Indirect scenario-based elicitation of response given covariates

the administrative minimum of 2 years, $X^* = \log(X - 2)$. These transformed times to submit were modelled using a normal distribution $\log X^* \sim \mathrm{N}(\mu, \sigma_n^2)$, where each expert was considered to have their own conceptual model for parameters μ_ℓ, σ_ℓ^2. Experts were asked to focus on a population comprising a cohort of about 100 PhD students, across all mathematical disciplines, including students who had enrolled during the last 5 years, and would enrol during the next 5 years. Only full-time students were considered.

In this situation the focus was on PhD students in a school of mathematical sciences, and the experts were research mathematicians, providing a rare situation where numerical cognitive biases could be considered negligible. Nonetheless, it was considered challenging to ask experts to estimate a mean and variance for transformed times to submit X^*. Elicitation applied the P and Q methods of elicitation (Spetzler and Staël von Holstein 1975) to ask experts for cumulative probabilities and quantiles of the distribution $p(X^*|\mu_\ell, \sigma_\ell^2)$. An example Q question was: *How long would it take the bottom 5% of students to submit their thesis?* and an example P question was *How many students would submit their thesis within 4 years?* In group 2, experts were asked P questions for quantiles $q \in \mathbf{Q} = \{3, 3.5, 4\}$ years, and Q questions for cumulative probabilities $p \in \mathbf{P} = \{0.025, 0.25, 0.75, 0.975\}$.

Consider responses from expert 1 in group 2 (Figure 3.2); see Albert *et al.* (In press) for details of the model designed to assess the consensus and diversity of opinions across experts. Conditioned on the expert's conceptual model μ_ℓ, σ_ℓ^2, we apply the normal measurement error model for *PQ* elicitations. Based on the expert's conceptual model, their 'true' quantiles were therefore $Q_p = \exp\left\{\Phi^{-1}(p|\mu_\ell, \sigma_\ell^2) + \delta\right\}$, and their true cumulative probabilities (on the logit-transformed scale) were $\mathrm{logit}(P_q) = \mathrm{logit}\left(\Phi(\log q - \delta|\mu_\ell, \sigma_\ell^2)\right)$.

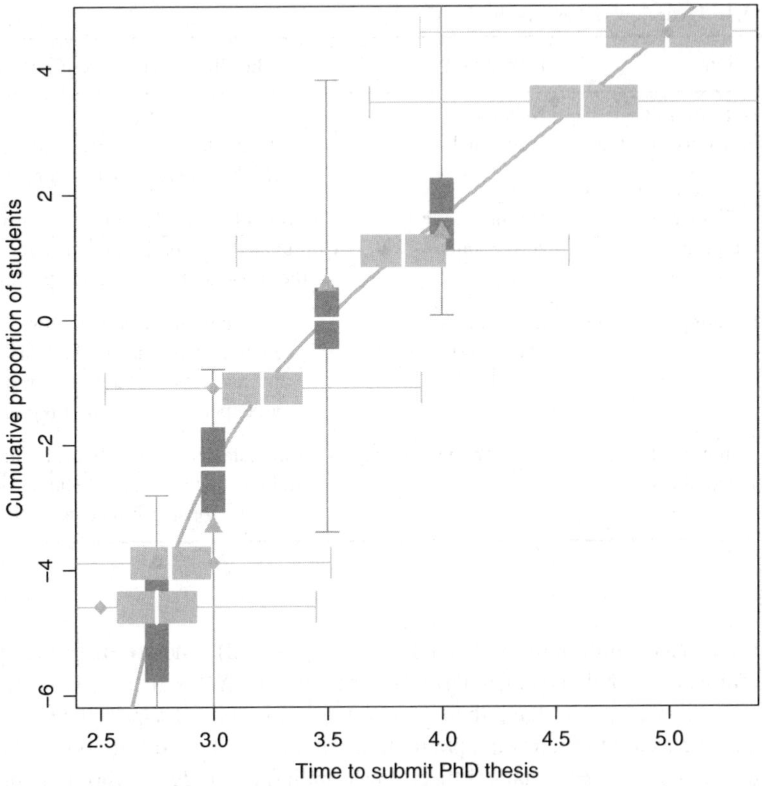

Figure 3.2 Case study 1: Gaussian elicitation error model (line) fit to P (vertical boxes) and Q elicitations (horizontal boxes) for expert 1, group 2. Boxes contain 50% intervals, and whiskers 95% intervals.

Contribution of prior and data

Here we suppose that $\sigma_n = 0.29$ is known. One researcher's assessment of their expected time to submit could be represented using a $N(0.41, 0.044^2)$ distribution (Albert *et al.* In press, expert 1, group 2). The data comprised 31 PhD theses submitted during the period 2003–2004 to the Faculty, by full-time students, with mean 0.4816 and standard deviation 0.4266. Then the relative weights governing the contribution of the prior and of the data (Section 3.2.1) were 75.20% and 24.80%. In this case the posterior average was very close to, and 'dominated' by, the prior average.

Sensitivity to information in the prior

If the analyst were to increase the uncertainty in the researcher's prior estimate, say by multiplying the prior standard deviation by 10, then the relative weights would become 2.94% and 97.06%, reversing the emphasis to be on the data, rather than on the prior. When combined with the data, this diluted prior yielded a posterior for

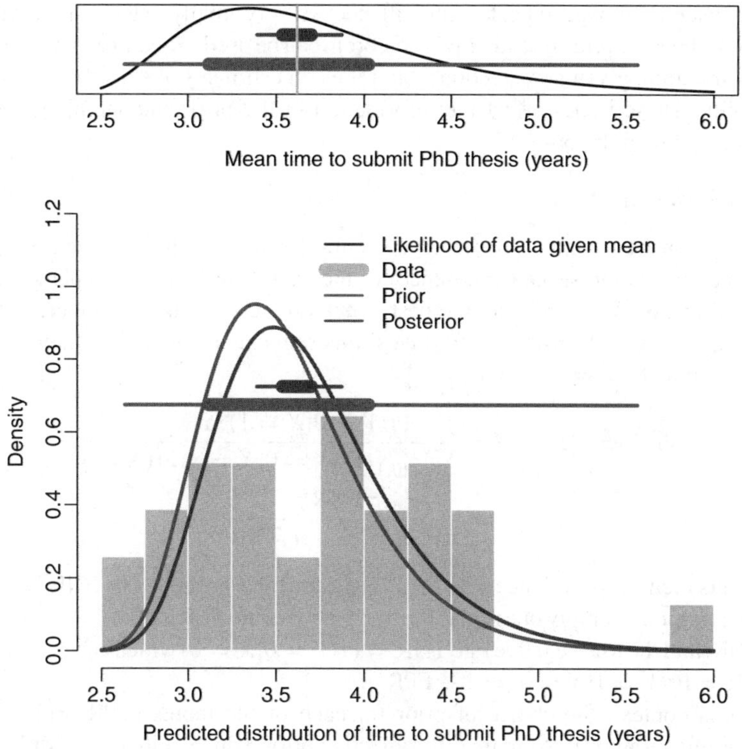

Figure 3.3 Time to submit PhD thesis: prior from one expert, data from one year. Boxes contain 50% Credible interval (CI), and whiskers 95% CI.

the mean $N(0.48, 0.075)$ (on the transformed scale). The time to submission varied widely between 3 and 4.5 years, so that the posterior average of between 3.5 and 3.7 years is dominated by the data (Figure 3.3).

3.4.2 Binomial likelihood: Surveillance for exotic plant pests

Surveillance for incursions of exotic pests on plants of agricultural or environmental significance poses two different inferential challenges, depending on whether the pest is detected or not. Any detection of an exotic pest will trigger an emergency response for more extensive surveillance to delineate the incursion and then inform a decision to eradicate or manage the pest. A variety of quantitative model-based design approaches are used to simulate the spatial and temporal dynamics of the pest incursion (see review in Anderson *et al.* accepted). These approaches typically focus on achieving a sensitivity of the surveillance system to detect the species when present. Often the false alarm rate (of reporting detections when the species is absent) is managed, rather than modelled, by escalating reports for confirmation by experts of increasing expertise. Conversely, a lack of detection does not necessarily imply the pest is absent, although repeated surveillance will eventually detect a pest when its symptoms become widespread. Bayes' theorem can be used to quantify the probability of area freedom, given no detections, but relies on estimates of sensitivity as well as specificity. These issues affect the information in the prior and sampling weights, examined in detail below.

Sampling weights

Let $Y = 1$ (or 0) denote an observed detection (or not) of the pest, and $X = 1$ (or 0) denote the true presence (or absence) of the pest. From a farmer's perspective, it is logical to ask 'If I don't find the pest, what's the chance that it is there?', which we recognize as $\Pr(X = 1|Y = 0)$. Bayes' theorem can estimate this in terms of the logical reverse $\Pr(Y = y|X = x)$:

$$\Pr(X = 1|Y = 0) = \frac{\Pr(Y = 0|X = 1)\Pr(X = 1)}{\sum_{x\in\{0,1\}} \Pr(Y = 0|X = x)\Pr(X = x)} = \frac{\pi \mathrm{FNR}}{\pi \mathrm{FNR} + (1-\pi)\mathrm{TNR}}. \tag{3.15}$$

The inputs include: prevalence $\pi = \Pr(X = 1)$, the false negative rate $\mathrm{FNR} = \Pr(Y = 0|X = 1)$ which is simply one minus the true positive rate $\mathrm{TPR} = \Pr(Y = 1|X = 1) = 1 - \mathrm{FNR}$, and the true negative rate $\mathrm{TNR} = \Pr(Y = 0|X = 0)$ which can be calculated as $\mathrm{TNR} = \Pr(Y = 1|X = 0) = 1 - \mathrm{FPR}$.

In this context, the choice of prior for each of the inputs (TPR, FPR, π) can have significant impact. Note that the objective prior, which coincides with Jeffreys' prior $p(\theta) \propto \theta^{-1/2}(1-\theta)^{-1/2}$, provides intervals that have excellent coverage and can outperform other classical confidence intervals (Example 1 in Berger 2006).

Information in the prior

Experts may estimate that moderately skilled inspectors, who have been trained in detecting the pest but are yet to encounter it, have a TPR of 67% and FPR of 10%. Of interest is the probability that the pest is in fact present even though nothing has been detected. As shown above, this depends on true prevalence, which can also be interpreted as a sampling weight for presence:

$$\frac{1}{3}\pi \Big/ \left(\frac{1}{3}\pi + \frac{9}{10}(1-\pi)\right) = \frac{10\pi}{27 - 17\pi}.$$

This simple model shows that the prior estimate of prevalence π will always exceed the posterior estimate that takes into account zero detections (Figure 3.4). For example, a prior estimate of 60% prevalence is scaled down to an estimate of just under 40% when nothing is detected.

Impact of objective priors when data comprise no detections

Several priors (Tuyl *et al.* 2008) are shown on the left of Figure 3.5. The impact of these priors on the posterior is compared, when the observed data comprise zero

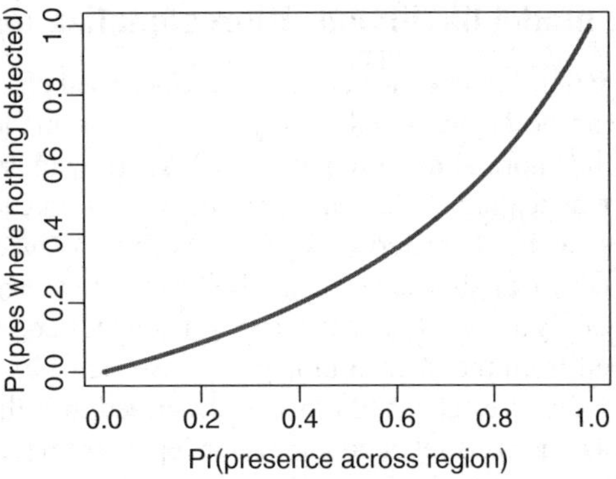

Figure 3.4 Case study 2: Posterior prevalence (y-axis) related to prior π (x-axis)

detections from 100 independent samples (right, Figure 3.5). The Bayes–Laplace prior and Zellner's maximal information prior provide nearly identical posteriors, on the range of $\theta \in [0, 0.10]$, despite clear differences in the priors. With no detections, the Haldane and Bayes–Laplace priors place the highest posterior density on a probability of zero, whereas Jeffreys' prior leads to a conclusion that the most likely value is 0.0098, which is very close to 1 in 100. Under Jeffreys' prior, there is zero posterior chance assigned to 0%, and higher posterior plausibility assigned to values over 4%, with the 95% highest probability density interval extending from 5 in 10 000 to 4.53 in 100.

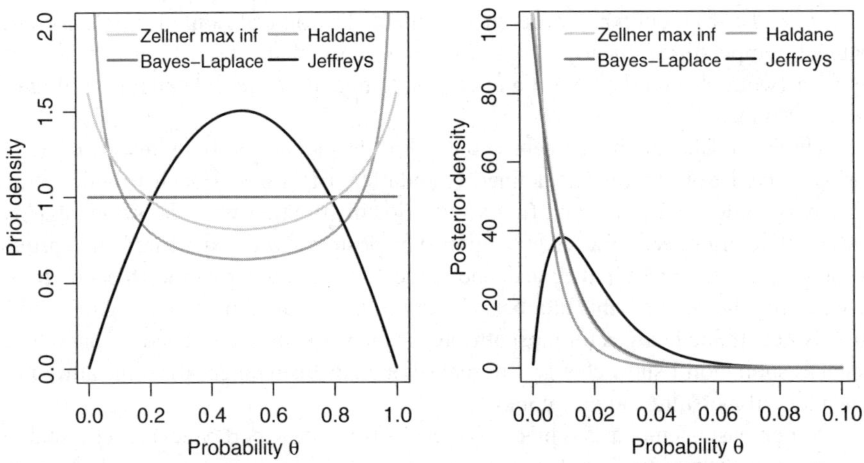

Figure 3.5 Case study 2: Reference priors for probabilities (left) and the corresponding posteriors (right) when observed data comprise zero detections from 100 independent samples.

3.4.3 Mixture model likelihood: Bioregionalization

Delineation of ecoregions can be viewed as an unsupervised classification problem. Prior information can be incorporated through a Bayesian model-based approach, based on a multivariate normal mixture model (MVMM), with an informative prior. The MVMM has an appealing ecological interpretation: (i) in each bioregion, environmental attributes can be characterized by their means, variances and covariances; and (ii) a latent variable can allocate each pixel on a map to a region.

In Australia, a readily available source of expert knowledge on ecoregional boundaries can be obtained from the interim bioregional assessments. An expert panel is consulted following a well-established process, described in Williams *et al.* (2011). The panel provides a consensus set of boundaries, adopted from relevant spatial layers (e.g. topography, soil classification, vegetation).

Indirect specification of prior via latent mixture membership

Direct elicitation would ask experts to perform the comprehensive task of estimating means, variances and covariances of the attributes. An indirect approach can utilize expert-informed boundaries. These boundaries can be used as initial estimates of regional membership $Z_i = k$ for each site i, and then enable estimation of mixture parameters μ_k, Σ_k, λ_k. This information is easily encoded into hierarchical priors.

In bioregionalization, the number of regions is already well established through the expert panel process, and is also well entrenched in legislation. What is of interest is how robust the boundaries between these 10 ecoregions might be to new Geographic Information System (GIS) data. Eight attributes were selected to reflect the key differences between regions governing differences at a subregional level, intermediate between bioregions and regional ecosystems. In the hierarchy of ecoregions, these are considered to be driven by: (a) topography, reflected by Compound Topographic Index (CTI); (b) bioclimatic differences, reflected by annual rainfall (rain), average annual temperatures (heat), and seasonal aspects characterized by the wettest quarter (wettest), and the hottest and coldest temperatures. For details see Williams *et al.* (2011).

Three models for the 10 ecoregions were developed by Williams *et al.*: (i) a prior derived from the expert-defined boundaries; (ii) a posterior distribution with the expert-informed prior; and (iii) a posterior distribution with the same vaguely informative prior used for each ecoregion. The posterior with expert-informed priors shows clear differences in the mean and range of attributes across the 10 ecoregions, confirming the need for this number of ecoregions (Figure 3.6). Some regions show tightly constrained ranges for most attributes (such as subregions 2 and 7), but others such as subregion 1 show clearly different (though diffuse) ranges for some attributes such as soil moisture and clay content.

As decreasing emphasis is placed on results from the expert panel process, spatial cohesion of the regions is reduced (Figure 3.7). Some regions remain well defined, such as northern region 10, southern region 1 and eastern region 9. Other regions, such as 2, 7 and 8, are shown to share similar attributes to regions not located nearby.

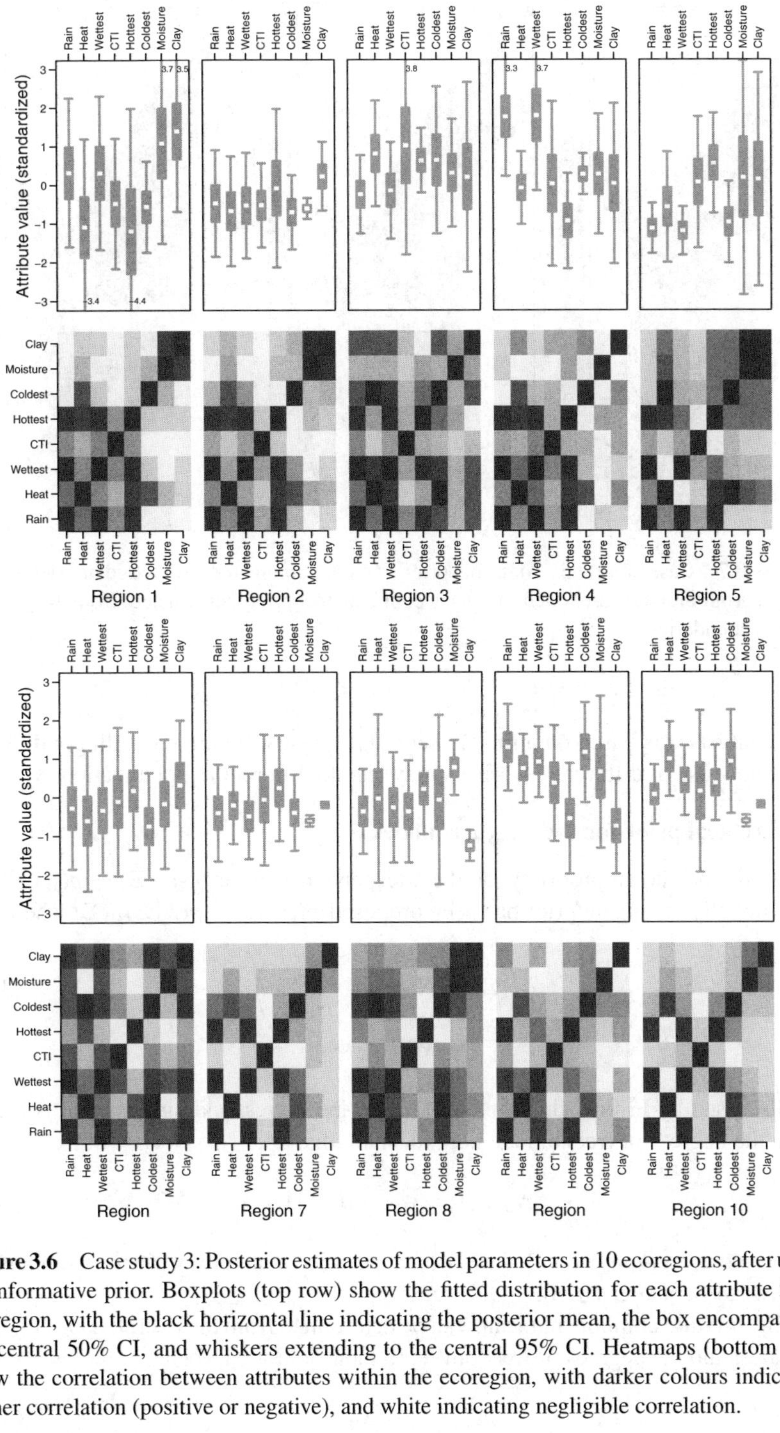

Figure 3.6 Case study 3: Posterior estimates of model parameters in 10 ecoregions, after using an informative prior. Boxplots (top row) show the fitted distribution for each attribute in an ecoregion, with the black horizontal line indicating the posterior mean, the box encompassing the central 50% CI, and whiskers extending to the central 95% CI. Heatmaps (bottom row) show the correlation between attributes within the ecoregion, with darker colours indicating higher correlation (positive or negative), and white indicating negligible correlation.

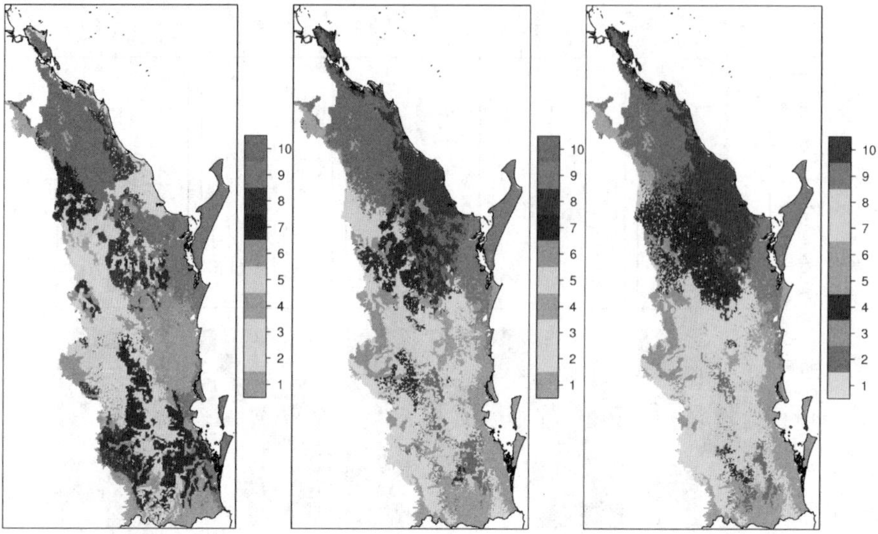

Figure 3.7 Case study 3: Map of prior regions dominated by the data (left), with reduced (middle) or as-elicited (right) weight on the prior constructed from the expert-defined boundaries.

This analysis was based on a limited set of GIS layers. Future modelling will investigate how use of different covariates may lead to different regionalizations.

Hierarchical prior for the mixture model

To overcome the impropriety of the independence prior $p(\mu, \Sigma) = p(\mu)p(\Sigma)$ a conditionally conjugate prior has been proposed $p(\mu, \Sigma) = p(\mu|\Sigma)p(\Sigma)$ where

$$\mu_k|\Sigma_k \sim \mathrm{N}_d\left(m_0, \frac{1}{N_0}\Sigma_k\right) \quad (3.16)$$

$$\Sigma_k^{-1} \sim \mathbf{W}_d(\nu_0, \Psi_0).$$

As discussed by Frühwirth-Schnatter (2006), the corresponding posterior is proper when the degrees of freedom are not too small, that is

$$N_0 > 1; \quad \nu_0 > (d-1)/2. \quad (3.17)$$

Furthermore, setting $\nu_0 > 2 + (d-1)/2$ permits non-zero variance components (Frühwirth-Schnatter 2006).

The mixture component means can be determined from the overall data summary statistics, and the degrees of freedom set to their minimal level

$$m_0 = \bar{y}, \quad N_0 = 1. \quad (3.18)$$

Then for the mixture component covariance matrices, setting $\Psi_0 = \xi S$ means that ξ describes the prior expected amount of heterogeneity (in means) between mixture components (Frühwirth-Schnatter 2006), as measured by

$$\mathrm{E}(R^2_{tr}) = 1 - \frac{\xi}{\nu_0 - (d+1)/2}. \tag{3.19}$$

With $\nu_0 = 2.5 + (d-1)/2$, just above its minimal level, then $\mathrm{E}(R^2_{tr}) = 1 - \xi/1.5$. Settings proposed by Robert (1996) are

$$\nu_0 = 3, \quad \Psi_0 = 0.75S \quad \text{so that } \xi = 50\% \tag{3.20}$$

and even more informative settings are (Bensmail *et al.* 1997)

$$\nu_0 = 2.5, \quad \Psi_0 = 0.5S \quad \text{so that } \xi = 67\%. \tag{3.21}$$

The bioregionalisation case study used the conditionally conjugate prior. However, a hierarchical prior (below) could have been used if the aim had been to describe how variable the diversity and correlation were across regions.

Hierarchical prior

Dependence can be introduced by centring the component covariances (Stephens 1997)

$$\prod_{k=1}^{K} p(\mu_k) \prod_{k=1}^{K} p(\Sigma_k|\Psi_0) \tag{3.22}$$

with the usual $\mu_k \sim \mathrm{N}_d(m_0, \Phi_0)$ and $\Sigma_k^{-1} \sim \mathbf{W}_d(\nu_0, \Psi_0)$. If the hyperparameters are scalar, then this is equivalent to the independence prior, otherwise $\Psi_0 \sim \mathbf{W}_r(g_0, G_0)$. Stephens (1997) chose

$$m_0 = \{m_k\}, \; \Phi_0 = \mathrm{diag}\{L_j\}, \; \nu_0 = 3, \; g_0 = 0.3, \; \Psi_0 = \mathrm{diag}\{100 g_0/(\nu_0 L_j^2)\} \tag{3.23}$$

where m_j and L_j are, respectively, the midpoints and the length of the range of the jth variable in the data. We note that strictly speaking it would be better to evaluate m_j and L_j using an independent data source.

3.4.4 Logistic regression likelihood: Mapping species distribution via habitat models

Logistic regression is a model-based approach to mapping species distribution. When applied within a Bayesian setting, logistic regression provides a useful platform for integrating expert knowledge, in the form of a prior, with empirical data. Here we focus on a recent study that aimed to determine habitat preferences, and use these to map the spatial distribution of the brush-tailed rock wallaby *Petrogale penicillatus*, a threatened species in Australia (Murray *et al.* 2009). At a landscape scale, the key habitat factors related to site occupancy were considered to be: geology, vegetation and landcover. At a finer scale, habitat factors that could be mapped within a GIS included slope, elevation and aspect (Murray *et al.* 2008). Other site-scale specific factors such as site complexity (indicative of the presence of boulders, crevices, etc.) could not be mapped at landscape scale.

Fieldwork collated data from over 200 sites in each of southern Queensland and northern New South Wales. Only areas falling within the species' physiological envelope were considered (Murray *et al.* 2008, e.g. omitting watercourses). Site access led to difficulty representing the full range of habitat preferences. Expert knowledge was therefore sought to help address these information gaps. Moreover, experts have the ability to translate their site-specific knowledge to broader scales in a way that is only roughly approximated by the selection of GIS attributes as 'proxies' for the desired habitat factors.

Scenario-based elicitation

Instead of asking experts to quantify the effect of each habitat variable on site occupancy, we asked them to assess the probability of site occupancy for specified combinations of habitat variables.

This was used to elicit expert knowledge for one training site, and 20 other sites, distributed across a region that no experts were known to have visited previously. Experts could explore the sites within their immediate surroundings (with scale constrained so that they could not identify the site's geographic location), together with all attributes within a purpose-built GIS. Installed on a portable computer, a software tool *Elicitator* was used to support elicitation using the scenario-based elicitation method described in James *et al.* (2010).

For each scenario represented by a mapped site, a visually hybridized *PQ* method was used to elicit quantiles, with feedback on cumulative probabilities, to define the distribution representing the proportion of sites (out of 100 matching the scenario) that the expert considered could be occupied by the rock wallaby. Experts elected to manipulate the boxplot for the first few sites, with some experts becoming proficient enough to feel confident about manipulating text or a beta probability density curve instead (Figure 3.8). An example script with the precise wording of questions is available (Supplementary material in Murray *et al.* 2009).

Experts were able to clearly differentiate low to high probabilities of site occupancy among the set of scenarios. Experts tended to show less uncertainty among sites (tighter elicited distributions) for scenarios considered uninhabitable.

Scenario-based elicitation for regression coefficients

Kadane *et al.* (1980) noted that experts found it natural to predict a response in a particular scenario, and to help ensure that all experts had the same starting point provided them with hypothetical data $(Y_H, X_H) = \{(y_h, x_h), h = 1, \ldots, H\}$. This amounted to eliciting a predictive posterior distribution for $[y_m^*|x_m, X_H, Y_H]$. From this the prior could be inferred. A simpler elicitation question posed by Bedrick *et al.* (1996) asked experts to describe the conditional mean $\mathbf{E}[y_m|x_m]$. If, in addition, experts could evaluate quantiles in the distribution $p(y_i|x_i)$ then these could be used to estimate $\mathbf{Var}[y_m|x_m]$. By applying a change of variables, the distribution of the conditional

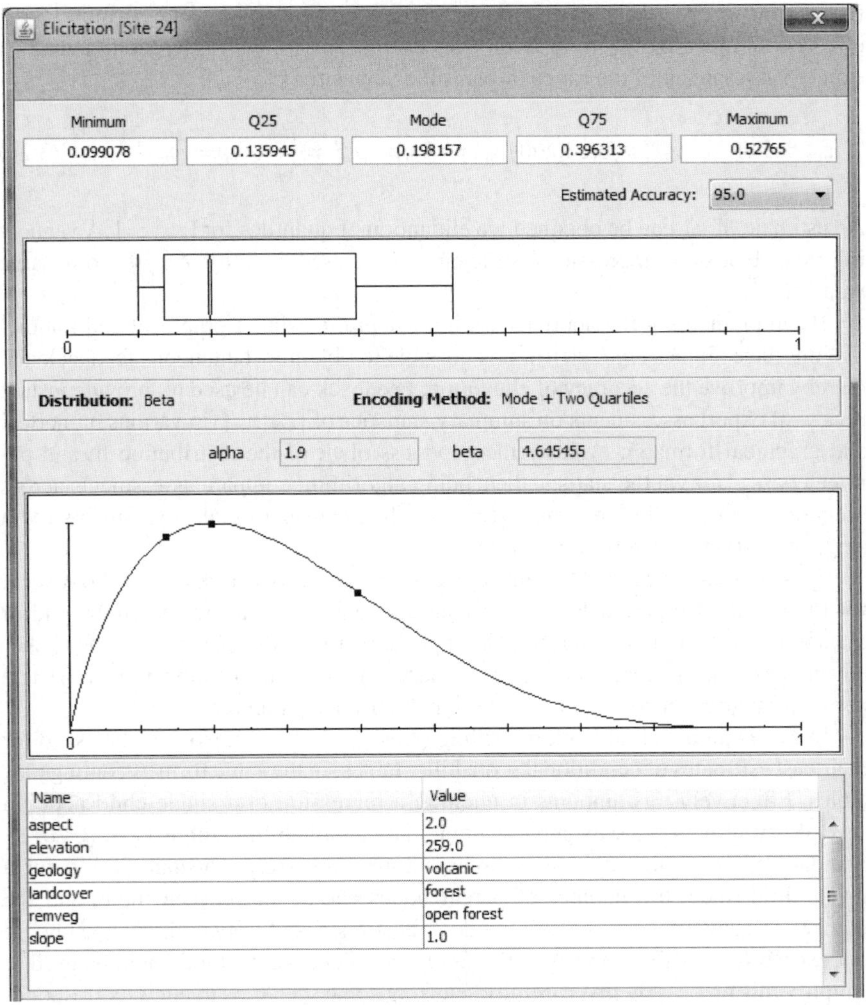

Figure 3.8 Case study 3: The Elicitation window in the *Elicitator* software package, for visually hybridized *PQ* elicitation of $p(y_m|x_m)$ of scenario m.

means could be used to infer the prior distribution of the regression coefficients. This deterministic solution required exactly the same number of elicitation scenarios as the number of covariates $M = J$.

Low Choy *et al.* (2010) extended this approach to the situation where more elicitation scenarios are available, $M > J$. With this approach expert knowledge z_m can be treated as elicited 'data' that can be related to covariates x_m, for each m via a beta regression, with the same explanatory variables X as the empirical data involved in the likelihood. In the usual form of a beta regression, the expected value μ_m may vary for each scenario m, but the effective sample size ν_m is assumed constant $\nu_m = \nu, \forall m$

(Branscum *et al.* 2007). By relaxing the second constraint, and permitting the effective sample size ν_m to also vary for each scenario, we may accommodate an expert's assessment of the range of plausible values for $[z_m|x_m]$:

$$z_m \sim \mathbf{Be}(\mu_m, \nu_m) \text{ where } \text{logit}(\mu_m) = X_m\beta,\ \mu_m = \frac{a_m}{\nu_m}; \nu_m = a_m + b_m. \quad (3.24)$$

An estimate of ν_m can be obtained via elicitation of quantiles for $[z_m|x_m]$. A vaguely informative prior for regression coefficients can be used $\beta \sim \text{N}(0, \sigma^2)$, for some large value of σ^2.

By implementing the approach within *Elicitator* using a graphical and interactive interface (James *et al.* 2010), it is possible to obtain instantaneous feedback and thereby improve the accuracy of elicitation. Feedback can be used to: provide instant review of expert assessments on summary statistics of $[z_m|x_m]$ (in various numerical and graphical formats); evaluate the goodness of fit of the distribution fit and parameters μ_m, ν_m via the statistical encoding algorithm; compare assessments across scenarios; and examine the implications β when considering all assessments (by a single expert) (Low Choy *et al.* 2011).

This approach is related to data augmentation priors (Bedrick *et al.* 1996) except for the use of two-parameters rather than a single parameter in the GLM, and of a statistical rather than deterministic approach to encoding. This approach takes advantage of the fact that elicitation of scenarios (more than the number of covariates) permits statistical encoding of the additional nuisance parameter.

In Bayesian belief network modelling (Uusitalo 2007), experts can be asked for their best estimates of conditional probability tables, in the same form as contingency tables. Effectively this amounts to eliciting a regression of a single child node on several parent nodes, for example defining habitat suitability (Johnson *et al.* 2011). Standard advice warns against complex sub-networks (e.g. Uusitalo 2007): for a given child node, the number of parent nodes should not exceed three, and the number of categories for the child node is best restricted to two–three, and should not exceed four or five. However, the *Elicitator* elicitation method could be used to simplify this process by first carefully selecting some scenarios to elicit (corresponding rows in the contingency table), and then inferring the underlying expert model (James *et al.* 2010).

Sensitivity to prior weight

Figure 3.9 shows the beta distributions $p(y_m|x_m)$ encoded as elicited from one of the four experts on the Queensland region. Two different encoded distributions are shown for each scenario, both utilizing the expert's best estimates for the number of occupied sites (out of 100 for that scenario). The first encoded distribution provided by *Elicitator* takes the elicited information on quantiles at face value. These resulted in fairly vague distributions. The second encoded distribution for each scenario shows the impacts on the posterior distribution, if the variability in the expert's estimates was decreased, by increasing the encoded effective sample size by a relatively modest factor of five. We see that the encoded distributions were much tighter.

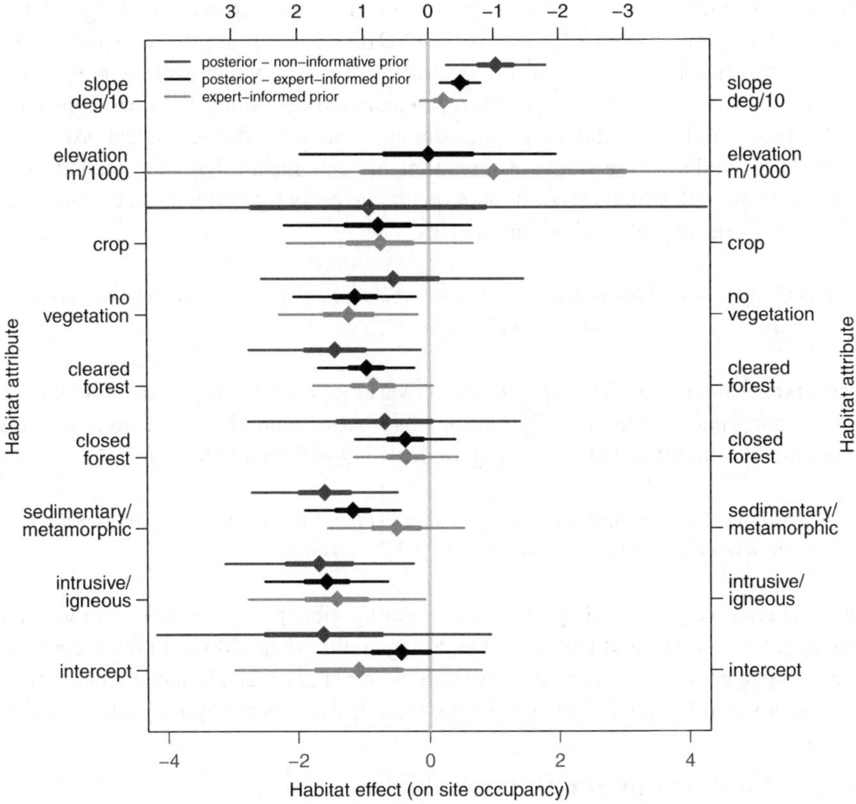

Figure 3.9 Case study 4: Impact of vaguely informative or expert-informed priors on posterior estimates.

3.5 Discussion

When implementing Bayesian analyses in practice, it can be useful to address priors within an overall framework of problem-solving and inference. Below we provide a checklist which may help develop this framework, particularly during collaboration between a statistical modeller and a researcher.

3.5.1 Limitations

At either end of the spectrum there are sometimes overwhelming reasons for selecting an objective or subjective prior.

> ***P1: Data-driven limitations.*** *Pragmatically, is there so much empirical data that the form and contents of the prior are unlikely to have any impact?*

In the case studies considered here, priors encoded from expert knowledge helped address major limitations in the data. In the PhD dissertation case study, expert knowledge was available at school level in contrast to the data, which were aggregated to faculty scale (Section 3.4.1.) In the bioregionalisation case study, data available at the desired scale and extent did not provide direct measures of the ecological processes. In the rock wallaby case study, observations of scats meant that sites needed to be accessible on foot, which was difficult if on private property or if in inaccessible areas of the hilly terrain preferred by this species.

> ***P2: Computational limitations.*** *Are simple mathematical forms for the prior preferred to ensure inference is computationally feasible?*

For mixture models, the Wishart priors on covariance matrices have been popular due to their conjugacy (Section 3.4.3). Normal distributions have been used as a basis for encoding several informative priors on regression coefficients (Section 3.4.4).

> ***P3: Philosophical limitations.*** *Determine how evidence will be used, and thus whether a subjective or objective approach (Section 3.1.2) suits the problem.*

In some contexts, a prevailing objective scientific view precludes that opinion enter into analyses (Hobbs and Hilborn 2006; Spiegelhalter *et al.* 2004). In other contexts such as species distribution modelling (Langhammer *et al.* 2007) and legal argument (Low Choy and Wilson 2009), experts are more trusted than empirical analyses.

3.5.2 Finding out about the problem

Information pertinent to priors will arise quite naturally when initially *discussing the problem*, to find out what the researcher knows, while trying to specify the statistical modelling challenge.

> ***P4: Find the sensible range.*** *Determine a range of parameter values considered sensible by the researcher.*

It is good practice for an experimenter in any discipline to have some idea of the expected range of values, so that they can detect errors. Hence it is usually possible to determine minimum and maximum values; values beyond this range would be considered in error or surprising. Such ranges can help construct the prior (e.g. Fisher *et al.* 2011; O'Leary *et al.* 2009).

> ***P5: Find out the current knowledge.*** *What is the starting state of knowledge about model parameters before analysis (Section 3.1.3)?*

In a pioneering study, prior knowledge may comprise a 'blank slate'. However, in most situations, ballpark estimates can be obtained by considering valid null and alternative hypotheses, or results in related contexts. For example, in the study on lifetimes of powerful owls, different sources of ecological prior information lead to different statistical priors, which may be used individually (McCarthy 2007) or in combination (Figure 3.1).

> ***P6: Find out about the design.*** *What information was used to design the study, especially sample size? This may help estimate uncertainty about parameter values, or point estimates of variance parameters.*

Spiegelhalter *et al.* (2004) provide a good introduction to this approach in the context of clinical studies.

3.5.3 Prior formulation

This phase focuses on modelling the prior information, including choice of the mathematical form of the prior distribution.

> ***P7: Nuisance parameters* versus *parameters with information.*** *Determine whether experts have knowledge relevant to at least some parameters in the model, particularly where empirical data are limited.*

This may be addressed formally via a sample size analysis to target collection or collation of relevant empirical data. Even if a subjective approach to Bayesian inference is selected, a baseline prior will still be required.

> ***P8: Baseline prior.*** *Select an appropriate baseline prior which will (a) provide a baseline for comparison with subjective priors, if necessary, and (b) have mathematical properties appropriate to the context. In particular consider objective priors, if they are available for the likelihood model of interest.*

There are several different mathematical criteria for formulating prior models (Section 3.2), including: propriety in mixture models (Section 3.4.3); and impact of prior knowledge on the posterior, with a binomial likelihood (Section 3.4.2).

> ***P9: Schools of thought.*** *Consider the research community, to whom the results of this analysis will be communicated. What schools of thought currently exist on the possible values of these parameters?*

Consider encoding different priors informed by different schools of thought. Is it important to determine how much evidence would convince, or change the mind of, the more extreme (sceptical versus enthusiastic) members of the community (Spiegelhalter *et al.* 2004)?

> ***P10: Prior pooling.*** *Determine whether sources of prior information need to be pooled, and whether this will be achieved using a more mathematical (i.e. empirical) or more psychological (i.e. subjective) approach.*

The options range, at one extreme, from objective to psychological pooling of prior information (Section 3.1.4).

> ***P11: Literature source.*** *In the literature, is there sufficient information to describe the parameters of the prior, and does this information need to be aggregated across studies?*

Will a meta-analysis of several studies provide useful information on a parameter's prior, delineating a range of potential values, and an estimate of variability based on variation among studies (and therefore experimental conditions or observational situations)?

> ***P12: Direct or indirect elicitation.*** *Are the experts sufficiently familiar with the likelihood model to define 'starting values' for the parameter?*

If the answer to the above question is 'no', then what information do experts know – that is relevant to the parameter, and can be comfortably elicited – and can used as a basis to infer the prior model, under an indirect approach to elicitation?

3.5.4 Communication

Finally, communication of results related to the prior can be key to the success of Bayesian inference, since this is an important point of difference compared with classical statistics.

> ***P13: Prior specification.*** *Key aspects of the design underlying prior modelling should be documented to ensure transparency and repeatability (Low Choy et al. 2010).*

These include:

1. The *purpose* or how prior information will be utilized.
2. The *goal* of prior specification (information being targeted).
3. The statistical *model* for the prior.
4. The method of *encoding* prior information into the model.
5. How inherent *variability* in an process of eliciting information (e.g. from experts or the literature) was managed.
6. A step-by-step *protocol* so that prior elicitation is repeatable.

As with all modelling, considerable insight can be gained by assessing sensitivity to prior modelling choices.

> ***P14: Prior sensitivity.*** *Compare different priors, for example in terms of location, informativeness and mathematical form.*

Some useful comparisons are:

Prior *vs* posterior Prior and posterior distributions should be compared on the scale of each parameter.

Community of priors The location of priors can be changed from sceptical (e.g. no change) to optimistic (e.g. large change).

Informativeness Sensitivity to the informativeness of the priors should be investigated. This may involve varying, for example, the effective sample size in a beta prior (Section 3.2.1), or the degrees of freedom in a Wishart prior on a covariance matrix (Section 3.4.3).

Mathematical form This may identify unintentionally over-constraining priors (Gelman 2006) and permit representation of qualitative information from experts relevant to the shape of the prior (Frühwirth-Schnatter and Wagner 2011).

Hypothesis testing More explicitly, hypotheses can be explicitly tested via Bayes factors.

3.5.5 Conclusion

Priors need not be the silent partners in Bayesian inference. Instead a more active consideration of their role and impact on Bayesian inference can lead to a better understanding of the model being fit, and to better insight into posterior results.

Acknowledgements

The author was supported by funding provided under project 90143 by the Cooperative Research Centre for National Plant Biosecurity, and would like to thank the experts involved in the examples used in this chapter: Chris Anderson, Robert Burdett, Nichole Hammond, Darryl Hardie, Frith Jarrad, James McGree, Ross McVinish, Lindsay Penrose, Chris Strickland, Mary Whitehouse, Therese Wilson, Ian Wood, Jin Xi, and others who contributed to Murray *et al.* (2009), Williams *et al.* (2011) and references therein. The examples owe a great deal to previous (cited) work undertaken with many collaborators: Clair Alston, Julian Caley, Rod Fensham, Rebecca Fisher, Allan James, Kerrie Mengersen, Justine Murray, Rebecca O'Leary, Kristen Williams, Therese Wilson.

References

Albert I, Donnet S, Guihenneuc-Jouyaux C, Low Choy S, Mengersen K and Rousseau J In press Combining expert opinions in prior elicitation. Bayesian Analysis.

Anderson C, Low Choy S, Dominiak B, Gillespie PS, Davis R, Gambley C, Loecker H, Pheloung P, Smith L, Taylor S and Whittle P accepted Biosecurity surveillance systems, plants. In *Biosecurity in Agriculture and the Environment* (ed. S McKirdy) CABI, Wallingford.

Bayarri MJ and Berger JO 2004 The interplay of Bayesian and Frequentist analysis. *Statistical Science* **19**, 58–80.

Bayes T and Price R 1763 An essay towards solving a problem in the doctrine of chances. *Philosophical Transactions of the Royal Society of London* **53**, 370–418.

Bedrick EJ, Christensen R and Johnson W 1996 A new perspective on priors for generalized linear models. *Journal of the American Statistical Association* **91**(436), 1450–1460.

Bensmail H, Celeux G, Raftery AE and Robert CP 1997 Inference in model-based cluster analysis. *Statistics and Computing* **7**, 1–10.

Berger J 2006 The case for objective Bayesian analysis. *Bayesian Analysis* **1**, 385–402.

Berger J and Bernardo JM 1992 On the development of reference priors (with discussion). In *Bayesian Statistics 4* (eds JM Bernado *et al.*), pp. 35–60. Oxford University Press, Oxford.

Bernardo J 2006 Objective Bayesian statistics: an introduction to Bayesian reference analysis. Postgraduate Tutorial Seminar, Valencia 8/ISBA 2006 World Meeting on Bayesian Statistics, Spain.

Bernardo JM 1979 Reference posterior distribution for Bayesian inference (with discussion). *Journal of the Royal Statistical Society, Series B* **41**, 113–147.

Box GEP and Tiao GC 1982 *Bayesian Inference in Statistical Analysis*. John Wiley & Sons, Inc., New York.

Branscum AJ, Johnson WO and Thurmond MC 2007 Bayesian beta regression: applications to household expenditure data and genetic distance between foot-and-mouth disease viruses. *Australian and New Zealand Journal of Statistics* **49**(3), 287–301.

Burgman M 2005 *Risks and Decisions for Conservation and Environmental Management*. Ecology, Biodiversity and Conservation. Cambridge University Press, Cambridge.

Chaloner KM and Duncan GT 1983 Assessment of a beta prior distribution: PM elicitation. *The Statistician* **32**(1/2), 174–180.

Congdon P 2001 *Bayesian Statistical Modelling*. John Wiley & Sons, Inc., New York.

Cooke RM and Goossens LLHJ 2008 TU Delft expert judgment data base. *Reliability Engineering and System Safety* **93**, 657–674.

Dawid AP 2004 Probability, causality and the empirical world: a Bayes-de Finetti-Popper-Borel synthesis. *Statistical Science* **19**(1), 44–57.

de la Place PS 1774 Mémoire sur la probabilité des causes par les événements. *Memoires de Mathematique et de Physique* **6**, 621–656.

Denham R and Mengersen K 2007 Geographically assisted elicitation of expert opinion for regression models. *Bayesian Analysis* **2**(1), 99–136.

El-Gamal M and Grether D 1995 Are people Bayesian? Uncovering behavioural strategies. *Journal of the American Statistical Association* **90**(432), 1137–1145.

Fienberg SE 2006 When did Bayesian inference become 'Bayesian'?. *Bayesian Analysis* **1**, 1–40.

Fisher R, O'Leary RA, Low-Choy S, Mengersen K and Caley MJ 2012 Elicit-n: New method and software for eliciting species richness, Environmental Modelling & Software.

French S 1986 Calibration and the expert problem. *Management Science* **32**(3), 315–321.

French S 2011 Aggregating expert judgement. *Revista de la Real Academia de Ciencias Exactas, Fisicas y Naturales* **105**, 181–206.

Frühwirth-Schnatter S 2006 *Finite Mixture and Markov Switching Models*. Springer Series in Statistics. Springer, Berlin.

Frühwirth-Schnatter S and Wagner H 2011 Bayesian variable selection for random intercept modelling of Gaussian and non-Gaussian data. In *Bayesian Statistics 9*, pp. 165–200, Oxford University Press, Oxford.

Funtowicz SO and Ravetz JR 1993 Science for the post-normal age. *Futures* **25**, 735–755.

Garthwaite PH and O'Hagan A 2000 Quantifying expert opinion in the UK water industry: an experimental study. *The Statistician* **49**, 455–477.

Gelman A 2006 Prior distributions for variance parameters in hierarchical models. *Bayesian Analysis* **1**(3), 515–533.

Gelman A, Carlin JB, Stern HS and Rubin DB 2004 *Bayesian Data Analysis*, 2nd edn. Chapman and Hall/CRC, Boca Raton, FL.

Genest C and Zidek JV 1986 Combining probability distributions: a critique and an annotated bibliography. *Statistical Science* **1**(1), 114–135.

Ghosh JK, Delampady M and Samanta T 2006 *An Introduction to Bayesian Analysis: Theory and Methods*. Springer, New York.

Gill J and Walker LD 2005 Elicited priors for Bayesian model specifications in political science research. *Journal of Politics* **67**(3), 841–872.

Goldstein M 2006 Subjective Bayesian analysis: principles and practice. *Bayesian Analysis* **1**, 403–420.

Goldstein M 2011 External Bayesian analysis for computer simulators. In *Bayesian Statistics 9* (eds JM Bernardo *et al.*),pp. 201–228. Oxford University Press, Oxford.

Hobbs NT and Hilborn R 2006 Alternatives to statistical hypothesis testing in ecology: a guide to self teaching. *Ecological Applications* **16**, 5–19.

Hora SC, Hora JA and Dodd NG 1992 Assessment of probability distributions for continuous random variables: a comparison of the bisection and fixed value methods. *Organizational Behaviour and Human Decision Processes* **51**, 135–155.

James A, Low Choy S, Murray J and Mengersen K 2010 Elicitator: an expert elicitation tool for regression in ecology. *Environmental Modelling & Software* **25**(1), 129–145.

Johnson S, Low Choy S and Mengersen K 2011 Integrating Bayesian networks and geographic information systems: good practice examples. *Integrated Environmental Assessment and Management*, published online 19 September, doi: 10.1002/ieam.262.

Kadane JB, Dickey JM, Winkler RL, Smith WS and Peters SC 1980 Interactive elicitation of opinion for a normal linear model. *Journal of the American Statistical Association* **75**, 845–854.

Kass RE and Greenhouse JB 1989 Comments on the paper by J. H. Ware. *Statistical Science* **4**(4), 310–317.

Langhammer PF, Bakarr MI, Bennun LA, Brooks TM, Clay RP, Darwall W, Silva ND, Edgar GJ, Eken G, Fishpool LDC, da Fonseca GAB, Foster MN, Knox DJ, Matiku P, Radford EA, Rodrigues ASL, Salaman P, Sechrest W and Tordoff AW 2007 Identification and gap analysis of key biodiversity areas: targets for comprehensive protected area systems. Best Practice Protected Area Guidelines Series no. 15, IUCN (The World Conservation Union), Gland, Switzerland. (accessed 9 May 2012).

Leal J, Wordsworth S, Legood R and Blair E 2007 Eliciting expert opinion for economic models: an applied example. *Value in Health* **10**(3), 195–203.

Lindley D 1983 Reconciliation of probability distributions. *Operations Research* **31**(5), 866–880.

Lindley DV 1956 On a measure of the information provided by an experiment. *Annals of Mathematical Statistics* **27**, 986–1005.

Lipscomb J, Parmigiani G and Hasselblad V 1998 Combining expert judgment by hierarchical modeling: an application to physician staffing. *Management Science* **44**, 149–161.

Low Choy S and Wilson T 2009 How do experts think about statistics? Hints for improving undergraduate and postgraduate training. *International Association for Statistics Education Satellite Conference Proceedings, Durban, South Africa.* (accessed 9 May 2012).

Low Choy S, James A, Murray J and Mengersen K 2011 *Elicitator*: a user-friendly, interactive tool to support the elicitation of expert knowledge. In *Expert Knowledge and Its Applications in Landscape Ecology* (eds AH Perera, CA Drew and CJ Johnson), pp. 39–67. Springer. New York

Low Choy S, Mengersen K and Rousseau J 2008 Encoding expert opinion on skewed nonnegative distributions. *Journal of Applied Probability and Statistics* **3**, 1–21.

Low Choy S, Murray J, James A and Mengersen K 2010 Indirect elicitation from ecological experts: from methods and software to habitat modelling and rock-wallabies. In *Handbook of Applied Bayesian Analysis* (eds A O'Hagan and M West). Oxford University Press, Oxford.

Marin JM and Robert CP 2007 *Bayesian Core: A practical approach to computational Bayesian statistics*. Springer, Berlin.

McCarthy M 2007 *Bayesian Methods for Ecology*. Cambridge University Press, Cambridge.

Morgan MG, Pitelka LF and Shevliakova E 2001 Elicitation of expert judgments of climate change impacts on forest ecosystems. *Climate Change* **49**, 279–307.

Murray J, Low Choy S, Possingham H and Goldizen A 2008 The importance of ecological scale for wildlife conservation in naturally fragmented environments: a case study of the brush-tailed rock-wallaby (*Petrogale penicillata). Biological Conservation* **141**, 7–22.

Murray JV, Goldizen AW, O'Leary RA, McAlpine CA, Possingham H and Low Choy S 2009 How useful is expert opinion for predicting the distribution of a species within and beyond the region of expertise? A case study using brush-tailed rock-wallabies *Petrogale penicillata*. *Journal of Applied Ecology* **46**, 842–851.

Oakley JE and O'Hagan A 2007 Uncertainty in prior elicitations: a nonparametric approach. *Biometrika* **94**, 427–441.

Oakley JE and O'Hagan A 2010 *SHELF: the Sheffield Elicitation Framework (version 2.0)*. School of Mathematics and Statistics, University of Sheffield.

O'Hagan A, Buck CE, Daneshkhah A, Eiser R, Garthwaite P, Jenkinson D, Oakley J and Rakow T 2006 *Uncertain Judgements: Eliciting Experts' Probabilities*. John Wiley & Sons, Inc., Hoboken, NJ.

O'Leary R, Low Choy S, Murray J, Kynn M, Denham R, Martin T and Mengersen K 2009 Comparison of three elicitation methods for logistic regression on predicting the presence of the threatened brush-tailed rock-wallaby *Petrogale penicillata*. *Environmetrics* **20**, 379–398.

O'Leary R, Murray J, Low Choy S and Mengersen K 2008 Expert elicitation for Bayesian classification trees. *Journal of Applied Probability and Statistics* **3**(1), 95–106.

Plous S 1993 *The Psychology of Judgment and Decision Making*. McGraw-Hill, New York.

Roback PJ and Givens GH 2001 Supra-Bayesian pooling of priors linked by a deterministic simulation model. *Communications in Statistics – Simulation and Computation* **30**(3), 447–476.

Robert CP 1996 Mixtures of distributions: inference and estimation. In *Markov Chain Monte Carlo in Practice* (eds WR Gilks, S Richardson and DJ Spiegelhalter), pp. 441–464. Chapman & Hall, London.

Speirs-Bridge A, Fidler F, McBride M, Flander L, Cumming G and Burgman M 2010 Reducing overconfidence in the interval judgments of experts. *Risk Analysis* **30**(3), 512–523.

Spetzler CS and Staël von Holstein CAS 1975 Probability encoding in decision analysis. *Management Science* **22**(3), 340–358.

Spiegelhalter DJ, Adams KR and Myles JP 2004 *Bayesian Approaches to Clinical Trials and Health-Care Evaluation*. Statistics in Practice. John Wiley & Sons, Ltd, Chichester.

Stephens M 1997 Bayesian methods for mixtures of normal distributions. PhD thesis. Faculty of Mathematical Sciences.

Stigler SM 1986 Laplace's 1774 memoir on inverse probability. *Statistical Science* **1**(3), 359–363.

Stone M 1961 The opinion pool. *Annals of Mathematical Statistics* **32**(4), 1339–1342.

Strauss H and Zeigler L 1975 The Delphi technique and its uses in social science research. *Journal of Creative Behaviour* **9**, 253–259.

Tebaldi C, Sanso B and Smith RL 2011 Characterizing uncertainty of future climate change projections using hierarchical Bayesian models. In *Bayesian Statistics 9, Proceedings of the Ninth Valencia International Meeting, 3–8 June 2010* (eds JM Bernardo *et al.*), pp. 639–658. Oxford University Press, Oxford.

Tuyl F, Gerlach R and Mengersen K 2008 A comparison of Bayes-Laplace, Jeffreys, and other priors: the case of zero events. *The American Statistician* **62**(1), 40–44.

Uusitalo L 2007 Advantages and challenges of Bayesian networks in environmental modelling. *Ecological Modelling* **203**, 312–318.

West M 1985 Generalized linear models: scale parameters, outlier accommodation and prior distributions. In *Bayesian Statistics 2* (eds JM Bernardo *et al.*). North-Holland, Amsterdam.

Williams K, Low Choy S, Rochester W and Alston C 2011 Using Bayesian mixture models that combine expert knowledge and GIS data to define ecoregions. In *Expert Knowledge and its Application in Landscape Ecology* (eds AH Perera, CA Drew and CJ Johnson), pp. 229–251. Springer, New York.

Zellner A and Siow A 1980 *Posterior Odds Ratios for Selected Regression Hypotheses (with discussion)*, pp. 585–603. Oxford University Press, Oxford.

4

Bayesian analysis of the normal linear regression model

Christopher M. Strickland and Clair L. Alston

Queensland University of Technology, Brisbane, Australia

4.1 Introduction

The normal linear regression model is a fundamental tool in statistical analysis. Bayesian analysis of this model is well developed; see for instance Zellner (1971), Judge *et al.* (1988), Koop (2003) and Marin and Robert (2007). The linear regression model relates k covariates, $x_{i,1}, x_{i,2}, \ldots, x_{i,k}$, to the ith observation y_i as follows:

$$y_i = x_{i,1}\beta_1 + x_{i,2}\beta_2 + \cdots + x_{i,k}\beta_k + \varepsilon_i, \tag{4.1}$$

for $i = 1, 2, \ldots, n$, where β_j, $j = 1, 2, \ldots, k$, are referred to as regression coefficients and ε_i is an error term, which throughout is assumed to be normally distributed.

This chapter aims to provide an introduction to the Bayesian analysis of the normal linear regression model, as given in Equation (4.1). We will use two case studies, presenting established data sets that have been widely analysed and are publicly available for researchers to use to develop their own models. These examples allow us to explore standard linear regression, variable selection, in this context, and to extend the standard model to account for serial correlation.

In this chapter, we analyse the basic linear regression model using natural conjugate priors, Jeffreys prior and Zellner's g-prior approach, illustrating the ease with which posterior estimation can be implemented. Sampling schemes to conduct linear regression with correlated errors are provided. The issue of variable selection is approached using the Bayesian lasso and stochastic search algorithms. These

Case Studies in Bayesian Statistical Modelling and Analysis, First Edition. Edited by Clair L. Alston, Kerrie L. Mengersen and Anthony N. Pettitt.

results are shown to be in close agreement with each other and the results of the g-prior analysis.

This chapter provides a foundation for the following chapters in this book by describing the mathematics underpinning the Bayesian analysis. As there are a number of software packages available for analysts to conduct Bayesian regression analysis, much of the mathematical detail can be skipped by readers who are only interested in the application results.

4.2 Case studies

4.2.1 Case study 1: Boston housing data set

The linear regression model is used to analyse the median value housing pricing in Boston, using a well-known data set, analysed in Harrison and Rubinfeld (1978) and Gilley and Pace (1996). The original data set, of sample size 506, is from a 1970 census and includes 13 dependent variables based on town, crime rate (CRIM), proportion of residential area zoned for large lots (ZN), proportion of non-retail business acres (INDUS), Charles River location (CHAS), nitric oxide concentration (NOX), average number of rooms (RM), proportion of pre-1940 dwellings (AGE), weighted distances to five employment centres (DIS), accessibility to highways (RAD), property tax level (TAX), pupil to teacher ratio (PTRATIO), racial mix (B) and percentage of low-income earners (LSTAT).

4.2.2 Case study 2: Production of cars and station wagons

The linear regression model is used to analyse monthly production of cars and station wagons (thousands of cars per month) in Australia from November 1969 to August 1995. The data set is a part of Rob Hyndman's time series library and can be found at `http://robjhyndman.com/tsdldata/data/cars&sw.dat`. The dependent variables for the analysis include a constant, a time trend, monthly dummy variables and a dummy variable for an outlier in November 1977.

Figure 4.1 plots monthly production of station wagons and cars (thousands per month) in Australia, from November 1969 to August 1995. It is clear that the data are seasonal.

4.3 Matrix notation and the likelihood

For mathematical convenience, the linear regression model in Equation (4.1) can be expressed, using matrix notation, as follows:

$$\boldsymbol{y} = X\boldsymbol{\beta} + \boldsymbol{\varepsilon}, \tag{4.2}$$

where $\boldsymbol{y} = \begin{bmatrix} y_1 \\ y_2 \\ \vdots \\ y_n \end{bmatrix}$, $X = \begin{bmatrix} x_{11} & x_{12} & \cdots & x_{1k} \\ x_{21} & x_{22} & \cdots & x_{2k} \\ \vdots & \vdots & \ddots & \vdots \\ x_{n1} & x_{n2} & \cdots & x_{nk} \end{bmatrix}$ and $\boldsymbol{\varepsilon} = \begin{bmatrix} \varepsilon_1 \\ \varepsilon_2 \\ \vdots \\ \varepsilon_n \end{bmatrix}$.

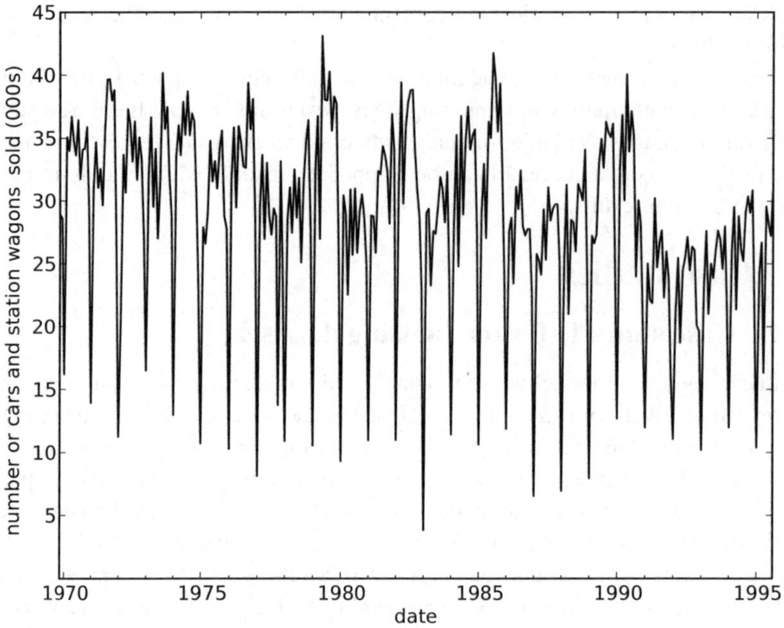

Figure 4.1 Monthly production of cars and station wagons in Australia (000s).

To define the likelihood function, an assumption needs to be made regarding the distribution of the errors in Equation (4.2). We begin with the assumption of spherical errors. That is, the errors are homoscedastic, serially uncorrelated and normally distributed, such that

$$\boldsymbol{\varepsilon} \sim \mathrm{N}\left(0, \sigma^2 \mathbf{I}\right). \tag{4.3}$$

Given Equation (4.2) and Equation (4.3), the likelihood $p(\boldsymbol{y}|\boldsymbol{X}, \boldsymbol{\beta}, \sigma)$ is a normally distributed $y \mid \boldsymbol{X}, \boldsymbol{\beta}, \sigma \sim \mathrm{N}(\boldsymbol{X\beta}, \sigma^2\mathbf{I})$. That is,

$$p\left(\boldsymbol{y} \mid \boldsymbol{X}, \boldsymbol{\beta}, \sigma\right) \propto \sigma^{-n} \exp\left\{-\frac{1}{2\sigma^2}\left(\boldsymbol{y} - \boldsymbol{X\beta}\right)^T\left(\boldsymbol{y} - \boldsymbol{X\beta}\right)\right\}. \tag{4.4}$$

4.4 Posterior inference

Posterior inference for the normal linear regression model, specified by Equation (4.2), is described in the following subsections. Initially, it is assumed that the error structure in the model is specified following Equation (4.3), and standard analytical results are presented for the natural conjugate prior, along with a brief description of Jeffreys prior and Zellner's g-prior. Proceeding this initial discussion, the normal linear regression model is generalized to account for non-spherical errors. A Markov chain Monte Carlo (MCMC) algorithm is detailed for estimation, with focus on the specific cases of heteroscedasticity and serial correlation. MCMC sampling schemes are described that can be used for variable selection for the normal linear regression model, with both spherical and non-spherical errors.

In the following sections, we use an underbar to denote parameters contained in prior distributions and an overbar to denote the parameters of posterior distributions.

4.4.1 Natural conjugate prior

The natural conjugate prior for the Bayesian analysis of the linear regression model, specified by Equation (4.4), is the normal inverted-gamma prior; see for instance Zellner (1971) and Judge *et al.* (1988). Under this scheme, the prior for $\boldsymbol{\beta} \mid \sigma$ is a normal distribution with mean $\underline{\boldsymbol{\beta}}$ and precision $\underline{\boldsymbol{V}}$, and can be specified as

$$p(\boldsymbol{\beta} \mid \sigma) \propto \sigma^{-k} \exp\left\{-\frac{1}{2\sigma^2}\left(\boldsymbol{\beta} - \underline{\boldsymbol{\beta}}\right)^T \underline{\boldsymbol{V}}\left(\boldsymbol{\beta} - \underline{\boldsymbol{\beta}}\right)\right\}. \tag{4.5}$$

where k is the number of regressors in the model.

The prior distribution for σ is an inverse-gamma distribution, with parameters $\underline{\nu}$ and $\underline{s}^2$, specified as

$$p(\sigma) \propto \sigma^{-(\underline{\nu}+1)} \exp\left\{-\frac{\underline{\nu s}^2}{2\sigma^2}\right\}, \tag{4.6}$$

Given the likelihood, Equation (4.4) and the prior described in Equations (4.5) and (4.6), we can derive the posterior distribution for the unknown parameters of interest, $\boldsymbol{\beta}$ and $\boldsymbol{\sigma}$. Our brief derivation begins with the foundation of Bayesian inference, which is that the posterior $\propto$ likelihood $\times$ prior. It follows that the joint posterior for $\boldsymbol{\beta}$ and σ can be expressed as

$$p(\boldsymbol{\beta}, \sigma \mid \boldsymbol{y}, \boldsymbol{X}) \propto p(\boldsymbol{y} \mid \boldsymbol{X}, \boldsymbol{\beta}, \sigma) \times p(\boldsymbol{\beta} \mid \sigma) \times p(\sigma)$$

Algebraically, this becomes

$$\begin{aligned} p(\boldsymbol{\beta}, \sigma \mid \boldsymbol{y}, \boldsymbol{X}) \propto\ & \sigma^{-n} \exp\left\{-\frac{1}{2\sigma^2}(\boldsymbol{y} - \boldsymbol{X\beta})^T(\boldsymbol{y} - \boldsymbol{X\beta})\right\} \\ & \times \sigma^{-k} \exp\left\{-\frac{1}{2\sigma^2}\left(\boldsymbol{\beta} - \underline{\boldsymbol{\beta}}\right)^T \underline{\boldsymbol{V}}\left(\boldsymbol{\beta} - \underline{\boldsymbol{\beta}}\right)\right\} \\ & \times \sigma^{-(\underline{\nu}+1)} \exp\left\{-\frac{\underline{\nu s}^2}{2\sigma^2}\right\}. \end{aligned}$$

Expanding each term in the equation above gives

$$\begin{aligned} p(\boldsymbol{\beta}, \sigma \mid \boldsymbol{y}, \boldsymbol{X}) \propto \sigma^{-n} \exp\left\{-\frac{1}{2\sigma^2}\left(\boldsymbol{y}^T\boldsymbol{y} + \boldsymbol{\beta}^T\boldsymbol{X}^T\boldsymbol{X}\boldsymbol{\beta} - 2\boldsymbol{\beta}^T\boldsymbol{X}^T\boldsymbol{y}\right)\right\} \\ \times \sigma^{-k} \exp\left\{-\frac{1}{2\sigma^2}\left(\boldsymbol{\beta}^T\underline{\boldsymbol{V}}\boldsymbol{\beta} + \underline{\boldsymbol{\beta}}^T\underline{\boldsymbol{V}}\underline{\boldsymbol{\beta}} - 2\boldsymbol{\beta}^T\underline{\boldsymbol{V}}\underline{\boldsymbol{\beta}}\right)\right\} \\ \times \sigma^{-(\underline{\nu}+1)} \exp\left\{-\frac{\underline{\nu s}^2}{2\sigma^2}\right\}. \end{aligned}$$

We can now combine these densities and obtain the jont posterior as

$$\begin{aligned} p(\boldsymbol{\beta}, \sigma \mid \boldsymbol{y}, \boldsymbol{X}) \propto \sigma^{-k} \exp\Bigg\{ -\frac{1}{2\sigma^2}\Big(\boldsymbol{y}^T\boldsymbol{y} + \boldsymbol{\beta}^T\left[\boldsymbol{X}^T\boldsymbol{X} + \underline{\boldsymbol{V}}\right]\boldsymbol{\beta} \\ - 2\boldsymbol{\beta}^T\left[\boldsymbol{X}^T\boldsymbol{y} + \underline{\boldsymbol{V}\boldsymbol{\beta}}\right]\Big) + \underline{\boldsymbol{\beta}}^T\underline{\boldsymbol{V}\boldsymbol{\beta}}\Bigg\}\sigma^{-(n+\underline{\nu}+1)} \exp\left\{-\frac{\underline{\nu s}^2}{2\sigma^2}\right\}. \end{aligned}$$

Completing the square gives the final posterior distribution,

$$p(\boldsymbol{\beta}, \sigma \mid \boldsymbol{y}, \boldsymbol{X}) \propto \sigma^{-k} \exp\left\{-\frac{1}{2\sigma^2}\left(\boldsymbol{\beta} - \overline{\boldsymbol{\beta}}\right)^T \overline{\boldsymbol{V}}\left(\boldsymbol{\beta} - \overline{\boldsymbol{\beta}}\right)\right\}\sigma^{-(\overline{\nu}+1)} \exp\left\{-\frac{\overline{\nu s}^2}{2\sigma^2}\right\} \tag{4.7}$$

where $\overline{\boldsymbol{\beta}}$ is the posterior mean and $\overline{\boldsymbol{V}}$ is the posterior precision. The posterior parameters for the inverse gamma distribution are $\overline{\nu}$ and $\overline{\nu s}^2$. They are defined as

$$\begin{aligned} \overline{\boldsymbol{\beta}} &= \overline{\boldsymbol{V}}^{-1}\left(\boldsymbol{X}^T\boldsymbol{y} + \underline{\boldsymbol{V}\boldsymbol{\beta}}\right), \\ \overline{\boldsymbol{V}} &= \left(\boldsymbol{X}^T\boldsymbol{X} + \underline{\boldsymbol{V}}\right), \\ \overline{\nu} &= n + \underline{\nu} \quad \text{and} \\ \overline{\nu s}^2 &= \underline{\nu s}^2 + \boldsymbol{y}^T\boldsymbol{y} - \overline{\boldsymbol{\beta}}^T\overline{\boldsymbol{V}\boldsymbol{\beta}} + \underline{\boldsymbol{\beta}}^T\underline{\boldsymbol{V}\boldsymbol{\beta}}. \end{aligned}$$

From this definition, we can see that the posterior parameters are combinations of the parameters attributable to the likelihood and the prior.

From the joint posterior density function for $\boldsymbol{\beta}$ and σ, we can factor the density as the conditional posterior density of $\boldsymbol{\beta}$ multiplied by the marginal posterior density of σ

$$p(\boldsymbol{\beta}, \sigma \mid \boldsymbol{y}, \boldsymbol{X}) = p(\boldsymbol{\beta} \mid \sigma, \boldsymbol{y}, \boldsymbol{X}) \times p(\sigma \mid \boldsymbol{y}),$$

where

$$\begin{aligned} \boldsymbol{\beta} \mid \sigma, \boldsymbol{y}, \boldsymbol{X} &\sim \mathrm{N}\left(\overline{\boldsymbol{\beta}}, \overline{\boldsymbol{V}}^{-1}\right) \\ \sigma \mid \boldsymbol{y} &\sim \mathrm{IG}\left(\frac{\overline{\nu}}{2}, \frac{\overline{\nu s}^2}{2}\right). \end{aligned}$$

Marginal posterior density

To obtain the marginal posterior density for $\boldsymbol{\beta}$, given the observations, $\boldsymbol{y}$, which we represent as $p(\boldsymbol{\beta} \mid \boldsymbol{y})$, we need to integrate σ out of the joint posterior probability density function (pdf) $p(\boldsymbol{\beta}, \sigma \mid \boldsymbol{y})$. When we do this we obtain the marginal posterior density of $\boldsymbol{\beta}$ as

$$\boldsymbol{\beta} \mid \boldsymbol{y} \sim \mathbf{T}\left(\overline{\nu}, \overline{\boldsymbol{\beta}}, \overline{s}^{-2}\overline{\boldsymbol{V}}^{-1}\right), \tag{4.8}$$

where $\mathbf{T}$ is a multivariate T-distribution.

Mathematically, we obtain this by integrating the parameter σ out of the joint conditional posterior distribution as follows:

$$\begin{aligned} p(\boldsymbol{\beta} \mid \boldsymbol{y}) &= \int_0^\infty p(\boldsymbol{\beta}, \sigma \mid \boldsymbol{y})\, d\sigma \\ &\propto \int_0^\infty \sigma^{-(\overline{\nu}+k+1)} \exp\left\{-\frac{1}{2\sigma^2}\left[\left(\boldsymbol{\beta}-\overline{\boldsymbol{\beta}}\right)^T \overline{\boldsymbol{V}}\left(\boldsymbol{\beta}-\overline{\boldsymbol{\beta}}\right)+\overline{\nu s}^2\right]\right\} d\sigma \\ &= \int_0^\infty \sigma^{-(\overline{\nu}+k+1)} \exp\left\{-\frac{a}{2\sigma^2}\right\} d\sigma, \end{aligned} \tag{4.9}$$

where $a = \left(\boldsymbol{\beta}-\overline{\boldsymbol{\beta}}\right)^T \overline{\boldsymbol{V}}\left(\boldsymbol{\beta}-\overline{\boldsymbol{\beta}}\right)+\overline{\nu s}^2$. Let $u = a/2\sigma^2$; then $\sigma = (-a/2u)^{1/2}$ and $d\sigma/du = 2^{-3/2}a^{1/2}u^{-3/2}$. Using integration by parts, it then follows that

$$p(\boldsymbol{\beta} \mid \boldsymbol{y}) \propto \left(\frac{a}{2}\right)^{-(\overline{\nu}+k)/2} \int_0^\infty u^{(\overline{\nu}+k-2)/2} \exp\{-u\}\, du. \tag{4.10}$$

From Equation (4.10) and using the definition of the gamma function, the marginal posterior density for $\boldsymbol{\beta}$ can be written as

$$\begin{aligned} p(\boldsymbol{\beta} \mid \boldsymbol{y}) &\propto a^{-(\overline{\nu}+k)/2}\Gamma\left(\frac{\overline{\nu}+k}{2}\right) \\ &\propto a^{-(\overline{\nu}+k)/2} \\ &= \left[\left(\boldsymbol{\beta}-\overline{\boldsymbol{\beta}}\right)^T \overline{\boldsymbol{V}}\left(\boldsymbol{\beta}-\overline{\boldsymbol{\beta}}\right)+\overline{\nu s}^2\right]^{-(\overline{\nu}+k)/2} \\ &= \left(1+\frac{\left(\boldsymbol{\beta}-\overline{\boldsymbol{\beta}}\right)^T \overline{\boldsymbol{V}}(\boldsymbol{\beta}-\overline{\boldsymbol{\beta}})}{\overline{\nu s}^2}\right)^{-(\overline{\nu}+k)/2}. \end{aligned} \tag{4.11}$$

Recognizing Equation (4.11) as the kernel for the multivariate t-distribution justifies Equation (4.8).

Marginal likelihood

The marginal likelihood, for the normal linear regression model in Equations (4.2) and (4.3), can also be expressed in closed form. The marginal likelihood is derived by integrating out the unknown parameters of interest, $\boldsymbol{\beta}$ and σ, from the joint pdf for $\boldsymbol{\beta}$, σ and $\boldsymbol{y}$, as follows:

$$\begin{aligned}
p(\boldsymbol{y} \mid \boldsymbol{X}) &= \int_0^{\infty} \int_{-\infty}^{\infty} p(\boldsymbol{\beta}, \sigma, \boldsymbol{y} \mid \boldsymbol{X})\, d\boldsymbol{\beta} d\sigma \\
&= \int_0^{\infty} \int_{-\infty}^{\infty} p(\boldsymbol{y} \mid \boldsymbol{X}, \boldsymbol{\beta}, \sigma) \times p(\boldsymbol{\beta} \mid \sigma) \times p(\sigma)\, d\boldsymbol{\beta} d\sigma \\
&= \int_0^{\infty} \int_{-\infty}^{\infty} (2\pi)^{-n/2} \sigma^{-n} \exp\left\{-\frac{1}{2\sigma^2} (\boldsymbol{y} - \boldsymbol{X\beta})^T (\boldsymbol{y} - \boldsymbol{X\beta})\right\} \\
&\times (2\pi)^{-k/2} \sigma^{-k} \mid \underline{\boldsymbol{V}} \mid^{1/2} \exp\left\{-\frac{1}{2\sigma^2} (\boldsymbol{\beta} - \underline{\boldsymbol{\beta}})^T \underline{\boldsymbol{V}} (\boldsymbol{\beta} - \underline{\boldsymbol{\beta}})\right\} \times \\
&\times \frac{2}{\Gamma(\nu/2)} \left(\frac{\underline{\nu s}^2}{2}\right)^{\underline{\nu}/2} \sigma^{-(\underline{\nu}+1)} \exp\left\{-\frac{\underline{\nu s}^2}{2\sigma^2}\right\} d\boldsymbol{\beta} d\sigma. \qquad (4.12)
\end{aligned}$$

Denoting $K = (2\pi)^{-(n+k)/2} \mid \underline{\boldsymbol{V}} \mid^{1/2} (2/\Gamma(\nu/2)) \left(\underline{\nu s}^2/2\right)^{\underline{\nu}/2}$, and taking note of the joint posterior in Equation 4.7, then Equation 4.12 can be simplified such that

$$\begin{aligned}
p(\boldsymbol{y} \mid \boldsymbol{X}) &= K \int_0^{\infty} \int_{-\infty}^{\infty} \exp\left\{-\frac{1}{2\sigma^2} (\boldsymbol{\beta} - \overline{\boldsymbol{\beta}})^T \overline{\boldsymbol{V}} (\boldsymbol{\beta} - \overline{\boldsymbol{\beta}})\right\} \\
&\times \sigma^{-(\overline{\nu}+k+1)} \exp\left\{-\frac{\overline{\nu s}^2}{2\sigma^2}\right\} d\boldsymbol{\beta} d\sigma \\
&= \int_0^{\infty} \sigma^{-(\overline{\nu}+k+1)} \exp\left\{-\frac{\overline{\nu s}^2}{2\sigma^2}\right\} \\
&\times \left[\int_{-\infty}^{\infty} \exp\left\{-\frac{1}{2\sigma^2} (\boldsymbol{\beta} - \overline{\boldsymbol{\beta}})^T \overline{\boldsymbol{V}} (\boldsymbol{\beta} - \overline{\boldsymbol{\beta}})\right\} d\boldsymbol{\beta}\right] d\sigma.
\end{aligned}$$

From the properties of the normal distribution, it follows that

$$\int_{-\infty}^{\infty} \exp\left\{-\frac{1}{2\sigma^2} (\boldsymbol{\beta} - \overline{\boldsymbol{\beta}})^T \overline{\boldsymbol{V}} (\boldsymbol{\beta} - \overline{\boldsymbol{\beta}})\right\} d\boldsymbol{\beta} = (2\pi)^{k/2} \sigma^k \mid \overline{\boldsymbol{V}} \mid^{-1/2}.$$

Defining $K_1 = (2\pi)^{-n/2} \left(\mid \underline{\boldsymbol{V}} \mid / \mid \overline{\boldsymbol{V}} \mid\right)^{1/2} (2/(\Gamma(\underline{\nu}/2)) \left(\underline{\nu s}^2/2\right)^{\underline{\nu}/2}$, then

$$\begin{aligned}
p(\boldsymbol{y} \mid \boldsymbol{X}) &= K_1 \int_0^{\infty} \sigma^{-(\overline{\nu}+1)} \exp\left\{-\frac{\overline{\nu s}^2}{2\sigma^2}\right\} d\sigma \\
&= \frac{1}{2} K_1 \Gamma\left(\frac{\overline{\nu}}{2}\right) \left(\frac{\overline{\nu s}}{2}\right)^{-\overline{\nu}/2} \\
&= (2\pi)^{-n/2} \left(\frac{\mid \underline{\boldsymbol{V}} \mid}{\mid \overline{\boldsymbol{V}} \mid}\right)^{1/2} \frac{\Gamma(\overline{\nu}/2)}{\Gamma(\underline{\nu}/2)} \frac{\left(\underline{\nu s}^2/2\right)^{\underline{\nu}/2}}{\left(\overline{\nu s}^2/2\right)^{\overline{\nu}/2}}. \qquad (4.13)
\end{aligned}$$

4.4.2 Alternative prior specifications

The normal–inverted-gamma prior discussed in Section 4.4.1 provides an attractive option for the researcher; however, there are other popular prior formulations, which can be used to formulate prior beliefs. In the following subsections, Jeffreys prior and Zellner's g-prior are briefly discussed.

Jeffreys prior

Jeffreys prior provides a way of assuming prior ignorance about the unknown parameters of interest, and is defined for an unknown parameter vector $\boldsymbol{\theta}$ as

$$p(\boldsymbol{\theta}) \propto |\,\mathbf{I}(\boldsymbol{\theta})\,|^{1/2}, \tag{4.14}$$

where $\mathbf{I}(\boldsymbol{\theta})$ is Fisher's information matrix, which is defined as

$$\mathbf{I}(\boldsymbol{\theta}) = -\mathbf{E}\left[\frac{\partial^2 \log p(\boldsymbol{y} \mid \boldsymbol{\theta})}{\partial\boldsymbol{\theta}\partial\boldsymbol{\theta}^T}\right],$$

where $\mathbf{E}[]$ is the mathematical expectation operator, and $\log p(\boldsymbol{y} \mid \boldsymbol{\theta})$ is the log-likelihood function. For the normal linear regression model in Equation (4.2) and Equation (4.3), and from the definition of Jeffreys prior in Equation (4.14), given the assumption of prior independence between $\boldsymbol{\beta}$ and σ, it can be shown that the joint prior distribution of the unknown parameters is

$$p(\boldsymbol{\beta}, \sigma) \propto \frac{1}{\sigma}. \tag{4.15}$$

Using a similar approach to the natural conjugate prior, the joint posterior distribution for $\boldsymbol{\beta}$ and σ based on Jeffreys prior (Equation 4.15) is defined by

$$\boldsymbol{\beta} \mid \boldsymbol{y} \sim \mathrm{N}\left(\hat{\boldsymbol{\beta}}, \overline{\boldsymbol{V}}^{-1}\right) \quad \text{and} \quad \sigma \mid \boldsymbol{y} \sim \mathrm{IG}\left(\frac{\overline{\nu}}{2}, \frac{\overline{\nu s}^2}{2}\right),$$

where $\hat{\boldsymbol{\beta}} = (\boldsymbol{X}^T\boldsymbol{X})^{-1}\boldsymbol{X}^T\boldsymbol{y}$, $\overline{\boldsymbol{V}} = \sigma^2(\boldsymbol{X}^T\boldsymbol{X})^{-1}$, $\overline{\nu} = n - k$ and $\overline{\nu s}^2 = \boldsymbol{e}^T\boldsymbol{e}$, and $\boldsymbol{e} = \boldsymbol{y} - \boldsymbol{X}\hat{\boldsymbol{\beta}}$.

Note that, as Jeffreys prior is improper, the marginal likelihood cannot be sensibly interpreted so other methods of model comparison must be used.

Zellner's g-prior

Zellner's g-prior, see Zellner (1986) and Marin and Robert (2007), is a special case of the normal–inverted-gamma prior in Equations (4.5) and (4.6), and is specified with $\underline{\boldsymbol{V}} = (\boldsymbol{X}^T\boldsymbol{X})/g$ and $\underline{\nu} = 0$, implying

$$\boldsymbol{\beta} \mid \sigma, \boldsymbol{X} \sim \mathrm{N}\left(\underline{\boldsymbol{\beta}}, g\sigma^2\left(\boldsymbol{X}^T\boldsymbol{X}\right)^{-1}\right) \quad \text{and} \quad p(\sigma) \propto \frac{1}{\sigma}, \tag{4.16}$$

where g is a scaling constant that allows the user to define the degree of uncertainty in the prior covariance structure; for example, a large g implies great uncertainty in the prior estimate of the mean, $\underline{\beta}$. A popular specification of Equation (4.16) is to assume $\underline{\beta} = \mathbf{0}$; then it follows from Section 4.4.1 that the posterior for β and σ is

$$\beta \mid \sigma, y, X \sim \mathrm{N}\left(\overline{\beta}, \overline{V}^{-1}\right) \quad \text{and} \quad \sigma \mid y, X \sim \mathrm{IG}\left(\frac{n}{2}, \frac{n\overline{s}^2}{2}\right)$$

where $\overline{\beta} = (g/(1+g))\hat{\beta}$ ($\hat{\beta}$ is as defined in the Jeffreys prior section). Also, ($\overline{V} = ((1+g)/g)(X^T X)$ and $n\overline{s}^2 = y^T y - (g/(g+1))y^T \hat{y}$, with $\hat{y} = X\hat{\beta}$. Note that, as $\underline{\nu} = 0$, then $\overline{\nu} = n$. Likewise, it follows that the marginal posterior density for β is defined such that

$$\beta \mid y, X \sim \mathbf{T}\left(n, \overline{\beta}, \overline{s^2 V}^{-1}\right).$$

Assuming that the constant of proportionality for the prior of σ is one, then the marginal distribution of y is

$$p(y \mid X) = \pi^{-n/2}(1+g)^{-k/2}\,\Gamma(n/2)\left(n\overline{s}^2\right)^{-n/2}. \tag{4.17}$$

However, the improper prior specification for σ in Equation (4.16) implies that Equation (4.17) can only be used for model comparison between models with the same σ.

4.4.3 Generalizations of the normal linear model

While the assumed error structure in Equation (4.3) is convenient in that closed form solutions are available for certain prior assumptions, it is somewhat naive for many data sets. In particular, for time series data it is common that serial correlation is present, while for cross-sectional data heteroscedasticity is fairly common. To handle both of these cases we can generalize the error structure in Equation (4.3) such that

$$\boldsymbol{\varepsilon} \sim \mathrm{N}\left(\mathbf{0}, \sigma^2 \boldsymbol{\Omega}^{-1}\right), \tag{4.18}$$

where the specification of $\boldsymbol{\Omega}$ is determined by the assumed structure of the heteroscedasticity or serial correlation in the data generating process and is assumed to be a function of a set of unknown parameters $\boldsymbol{\theta}$. Given Equations (4.2) and (4.18), the joint density function for the complete set of observations, y, is given by

$$p(y \mid X, \beta, \sigma, \boldsymbol{\Omega}) \propto \sigma^{-n} \mid \boldsymbol{\Omega} \mid^{1/2} \exp\left\{-\frac{1}{2}(y - X\beta)^T \boldsymbol{\Omega}(y - X\beta)\right\}. \tag{4.19}$$

Given a suitable transformation W, which is a function of $\boldsymbol{\theta}$ and is defined such that

$$\tilde{\boldsymbol{\varepsilon}} \sim N\left(0, \sigma^2 \boldsymbol{I}\right),$$

where $\tilde{\boldsymbol{\varepsilon}} = W\boldsymbol{\varepsilon}$, then defining the transformed variables $\tilde{y} = Wy$ and $\tilde{X} = WX$ allows the use of the standard results presented in Section 4.4.1 in developing a simple

MCMC sampling scheme. In particular, it is clear that

$$p\left(\boldsymbol{\beta}, \sigma \mid \tilde{\boldsymbol{y}}, \tilde{\boldsymbol{X}}\right) = p\left(\boldsymbol{\beta}, \sigma \mid \boldsymbol{y}, \boldsymbol{X}, \boldsymbol{\Omega}\right). \tag{4.20}$$

Further, it is convenient to note that

$$p\left(\boldsymbol{y} \mid \boldsymbol{X}, \boldsymbol{\beta}, \sigma, \boldsymbol{\Omega}\right) \propto \sigma^{-n} \mid \boldsymbol{\Omega} \mid^{1/2} \exp\left\{-\frac{1}{2}\left(\tilde{\boldsymbol{y}} - \tilde{\boldsymbol{X}}\boldsymbol{\beta}\right)^T \left(\tilde{\boldsymbol{y}} - \tilde{\boldsymbol{X}}\boldsymbol{\beta}\right)\right\}. \tag{4.21}$$

A general MCMC scheme, which can be used in making inference about the joint posterior for $\boldsymbol{\theta}$, $\boldsymbol{\beta}$ and $\boldsymbol{\sigma}$, is defined at iteration j in Algorithm 1).

Algorithm 1: MCMC sampling scheme for the normal linear regression model with non-spherical errors

---

1. Use $\tilde{\boldsymbol{\theta}}^{(j-1)}$ to calculate $\tilde{\boldsymbol{y}} = \boldsymbol{W}\boldsymbol{y}$ and $\tilde{\boldsymbol{X}} = \boldsymbol{W}\boldsymbol{X}$.
2. Sample $\boldsymbol{\beta}^{(j)}$ and $\sigma^{(j)}$ from $p\left(\boldsymbol{\beta}, \sigma \mid \tilde{\boldsymbol{y}}, \tilde{\boldsymbol{X}}\right)$ as per Equation (4.20).
3. Sample $\boldsymbol{\theta}^{(j)}$ from $p\left(\boldsymbol{\theta} \mid \boldsymbol{y}, \boldsymbol{X}, \boldsymbol{\beta}^{(j)}, \sigma^{(j)}\right)$.

The first step of the algorithm forms the weighted linear model. The second step, which is to sample $\boldsymbol{\beta}$ and σ jointly, has a closed form solution. For the third step, provided the dimension of $\boldsymbol{\theta}$ is small, a random walk Metropolis–Hastings (RWMH), see Robert and Casella (1999), may be used to sample $\boldsymbol{\theta}$ from its posterior distribution.

The marginal likelihood, for the linear regression model defined by Equations (4.2) and (4.18), is given by

$$p\left(\boldsymbol{y} \mid \boldsymbol{X}\right) = \iiint p\left(\boldsymbol{y} \mid \boldsymbol{X}, \boldsymbol{\beta}, \sigma, \boldsymbol{\Omega}\right) p\left(\boldsymbol{\beta}, \sigma, \boldsymbol{\Omega}\right) d\boldsymbol{\beta} d\sigma d\boldsymbol{\Omega},$$

which is not available in closed form; however, it can be estimated numerically. A generic importance sampling approach, see for example Robert and Casella (1999) and Marin and Robert (2007), is constructed with a normally distributed importance density, denoted as $g\left(\boldsymbol{\beta}, \sigma, \boldsymbol{\Omega}\right)$, which has a mean $\hat{\boldsymbol{m}}$ and a covariance $2\hat{\boldsymbol{\Sigma}}$, where $\hat{\boldsymbol{m}}$ and $\hat{\boldsymbol{\Sigma}}$ are the sample mean and covariance respectively, of the MCMC iterates for $(\boldsymbol{\beta}^T, \sigma, \boldsymbol{\theta}^T)^T$ that are generated using Algorithm 1. Given M draws of $(\boldsymbol{\beta}^T, \sigma, \boldsymbol{\theta}^T)^T$ from $g(\boldsymbol{\beta}, \sigma, \boldsymbol{\Omega})$ the marginal likelihood may be estimated as follows:

$$\widehat{p\left(\boldsymbol{y} \mid \boldsymbol{X}\right)} = \frac{1}{M}\sum_{i=1}^{M} \frac{p\left(\boldsymbol{y} \mid \boldsymbol{X}, \boldsymbol{\beta}^{(i)}, \sigma^{(i)}, \boldsymbol{\Omega}^{(i)}\right) p\left(\boldsymbol{\beta}^{(i)}, \sigma^{(i)}, \boldsymbol{\Omega}^{(i)}\right)}{g\left(\boldsymbol{\beta}^{(i)}, \sigma^{(i)}, \boldsymbol{\Omega}^{(i)}\right)}.$$

Specific details for modifying the linear regression model to allow for either heteroscedasticity or first-order serial correlation are given in the following subsections.

Heteroscedasticity

Modification of the linear regression model in Equation (4.1) to account for heteroscedasticity is achieved by defining $\boldsymbol{\Omega} = \text{diag}\left(\tau_1^2, \tau_2^2, \ldots, \tau_n^2\right)$, where, for $i = 1, 2, \ldots, n$,

$$\tau_i^2 = \exp\left(\alpha_1 z_{1,i} + \alpha_2 z_{2,i} + \cdots + \alpha_s z_{s,i}\right), \tag{4.22}$$

where $z_{1,i}$ are exogenous variables; see for instance Judge *et al.* (1988).

To implement Algorithm 1, simply define $\boldsymbol{W} = \text{diag}\left(w_1, w_2, \ldots, w_n\right)$, with $w_i = 1/\tau_i$. In sampling $\boldsymbol{\theta}$, where for the heteroscedasticity case $\boldsymbol{\theta} = (\alpha_1, \alpha_2, \ldots, \alpha_s)$, it is convenient to note that the diagonal structure of $\boldsymbol{W}$ implies that $\mid \boldsymbol{\Omega} \mid = \prod_{i=1}^n \tau_i^2$, resulting in a cheap and simple way of calculating the determinant in Equation (4.21).

Serial correlation

The linear regression model in Equation (4.1) can be modified to account for normally distributed errors, with first-order serial correlation, by assuming, for $i = 1, 2, \ldots, n$,

$$\varepsilon_i = \rho\varepsilon_{i-1} + \nu_i, \quad \nu_i \sim \text{N}\left(0, \sigma^2\right), \tag{4.23}$$

where ρ is a parameter that models spatial dependence and ν_i is an independent identically distributed (iid) random variable that is normally distributed with a mean of 0 and a variance σ^2; see for example Zellner (1971). It is further assumed that the underlying process in Equation (4.23) is stationary, which implies $\mid \rho \mid < 1$. Assuming that the process in Equation (4.23) has been running for n time steps, then it follows that

$$\begin{aligned} p(\boldsymbol{\varepsilon}) &= p(\varepsilon_1) \prod_{i=2}^{n} p(\varepsilon_i) \\ &\propto \sigma^{-n} \exp\left\{-\frac{1}{2\sigma^2}\left[\upsilon_1 \varepsilon_1^2 + \sum_{i=2}^{n} (\varepsilon_i - \rho\varepsilon_{i-1})^2\right]\right\}, \end{aligned}$$

where $\upsilon_1 = \left(1 - \rho^2\right)$. It follows that

$$\begin{aligned} \sum_{i=2}^{n} (\varepsilon_i - \rho\varepsilon_{i-1})^2 &= \begin{bmatrix} 1 & -\rho & 0 & \cdots & 0 \\ 0 & 1 & -\rho & \ddots & \vdots \\ \vdots & \ddots & \ddots & \ddots & 0 \\ 0 & & 0 & 1 & -\rho \end{bmatrix} \begin{bmatrix} \varepsilon_1 \\ \varepsilon_2 \\ \vdots \\ \varepsilon_n \end{bmatrix} \\ &= \boldsymbol{R\varepsilon}. \end{aligned} \tag{4.24}$$

Equation (4.18) and Equation (4.24) imply

$$p(\boldsymbol{\varepsilon}) \propto \sigma^{-n} \exp\left\{-\frac{1}{2\sigma^2}\boldsymbol{\varepsilon}^T \boldsymbol{\Omega}\boldsymbol{\varepsilon}\right\},$$

and it is then clear that $\boldsymbol{\Omega} = \boldsymbol{R}^T \boldsymbol{R} + \boldsymbol{D}$, where

$$\boldsymbol{D} = \begin{bmatrix} v_1 & 0 & \cdots & 0 \\ 0 & 0 & \ddots & \vdots \\ \vdots & \ddots & \ddots & \vdots \\ 0 & \cdots & \cdots & 0 \end{bmatrix}.$$

This implies that

$$\boldsymbol{\Omega} = \begin{bmatrix} 1 & -\rho & 0 & 0 & \cdots & 0 \\ -\rho & 1+\rho^2 & -\rho & 0 & \ddots & 0 \\ 0 & -\rho & 1+\rho^2 & \ddots & \ddots & \vdots \\ 0 & 0 & \ddots & \ddots & -\rho & 0 \\ \vdots & \vdots & \ddots & -\rho & 1+\rho^2 & -\rho \\ 0 & 0 & \cdots & 0 & -\rho & 1 \end{bmatrix}.$$

To implement Algorithm 1, step 1 requires the weight matrix $\boldsymbol{W}$. One possibility is take $\boldsymbol{W}$ as the upper Cholesky triangle in the Cholesky factorization of $\boldsymbol{\Omega}$, that is $\boldsymbol{\Omega} = \boldsymbol{W}^T \boldsymbol{W}$. It can be shown that

$$\boldsymbol{W} = \begin{bmatrix} 1 & -\rho & 0 & 0 & \cdots & 0 \\ 0 & 1 & -\rho & \ddots & \ddots & 0 \\ 0 & 0 & \ddots & \ddots & \ddots & \vdots \\ \vdots & \ddots & \ddots & \ddots & \ddots & 0 \\ \vdots & \ddots & \ddots & \ddots & 1 & -\rho \\ 0 & 0 & \cdots & \cdots & 0 & \sqrt{1-\rho^2} \end{bmatrix}. \tag{4.25}$$

Note that a practical implementation of Algorithm 1 should take advantage of the sparse structure $\boldsymbol{W}$. For $n \gg k$, this implies that step 1 can be computed in $\mathbf{O}(n)$ rather that $\mathbf{O}\left(n^2\right)$ operations. For the practitioner, this means that the estimation for a data set 10 times the size will take 10 times as long, rather than 100 times as long.

To implement step 3 of Algorithm 1, RWMH requires the evaluation of Equation (4.21). Given the form of $\boldsymbol{W}$ in Equation (4.25), it follows that $\mid \boldsymbol{\Omega} \mid^{-1/2} = \sqrt{1-\rho^2}$ and as a consequence Equation (4.21), in this case, has the following simple form:

$$p(\boldsymbol{y} \mid \boldsymbol{X}, \boldsymbol{\beta}, \sigma, \boldsymbol{\Omega}) \propto \frac{\sqrt{1-\rho^2}}{\sigma^n} \exp\left\{-\frac{1}{2}\left(\tilde{\boldsymbol{y}} - \tilde{\boldsymbol{X}}\boldsymbol{\beta}\right)^T \left(\tilde{\boldsymbol{y}} - \tilde{\boldsymbol{X}}\boldsymbol{\beta}\right)\right\}.$$

4.4.4 Variable selection

The variable selection problem in the multiple linear regression framework is one of choosing the 'best' subset, made up of k_γ predictors, given k possible explanatory variables. The difficulty in this problem is that there are 2^k competing models, so for large k traditional methods become computationally infeasible. Two alternative approaches that make use of MCMC machinery, to effectively search the parameter space of possible models in order to find the subset of most probable models, are described. Specifically, the stochastic search variable selection (SSVS) and the Bayesian lasso are detailed.

Stochastic search

SSVS was first proposed in George and McCulloch (1993) and further refined in Geweke (1996) and George and McCulloch (1997), among others. The idea essentially involves augmenting the standard multiple linear regression model with a set of k binary indicator variables that take the value 1 if the corresponding regressor is to be included and the value 0 otherwise. The k indicator variables are treated as unknown and estimated as part of an MCMC sampling scheme.

Denoting the ith indicator variable, for $i = 1, 2, \ldots, k$, as γ_i, then the conditional distribution for β_i can be specified as

$$\beta_i \mid \sigma, \gamma_i \sim (1 - \gamma_i)\,\mathrm{N}\left(0, \kappa_0^{-1}\sigma^2\right) + \gamma_i \mathrm{N}\left(0, \kappa_1^{-1}\sigma^2\right), \tag{4.26}$$

where κ_0 is a precision parameter that is specified to be relatively large (indicating a tight distribution centred about zero) and κ_1 is a precision parameter that is specified to be relatively small. The intuition is that when β_i is small enough to warrant the exclusion of X_i, then γ_i should equal zero, while when β_i is relatively large and should be included, then γ_i should equal one. A Bernoulli prior is assumed for γ_i, such that

$$p(\gamma_i) = p_i^{\gamma_i} (1 - p_i)^{1-\gamma_i}, \tag{4.27}$$

where p_i is the prior probability that $\gamma_i = 1$. Assuming an inverted-gamma prior for σ, following Equation (4.7), and combining with the prior for $\boldsymbol{\beta}$ in Equation (4.26), it is clear that, conditional on $\boldsymbol{\gamma} = (\gamma_1, \gamma_2, \ldots, \gamma_k)^T$, the joint prior is $\boldsymbol{\beta}$ and σ is a normal–inverted-gamma prior, with the prior precision $\underline{\boldsymbol{V}}$ which can be parameterized as

$$\underline{\boldsymbol{V}} = \boldsymbol{D}\boldsymbol{R}^{-1}\boldsymbol{D}. \tag{4.28}$$

Here $\boldsymbol{R}$ is the prior correlation matrix for $\boldsymbol{\beta}$ and $\boldsymbol{D}$ is a diagonal matrix, defined such that the ith diagonal element is as follows:

$$\boldsymbol{D}_{ii}^2 = \begin{cases} \kappa_{0,i} & \text{if } \gamma_i = 0 \\ \kappa_{1,i} & \text{if } \gamma_i = 1. \end{cases}$$

If variable selection is of primary interest, and assuming $\kappa_{0,i} > 0$ and $\kappa_{1,i} > 0$, for $i = 1, 2, \ldots, k$, then the results in Section 4.4.1 can be used in implementing a simple Gibbs sampling scheme. In particular, the jth iteration of a Gibbs sampling scheme that can be used in making inference about the posterior distribution for $\boldsymbol{\gamma}$ is as follows in Algorithm 2.

Algorithm 2: Stochastic search variable selection for the normal linear regression model with spherical errors

---

1. $\gamma_1^{(j)}$ from $p\left(\gamma_1 \mid \mathbf{y}, \mathbf{X}, \gamma_2^{(j-1)}, \gamma_3^{(j-1)}, \ldots, \gamma_k^{(j-1)}\right)$.
2. Sample $\gamma_2^{(j)}$ from $p\left(\gamma_2 \mid \mathbf{y}, \mathbf{X}, \gamma_1^{(j)}, \gamma_3^{(j-1)}, \gamma_4^{(j-1)}, \ldots, \gamma_k^{(j-1)}\right)$.
3. $\vdots \quad \vdots$
4. Sample $\gamma_k^{(j)}$ from $p\left(\gamma_k \mid \mathbf{y}, \mathbf{X}, \gamma_1^{(j)}, \gamma_2^{(j)}, \ldots, \gamma_{k-1}^{(j)}\right)$.

To sample from $p(\gamma_i \mid \mathbf{y}, \mathbf{X}, \boldsymbol{\gamma}_{\setminus i})$, where $\boldsymbol{\gamma}_{\setminus i}$ is the complete set of $\boldsymbol{\gamma}$ excluding γ_i, note that

$$p\left(\gamma_i \mid \mathbf{y}, \mathbf{X}, \boldsymbol{\gamma}_{\setminus i}\right) \propto p\left(\boldsymbol{\gamma} \mid \mathbf{y}, \mathbf{X}\right),$$

so using Equation (4.13) it is clear that

$$p(\gamma_i \mid \mathbf{y}, \mathbf{X}, \boldsymbol{\gamma}_{\setminus i}) \propto \left(\frac{|\mathbf{V}|}{|\overline{\mathbf{V}}|}\right)^{1/2} (\overline{\nu s}^2/2)^{-\overline{\nu}/2}. \tag{4.29}$$

As γ_i is a binary variable, it is straightforward to evaluate $p\left(\boldsymbol{\gamma}_i = 1 \mid \mathbf{y}, \mathbf{X}, \gamma_{\setminus i}\right)$ as

$$p\left(\gamma_i = 1 \mid \mathbf{y}, \mathbf{X}, \boldsymbol{\gamma}_{\setminus i}\right) = \frac{a}{a+b},$$

where a is simply calculated by evaluating Equation (4.29), with $\gamma_i = 1$, and b is calculated by evaluating Equation (4.29), with $\gamma_i = 0$.

To specify $\mathbf{R}$ and $\mathbf{D}$, George and McCulloch (1997) suggest for $\kappa_{0,i} > 0$ and $\kappa_{1,i} > 0$, for $i = 1, 2, \ldots, k$, that both $\mathbf{R} = \mathbf{I}_k$ and $\mathbf{R} = (\mathbf{X}^T\mathbf{X})/g$ are good choices. An alternative choice, also suggested in George and McCulloch (1997), that is commonly implemented is to set $\kappa_{0,i}^{-1} = 0$, and $\kappa_{1,i} > 0$, for $i = 1, 2, \ldots, k$, and to set $\mathbf{R} = (\mathbf{X}^T\mathbf{X})/g$, with $\underline{\boldsymbol{\beta}} = \mathbf{0}$. Under this specification, Equation (4.17) implies that Algorithm 2 is modified such that

$$p(\gamma_i \mid \mathbf{y}, \mathbf{X}, \boldsymbol{\gamma}_{\setminus i}) \propto (1+g)^{-k_\gamma/2}(n\overline{s}^2)^{-n/2} \mid_{\mathbf{X}=\mathbf{X}_\gamma},$$

where $\mathbf{X}_\gamma$ comprises the columns of $\mathbf{X}$, so that column i is included only if $\gamma_i = 1$, for $i = 1, 2, \ldots, k$ and $k_\gamma = \sum_{i=1}^{k} \gamma_i$.

For the normal linear regression model in Equation (4.2), where the error distribution is specified with the general covariance matrix in Equation (4.18), the marginal likelihood is not available in closed form. However, it is possible to design an MCMC algorithm to make inference about the posterior distribution for the complete set of unknowns, $(\boldsymbol{\beta}, \sigma, \boldsymbol{\Omega}, \boldsymbol{\gamma})$, and use standard MCMC results to draw conclusions about the marginal posterior distribution for $\boldsymbol{\gamma}$. Specifically, Algorithm 3 assumes $\kappa_{0,i} > 0$ and $\kappa_{1,i} > 0$, for $i = 1, 2, \ldots, k$, at the jth iteration of the MCMC scheme.

Algorithm 3: Stochastic search variable selection for the normal linear regression model with non-spherical errors

1. Use $\tilde{\boldsymbol{\theta}}^{(j-1)}$ to calculate $\tilde{\boldsymbol{y}} = \boldsymbol{W}\boldsymbol{y}$ and $\tilde{\boldsymbol{X}} = \boldsymbol{W}\boldsymbol{X}$.
2. Sample $\boldsymbol{\beta}^{(j)}$ and $\sigma^{(j)}$ from $p\left(\boldsymbol{\beta}, \sigma \mid \tilde{\boldsymbol{y}}, \tilde{\boldsymbol{X}}, \boldsymbol{\gamma}^{(j-1)}\right)$.
3. Sample $\boldsymbol{\theta}^{(j)}$ from $p\left(\boldsymbol{\theta} \mid \boldsymbol{y}, \boldsymbol{X}, \boldsymbol{\beta}^{(j)}, \sigma^{(j)}, \boldsymbol{\gamma}^{(j-1)}\right)$.
4. Sample $\gamma_1^{(j)}$from $p\left(\gamma_1 \mid \boldsymbol{y}, \boldsymbol{X}, \boldsymbol{\beta}^{(j)}, \sigma^{(j)}, \boldsymbol{\theta}^{(j)}, \gamma_2^{(j-1)}, \gamma_3^{(j-1)}, \ldots, \gamma_k^{(j-1)}\right)$.
5. Sample $\gamma_2^{(j)}$ from $p\left(\gamma_2 \mid \boldsymbol{y}, \boldsymbol{X}, \boldsymbol{\beta}^{(j)}, \sigma^{(j)}, \boldsymbol{\theta}^{(j)}, \gamma_1^{(j)}, \gamma_3^{(j-1)}, \gamma_4^{(j-1)}, \ldots, \gamma_k^{(j-1)}\right)$.
6. $\vdots \qquad \vdots$
7. Sample $\gamma_k^{(j)}$ from $p\left(\gamma_k \mid \boldsymbol{y}, \boldsymbol{X}, \boldsymbol{\beta}^{(j)}, \sigma^{(j)}, \boldsymbol{\theta}^{(j)}, \gamma_1^{(j)}, \gamma_2^{(j)}, \ldots, \gamma_{k-1}^{(j)}\right)$.

Steps 1, 2 and 3 in Algorithm 3 are virtually identical to the steps in Algorithm 1. To sample from the posterior distributions for $\boldsymbol{\gamma}_i$, $\boldsymbol{y}$, $\boldsymbol{X}$, $\boldsymbol{\beta}$, σ, $\boldsymbol{\theta}$ and $\boldsymbol{\gamma}_{\backslash i}$ note that the form of the hierarchical prior in Equation (4.26) implies

$$p\left(\gamma_i \mid \boldsymbol{y}, \boldsymbol{X}, \boldsymbol{\beta}, \sigma, \boldsymbol{\theta}, \boldsymbol{\gamma}_{\backslash i}\right) \propto p\left(\boldsymbol{\beta} \mid \boldsymbol{\gamma}\right) \times p\left(\gamma_i\right) \tag{4.30}$$

where $p(\boldsymbol{\beta} \mid \boldsymbol{\gamma})$ is defined by Equation (4.26), and $p(\gamma_i)$ is defined in Equation (4.27). As for the case with spherical errors, the binary nature of $\boldsymbol{\gamma}$ means that $p(\gamma_i = 1 \mid \boldsymbol{y}, \boldsymbol{X}, \boldsymbol{\beta}, \sigma, \boldsymbol{\theta}, \boldsymbol{\gamma}_{\backslash i})$ is straightforward to evaluate, and in particular

$$p\left(\gamma_i = 1 \mid \boldsymbol{y}, \boldsymbol{X}, \boldsymbol{\beta}, \sigma, \boldsymbol{\theta}, \boldsymbol{\gamma}_{\backslash i}\right) = \frac{c}{c+d},$$

where c is Equation (4.30) evaluated at 1, and d is Equation (4.30) evaluated at 0.

Bayesian lasso

Let us represent the linear regression model as

$$\boldsymbol{y} = \alpha \boldsymbol{1}_N + \boldsymbol{X}\boldsymbol{\beta} + \boldsymbol{\varepsilon}$$

where the constant, also known as the overall mean, is represented as α, $\boldsymbol{X}$ is an $N \times p$ matrix of 'standardized' regressors, and $\boldsymbol{\varepsilon}$ are the usual normal errors.

Variable selection is achieved using the Bayesian lasso approach by

$$\min_{\boldsymbol{\beta}}(\tilde{\boldsymbol{y}} - \boldsymbol{X}\boldsymbol{\beta})^T(\tilde{\boldsymbol{y}} - \boldsymbol{X}\boldsymbol{\beta}) + \lambda \sum_{j=1}^{p} \mid \beta_j \mid$$

for some penalty parameter, λ. Here, $\tilde{\boldsymbol{y}} = \boldsymbol{y} - \alpha \boldsymbol{1}_N$.

As noted by Tibshirani (1996) and implemented by Park and Casella (2008), the lasso estimates can be interpreted as estimated posterior modes when the parameters,

$\boldsymbol{\beta}$, have independent and identical Laplace priors. The fully Bayesian analysis of Park and Casella (2008) used a conditional Laplace prior for the regression coefficients

$$p(\boldsymbol{\beta} \mid \sigma^2) = \prod_{j=1}^{p} \frac{\lambda}{2\sigma} \exp\left\{\frac{-\lambda \mid \beta_j \mid}{\sigma}\right\}.$$

Conditioning $\boldsymbol{\beta}$ on σ^2 guarantees a unimodal full posterior. This is important to the modelling process, as a lack of unimodality tends to slow convergence of MCMC samples and results in point estimates that have wide credible intervals.

The marginal prior for the error variance σ^2 is taken as

$$p(\sigma^2) \propto \sigma^{-2}$$

or alternatively as

$$\sigma^2 \sim \text{IG}(a, b). \tag{4.31}$$

Advantage is then taken of the idea that the Laplace distribution can be represented as a scale mixture of normals, so that the prior for the standardized regression coefficients is taken as a p-dimensional normal distribution

$$\boldsymbol{\beta} \mid \sigma^2, \tau_1^2, \tau_2^2, \ldots, \tau_p^2 \sim \text{N}\left(\mathbf{0}_p, \sigma^2 \mathbf{D}_\tau\right)$$

where $\mathbf{D}_\tau = \text{diag}(\tau_1^1, \tau_2^2, \ldots, \tau_p^2)$.

A joint prior for the variances is taken as

$$p(\sigma^2, \tau_1^2, \tau_2^2, \ldots, \tau_p^2) \propto \sigma^{-2} d\sigma^2 \prod_{j=1}^{p} \frac{\lambda^2}{2} \exp\left(\frac{-\lambda^2 \tau_j^2}{2}\right) d\tau_j^2.$$

A gamma prior is taken on the square of the penalty term such that

$$p(\lambda^2) \sim \text{Gamma}(r, \delta) \tag{4.32}$$

where $r > 0$ and $\delta > 0$.

This prior assumption is convenient in that it is a conjugate prior, and consequently fits naturally in the Gibbs sampling framework given in Algorithm 4.

4.5 Analysis

4.5.1 Case study 1: Boston housing data set

g-prior analysis

Initially, an analysis of the Boston housing data set described in Section 4.2 using a g-prior, see Section 4.4.2, is conducted, as specified by Equations (4.2), (4.3) and (4.16). The parameters $\underline{\boldsymbol{\beta}}$ and g were set to 0 and 1000, respectively, which implies that we believe the prior mean estimate for $\boldsymbol{\beta}$ is not very precise.

Algorithm 4: Gibbs sampler for estimation of parameters in Bayesian lasso model

1. Update $\boldsymbol{\beta} \sim \text{N}\left((\boldsymbol{X}^T\boldsymbol{X} + \boldsymbol{D}_\tau^{-1})\boldsymbol{X}^T\tilde{\boldsymbol{y}}, \sigma^2(\boldsymbol{X}^T\boldsymbol{X} + \boldsymbol{D}_\tau^{-1})\right).^\dagger$
2. Update $\sigma^2 \sim \text{IG}\left(\frac{n-1+p}{2}, \frac{(\tilde{\boldsymbol{y}}-\boldsymbol{X}\boldsymbol{\beta})^T(\tilde{\boldsymbol{y}}-\boldsymbol{X}\boldsymbol{\beta})}{2} + \frac{\boldsymbol{\beta}_j^T\boldsymbol{D}_j^{-1}\boldsymbol{\beta}}{2}\right).^*$
3. Update $\tau_j^{-2} \sim$ Inverse Gaussian $\left(\sqrt{\frac{\lambda^2\sigma^2}{\beta_j^2}}, \lambda^2\right),^\dagger$

   where $f(x) = \sqrt{\frac{\lambda^2}{2\pi}}x^{-3/2}\exp\left[\frac{-\lambda^2\left(x-\sqrt{\lambda^2\sigma^2/\beta_j^2}\right)^2}{2x\lambda^2\sigma^2/\beta_j^2}\right].$
4. Update $\lambda^2 \sim \text{Gamma}\left(p + r, \sum_{j=1}^p \tau_j^2 + \delta\right).$

   † Perform by block updating for efficiency.
   $*$ Based on $p(\sigma^2) \propto \sigma^{-2}$.

Table 4.1 contains output from this analysis. The table reports the variable names in the first column, and then the marginal posterior mean, the posterior standard deviation, and the lower and upper 95% highest posterior density (HPD) intervals, respectively. It is clear that zero is contained in several of the 95% HPD intervals, indicating the possible inclusion of unnecessary variables. Specifically, the HPD intervals for ZN, INDUS, CHAS, AGE, TAX and B all contain zero.

Stochastic search analysis

The output in Table 4.2 is produced using 20 000 iterations of Algorithm 2, where the first 5000 iterations are discarded. The prior hyperparameters were $\kappa_{0,i}^{-1} = 0$

Table 4.1 Output from the analysis of the Boston housing data set.

Variable name	Mean	SD	2.5%	97.5%
CONSTANT	2.8460	0.1673	2.5170	3.1740
CRIM	−0.0118	0.0014	−0.0146	−0.0091
ZN	0.0001	0.0006	−0.0010	0.0012
INDUS	0.0002	0.0026	−0.0050	0.0054
CHAS	0.0913	0.0371	0.0183	0.1643
NOX^2	−0.6374	0.1266	−0.8861	−0.3887
RM^2	0.0063	0.0015	0.0034	0.0092
AGE	0.0001	0.0006	−0.0011	0.0012
log(DIS)	−0.1911	0.0374	−0.2645	−0.1177
log(RAD)	0.0956	0.0214	0.0536	0.1377
TAX	−0.0004	0.0001	−0.0007	−0.0002
PTRATIO	−0.0311	0.0056	−0.0421	−0.0201
B	0.3633	0.1154	0.1367	0.5900
LSTAT	−0.3708	0.0280	−0.4258	−0.3158
σ	0.2045	0.0064	NA	NA

Table 4.2 Output from the analysis of the Boston housing data set using the stochastic search algorithm.

Probability	Model
0.419	0, 1, 5, 6, 8, 9, 10, 11, 12, 13
0.291	0, 1, 4, 5, 6, 8, 9, 10, 11, 12, 13
0.070	0, 1, 5, 6, 8, 9, 10, 11, 13

and $\kappa_{1,i} = 1$, for $i = 1, 2, \ldots, k$. Further, we set $\underline{\beta} = \mathbf{0}$ and $\boldsymbol{R} = (\boldsymbol{X}^T\boldsymbol{X}/g$, with $g = 1000$. The table lists the model probability on the left and the index of the selected variable on the right.

Table 4.3 presents output from the regression analysis of the preferred model from the stochastic search algorithm. As for Table 4.1, the variable names are in the first column and then the marginal posterior mean, marginal posterior standard deviation, and the lower and upper 95% HPD intervals, respectively. Unlike the initial estimates in Table 4.3, none of the HPD intervals contain zero.

Bayesian lasso analysis

The Bayesian lasso analysis is run using the R library *monomvn* (Grammacy 2011). The Gibbs sampler of Algorithm 4 was run for 110 000 iterations, with a burn-in of 10 000 iterations and a thinning of 100 iterations.

The hyperparameters for the prior of the square of the lasso parameter (λ^2) in Equation (4.32) were taken as $r = 0.025$ and $\delta = 0.1$, as this provides a distribution with $E(\lambda^2) = 0.25$ and a variance of 10 times this amount, which is adequate to cover the likely values of this parameter. In this analysis we use the inverse-gamma prior for the model variance, σ^2, defined in Equation (4.31), setting $a = 4$ and $b = 60$, which once again adequately allows for all likely values of σ^2.

Table 4.3 Regression analysis of the preferred model from the stochastic search analysis.

Variable names	Mean	SD	2.5%	97.5%
CONSTANT	2.8600	0.1565	2.5530	3.1680
CRIM	−0.0120	0.0014	−0.0147	−0.0093
NOX^2	−0.6056	0.1231	−0.8475	−0.3637
RM^2	0.0065	0.0014	0.0037	0.0092
log(DIS)	−0.1977	0.0296	−0.2558	−0.1396
log(RAD)	0.0988	0.0205	0.0586	0.1390
TAX	−0.0004	0.0001	−0.0007	−0.0002
PTRATIO	−0.0319	0.0053	−0.0423	−0.0215
B	0.3772	0.1157	0.1499	0.6044
LSTAT	−0.3731	0.0255	−0.4231	−0.3231
σ	0.2058	0.0065	NA	NA

Table 4.4 Standardized coefficient estimates using the Bayesian lasso technique. The p-value represents the proportion of MCMC iterates which are of opposing sign to the mean. The reversible jump (RJ) probability of a coefficient not being equal to 0 is also given.

Variable names	Mean	2.5%	97.5	p-value	RJ probability
CONSTANT	34.6412	24.4827	44.1508	0.0000	1.0000
CRIM	−0.0970	−0.1594	−0.0321	0.0010	0.9425
ZN	0.0413	0.0152	0.0684	0.0030	0.9536
INDUS	0.0005	−0.1084	0.1157	0.4955	0.1853
CHAS	2.7035	0.9523	4.3540	0.0020	0.9519
NOX^2	−16.4163	−23.8022	−9.0102	0.0000	0.9998
RM^2	3.8533	3.0347	4.6269	0.0000	0.9999
AGE	−0.0008	−0.0230	0.0235	0.4685	0.1638
log(DIS)	−1.3978	−1.7854	−0.9880	0.0000	0.9999
log(RAD)	0.2539	0.1196	0.3855	0.0000	0.9928
TAX	−0.0099	−0.0170	−0.0025	0.0030	0.9550
PTRATIO	−0.9335	−1.1969	−0.6801	0.0000	0.9999
B	0.0091	0.0032	0.0145	0.0010	0.9807
LSTAT	−0.5207	−0.6167	−0.4194	0.0000	0.9999

Table 4.4 shows the point estimates and HPD intervals for the regression coefficients using the Bayesian lasso technique. From this table we observe a close agreement in the influence of parameter estimates with that of the stochastic search analysis; however, the variables of ZN and CHAS appear to be more important than in the stochastic search analysis. Both analyses agree that INDUS and AGE are not important variables in the linear regression, as their 95% HPD intervals contain zero.

The *monomvn* library also contains an option to fit the Bayesian lasso model using the reversible jump algorithm. This is an algorithm that allows moves between models of different dimensions. In the case of linear regression, a new model is proposed, with some probability, at each iteration, which will involve either adding or removing a randomly chosen variable. This move is then accepted or rejected according to a specific probabilistic rule; for further details see Robert and Casella (1999). When the reversible jump probabilities are considered, it is highlighted that INDUS and AGE are relatively unimportant, and that ZN and CHAS are the least important of the remaining variables (Table 4.4). As such, the results of the Bayesian lasso regression are in close agreement with the stochastic search analysis.

The coefficients and their 95% HPD intervals are graphed in Figure 4.2 to illustrate the relative size of the coefficients in the model.

Figure 4.3 illustrates the proportion of time the model spends in each state while conducting the Bayesian lasso analysis. It can be seen that a model with 11 regressors is favoured.

Table 4.5 gives the probability of each set of regressors forming the model using the reversible jump algorithm. The favoured model has 11 regressors (plus a constant), and corresponds to the model that excludes the regressors INDUS and AGE. The second favoured model has 12 regressors, and this additional variable is

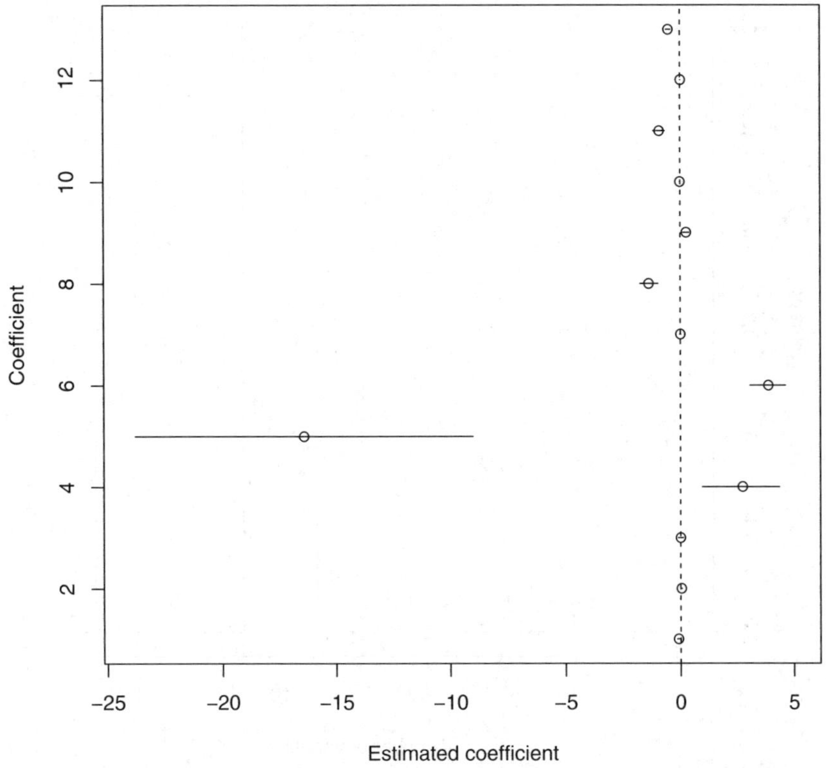

Figure 4.2 Estimated standardized coefficients for Boston housing data regression analysis using the Bayesian lasso. Constant not included in graph.

interchangeable between INDUS and AGE, with neither variable being included more often than the other.

4.5.2 Case study 2: Car production data set

To analyse the car production data set several different model specifications are assumed. The design matrix is constructed to include a constant, a time trend and 11 dummy variables, to account for the differences in the level of production for each month and the base month (October), which is captured via the constant. Further, a dummy variable is included to account for an outlier in November 1977. The models are specified as follows.

Model 1: The normal linear regression model is assumed, with spherical errors. A normal–inverted-gamma prior is assumed, with $\underline{\nu} = 10$, $\underline{\nu s^2} = 0.01$, $\underline{\boldsymbol{\beta}} = \mathbf{0}$ and $\underline{V} = 0.01 \times \boldsymbol{I}_k$.

Model 2: The normal linear regression model is assumed, with non-spherical errors. Specifically, it is assumed there is first-order serial correlation in the residuals.

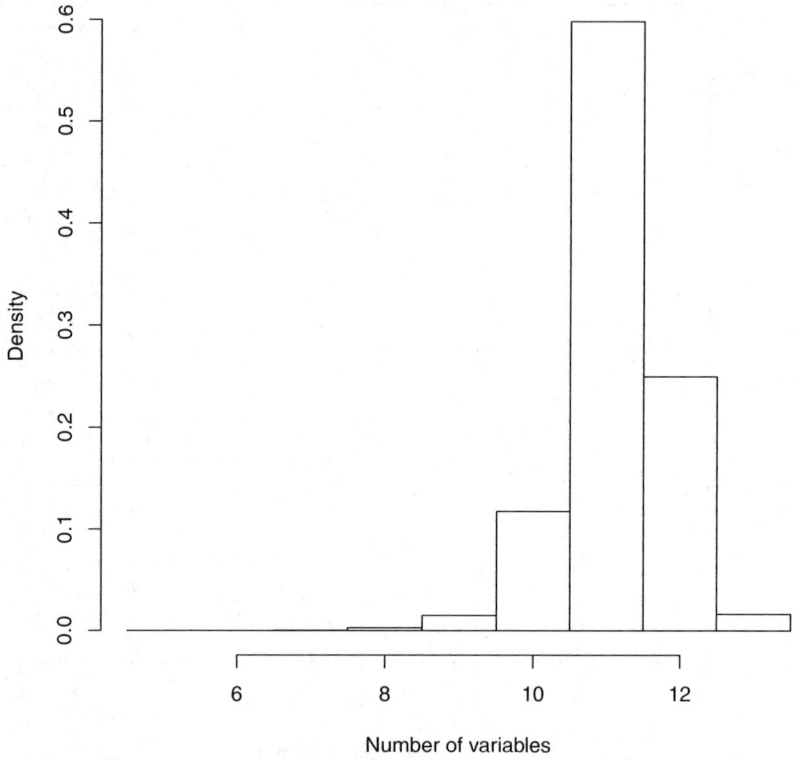

Figure 4.3 Histogram of the number of parameters in the regression model using the reversible jump algorithm.

Table 4.5 Probability of the number of coefficients in the linear regression model using the Bayesian lasso technique and reversible jump. The constant is not included.

Number of variables	Model probability
5	0.000 01
6	0.000 01
7	0.000 32
8	0.00281
9	0.014 90
10	0.117 48
11	0.598 1
12	0.249 7
13	0.016 62

For $\boldsymbol{\beta}$ and σ a normal–inverted-gamma prior is assumed, with $\underline{\nu} = 10$, $\underline{\nu s^2} = 0.01$, $\underline{\boldsymbol{\beta}} = \mathbf{0}$ and $\underline{V} = 0.01 \times \boldsymbol{I}_k$. A uniform prior over the zero one interval is assumed for ρ.

Model 3: Algorithm 2 is applied to the normal linear regression model, with spherical errors, to select a reduced subset of explanatory variables. A normal–inverted-gamma prior is specified, with $\underline{\boldsymbol{V}}$ defined following Equation (4.28), where $\boldsymbol{R} = \boldsymbol{I}_k$, $\kappa_{0,1} = \kappa_{0,2} = \cdots = \kappa_{0,k} = 3.0$, $\kappa_{1,1} = \kappa_{1,2} = \cdots = \kappa_{1,k} = 0.03$, $\nu = 10$ and $\nu s^2 = 0.01$. The normal linear regression model, with spherical errors, is used for estimation, with the reduced set of explanatory variables from the prior analysis. As in Model 1 a normal–inverted-gamma prior is assumed, with $\underline{\nu} = 10$, $\underline{\nu s^2} = 0.01$, $\underline{\boldsymbol{\beta}} = \mathbf{0}$ and $\underline{V} = 0.01 \times \boldsymbol{I}_q$, where q is the number of explanatory variables.

Model 4: Algorithm 2 is applied to the normal linear regression model, with spherical errors, to select a reduced subset of explanatory variables. A normal–inverted-gamma prior is specified, with $\underline{\boldsymbol{V}}$ defined following Equation (4.28), where $\boldsymbol{R} = \boldsymbol{I}_k$, $\kappa_{0,1} = \kappa_{0,2} = \cdots = \kappa_{0,k} = 6.0$, $\kappa_{1,1} = \kappa_{1,2} = \cdots = \kappa_{1,k} = 0.06$, $\nu = 10$ and $\nu s^2 = 0.01$. The normal linear regression model, with first-order serial correlation in the errors is used for estimation, with the reduced set of explanatory variables from the prior analysis. As in Model 2 a normal–inverted-gamma prior is assumed, with $\underline{\nu} = 10$, $\underline{\nu s^2} = 0.01$, $\underline{\boldsymbol{\beta}} = \mathbf{0}$ and $\underline{V} = 0.01 \times \boldsymbol{I}_q$, where q is the number of explanatory variables and a uniform prior over the interval between zero and one is assumed for ρ.

Model 5: Algorithm 3 is applied to the normal linear regression model, with first-order serial correlation in the residuals. A normal–inverted-gamma prior is specified, with $\underline{\boldsymbol{V}}$ defined following Equation (4.28), where $\boldsymbol{R} = \boldsymbol{I}_k$, $\kappa_{0,1} = \kappa_{0,2} = \cdots = \kappa_{0,k} = 6.0$, $\kappa_{1,1} = \kappa_{1,2} = \cdots = \kappa_{1,k} = 0.06$, $\nu = 10$ and $\nu s^2 = 0.01$. The normal linear regression model, with first-order serial correlation in the errors is used for estimation, with the reduced set of explanatory variables from the prior analysis. As in Model 2, a normal–inverted-gamma prior is assumed, with $\underline{\nu} = 10$, $\underline{\nu s^2} = 0.01$, $\underline{\boldsymbol{\beta}} = \mathbf{0}$ and $\underline{V} = 0.01 \times \boldsymbol{I}_q$, where q is the number of explanatory variables and a uniform prior over the interval between zero and one is assumed for ρ.

The analyses undertaken using MCMC are run for 20 000 iterations, with the first 5000 iterations discarded as burn-in.

Table 4.6 reports the estimated model probabilities for each of the five models used in the analysis. The model probabilities are calculated, with the a priori assumption that each of the five models is equally probable. Given the prior assumptions, Model 5 is clearly preferred over the other models, indicating the importance of accounting for serial correlation in the analysis. The explanatory variables included in the analysis in

Table 4.6 Model probabilities for each of the five models used in the estimation of the car production data set.

Model	Model 1	Model 2	Model 3	Model 4	Model 5
Model probability	0.0000	0.0004	0.0000	0.0042	0.9954

Table 4.7 Output from the estimation of Model 5 on the car production data set.

Variable	Mean	SD	2.5%	97.5%	IF
Constant	3.20	0.06	3.09	3.31	3.6
December	−0.69	0.06	−0.81	−0.57	3.8
January	−2.09	0.07	−2.23	−1.95	3.8
February	−0.32	0.06	−0.44	−0.20	3.6
April	−0.60	0.05	−0.70	−0.49	3.6
June	−0.22	0.05	−0.31	−0.11	3.8
Outlier	−1.73	0.27	−2.24	−1.18	3.6
ρ	0.66	0.66	0.57	0.74	3.5
σ	0.32	0.01	0.30	0.35	3.8

the final analysis using Model 5 are a constant and dummy variables for the months of December, January, February, April and June and the dummy variable for the outlier on November 1977.

Table 4.7 reports output from the estimation of Model 5 on the cars and station wagons data set. The first column of the table reports the variable names, while the subsequent columns report the marginal posterior mean, the marginal posterior standard deviation, the lower and upper %95 HPD intervals and the inefficiency factor. The constant could be interpreted as the marginal posterior mean level of production over the time of the analysis for the month of October. The estimated values for the summer months of December, January and February show a drop in production relative to October, which is possibly partially explained by the national holidays during that period. The analysis also finds the modelled outlier is statistically important and indicates a large drop in production for the month of November 1977. It is clear that none of the explanatory variables included in the model contain zero in the estimated 95% HPD interval. The inefficiency factors are all low, indicating that the MCMC algorithm mixes well.

References

George EI and McCulloch RE 1993 Variable selection via Gibbs sampling. *Journal of the American Statistical Association* **88**, 881–889.

George EI and McCulloch RE 1997 Approaches for Bayesian variable selection. *Statistica Sinica* **7**, 339–373.

Geweke J 1996 Variable selection and model comparison in regression. In *Bayesian Statistics 5*, pp. 609–620. Oxford University Press, Oxford.

Gilley OW and Pace RK 1996 On the Harrison and Rubinfeld data. *Journal of Environmental Economics and Management* **31**, 403–405.

Grammacy R 2011 Package ‘monomvn’.

Harrison D and Rubinfeld DL 1978 Hedonic housing prices and the demand for clean air. *Journal of Environmental Economics and Management* **5**, 81–102.

Judge GG, Hill RC, Griffiths WE, Lütkepohl H and Lee TC 1988 *Introduction to Theory and Practice of Econometrics*. John Wiley & Sons, Ltd, Chichester.

Koop G 2003 *Bayesian Econometrics*. John Wiley & Sons, Ltd, Chichester.

Marin JM and Robert CP 2007 *Bayesian Core*. Springer, Berlin.

Park T and Casella G 2008 The Bayesian lasso. *Journal of the American Statistical Association* **103**(482), 681–686.

Robert CP and Casella G 1999 *Monte Carlo Statistical Methods*. Springer, New York.

Tibshirani R 1996 Regression shrinkage and selection via the lasso. *Journal of the Royal Statistical Society, Series B* **583**, 267–288.

Zellner A 1971 *An Introduction to Bayesian Inference in Econometrics*. John Wiley & Sons, Inc., New York.

Zellner A 1986 On assessing prior distributions and Bayesian regression analysis with g-prior distributions. In *Bayesian Inference and Decision Techniques: Essays in Honor of Bruno de Finetti*. Elsevier, Amsterdam.

5

Adapting ICU mortality models for local data: A Bayesian approach

Petra L. Graham,[1] Kerrie L. Mengersen[2] and David A. Cook[3]

[1]*Macquarie University, North Ryde, Australia*
[2]*Queensland University of Technology, Brisbane, Australia*
[3]*Princess Alexandra Hospital, Brisbane, Australia*

5.1 Introduction

Risk adjustment is a method of controlling for casemix and severity of illness in a sample of patients using a statistical model to estimate the probability of death. For example, risk adjustment may be performed within a hospital to facilitate comparable care across a patient mix, or so that the quality of care given by different hospitals or units may be compared on an equal footing. In the intensive care setting models adjust for age, severity of illness, chronic health and reason for admission. Such variables are generally objective characteristics of the patient and not characteristics attributable to the unit involved (Iezzoni 1997; Zaslavsky 2001).

Case Studies in Bayesian Statistical Modelling and Analysis, First Edition. Edited by Clair L. Alston, Kerrie L. Mengersen and Anthony N. Pettitt.

5.2 Case study: Updating a known risk-adjustment model for local use

The Acute Physiology and Chronic Health Evaluation III (APACHE® III) system[1] (Knaus *et al.* 1991, 1993; Wagner *et al.* 1994; Zimmerman 1989) is one of several available risk-adjustment tools and is used by medical practitioners to predict hospital or intensive care unit (ICU) mortality and to assess the severity of illness of their patients. It is a logistic regression model developed from a database of over 17 000 patients from 40 US hospitals. The APACHE III system is widely considered to be a 'gold-standard' mortality prediction system. Validation on independent US (Sirio *et al.* 1999; Zimmerman *et al.* 1998) and Australian (Cook 2000) data showed that the model performed well in terms of goodness of fit of the model (calibration) and discrimination. However, risk-adjustment tools in general may not perform well in data sets beyond those for which the tools were developed (Glance *et al.* 2002; Pillai *et al.* 1999). This argument is supported by several international studies (Bastos *et al.* 1996; Beck *et al.* 1997; Markgraf *et al.* 2000; Pappachan *et al.* 1999) in which the APACHE III system was found to have poor model calibration, with observed hospital mortality higher than predicted.

Risk-adjustment models exhibit poor calibration for several reasons. The hospitals involved could have differences in the quality of care or the risk-adjustment tool does not account adequately for features in the local casemix that differ from the model development data (Glance *et al.* 2002; Iezzoni 1997). In intensive care, a typical response to poor model calibration is to examine the process of care, data collection and the model. As a result of this investigation, the medical practitioner may place less trust in risk-adjustment models or may search among the available risk-adjustment tools to find one that better meets expectations or need. Alternatively the practitioner may build a new risk-adjustment tool *de novo*, ignoring the substantial information available in models already at hand. However, building new models poses special problems whereby small sample sizes, data quality, variable selection, rare events and other issues must be considered carefully. The new model must pass tests of generalizability, validation and acceptance by the wider community.

An alternative to building entirely new models is to update the original model in light of local data through Bayesian logistic regression. We propose that a model may be adapted for use in local prediction through Bayesian logistic regression with coefficients from the APACHE III model used as informative priors. Bayesian techniques are well established for modelling the uncertainty inherent in most data sets (see Gelman *et al.* (2003) or Congdon (2010), for example). Parmigiani (2002) provides a useful guide to Bayesian approaches in medical decision making. Bedrick *et al.* (1997) used a Bayesian logistic model to predict survival at a trauma centre with prior probabilities relating to patient condition elicited from a trauma surgeon. Pliam et al. (1997) compared several mortality prediction tools for coronary bypass surgery patients with locally built Bayesian and frequentist models. Although not compared

[1] APACHE is a registered trademark of Cerner Corporation, Kansas, Missouri, USA.

with respect to goodness of fit, all models discriminated well but demonstrated slight differences with respect to their predictions of overall mean mortality. Tekkis *et al.* (2002) developed an empirical Bayesian and a frequentist logistic regression tool for predicting outcome after oesophagogastric surgery. Both models discriminated reasonably well but the calibration of the Bayesian model was poor. The purpose of this study is to examine the performance of an updated model using informative priors based on a published risk-adjustment model.

5.3 Models and methods

The data used in this study comprise simulated admissions to an Australian Hospital over a 5 year period. The data were simulated to be representative of real data from an Australian hospital. As such, data for 5278 elective, surgical and non-surgical adult emergency admissions were simulated according to the rules of APACHE III (Knaus *et al.* 1991; Zimmerman 1989) and included all variables necessary for calculation of the unadjusted APACHE III hospital mortality predictions. Included in this data set were variables associated with patient demographics, time from hospital admission until ICU admission, initial diagnosis or disease group and the outcome on discharge from hospital. No patients were simulated to have readmissions.

The APACHE III model has been regularly updated with revised coefficients since its original publication. The coefficients used in this study come from the APACHE® III-J version. The coefficients and details of the model were available at initiation of this study from the Cerner APACHE website (www.apache-msi.com) but are no longer available. From this point forward reference to the APACHE III model implies APACHE® III-J.

For each simulated patient in this study, APACHE III mortality predictions were calculated based on the APACHE III coefficients. These coefficients may be seen in the second column of Table 5.1. The logistic equation comprises coefficients for a categorical age variable, an acute physiology score and a four-level variable derived from this score, the source of the patient admission (i.e. from another hospital, from the emergency room and so on), the pre-ICU length of stay, the admitting diagnosis and whether or not the admission was for emergency surgery. Also, for all emergency (surgical and non-surgical) patients a coefficient representing their most serious comorbid or chronic health item (where, except for *none* the chronic health items are ordered in Table 5.1 from most serious to least serious).

For these calculations, the admitting diagnosis and chronic health items presented minor problems since some of the required information was not available for local simulated patients. With the chronic health items it was not possible to tell the difference between immunosuppression and leukaemia/ multiple myeloma, since these two outcomes share the same code in Australian medical data management systems. Thus for patients in the simulated data set with these chronic health items, an average of the two coefficients was used as the best guess of a coefficient for these combined items. Similarly for patients with a simulated admitting diagnosis of either acute myocardial infarction or trauma, additional information about their treatment

Table 5.1 Coefficients of the logistic regression models.

Variable		APACHE III-J[a] coefficient	Posterior mean[b] (95% credible interval)	MLE[c]
Intercept		−6.413	−6.167 (−6.538, −5.792)	−5.652
Age	≤44	0	0	0
	45–54	0.343	0.245 (-0.029, 0.452)	0.277
	55–59	0.321	0.305 (-0.009, 0.613)	0.337
	60–64	0.613	0.602 (0.300, 0.901)	0.624
	65–69	0.757	0.793 (0.524, 1.059)	0.823
	70–74	1.006	0.925 (0.645, 1.203)	0.857
	75-84	1.127	1.223 (0.964, 1.484)	1.256
	≥85	1.495	1.694 (1.329, 2.055)	2.056
Comorbidity[d]	None	0	0	0
	AIDS	1.024	1.024 (0.586, 1.463)	NA
	Hepatic failure	1.133	0.833 (0.455, 1.211)	−0.029
	Lymphoma	1.005	0.929 (0.516, 1.343)	0.197
	Metastatic cancer	0.886	0.724 (0.364, 1.082)	0.329
	Leukaemia/immun.[e]	0.539	0.412 (0.108, 0.710)	0.210
	Cirrhosis	0.860	0.702 (0.340, 1.060)	0.366
Acute Physiology Score (APS)		0.088	0.087 (0.073, 0.100)	0.079
Splined APS	1	−0.309	−0.402 (−0.540, −0.259)	−0.434
	2	0.513	0.711 (0.453, 0.960)	0.819
	3	−0.353	−0.502 (−0.739, −0.267)	−0.681
	4	0.451	0.355 (−0.047, 0.758)	0.431
Admission source	ER/ Direct admit	0	0	0
	OR/RR	−0.238	−0.474 (−0.717, −0.232)	−0.707
	Floor	0.048	0.209 (−0.081, 0.497)	0.133
	Other hospital	0.206	0.181 (−0.091, 0.452)	0.298
Emergency surgery=True		0.079	0.375 (0.098, 0.648)	0.614
Pre-ICU length of stay[f]		0.141	0.121 (0.005, 0.239)	0.118
Diagnosis		Not shown[g]	NA[h]	NA[h]
Disease group risk category	Low	—[h]	0	0
	Neutral	0.189[h]	0.252 (−0.012, 0.516)	0.273
	High	1.118[h]	1.081 (0.862, 1.302)	1.008

[a]Based on external population only. Coefficients as reported on www.apache-msi.com.
[b]Based on Bayesian analysis of the simulated local population using APACHE III-J estimates as priors. All variances, σ_j^2, are set equal to 0.05.
[c]Based on frequentist analysis of the simulated data only.
[d]Coefficient was applied to surgical and non-surgical emergency patients only.
[e]The average of the leukaemia/multiple myeloma and immunosuppression coefficients was used as a best guess for this coefficient.
[f]This coefficient was multiplied by the square root of the number of days between hospital and ICU admission.
[g]There are too many diagnosis-related groups to be shown here. Coefficients for these were used in calculating the APACHE III hospital mortality predictions.
[h]This coefficient is based on Graham and Cook (2004). It has not been obtained from APACHE.

was needed to be able to apply appropriate coefficients. Since this information was not available in the data set, an average of the relevant coefficients was used to give approximate APACHE III mortality predictions.

The data were then split into a training set containing 3708 randomly selected cases (approximately 70% of the data) and a test set containing the remaining 1570 cases. The training set was then used to build a Bayesian logistic regression equation in which the APACHE III coefficients, described in the previous paragraphs, were employed as parameters for the means in priors for the coefficients of the new model. The only exception to this is that alternative means were used in priors for the admitting diagnoses in order to simplify the modelling. This involved replacing the coefficients for about 80 diagnostic categories with coefficients based on a simpler, three-level categorical variable. This variable, developed from the data in Table 3 of Knaus *et al.* (1991) and known as disease group risk category, is described in detail in Graham and Cook (2004). Means for these priors are based on the second model that Graham and Cook describe. The test set was used to validate the Bayesian model. This validation provides an assessment of model reproducibility (Justice *et al.* 1999).

The Bayesian logistic model used here has the familiar form

$$\begin{aligned} y_i | p_i &\sim \text{Bernoulli}(p_i), i = 1, \ldots, n \\ \text{logit}(p_i) &= \beta_0 + \beta_1 x_1 + \cdots + \beta_m x_m \\ \beta_j &\sim \text{Normal}(c_j, \sigma_j^2), j = 0, \ldots, m \end{aligned} \tag{5.1}$$

where y_i is the binary outcome (death or survival) simulated for each patient at discharge from hospital, $\text{logit}(p_i) = \log(p_i/(1 - p_i))$, c_j is the jth ($j = 0, \ldots, m$) APACHE III coefficient and σ_j^2 is the specified variance. Note that these variances were not available in the public information accessed. As such, for each of the coefficients a variance of 0.05 was chosen. This choice reflects our confidence that although local variation might be seen, the coefficients for our model should not differ greatly from the original APACHE model. This choice also acknowledges the large size of the APACHE III database.

The Bayesian analysis using Gibbs sampling (WinBUGS, Spiegelhalter *et al.* (2003)) package was used to perform the Markov chain Monte Carlo simulations required for solving this model. Three chains were run and the initial 10 000 iterations were discarded as burn-in and a further 10 000 iterations were used to generate approximate posterior distributions for these coefficients. Convergence was assessed using the Brooks, Gelman and Rubin multivariate potential scale reduction factor and passed this test unless otherwise stated. The posterior distributions of the probability of death, p_i, for each patient were also calculated using 10 000 iterations after burn-in.

The APACHE III and updated Bayesian models were initially assessed using frequentist calibration and discrimination methods. Calibration or model fit was assessed using Hosmer–Lemeshow $\hat{C}$ statistics (Hosmer and Lemeshow 2000). Here the admissions were divided into 10 groups using equally spaced cut points based on the predicted risk of death (we call these ROD groups). The agreement between predicted and observed mortality rates in these groups was assessed by calculating the Pearson chi-squared statistic from the 2×10 table of observed and estimated expected frequency of deaths and survivals. All of the data were used in assessments of the APACHE III model and the training and test sets were assessed separately for the updated Bayesian model. For the $\hat{C}$ statistic, 8 degrees of freedom were used for the training data and 10 degrees of freedom were used for both the APACHE III validation and the updated Bayesian model test data (Hosmer and Lemeshow 2000). The null hypothesis of no difference between observed and expected mortality rates was rejected if the p-value was less than 0.05.

Discrimination was assessed by calculating the area under the receiver operating characteristic (ROC) curves and is presented with 95% confidence intervals (Hanley and McNeil 1982). The area under the ROC curve estimates the probability that a randomly selected mortality will be given a higher risk of death estimate than a randomly selected survivor. It is a global measure of the ability of the model to assign a higher risk of death to patients who actually die (Justice *et al.* 1999).

In the Bayesian framework, alternative diagnostic methods are preferable for assessing the goodness of fit of the models. These diagnostics involve posterior predictive checks such as those suggested by Gelman *et al.* (2000, 2003). These checks involve predicting the unknown but observable outcome, $\tilde{y}$, given the known outcome y using new data or predicting the result (when new data are unavailable) in replicated data $y^{\text{rep}} = y_1^{\text{rep}}, \ldots, y_n^{\text{rep}}$ and comparing the results with that of the observed data. Replicated data are literally replications of the data used to build the model. Replicated data are not ideal since they are likely to produce optimistic results. This is because the combinations of coefficients that the predictions are based on are already known to the model. For the new data we can write the predictive distribution as (Gelman *et al.* 2003)

$$P(\tilde{y}|y) = \int P(\tilde{y}|\beta)P(\beta|y)d\beta \tag{5.2}$$

where $y = (y_1, \ldots, y_n)$ are the discrete observed data and β is a vector of the parameters of the logistic model. For the replicated data the $\tilde{y}$ in Equation (5.2) is replaced by y^{rep}.

Naturally the test data are used as new data in order to independently assess the goodness of fit of the model. For completeness, we also examined the goodness of fit in the training set using replicated data.

Here, in each of $l = 1, \ldots, 10\,000$ draws from the posterior predictive distribution,

$$I(y_i^{\text{rep}\,l}) = \begin{cases} 1 \text{ if } y_i^{\text{rep}\,l} = y_i \\ 0 \text{ otherwise} \end{cases}$$

was calculated for each patient. Thus the probability of correct prediction, d, for each patient i was equal to the average number of times that the outcome is correctly predicted, that is

$$d_i = \frac{\sum_{l=1}^{10\,000} I(y_i^{\text{rep}\,l})}{10\,000}. \tag{5.3}$$

Similar summaries were calculated for the test data, $\tilde{y}$.

Lastly, in practice it may be preferable to update the model on a more regular basis so that the model consistently reflects changes in the patient mix. To determine the feasibility of updating the model every 12 months or so, the Bayesian modelling was repeated for each of the 5 years of simulated data using all of the data in each year to update the model. Parameters for the means of priors for the first year were based on the APACHE III model and for subsequent years parameters for the means and variances of priors for coefficients were based on the model from the previous year. This provides a dynamically updating model in which the changing coefficients can be monitored to determine how the mix of cases is varying over time. Since all of the data are used in each year, Bayesian model checking could only be performed on the replicated data. As previously described, posterior summaries are based on 10 000 iterations after burn-in.

5.4 Data analysis and results

5.4.1 Updating using the training data

A summary of the simulated data is shown in Table 5.2. The percentages in both the training and test sets were similar for the characteristics shown.

Table 5.1 gives the posterior means and 95% credible intervals for the coefficients defined using a Bayesian logistic analysis of the local (training) data with priors informed by the APACHE III model (third column). For comparison, the APACHE III coefficients, as publicly reported, are also given in Table 5.1 (second column).

Table 5.2 Summary characteristics of the simulated data set.

Variable	Category	Developmental data ($n = 3708$)	Validation data ($n = 1570$)
Sex	Male	2312 (62.4)	981 (62.5)
Admission source	ER/direct admit	822 (22.2)	280 (17.8)
	OR/RR	2056 (55.4)	920 (58.6)
	Floor	359 (9.7)	147 (9.4)
	Other hospital	471 (12.7)	223 (14.2)
Admission type	Elective	1576 (42.5)	677 (43.1)
	Surgical emergency	480 (12.9)	243 (15.5)
	Non-surgical emergency	1652 (44.6)	650 (41.4)
Disease group risk category	Low	2058 (55.5)	891 (56.8)
	Neutral	608 (16.4)	262 (16.7)
	High	1042 (28.1)	417 (26.6)
Age	≤44	1227 (33.1)	519 (33.1)
	45–54	537 (14.5)	260 (16.6)
	55–59	307 (8.3)	129 (8.2)
	60–64	337 (9.1)	140 (8.9)
	65–69	438 (11.8)	177 (11.3)
	70–74	368 (8.8)	160 (10.2)
	75–84	427 (10.3)	161 (10.3)
	≥85	67 (1.2)	24 (1.5)
Hospital outcome	Dead	576 (15.5)	204 (13.0)

Values in parentheses are percentages.

Confidence intervals or variances of these estimates were not made public. Also provided are coefficients based on a standard (maximum likelihood) logistic regression model (fourth column) using only the simulated local training data. Note that both the APACHE III and the coefficients in the fourth column are maximum likelihood estimates.

The values in Table 5.1 show that the posterior means resulting from inclusion of local data are not a lot different compared with the APACHE III coefficients for most of the variables. For example, the locally updated model has coefficients for the comorbidities (except for AIDS) that are a little smaller than that of APACHE III although the credible intervals still contain the original APACHE III values. On the other hand, the coefficient for patients who received emergency surgery is somewhat larger in the locally updated Bayesian model with a credible interval that suggested a 95% probability that the coefficient is between 0.098 and 0.648.

In contrast, the coefficients from the maximum likelihood model (fourth column of Table 5.1) sometimes differ quite a lot from the APACHE III coefficients. For example, the coefficients for the oldest age group and the emergency surgery patients are substantially larger in the maximum likelihood model compared with the APACHE III model.

Frequentist analysis of the goodness of fit or calibration of the logistic models is shown in the second column of Table 5.3. The values suggest that the original APACHE III model calibrated poorly to the simulated data since the p-value for the Hosmer–Lemeshow test was much smaller than 0.05. Moreover, comparison of the observed and predicted values showed that the APACHE III model underpredicted deaths in almost every ROD group. In contrast, the results for both the developmental

Table 5.3 Calibration and discrimination of the logistic models.

	APACHE III ($n = 5278$)		Bayesian developmental ($n = 3708$)		Bayesian test ($n = 1570$)	
Calibration ROD group	Observed mortality	Predicted mortality	Observed mortality	Predicted mortality	Observed mortality	Predicted mortality
0-0.1	161	104.79	78	81.99	35	36.61
0.1–0.2	107	76.65	76	67.99	27	30.65
0.2–0.3	95	65.85	65	58.78	16	20.26
0.3–0.4	88	64.63	48	49.85	18	20.35
0.4–0.5	51	47.15	57	55.03	17	18.75
0.5–0.6	60	54.39	68	56.86	16	17.27
0.6–0.7	63	64.13	35	40.92	16	19.73
0.7–0.8	67	59.98	62	65.94	14	15.01
0.8–0.9	49	44.52	43	48.30	25	27.42
0.9–1.0	39	39.50	44	45.22	20	21.63
$\hat{C}$ (p-value)	83.1 (0.000)		14.9 (0.06)		8.6 (0.57)	
Discrimination	APACHE III		Bayesian developmental		Bayesian test	
ROC curve area (95% CI)	0.886 (0.874, 0.889)		0.879 (0.864, 0.894)		0.863 (0.834, 0.893)	

Table 5.4 Summary of the Bayesian predictive probabilities.

	Training (replicated) data		Test (new) data	
ROD group	Survivors n (% $d_i > 0.5$)	Non-survivors n (% $d_i > 0.5$)	Survivors n (% $d_i > 0.5$)	Non-survivors n (% $d_i > 0.5$)
0.0–0.1	2280 (100.0)	78 (0)	1000 (100)	35 (0)
0.1–0.2	406 (100.0)	76 (0)	189 (100)	27 (0)
0.2–0.3	175 (100.0)	65 (0)	66 (100)	16 (0)
0.3–0.4	97 (100.0)	48 (0)	40 (100)	18 (0)
0.4–0.5	65 (96.9)	57 (3.5)	26 (100)	17 (0)
0.5–0.6	37 (0)	68 (98.5)	15 (0)	16 (100)
0.6–0.7	28 (0)	35 (100)	14 (0)	16 (100)
0.7–0.8	26 (0)	62 (100)	6 (0)	14 (100)
0.8–0.9	14 (0)	43 (100)	7 (0)	25 (100)
0.9–1.0	4 (0)	44 (100)	3 (0)	20 (100)

n = total number of patients in each ROD group

and test data suggest that the updated model (using both APACHE III and local data) calibrated well (since both p-values are larger than 0.05), although the developmental set shows some evidence of lack of fit since the p-value is only 0.06. The updated model slightly overpredicted mortality in the upper ROD groups although this discrepancy is not statistically significant at the 5% level. This indicates that the Bayesian updating has, in effect, recalibrated the APACHE III model so that it better fits the local data.

The ROC curve analysis, shown in the last three rows of Table 5.3, indicates that both models have excellent discrimination. The width of the intervals for the training set and test sets were wider than the APACHE III model providing an appropriate reflection of the uncertainty in the local data.

A summary of the Bayesian predictive probabilities is shown in Table 5.4. Here the number of patients in each ROD group in the Bayesian logistic model for survivors and non-survivors together with the percentage of these patients correctly predicted at least 50% of the time is shown. The table shows for both the training and test data that all of the survivors were correctly predicted to survive more than 50% of the time in the lower probability groups but none of the survivors were correctly predicted more than 50% of the time in the higher probability groups. Similarly, none of the non-survivors were correctly predicted to survive more than 50% of the time in the lower ROD groups but all of the non-survivors were correctly predicted more than 50% of the time in the higher probability groups.

5.4.2 Updating the model yearly

Calibration and discrimination results for the yearly updates may be seen in Table 5.5. Results indicate that calibration for each year was good although the first and last years indicate weak evidence of a lack of calibration. The ROC curve area

Table 5.5 Summary of the calibration and discrimination for each year.

Year	Calibration $\hat{C}$ (p-value)	Discrimination ROC curve area (95% CI)
1	15.41 (0.05)	0.78 (0.74, 0.83)
2	3.99 (0.86)	0.71 (0.67, 0.75)
3	3.45 (0.88)	0.57 (0.52, 0.61)
4	7.99 (0.43)	0.78 (0.74, 0.83)
5	14.22 (0.07)	0.83 (0.79, 0.88)

indicates reasonable discrimination except for the third year where the discrimination is quite poor.

A plot of 95% credible intervals for a selection of the coefficients over time is shown in Figure 5.1. Here the lowermost credible interval for each coefficient was for the first year and the uppermost credible interval was for the final year. The posterior distributions for many coefficients did not vary a great deal from year to year and are not presented. For the coefficients presented, point estimates moved away from the original parameter estimate by a small consistent amount over time.

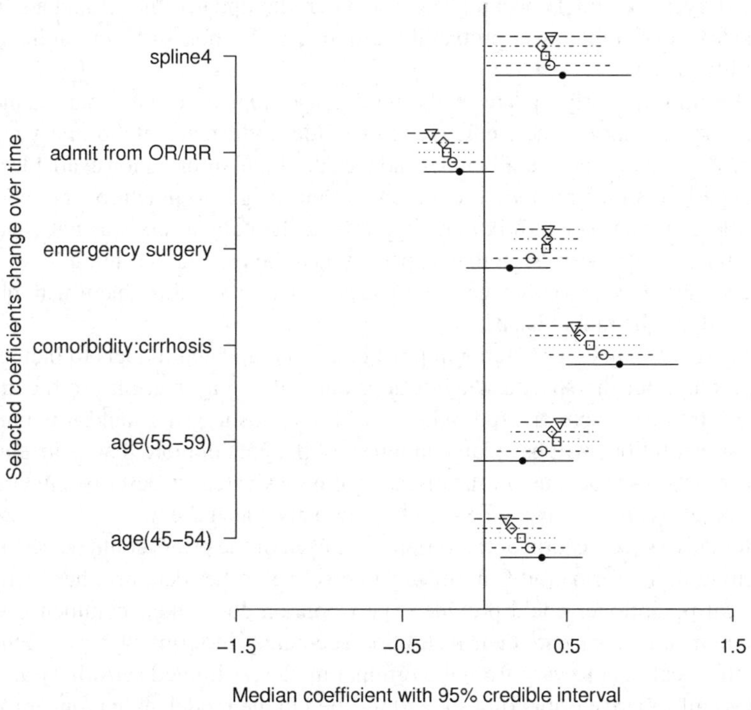

Figure 5.1 The 95% credible intervals for selected coefficients over time. Lowermost interval for each coefficient is for first year, uppermost interval is for final year.

Examination of the performance of the updated models in each year through the Bayesian predictive probabilities (results not shown) indicated that the Bayesian goodness of fit was approximately the same as the overall model for most years. An exception to this was in the third year where at least 60% of patients were correctly predicted at least 50% of the time in most of the deciles of risk.

Tips and Tricks: WinBUGS can produce errors (Trap 66) when running these models. One possible fix is to use a larger value for the precision or to ensure that initial values are chosen so that they are not in the tails of the relevant distributions from which they were generated.

5.5 Discussion

This chapter has presented a method for Bayesian updating of the original APACHE III model by using its coefficients as parameters for priors in a new model. This has resulted in an improved discriminatory model for local use. Frequentist model checks (for the model using all of the years together) suggested that the updated model has reasonable or good calibration where the original model alone calibrated poorly. The discrimination of both the APACHE III and Bayesian models was competitive. Bayesian model checking suggested that the updated model had a good fit for the survivors in the lower probability groups and for non-survivors in the upper probability groups.

Performing a yearly update of the model also appears to be a reasonable approach. The summary of the predictive probabilities indicated that the yearly models performed as well as the overall model and the additional information available about the change in coefficients may provide important insight to practitioners about the constantly evolving patient mix in their wards. If the patient mix was not changing and yet more deaths were seen than expected then the practitioner might have more confidence that the reason for the discrepancy is due to a real problem with quality of care and not the model itself.

Although it may be preferable in practice for the updated models to include the information about the specific diagnostic groups, this may present a difficulty for many institutions. There are approximately 80 diagnostic groups and it is unlikely that most institutions will have large numbers of patients in more than a dozen or so of these groups. As such, use of the disease group risk category described in Graham and Cook (2004) may prove to be a useful alternative, as it did here.

Alternative specification of the variance for each of the priors could be performed by eliciting this information from medical personnel, other data or other literature. A medical practitioner could provide expert opinion about their confidence that a given coefficient represents their institution accurately in terms of how much they expect the local data to vary from the original model. A limited sensitivity analysis using less informative priors (not shown) resulted in the model taking longer to converge although the resulting coefficients were similar and the predictive probabilities appeared to be close to that of the model that used more informative priors.

An additional approach for specifying informative priors for regression coefficients is described by Bedrick *et al.* (1996) for generalized linear models including logistic models. Their method involves specifying a prior distribution for the mean of potential observations at different locations of the predictor space and inferring the required priors on the model coefficients. This approach might be preferable if particular combinations of predictors are perhaps poorly accounted for in the original model as compared with the local data.

Future examination of the updating approach by using coefficients from other published risk-adjustment models and a more comprehensive sensitivity analysis would also be useful to determine whether similar results are obtained.

The methods used here are only appropriate if the aim of the analysis was to predict mortality in the hospital in which the updating has been performed. If interest was in comparing the local hospital with an international gold standard then recalibration will mask any differences therein.

References

Bastos P, Sun X, Wagner D, Zimmerman J, Gomes D, Coelho C, deSouza P, Lima R, Passos J, Livianu J, Dias M, Terzi R, Rocha M and Vieira S 1996 Application of the APACHE III prognostic system in Brazilian intensive care units: a prospective multicenter study. *Intensive Care Medicine* **22**, 564–570.

Beck D, Taylor B, Millar B and Smith G 1997 Prediction of outcome from intensive care: a prospective cohort study comparing APACHE II and III prognostic systems in a United Kingdom intensive care unit. *Critical Care Medicine* **25**, 9–15.

Bedrick E, Christensen R and Johnson W 1996 A new perspective on priors for generalized linear models.. *Journal of the American Statistical Association* **91**, 1450–1460.

Bedrick E, Christensen R and Johnson W 1997 Bayesian binomial regression: predicting survival at a trauma center. *The American Statistician* **51**, 211–218.

Congdon P 2010 *Applied Bayesian Hierarchical Methods*. Chapman & Hall/CRC Press, Boca Raton, FL.

Cook D 2000 Performance of APACHE III models in an Australian ICU. *Chest* **118**, 1732–1738.

Gelman A, Goegebeur Y, Tuerlinckx F and Mechelen IV 2000 Diagnostic checks for discrete-data regression models using posterior predictive simulations. *Journal of the Royal Statistical Society, Series C – Applied Statistics* **49**, 247–268.

Gelman A, Carlin J, Stern H and Rubin D 2003 *Bayesian Data Analysis*, 2nd edn. Chapman & Hall/CRC Press, Boca Raton, FL.

Glance L, Osler T and Dick A 2002 Rating the quality of intensive care units: is it a function of the intensive care unit scoring system? *Critical Care Medicine* **30**, 1976–1982.

Graham P and Cook D 2004 Prediction of risk of death using thirty-day outcome: a practical endpoint for quality auditing in intensive care. *Chest* **125**, 1458–1466.

Hanley J and McNeil B 1982 The meaning and use of a ROC curve. *Radiology* **143**, 29–36.

Hosmer D and Lemeshow S 2000 *Applied Logistic Regression*, 2nd edn. John Wiley & Sons, Inc., New York.

Iezzoni L 1997 The risks of risk adjustment. *Journal of the American Medical Association* **278**, 1600–1607.

Justice A, Covinsky K and Berlin J 1999 Assessing the generalizability of prognostic information. *Annals of Internal Medicine* **130**, 515–524.

Knaus W, Wagner D, Draper E, Zimmerman J, Bergner M, Bastos P, Sirio C, Murphy D, Lotring T, Damiano A and Harrell F 1991 The APACHE III prognostic system: risk prediction of hospital mortality for critically ill hospitalized adults. *Chest* **100**, 1619–1636.

Knaus W, Wagner D, Zimmerman J and Draper E 1993 Variations in hospital mortality and length of stay from intensive care. *Annals of Internal Medicine* **118**, 753–761.

Markgraf R, Deutschinoff G, Pientka L and Scholten T 2000 Comparison of acute physiology and chronic health evaluations II and III and simplified acute physiology score II: a prospective cohort study evaluating these methods to predict outcome in a German interdisciplinary intensive care unit. *Critical Care Medicine* **28**, 26–33.

Pappachan J, Millar B, Bennett D and Smith G 1999 Comparison of outcome from intensive care admission after adjustment for case mix by the APACHE III prognostic system. *Chest* **115**, 802–810.

Parmigiani G 2002 *Modeling in Medical Decision Making: A Bayesian Approach*. John Wiley & Sons, Ltd, Chichester.

Pillai S, van Rij A, Williams S, Thomson I, Putterill M and Greig S 1999 Complexity- and risk-adjusted model for measuring surgical outcome. *British Journal of Surgery* **86**, 1567–1572.

Pliam M, Shaw R and Zapolanski A 1997 Comparative analysis of coronary surgery risk stratification models. *Journal of Invasive Cardiology* **9**, 203–222.

Sirio C, Shepardson L, Rotondi A, Cooper G, Angus D, Harper D and Rosenthal G 1999 Community wide assessment of intensive care outcomes using a physiologically based prognostic measure - implications for critical care delivery from Cleveland health quality choice. *Chest* **115**, 793–801.

Spiegelhalter D, Thomas A, Best N and Lunn D 2003 *WinBUGS Version 1.4 User Manual*. MRC Biostatistics Unit, Cambridge, and Imperial College, London.

Tekkis P, Kocher H, Kessaris N, Poloniecki J, Prytherch D, Somers S, McCulloch P, Ellut J and Steger A 2002 Risk adjustment in oesophagogastric surgery: a comparison of Bayesian analysis and logistic regression. *British Journal of Surgery* **89**, 381.

Wagner D, Knaus W, Harrell F, Zimmerman J and Watts C 1994 Daily prognostic estimates for critically ill adults in intensive care units: results from a prospective, multicenter, inception cohort analysis. *Critical Care Medicine* **22**, 1359–1372.

Zaslavsky A 2001 Statistical issues in reporting quality data: small samples and casemix variation. *International Journal for Quality in Health Care* **13**, 481–488.

Zimmerman J 1989 The APACHE III study design: analytic plan for evaluation of severity and outcome. *Critical Care Medicine* **17**, S169–S221.

Zimmerman J, Wagner D, Draper E, Wright L, Alzola C and Knaus W 1998 Evaluation of acute physiology and chronic health evaluation III predictions of hospital mortality in an independent database. *Critical Care Medicine* **26**, 1317–1326.

6

A Bayesian regression model with variable selection for genome-wide association studies

Carla Chen[1], Kerrie L. Mengersen[1], Katja Ickstadt[2] and Jonathan M. Keith[1,3]

[1]*Queensland University of Technology, Brisbane, Australia*
[2]*TU Dortmund University, Germany*
[3]*Monash University, Melbourne, Australia*

6.1 Introduction

In the previous chapter, we have seen the use of stochastic searching for variable selection in a simple linear model setting. In this chapter, we extend the concept for a specific application, namely genome-wide association studies (GWAs). GWAs aim to identify, from among a large number of marker loci drawn from across the genome, those markers that are in linkage disequilibrium with a locus associated with some disease or phenotypic trait. Due to increasing knowledge of common variations in the human genome, advancements in genotyping technologies and in particular reduction in the cost of gene chips, GWAs have become more prevalent. One of the current challenges faced in GWAs is to find an adequate and efficient statistical method for analysing large single nucleotide polymorphism (SNP) data sets.

Case Studies in Bayesian Statistical Modelling and Analysis, First Edition. Edited by Clair L. Alston, Kerrie L. Mengersen and Anthony N. Pettitt.

In this chapter, we propose to regress the trait or disease status of interest against all SNPs simultaneously with the stochastic variable selection algorithm. Excellent methods for the variable selection problem have been developed within a Bayesian context. The issue of multiple comparisons is also handled simply and effectively in a Bayesian context (Berry and Hochberg 1999). Mitchell and Beauchamp (1988) introduced model selection via an assignment of prior probabilities to the various models, and subsequent updating of those probabilities in accordance with Bayes' rule.

The method proposed here is able to detect both the additive and the multiplicative effects. The variable selection adopted in our model is allied with that in George and McCulloch (1993), which introduced the use of a latent variable for the identification of promising subsets of variables as shown in the previous chapter.

In this chapter, we introduce two Bayesian models, the first for continuous traits and the second for dichotomous traits. These models are initially described in the context of main effects only and then further extended for the detection of gene–gene interaction effects, that is multiplicative effects.

6.2 Case study: Case–control of Type 1 diabetes

We tested the performance of the proposed model using a Type 1 diabetes (TID) data set. The data were obtained from the Wellcome Trust Case Control consortium (WTCCC, http://www.wtccc.org.uk). In its study, the WTCCC collected 14 000 cases and 3000 shared controls for seven different familial diseases. Here we focus on TID.

Individuals involved in this study are self-identified white Europeans who live in Great Britain. The controls are recruited from two sources: 1500 are from the 1958 British Birth Cohort and the remainder are blood donors recruited for the WTCCC project.

TID cases are recruited from three sources. The first is from approximately 8000 individuals who attend the paediatric and adult diabetes clinics of 150 National Health Service hospitals across mainland UK. The second source of cases is voluntary members of the British Society for Paediatric Endocrinology and Diabetes. The rest are from the peripatetic nurses employed by the JDRF/WT GRID project (http://www-gene.cimr.cam.ac.uk/todd/).

Diagnosis of the TID cases is based on the participants' age of diagnosis and their insulin dependency. The cases of the TID study are required to be diagnosed with TID at age less than 17 and have been insulin dependent for more than 6 months. Individuals with other forms of diabetes, such as maturity onset diabetes of the young, are excluded from the data set.

Both cases and controls were genotyped with the GeneChip 500K Mapping array (Affymetrix Chip) with 500 568 SNPs. After filtration, there was a total of 469 557 SNPs. Details of the WTCCC experimental design, data collection, data filtration and more are given in The Wellcome Trust Case Control Consortium (2007).

Previously published results of single locus analysis indicated strong signal association of Chromosome 6 (The Wellcome Trust Case Control Consortium 2007). In light of this, we used only the SNP data on Chromosome 6 for this study.

In addition to the filtration methods and exclusion genotypes recommended by The Wellcome Trust Case Control Consortium (2007), we set CHIAMO calls with a score less than 0.9 to missing and removed all SNPs with one or more missing values to speed up the computation time. This leads to a total of 26 291 SNPs in the TID data.

6.3 Case study: GENICA

GENICA is an interdisciplinary study group on Gene ENvironmental Interaction and breast CAncer in Germany, with its main focus on the identification of both genetic and environmental effects on sporadic breast cancer. The data were collected between August 2000 and October 2002 on incident breast cancer cases and population-based controls in the Bonn region in Germany. Among the cases, 688 were first-time diagnoses of primary breast cancer, and were later histologically confirmed. There were 724 controls, matched within 5 year age classes. Samples contain only Caucasian females younger than 80 years old.

Each SNP genotype can take one of three forms: homozygous reference genotype, heterozygous variant genotype and homozygous variant genotype. The homozygous reference genotype is taken to be the genotype which has both alleles being the most frequent variant. The heterozygous variant genotype occurs when one of the base pairs is more frequent while the other base is less frequent, and the homozygous variant genotype is when both members of the pair are less frequent.

Not all genotype data are used in this study. The subset of SNPs which are related to oestrogen, DNA repair or control of cell cycle pathway is tested here, with a total of 39 SNPs. From a total of 1234 females, including 609 cases and 625 controls, individuals with more than three genotypes missing were excluded from the analysis. The final data therefore included 1199 women and were composed of 592 cases and 607 controls. Other missing genotypes were imputed using the k-nearest neighbour method (Schwender and Ickstadt 2008b). Details of data collection and genotyping procedure are in Justenhoven *et al.* (2004).

Although factors such as smoking history, family history of breast cancer and menopausal status were collected in the GENICA study, these variables were not available at the time of our study.

6.4 Models and methods

6.4.1 Main effect models

Continuous trait model

Let y_i be the observed value or realization of the dependent variable (continuous trait) for individual i, $i = 1, \ldots, n$. We model y_i as in Equation (6.1) below, dependent on a constant term μ_i, on n_c continuous-valued covariates, n_d discrete-valued covariates, and up to n_s SNPs. Let the jth continuous-valued covariate for individual i be x_{ji}. For each continuous-valued covariate, we introduce a regression parameter β_j. For the jth discrete-valued covariate, let L_j be the number of levels and let h_{jki} be 1 if the covariate has level k for individual i and 0 otherwise, for $k = 1, \ldots, L_j$. For each discrete-valued covariate and each level of that covariate we introduce a regression parameter ω_{jk}. Let z_s be an indicator variable for SNP s, taking the value 1 or 0 depending on whether SNP s is included in the model or not. Let g_{sli} be an indicator variable taking the value 1 or 0 depending on whether individual i has genotype l (where $l = 0, 1, 2$) at SNP s or not. Let ν_{sl} be the contribution to the dependent variable made by genotype l at SNP s. Let ε_i be the residual. Then

$$y_i = \mu + \sum_{j=1}^{n_c} \beta_j x_{ij} + \sum_{j=1}^{n_d} \sum_{k=1}^{L_j} \omega_{jk} h_{jki} + \sum_{s=1}^{n_s} z_s \sum_{l=0}^{2} \nu_{sl} g_{sli} + \varepsilon_i. \tag{6.1}$$

Because the SNPs are categorical variables, we arbitrarily assign the value $\nu_{s2} = 0$ for all SNPs s.

Case–control model (logistic model)

For case–control data, y_i is the presence/absence of the phenotypic trait, and takes the value 1 when the phenotype is present, else 0. The model proposed in Equation (6.1) can be simply modified by introducing a logit link function $y_i = \log(q_i/(1 - q_i))$, where q_i is the probability that individual i has the trait of interest. Then Equation (6.1) follows with the same notation for the model parameters.

Prior distributions

As part of the Bayesian approach, a prior distribution is required for each of the model parameters. In our two-case studies, no prior information is available, therefore non-informative priors are considered here. Moreover, because the indicator variable Z, $Z = (z_1, \ldots, z_S)$, is not directly observed, we adopted a hierarchical approach. Details on the priors used in our case studies are described in the examples.

Parameter estimation

Model parameters are estimated using Markov chain Monte Carlo (MCMC). The Gibbs sampler involves sampling from one-dimensional conditional distributions given other parameters and this is used for the estimation of all variables with one exception which we discuss below. Except for z_s, all other parameters possess non-standard conditional distributions; thus we used the slice sampler (Neal 2003) to draw from these.

Instead of sampling from the distribution function, the slice sampler samples from the area under the density function. Despite the complexity of using the slice sampler for multivariate distributions, it is relatively simple to implement for updating a single variable. Let x denote a model parameter and x_0 and x_1 be the current and new values of x, respectively. The procedure for updating x involves three steps. Firstly, draw a real value y uniformly from $(0, f(x))$ and consider the horizontal 'slice' $S = \{x : y < f(x)\}$. Next, establish an interval, $I = (L, R)$, around x_0 which contains this slice. A new value is then drawn uniformly from the interval and becomes x_1 if it is within S, else it is iteratively redrawn.

For simplicity, we used an initial interval of $(-1000, 1000)$ and used the shrinkage procedure (Neal 2003) for sampling from the interval.

The estimation procedure for z_s is described in the following.

Variable selection

Variable selection is an important element of the new models, which utilize the variable inclusion indicator (z_s) to determine the importance of SNP s. At each MCMC iteration, the value of z_s depends on the ratio of the conditional posterior probabilities of including and excluding SNP s. At the first iteration, start with a randomly generated vector of length n_S, comprising 0s and 1s, denoted $z^0 = (z_1^0, \ldots, z_{n_S}^0)$. Let t denote the MCMC iteration $t = 1, \ldots, T$, where T is the total number of iterations. Let Θ^t be a vector containing all parameters other than z at iteration t. At each t, SNP s is randomly selected from all SNPs and z_s is updated as follows:

1. Estimate the conditional posterior probability with $z_s = z_s^{t-1}$, that is $P(z_s^{t-1}|\Theta^t, Y, z_{-s})$.
2. Estimate the conditional posterior probability with the complementary value, $P(z_s^{'}|\Theta^t, Y, z_{-s})$, $z_s^{'} = 1 - z_s^{t-1}$.
3. Determine the ratio of the values computed in steps 2 and 1.
4. Accept the proposed $z_s^{'}$ if the value of step 2 is greater than a value randomly generated from a uniform distribution with minimum 0 and maximum 1; else retain z_s^{t-1}.

After SNP s is updated, the procedure is repeated for another SNP drawn randomly from the remaining SNPs. This continues until all SNPs are updated. The probability that SNP s is associated with the trait of interest is then estimated as the number of times SNP s is included in the model over the total number of iterations after burn-in.

Example 1: Case–control of TID

As only the genotype information is presented in the data set thus obtained, the logistic regression model is simply

$$y_i = \log\left(\frac{q_i}{1 - q_i}\right) = \mu + \sum_{s=1}^{n_s} z_s \sum_{l=0}^{2} \nu_{sl} g_{sli} + \varepsilon_i \tag{6.2}$$

where $i = 1, \ldots, 4857$ and $s = 1, \ldots, 26\,291$ and we arbitrarily assigned $\nu_{s2} = 0$.

Non-informative priors are used for this model as follows. The prior probability distributions for both overall mean (μ) and the contribution of level l of SNP s are assumed to be normally distributed with mean 0 and precision 1. The prior distribution for the residual, ε, is assumed to be a normal distribution with mean 0 and precision τ, and the prior for τ is assumed to be a gamma distribution, with parameters set to 0.05 ($\alpha = \beta = 0.05$). For z_s, we adopted a hierarchical approach, and let the probability that $z_s = 1$ be p_z, where p_z is a hyperparameter. We assumed the prior probability of z_s follows a Bernoulli distribution.

Five independent MCMC chains were generated with 100 000 iterations each. The first 50 000 iterations of each were considered as burn-in and the remaining ones were extracted for building the posterior marginal distributions. The algorithm was implemented in C.

6.4.2 Main effects and interactions

The model introduced in Equation (6.1) includes the main effects only. This can be extended for detecting SNP interaction effects as follows. Using the same notation as before, let η_{jk} be the indicator parameter, $\eta_{jk} = 1$ if the interaction of SNPs j and k is included in the model, else 0, and let $\gamma_{jl_jkl_k}$ be the coefficient of the interaction between the genotype l_j of SNP j and the genotype l_k of SNP k ($l_j = 0, 1, 2$; $l_k = 0, 1, 2$ and $j \neq k$). Then the model with two-way interactions is as follows:

$$y_i = \mu + \sum_{j=1}^{n_c} \beta_j x_{ij} + \sum_{j=1}^{n_d} \sum_{k=1}^{L_j} \omega_{jk} h_{jki} + \sum_{s=1}^{n_s} z_s \sum_{l=0}^{2} \nu_{sl} g_{sli} + \sum_{j=1}^{n_s} \sum_{k=1, j \neq k}^{n_s} \eta_{jk} \sum_{l_j=0}^{n_1} \sum_{l_k=0}^{n_2} \gamma_{jl_jkl_k} g_{jl_jkl_ki} + \varepsilon_i. \tag{6.3}$$

This model can be extended in an obvious manner to include multi-way interactions. By introducing a logit link function, this model can be implemented for the case–control study.

Typically, when an interaction effect and the two corresponding main effects are included in a model, then the number of levels for the interaction is $(n_1 - 1)(n_2 - 1)$, where n_1 and n_2 are the number of levels for each of the main effects (the maximum number of levels is nine). However, here we have chosen to include $n_1 * n_2 - 1$ levels for the interaction, because one or both of the main effects may not be included in the model (i.e. $z_1 = 0$ and/or $z_2 = 0$).

Parameters of this model are estimated following the same procedure as described earlier. The combination of Gibbs and slice samplers was implemented for sampling from the conditional posterior distributions. Likewise, variable η_{jk} was updated following the same procedure as for z_s.

Example 2: GENICA We illustrate this expanded model using the GENICA data set. Let θ denote the parameter space. The parameters for the GENICA data are thus

$$\theta = \{z_s, \nu_{sl}, \gamma_{jl_jkl_k}, \tau\}$$

where $s, j, k = 1, \ldots, 39, l, l_j, l_k = 1, 2, 3$ and z_s and η_{jk} are independent. The priors for model parameters were similar to the ones used in Example 1, as follows:

$$\varepsilon \sim N(0, \tau), \qquad \mu \sim N(0, 1), \qquad \tau \sim Ga(0.01, 0.01)$$
$$\nu_{sl}, \gamma_{jl_jkl_k} \sim N(0, 10), \qquad z_s \sim Bern(p_z), \qquad \eta_{jk} \sim Bern(p_\eta).$$

Ten MCMC chains were generated with 300 000 iterations each. Of these, the first 250 000 iterations were considered burn-in, and the remaining 50 000 cases were extracted for the construction of the marginal posterior distribution of θ. The computational algorithm was implemented in C.

6.5 Data analysis and results

6.5.1 WTCCC TID

The results of the MCMC runs for the WTCCC TID data indicated multiple modes in the posterior distribution. No prominent model was identified across all five chains. At each MCMC run, at least 13 000 unique models were tested, with the five most common models occupying only 1.25% to 4.5% of the post burn-in iterations. These models identified 17 to 24 SNPs of the total 26 291 SNPs, with some SNPs commonly found among many models (Table 6.1). These include SNPs 1576 (rs10901001), 4073 (rs874448), 4887 (rs950877) and 6222 (rs9272723). Five additional chains were generated using SNPs listed in Table 6.1. The posterior log-likelihood was well mixed after 150 000 iterations and with log-likelihood value between −2012 and −2052.

Although all SNPs on Chromosome 6 had the opportunity to enter the model at each of the MCMC iterations in the analysis, more than half (51%) of the SNPs were not selected in any of the 250 000 iterations (50 000 iterations, five chains). In contrast, 4% of SNPs (1143 SNPs) were included at least once in the iterations of all five chains. Of these 1143 SNPs, all five chains selected SNPs 1576 (rs10901001) and 4073 (rs874448) in nearly all iterations ($\approx 97\%$), followed by SNP 4887 (rs950877, 76%), which is also included in the five optimal models.

The results of the MCMC runs also identified a group of SNPs with highly variant probability of inclusion across the chains. For instance, SNP 6051(rs3131631) had a high probability of inclusion for chains 1, 3 and 4, but was selected in less than 1% of iterations in chains 2 and 5. This indicated that the inclusion of an SNP from this group depends on other SNPs already present in the model during the variable selection procedure. This was also observed for SNPs 6232 (rs9275418) and 6233 (rs9275523). SNP 6232 was selected in nearly 100% of iterations for chains 1, 2 and 4, but was not

Table 6.1 SNPs included in the most common models from each of the five chains.

Chain	No SNPs	SNPs ID
1	20	**1576**, **4073**, **4887**, 5587, 5638, 5663, *5919, 5969, 6051, 6110, 6122, 6158, 6195, 6205, 6211, 6217, 6221, 6222, 6232*, 8390
2	18	1112, **1576**, **4073**, **4887**, 5545, 5661, *5957, 6025, 6073, 6087, 6156, 6157, 6180, 6217, 6221, 6222, 6228, 6232*
3	24	1112, **1576**, **4073**, **4887**,5447, 5577, 5587, 5588, *5802, 5947, 6051, 6110, 6156, 6160, 6172, 6174, 6177, 6180, 6217, 6221, 6222, 6233*, 8169, 21883
4	17	**1576**, **4073**, **4887**, 5566, *5919, 5969, 6051, 6117, 6179, 6189, 6214, 6222, 6225, 6232, 6382, 6385*, 26289
5	23	**1576**, 3302, **4073**, **4887**,5553, 5571, *5932, 6025, 6043, 6121, 6149, 6154, 6173, 6180, 6191, 6205, 6219, 6227, 6233*, 12097, 17510, 22015, 24454

1. The reference codes of these SNPs are supplied in the Table 6.3.
2. SNPs in **bold** are the common SNPs identified across chains.
3. SNPs in *italics* are the SNPs from the major histocompatibility complex region.

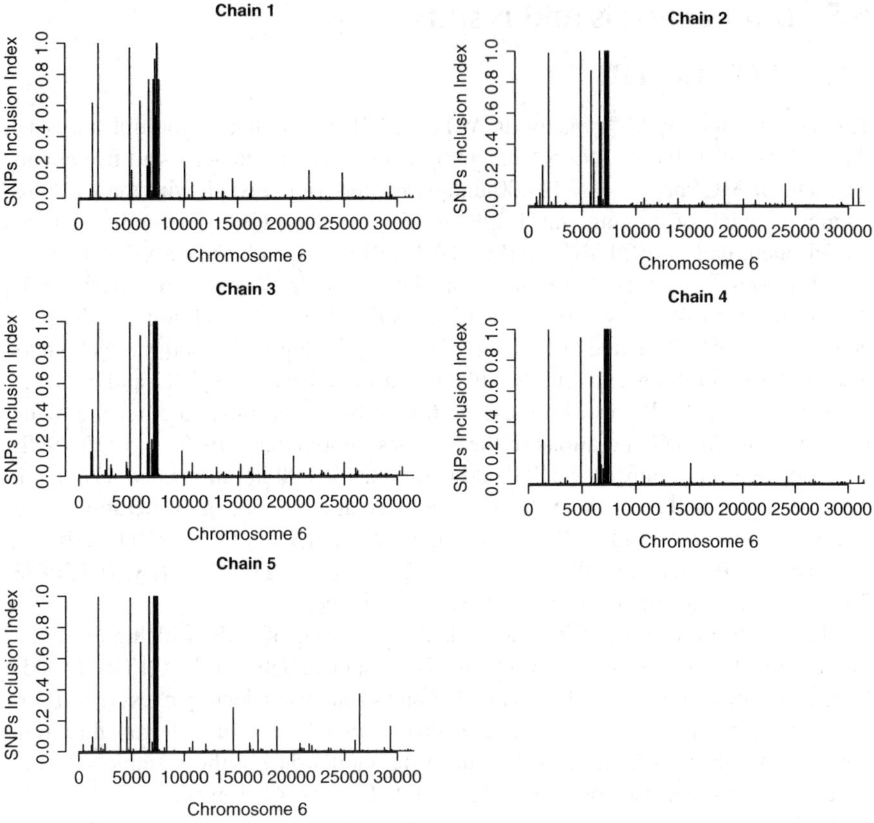

Figure 6.1 The contribution of individual SNPs on Chromosome 6 to TID across five chains.

selected for chains 3 and 5; in contrast, SNP 6232 was included in nearly all iterations for chains 3 and 5, but was never included for chains 1, 2 and 4. Since these two SNPs are physically nearby, they may be in linkage disequilibrium.

Figure 6.1 shows the ranking of SNPs across Chromosome 6 for the five chains. The first two peaks correspond to SNPs 1576 and 4073. This figure also shows a strong association with TID on a region of the shorter arm of Chromosome 6 which is the major histocompatibility complexity (MHC) region (SNP 5802 to SNP 6358).

6.5.2 GENICA

As in the previous case study, the results of the MCMC runs for the GENICA breast cancer data indicated that the posterior distribution has multiple modes. Table 6.2 lists the most frequently selected model in each of the 10 MCMC chains. These models were selected in each chain for at least 61% of the 50 000 post-burn-in iterations. Of the 10 chains, 6 converged to the same model (chains 2, 4, 5, 8, 9 and 10), which contains only two SNPs – SNP 20 and 21 – and both are fitted as main effects. In contrast, the remaining four models indicated the presence of interaction effects.

Table 6.2 Unique best models in 10 chains.

Model	Parameters	Chains	Frequency (%)
1	μ, SNP 20, SNP 21×SNP 23, SNP 3 ×SNP 28, SNP 4×SNP 28	1	61.4
2	μ, SNP 20, SNP 21	2, 4, 5, 8, 9, 10	88.6–93.1
3	μ, SNP 20, SNP 6×SNP 21	3	90.5
4	μ, SNP 20, SNP 14×SNP 21, SNP 2×SNP 14, SNP 3×SNP 14	6	89.5
5	μ, SNP 20, SNP 21, SNP 2×SNP 37, SNP 3×SNP 37, SNP 4×SNP 37	7	68.6

SNP 20 is the most prominent main effect and is observed in all models (Table 6.2). In contrast, SNP 21 is included in a model as either a main or an interaction effect. When SNP 21 is selected as an interaction effect, it interacts with a different SNP in different models. For instance, SNP 21 interacts with only SNP 23 in Model 1. Besides these two SNPs, other possible SNPs and interactions are also identified as indicated in Table 6.2.

The estimated coefficient of SNP 20 is fairly consistent across models and ranges between -1.17 and -1.12 for the homozygous reference variants (level 0) and between -0.57 and -0.52 for the heterozygous genotype variants (level 1). This indicates that individuals with a homozygous variant genotype (level 2) at SNP 20 associated with a higher chance of having breast cancer, followed by individuals having a heterozygous variant genotype (level 1) and homozygous reference variants (level 0) at SNP 20. In two of the models (models 2 and 5), SNP 21 is included as a main effect and the posterior estimates of the coefficient for genotype variants are also consistent for both models. The coefficients indicate that having homozygous reference variants at SNP 21 is associated with a higher probability of breast cancer than the other two genotype variants. However, these two SNPs needed to be considered conjointly to estimate the probability of sporadic breast cancer (Model 2).

Considering SNPs 20 and 21 as additive effects, the highest chance of sporadic breast cancer occurs when individuals have homozygous genotype variant (level 2) at SNP 20 and homozygous reference genotype variant (level 0) at SNP 21, with an odds ratio of 4.17 (CI: 2.63–6.67) compared with individuals with homozygous genotype variants (level 2) at both SNP 20 and SNP 21. The next highest probability occurs for individuals with heterozygous genotype variants (level 1) and homozygous reference variant (level 0) at SNPs 20 and 21 respectively; these individuals have an odds ratio of 2.37(CI: 1.01–5.58). The lowest chance of sporadic breast cancer is for subjects with homozygous reference variants at SNP 20 and homozygous variants (level 2) at SNP 21.

In other models, where SNP 21 is selected as part of an interaction effect, the effect of genotype variants at this SNP becomes more complicated. Figure 6.2 shows

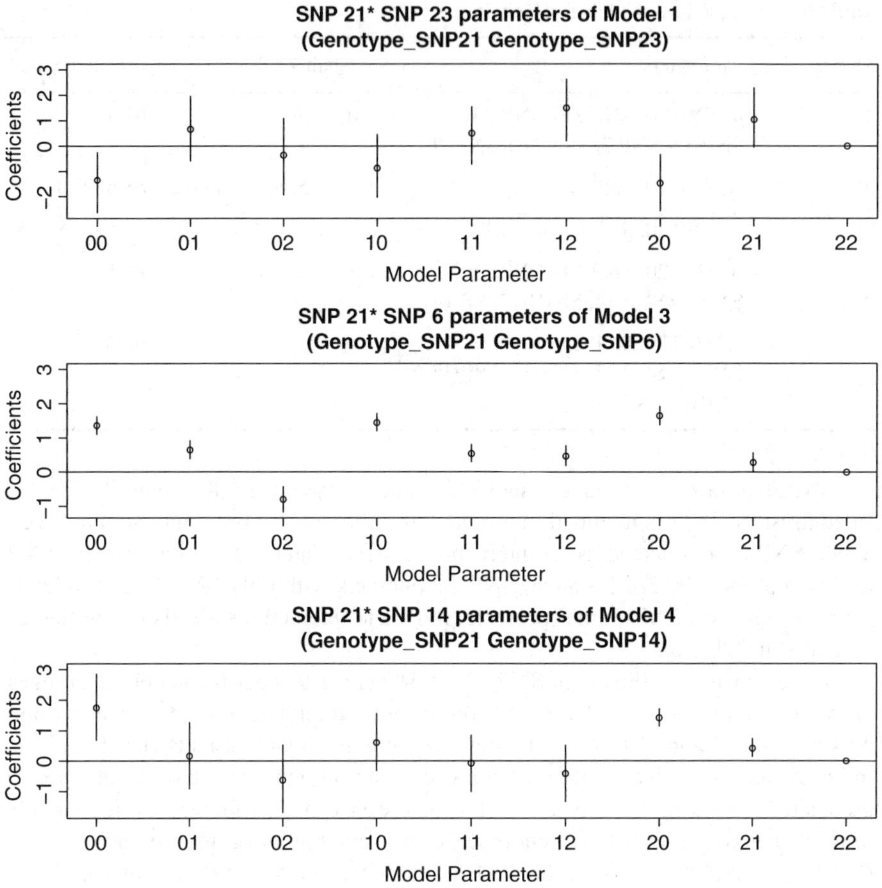

Figure 6.2 Coefficients of interaction terms with SNP 21 with credible intervals.

the posterior mean and credible intervals of the interaction terms of Models 1, 3 and 4, which all involve SNP 21. In Model 1, SNP 21 contributed to the probability of breast cancer by associating with SNP 23 and the genotype variants of SNP 21 in this combination are quite different from the genotype variants of SNP 21 combinations in Models 3 and 4 (SNP 21*SNP 6, SNP 21*SNP 14, respectively), but some similarities are found in Models 3 and 4.

6.6 Discussion

Here we introduced a simple regression model with stochastic search algorithm into GWA analysis which is an alternative to the commonly used single locus analysis. Estimation was undertaken using a novel algorithm. The model is capable of identifying a group of SNPs that contributed to the genetic causes of the disease status through additive or interaction effects.

The use of Bayesian regression for identifying important loci is not novel. The fairly recent paper by Hoggart *et al.* (2008) identified subsets of important SNPs using Bayesian inspired penalized maximum likelihood. They assigned a sharp prior mode at zero to the regression coefficients and SNPs with non-zero coefficient estimates were said to have some signal of association. Apart from this, most studies to date are focused on the analysis of quantitative trait loci (QTL) data (Xu 2003, 2007; Xu and Jia 2007; Yi and Xu 2000; Yi *et al.* 2003, 2005). The model of Xu (2003) was initially developed for detecting single locus effects simultaneously, and was later developed for detecting epistasis effects (Xu 2007). In Xu (2007), the empirical Bayes approach is implemented to estimate genetic epistasis effects without using variable selection, and the relative importance of effects is based on the ratio of variances.

In contrast, our model is more closely allied to Yi *et al.* (2005), but our method differs in a number of aspects. Firstly, Yi *et al.* (2005) partition the genome into a fixed number of loci and assume that the QTL occur at one of these sites. This partitioning is required to be specified prior to the analysis. In contrast, the SNP data can be directly utilized in our model and the number of potential causal loci is estimated directly from the data set without boundaries. Secondly, because the model by Yi *et al.* (2005) is for a QTL study, a design matrix can be employed; although this is an ideal approach, it is not feasible for population studies.

The results of the WTCCC analysis illustrated the ability of our model to search for main additive effects in a relatively large data set. The model indicated that more than 50% of SNPs on Chromosome 6 do not contribute substantively to the determination of the phenotype and only less than 4% of SNPs on this chromosome are strongly informative. Of these, 17 to 24 SNPs were selected to best describe the genetic association with TID. All four MCMC chains identified the three SNPs rs10901001, rs874448 and rs950877, which are outside the major histocompatibility complex (MHC) region, and 12 to 13 SNPs from the MHC region. The three SNPs that are outside the MHC region all show a strong signal of association with TID. These are novel SNPs which have not been identified by other studies. In contrast, the MHC region is known to be associated with a large number of infectious and autoimmune diseases (de Bakker *et al.* 2006). The association between this region and TID has also been published (The Wellcome Trust Case Control Consortium 2007) and is successfully replicated in our study.

We repeated the analysis of WTCCC data using the common SNP by SNP search algorithm. As expected, the SNP by SNP search algorithm identified strong association signals in the MHC region; however, three novel SNPs identified by our models (rs10901001, rs874448 and rs950877) have little association to TID when tested in this manner. The unadjusted p-values for these three SNPs are 10 to the power of -6, -12 and -5, respectively. It is possible that these SNPs cannot be detected in isolation, but interact with SNPs in the MHC region. Further investigation of this effect is needed.

The aim of the analysis of the GENICA data is to exemplify the ability to detect the combination of the main and interaction effects by the described model. This study is designed for targeted gene search studies rather then running GWAs. Ten MCMC runs revealed five different models, with the most frequently identified model composed of

SNPs 20 and 21 as main effects. Among the models, SNP 20 is consistently selected as a main effect, but SNP 21 appears to be associated with sporadic breast cancer as either a main or an interaction effect. Analysis of the same data set using two different types of logic regression (Schwender and Ickstadt 2008a) also revealed the importance of these two SNPs.

In general, according to the results from both WTCCC and GENICA analyses, SNPs identified by the models may potentially be separated into two major categories: those present in all models and those implicated in only some of the models. The second category may be the result of correlation between SNPs entering a model and those SNPs that are already in the model during an iteration (linkage disequilibrium (LD) between SNPs). In the WTCCC case study, the SNPs that fall into the second category are mainly from the MHC region of Chromosome 6. Given that the LD structure of this region of the genome is longer and more complicated than other regions of the genome (de Bakker *et al.* 2006; Walsh *et al.* 2003), the SNPs of the MHC region identified by one MCMC run are potentially in linkage disequilibrium with the same region of SNPs identified by other runs. However, a more complete understanding of the effect of SNPs in LD on the model requires further research.

The use of a Bayesian framework overcomes the problem raised by Lin *et al.* (2008), who listed three drawbacks of using the logistic model in conjunction with variable selection (i.e. Akaike information criterion (AIC), Schwarz criterion (SC)) under the frequentist framework. Firstly, the empty cell effect, which occurs when there is a low frequency of some genotype or genotype combination, can make the interpretation of the logistic regression result invalid. In our model, these empty cells are filtered out during the updating procedures. Although the effect of these empty cells is inconclusive, the results are not affected by the empty cells. The second and third concerns raised by Lin *et al.* (2008) are that the frequentist logistic regression demonstrated weak power for variable selection due to the correlation between variables and the problem due to the genetic heterogeneity. Although this may be true when explaining the genetic make-up with only the most prominent model, these two problems can be simply overcome by allowing multiple chains or incorporating the technique of model averaging (Yi and Xu 2000).

Although it is not illustrated here with the case studies, our model has the ability to accommodate missing data by introducing extra parameters. When there are missing genotypes, the model is modified as follows. We treat the values of g_{sli} at each SNP s as unknown parameters of the model and introduce the observed genotypes g^o_{sli}. Here l can take one of four values: 0, 1, 2 or missing. For each SNP, and each of the three possible true genotypes, we let the probability that the data are missing be ϕ, a real value in the interval (0,1). The value of ϕ is then estimated as a model parameter via the hierarchical approach.

Although the model proposed in this chapter is relatively simple conceptually, there are some drawbacks. The first is the indecisive nature of the variable selection in each chain, indicated by the moderate contribution by various SNPs. The challenge that is admittedly only partially addressed here is how to optimally combine this information.

The second drawback is the computational burden. This problem may be overcome by adopting different MCMC algorithms, such as the reversible jump MCMC (Green 1995), simulated tempering (Marinari and Parisi 1992), variational approaches (Jordan *et al.* 1999) or population MCMC (Cappe *et al.* 2004).

Despite the above drawbacks, the proposed model is able to detect the relevant SNPs for both TID and sporadic breast cancer. It is hoped that such investigations of alternative ways of exploring and describing the role of SNPs and their interaction in GWA studies can facilitate a better understanding of the genetics of complex disease.

Acknowledgements

This study is funded by the National Health and Medical Research Council capacity Building grant (grant number 389892). The authors would also like to thank all partners within the GENICA research network for their cooperation. This study makes use of data generated by the Wellcome Trust Case Control Consortium. A full list of the investigators who contributed to the generation of the data is available from www.wtccc.org.uk. Funding for the project was provided by the Wellcome Trust under award 076113.

The authors would also like to thank the high-performance computing team of Queensland University of Technology for their assistance in provision of computing resources.

References

Berry DA and Hochberg Y 1999 Bayesian perspectives on multiple comparisons. *Journal of Statistical Planning and Inference* **82**, 215–227.

Cappe O, Guillin A, Marin JM and Robert CP 2004 Population Monte Carlo. *Journal of Computational and Graphical Statistics* **13**(4), 907–929.

de Bakker PIW, McVean G, Sabeti PC, Miretti MM, Green T, Marchini J, Ke X, Monsuur AJ, Whittaker P and Delgado M 2006 A high-resolution HLA and SNP haplotype map for disease association studies in the extended human MHC. *Nature Genetics* **38**(10), 1166–1172.

George EI and McCulloch RE 1993 Variable selection via Gibbs sampling. *Journal of the American Statistical Association* **88**(423), 881–889.

Green PJ 1995 Reversible jump Markov chain Monte Carlo computation and Bayesian model determination. *Biometrika* **82**(4), 711–732.

Hoggart CJ, Whittaker JC, De Iorio M and Balding DJ 2008 Simultaneous analysis of all SNPs in genome-wide and re-sequencing association studies. *PLoS Genetics*. http://www.plosgenetics.org/article/info

Jordan MI, Ghahramani Z, Jaakkola TS and Saul LK 1999 An introduction to variational methods for graphical models. *Machine Learning* **37**(2), 183–233.

Justenhoven C, Hamann U, Pesch B, Harth V, Rabstein S, Baisch C, Vollmert C, Illig T, Ko YD, Bruning T, Brauch H, for the Interdisciplinary Study Group on Gene Environment Interactions and Breast Cancer in Germany 2004 ERCC2 genotypes and a corresponding haplotype are linked with breast cancer risk in a German population. *Cancer Epidemiology, Biomarkers & Prevention* **13**(12), 2059–2064.

Lin HY, Desmond R, Louis Bridges S and Soong S 2008 Variable selection in logistic regression for detecting SNP–SNP interactions: the rheumatoid arthritis example. *European Journal of Human Genetics* **16**(6), 735–741.

Marinari E and Parisi G 1992 Simulated tempering: a new Monte Carlo scheme. *Europhysics Letters* **19**(6), 451–458.

Mitchell TJ and Beauchamp JJ 1988 Bayesian variable selection in linear regression. *Journal of the American Statistical Association* **83**(404), 1023–1032.

Neal RM 2003 Slice sampling. *Annals of Statistics* **31**(3), 705–767.

Schwender H and Ickstadt K 2008a Identification of SNP interactions using logic regression. *Biostatistics* **9**(1), 187.

Schwender H and Ickstadt K 2008b Imputing missing genotypes with k nearest neighbors. Technical report, Collaborative Research Center 475, Department of Statistics, University of Dortmund.

The Wellcome Trust Case Control Consortium 2007 Genome-wide association study of 14000 cases of seven common diseases and 3000 shared controls. *Nature* **447**(7), 661–590.

Walsh EC, Mather KA, Schaffner SF, Farwell L, Daly MJ, Patterson N, Cullen M, Carrington M, Bugawan TL and Erlich H 2003 An integrated haplotype map of the human major histocompatibility complex. *American Journal of Human Genetics* **73**(3), 580–590.

Xu S 2003 Estimating polygenic effects using markers of the entire genome. *Genetics* **163**, 789–801.

Xu S 2007 An empirical Bayes method for estimating epistatic effects of quantitative trait loci. *Biometrics* **63**(2), 513–521.

Xu S and Jia Z 2007 Genomewide analysis of epistatic effects for quantitative traits in barley. *Genetics* **175**, 1955–1963.

Yi N and Xu S 2000 Bayesian mapping of quantitative trait loci for complex binary traits. *Genetics* **155**(3), 1391–1403.

Yi N, George V and Allison DB 2003 Stochastic search variable selection for identifying multiple quantitative trait loci. *Genetics* **164**, 1129–1138.

Yi N, Yandell BS, Churchill GA, Allison DB, Eisen EJ and Pomp D 2005 Bayesian model selection for genome-wide epistatic quantitative trait loci analysis. *Genetics* **170**(3), 1333–1344.

6.A Appendix: SNP IDs

Table 6.3 The SNP IDs referenced in this study.

SNPID	SNP names	SNPID	SNP names
1112	rs4959334	6157	rs9268403
1576	rs10901001	6158	rs12201454
3302	rs7749556	6160	rs12528797
4073	rs874448	6172	rs3806156
4887	rs950877	6173	rs3763307
5447	rs16894900	6174	rs3763308
5545	rs9258205	6177	rs2001097
5553	rs9258223	6179	rs3135378
5566	rs1633030	6180	rs3135377
5571	rs1632973	6189	rs9268560
5577	rs9258466	6191	rs3135342
5587	rs1233320	6195	rs9268645
5588	rs16896081	6205	rs9268858
5638	rs1150743	6211	rs9268877
5661	rs9261389	6214	rs9270986
5663	rs9261394	6217	rs4530903
5802	rs2394390	6219	rs9272219
5919	rs9263702	6221	rs9272723
5932	rs2073724	6222	rs9273363
5947	rs3095238	6225	rs7775228
5957	rs3130531	6227	rs6457617
5969	rs7382297	6228	rs6457620
6025	rs16899646	6232	rs9275418
6043	rs2523650	6233	rs9275523
6051	rs3131631	6382	rs3129207
6073	rs2242655	6385	rs7382464
6087	rs480092	8169	rs16872971
6110	rs408359	8390	rs2028542
6117	rs438475	12097	rs9343272
6121	SNP_A.2064274	17510	rs6938123
6122	rs377763	21883	rs9497148
6149	rs9268302	22015	rs3763239
6154	rs9268402	24454	rs16891392
6156	rs9391858	26289	rs16901461

7

Bayesian meta-analysis

Jegar O. Pitchforth and Kerrie L. Mengersen

Queensland University of Technology, Brisbane, Australia

7.1 Introduction

Meta-analysis is the process of quantitatively combining a set of estimates about an outcome of interest. Typically, these estimates are obtained from a set of studies identified through a systematic review of relevant literature on the topic. The combined analysis reduces dependence on any one finding or narrative and 'borrows strength' across studies which may each provide inconclusive, or conclusive but conflicting evidence. Bayesian meta-analysis has advantages over frequentist approaches in that it provides a more flexible modelling framework, allows more appropriate quantification of the uncertainty around effect estimates, and facilitates a clearer understanding of sources of variation within and between studies (Sutton and Abrams 2001).

In this chapter, we describe Bayesian meta-analysis models and their attributes through two substantive case studies.

The first study is illustrative of a very wide range of univariate meta-analyses that can be described using a standard random effects model. Extensions to this model are also described. These include the extension of the hierarchical structure of the model to allow for dependencies between the studies, incorporation of covariates via meta-regression, and analysis of multiple effects via a full multivariate meta-analysis model. Although not all of these extensions are employed in the case study, it is hoped that by describing the various models in a unified manner the reader can more easily see how to develop a suitable model for their particular problem.

The second case study illustrates a range of problems requiring the synthesis of results from time series or repeated measures studies. The repeated measurements

Case Studies in Bayesian Statistical Modelling and Analysis, First Edition. Edited by Clair L. Alston, Kerrie L. Mengersen and Anthony N. Pettitt.

might be taken on the same individual, or they might represent separate samples of unique sets of individuals measured at different time periods within a study. By fitting a parametric (linear, quadratic or higher order polynomial) trend model for each study, the aims of the meta-analysis are now to evaluate trends within studies, describe the heterogeneity among study-specific regression parameters and, if it is sensible, to combine these values to obtain an overall trend estimate.

7.2 Case study 1: Association between red meat consumption and breast cancer

7.2.1 Background

Many factors have been implicated in the occurrence of breast cancer, including smoking (Terry *et al.* 2002), hormone replacement therapy (Beral 2003), family history (Kingsmore *et al.* 2003), dietary fat intake (Boyd *et al.* 2003), menopausal status (Byrne *et al.* 1995) and body mass index (Vatten and Kvinnsland 1990), among others. The question of whether there is an association between red meat consumption and the disease has been actively debated in both scientific and popular literature. Formal studies of the association have reported conflicting results: from no association between any level of red meat consumption (van der Hel *et al.* 2004) to a significantly raised relative risk (Taylor *et al.* 2009).

The focal variable for this meta-analysis is the iron content of red meat. Where data allow, we are also interested in iron levels in blood. Iron is proposed to promote the carcinogenic effects of meat mutagens and interact with catechol oestrogen metabolites producing hydroxyl radicals (Ferrucci *et al.* 2009). High iron levels are also hypothesized to result in mammary carcinogenesis as a function of oxidative stress (Bae *et al.* 2009).

In the studies included in this analysis, levels of red meat and iron intake were defined by the sample population. The sample was split generally into quintiles based on their intake. The thresholds of group membership thus varied by iron intake quantity. Only two studies included in the final analysis measured both red meat and dietary iron (Ferrucci *et al.* 2009; Kabat *et al.* 2007).

To our knowledge, there are currently no systematic reviews of the association between red meat consumption and breast cancer. Based on our search of the literature, the study that is most similar to the review described in this chapter was conducted a decade ago by Missmer *et al.* (2001), whose meta-analysis of meat and dairy food consumption and breast cancer found no significant association between red meat intake and the disease. However, most of the studies in Missmer *et al.*'s review focused on the effect of heterocyclic amines in red meat as the primary factor, and a number of studies have since been published on the subject. A similar meta-analysis was conducted by Taylor *et al.* (2009), but this included studies focusing on dietary fat with red meat as a subset of included foods. There is also literature on related associations: for example, Cho *et al.* (2006) found an association between red meat and hormone receptor-positive cancer but no association with breast cancer, while Stevens *et al.* (1988) reported a positive association between iron and other types of cancer.

We conducted a systematic search of the literature using Google Scholar, PubMed, ProQuest, UQ (University of Queensland) Summons and QUT (Queensland University of Technology) Library (all year ranges). Key search terms were breast, breast tissue, breast cancer, dietary, iron, red meat, carcinogen, epidemiology and mammary. We ran searches on individual terms, then combinations of terms. Grey literature was excluded from the search. Two reviewers independently verified eligible articles for which abstracts were available and the full article text was either downloaded or linked to the reference list for future assessment.

The inclusion criteria were as follows: (i) original studies and articles reporting on human females, written in English and published in peer-reviewed journals; and (ii) observational studies including the consumption or non-consumption of red meat as factors and risk of breast cancer as the outcome. The exclusion criteria were as follows: (i) review articles; and (ii) articles focusing on a mediating effect of a third variable, such as genetics, in the association between red meat and breast cancer. The term red meat in this study includes both processed and unprocessed dark animal meat from mammals, such as cows, sheep and goats but excluding pig products. All studies included in this review excluded pig products from the analysis.

For each study we extracted data including details of study design, age range, population type, red meat consumption measurement, iron intake measurement, control type, sample size. If more than one article with the same data set was identified, we only used the article containing the most complete data. The two reviewers independently checked the data for consistency in extraction and interpretation.

The difference in risk of breast cancer participants who consumed red meat compared with participants who did not consume red meat was expressed in each study as either an odds ratio, hazard ratio or risk ratio. Despite their mathematical difference, for the purposes of this meta-analysis these statistics were considered to be sufficiently equivalent to justify the calculation of an overall effect.

An initial search yielded over 2 million articles written in English with contents relating to red meat consumption and breast cancer, of which 28 were potentially relevant articles and were subsequently retrieved. Eleven articles were excluded because they were reviews, did not report complete data, were primarily concerned with examining a possible moderating factor in the relationship, or did not report breast cancer risk specifically. An additional two articles were excluded as they used secondary analysis of previously reported data. As a result, 15 studies were included in the final analysis.

The 15 studies included 287 492 individual participants aged between 20 and 85 years. Thirteen studies were conducted in primarily Caucasian populations, and two studies were conducted in primarily Asian populations. Nine studies had a prospective cohort design and eight were case-matched, cross-sectional studies (contact authors for further details).

The aims of the meta-analysis were as follows: (i) to obtain an overall estimate of the association between red meat intake and breast cancer; (ii) to assess whether the association differed for pre-menopausal and post-menopausal women; (iii) to assess whether an exposure–response relationship exists. Geographic and/or cultural differences were also of interest, but there were too few studies to undertake a valid assessment of the association by continent.

7.2.2 Meta-analysis models

Aim 1

For the first aim, we consider a random effects model that assumes exchangeability between studies and allows for variation within and between studies. In this case study, the effect of interest is the log odds ratio. The model assumes that the effect estimate of the log odds ratio T_i from the ith study is normally distributed around the true effect θ_i for that study, and that these effects are normally distributed around an overall effect μ_0. Thus

$$T_i \sim N\left(\theta_i, \sigma_i^2\right)$$
$$\theta_i \sim N\left(\mu_0, \tau^2\right).$$

Prior distributions are required for σ_i^2, μ_0 and τ^2. Accepted practice in meta-analysis is to take the reported variances for T_i (S_i^2, say) as known (i.e. $S_i^2 = \sigma_i^2$). The argument is that this source of variation is minor compared with the within- and between-study variation. A more complete representation can be obtained by the prior $\sigma_i^2 \sim \nu S_i^2/\chi_\nu^2$ where ν is representative of the overall size of the studies (e.g. the number of cases or the number of exposed cases) or set as small to indicate non-informativeness (DuMouchel 1990). Non-informative prior distributions were set for μ_0 (normal with mean 0 and variance 10 000) and τ (uniform over range 0 to 100). The question of whether there is an association between red meat consumption and breast cancer is addressed primarily by inspection of the posterior distribution for μ_0. In addition to posterior credible intervals for this overall effect, it is also useful to compute the probability that the effect is greater than zero (equivalent to no association on the log scale) or, even more usefully, greater than a threshold that is considered to be clinically relevant. Note that computation of these probabilities is very straightforward as part of the MCMC analysis of the model using a Metropolis–Hastings sampling algorithm. (At each MCMC iteration, update an indicator variable if μ_0 exceeds the nominated value; the posterior estimate of this variable is the probability of exceedance.)

This model returns not only a posterior distribution for the overall effect (log odds ratio) μ_0, but also posterior estimates of study-specific effects θ_i. The θ_i are commonly known as 'shrinkage estimates' since they tend to be pulled away from the observed estimate T_i towards the overall estimate μ_0, with the degree of shrinkage depending on the degree of departure from the overall estimate and the precision of the study-specific estimates. The model also provides important information about the heterogeneity of estimates between studies, through the posterior distribution for τ^2. If this is relatively large, the analyst should consider whether a combined effect is meaningful and investigate possible explanations for this between-study variability.

As an aside, a fixed effects model can be obtained in two ways: by allowing the between-study variance to tend to infinity, in which case the studies are not able to be combined; or (more appropriately for meta-analysis) by taking $\tau^2 = 0$. In some (rare) situations, studies are considered sufficiently homogeneous to warrant this reduced model. A common practice is to fit this model if a preliminary analysis of the between-study heterogeneity returns a 'non-significant' outcome. This is not recommended, since 'non-significance' does not indicate a complete lack of heterogeneity, and even small sources of variance such as this can influence posterior estimates and inferences. Small between-study variation can be easily accommodated in the random effects model. However, it is acknowledged that τ^2 is poorly estimated if the number of studies is small; in this case, the analyst could fit both models and evaluate the comparability and stability of the parameter and effect estimates.

Aim 2

The second aim of the analysis was to evaluate whether there is a difference in risk for pre-menopausal and post-menopausal women. There are a number of ways that this could be addressed. We could fit the random effects model described above separately to the reported estimates for pre-menopausal women and for post-menopausal women, and compare the posterior distributions for the overall effect estimates in the two groups. Alternatively, a meta-regression model could be fitted, with an indicator X_i for menopausal status as the explanatory variable. Thus, the study-specific reported effect estimates (log odds ratios) T_i are assumed to be normally distributed around a true study-specific effect θ_i, and these effects are fitted by a regression $\theta_i = \beta_0 + \beta_1 X_i$. The model can be formulated as follows:

$$T_i \sim N(\theta_i, \sigma_i^2)$$

$$\theta_i \sim N(\mu_i, \tau^2); \qquad \mu_i = \beta_0 + \beta_1 X_i$$

where (as before) σ_i^2 represents the sampling variation associated with T_i and τ_i^2 represents the residual variance associated with the regression of the θ_i. Hence if $X_i = 0$ indicates pre-menopausal status and $X_i = 1$ indicates post-menopausal status, the overall effects for the two groups are β_0 and $(\beta_0 + \beta_1)$, respectively. The focus then is on the posterior distributions of these two terms. Posterior distributions are also returned for the study-specific effects θ_i, and between-study heterogeneity, after accounting for menopausal status, is indicated by τ^2.

A non-informative prior distribution was set for τ (uniform over range 0 to 100). Non-informative prior distributions were also set for β_0 and β_1 to ensure more conservative posterior distributions than necessary and demonstrate a basic model only. These were first specified independently as univariate normal with mean 0 and variance 10 000. In a subsequent analysis, the non-independence of these two parameters was accommodated by fitting a bivariate normal prior for (β_0, β_1), with mean vector $(0, 0)$ and covariance matrix set with diagonal terms equal to 10 000 and off-diagonal terms equal to 100.

The final model considered for this aim is a multi-level or hierarchical meta-analysis, in which an additional hierarchy is added to the meta-analysis model, with pre- and post-menopausal women considered as separate groups. Hence the study-level estimates would be considered to be drawn from group-specific distributions, and the group-level effects would be considered to be drawn from an overall distribution. Letting T_{kj} represent the jth estimate (reported log odds ratio) in the kth group ($k = 1, 2$), the model becomes

$$T_{kj} \sim N(\theta_{kj}, \sigma_{kj}^2)$$

$$\theta_{kj} \sim N(\mu_k, \gamma_k^2); \qquad \mu_k \sim N(\mu_0, \omega^2)$$

where θ_{kj} represents the study-specific effect, μ_k represents the overall effect for the ith group and μ_0 represents the overall effect. (Recall that, for the case study, this is the overall log odds ratio for the association between red meat and breast cancer.) Note that there are now three sources of variation: sampling variation for each effect estimate (σ_{kj}^2), variation between effect estimates within groups (γ_k^2) and variation between groups (ω^2). Note that the studies are now considered exchangeable within groups, but not exchangeable between groups. To address the study aim, a posterior distribution for the difference between the group effects, $\mu_1 - \mu_2$, can be computed as part of the analysis using the Metropolis–Hastings sampling algorithm. The difficulty with applying this model to the case study, however, is that the between-group variance is very poorly estimated with only two groups. Hence we do not pursue the analysis using this approach.

Aim 3

An exposure–response model was based on the reported estimates of log odds ratios (and corresponding variances) at each of a study-specific set of quantiles of red meat consumption (expressed as unexposed/exposed; or tertiles, quartiles or quintiles of exposure). Due to the paucity of data, a linear response was fitted to the estimates within each study, assuming equal spacing between the exposure levels L_{ij} in the study. The model is thus represented as

$$\theta_{ij} \sim N(\mu_{ij}, \tau_i^2); \qquad T_{ij} \sim N(\theta_{ij}, \sigma_{ij}^2)$$

$$\beta_{1i} \sim N(\beta_0, \tau_0^2); \qquad \mu_{ij} = \beta_{1i} L_{ij}$$

where σ_i^2 represents the sampling variation associated with T_{ij}, τ_i^2 represents the residual variance associated with the regression of the θ_{ij} within the ith study, the coefficient β_{1i} represents the exposure–response relationship in the ith study, and β_0 represents the overall exposure-response relationship.

As in the previous models the Metropolis–Hastings sampling algorithm was used. Non-informative normal prior distributions were set for the regression parameters and non-informative uniform priors were set for the standard deviations.

Model extensions

The above model is easily extensible to a very wide range of meta-analysis situations. A full multivariate meta-analysis that allows for possible multiple effect estimates, non-independence between effects within groups, and between effects between groups can be described as follows. Letting T_i denote the vector of effect estimates T_{ij},

$$\begin{aligned} T_i &\sim MVN(\theta_i, \Sigma_i) \\ \theta_i &\sim MVN(\mu_i, \Gamma_i) \\ \mu_i &\sim MVN(\mu_0, \Omega). \end{aligned}$$

Here Σ_i is a matrix that describes the variability of the effect estimates around the true effects in the ith study, with diagonal terms σ_{ij}^2 that describe the variances of $T_{i1}, T_{i2} \ldots, T_{iPi}$ and the off-diagonal terms σ_{ijk} describe the covariances between the jth and kth estimates, T_{ij} and T_{ij}, in the ith study. The corresponding matrix of estimated variances and covariances is denoted by C_i. Hence

$$C_i = \begin{bmatrix} S_{i1}^2 & S_{i2} & \ldots & S_{iR} \\ \ldots & S_{i2}^2 & \ldots & \ldots \\ \vdots & \vdots & \ddots & \vdots \\ S_{i1P_1} & \ldots & \ldots & S_{iP_I}^2 \end{bmatrix}, \quad \Sigma_i = \begin{bmatrix} \sigma_{i1}^2 & \sigma_{i2} & \ldots & \sigma_{iR} \\ \ldots & \sigma_{i2}^2 & \ldots & \ldots \\ \vdots & \vdots & \ddots & \vdots \\ \sigma_{i1P_1} & \ldots & \ldots & \sigma_{iP_I}^2 \end{bmatrix}.$$

Similarly, for the ith study, Γ_i describes the variability of the true effects around the overall group effect μ_i. Thus the diagonal terms describe the variances of $\theta_1, \theta_2, \ldots, \theta_{Pi}$ and the off-diagonal terms describe the covariances (relationships) between the effects θ_j and θ_k for the ith study:

$$\Gamma_i = \begin{bmatrix} \gamma_{i1}^2 & \gamma_{i12} & \ldots & \gamma_{i1I} \\ \ldots & \gamma_{i2}^2 & \ldots & \ldots \\ \vdots & \vdots & \ddots & \vdots \\ \gamma_{i12} & \ldots & \ldots & \gamma_{iI}^2 \end{bmatrix}.$$

Finally, the matrix Ω describes the variability of the overall group effects μ_i around the global effect μ_0. The diagonal terms of this matrix represent the variances of $\mu_1, \mu_2, \ldots, \mu_I$ and the off-diagonal terms describe the covariances between the effects:

$$\Omega = \begin{bmatrix} \omega_1^2 & \omega_{12} & \ldots & \omega_{1I} \\ \ldots & \omega_2^2 & \ldots & \ldots \\ \vdots & \vdots & \ddots & \vdots \\ \omega_{i1} & \ldots & \ldots & \omega_I^2 \end{bmatrix}.$$

For exposition, we describe this general model more explicitly for the situation in which the meta-analysis contains two effects with no covariates. In this case, the model expands as follows:

$$\begin{bmatrix} T_{i1} \\ T_{i2} \end{bmatrix} \sim N\left(\begin{bmatrix} \theta_{i1} \\ \theta_{i2} \end{bmatrix}, \Sigma_i\right); \qquad \Sigma_i = \begin{bmatrix} \sigma_{i1}^2 & \sigma_{i12} \\ \sigma_{i12} & \sigma_{i2}^2 \end{bmatrix}$$

$$\begin{bmatrix} \theta_{i1} \\ \theta_{i2} \end{bmatrix} \sim N\left(\begin{bmatrix} \mu_1 \\ \mu_2 \end{bmatrix}, \Gamma\right); \qquad \Gamma = \begin{bmatrix} \gamma_1^2 & \gamma_{12} \\ \gamma_{12} & \gamma_2^2 \end{bmatrix}.$$

Note that, as before, Σ_i is estimated by C_i. Note also that sometimes it is easier to describe the model in terms of correlations. For example, if r_{i12} is the observed correlation between the first and second effect estimates, then $S_{i12} = r_{i12}S_{i1}S_{i2}$, similarly, if ρ_{12} is the correlation between the overall effects $\mu_1$and μ_2, then $\gamma_{12} = \rho_{12}\gamma_1\gamma_2$.

By merging the levels of the model, the observed effects are related to the overall effects as follows:

$$\begin{bmatrix} T_{i1} \\ T_{i2} \end{bmatrix} \sim N\left(\begin{bmatrix} \mu_1 \\ \mu_2 \end{bmatrix}, V_i\right);$$

$$V_i = \begin{bmatrix} \sigma_{i1}^2 + \gamma_1^2 & \sigma_{i12} + \gamma_{12} \\ \sigma_{i12} + \gamma_{12} & \sigma_{i2}^2 + \gamma_2^2 \end{bmatrix}.$$

Thus the observed effects are centred around the overall effects (μ_1, μ_2) and (as for the simple random effects model) the combined variance is the sum of the within-study variances (given by the σs) and the between-study variances (given by the γs).

7.2.3 Computation

All analyses were undertaken using WinBUGS. For each analysis, two chains of 200 000 iterations were run, with 50 000 iterations discarded as burn-in. Convergence was assessed using the diagnostics available in WinBUGS, namely inspection of trace and density plots and evaluation of the Brooks–Gelman–Rubin (BGR) statistic.

7.2.4 Results

Aim 1

The maximum odds ratios and associated 95% confidence intervals obtained from the 15 studies are reported in Table 7.1, along with their posterior estimates gained from the meta-analysis. Figure 7.1 shows a forest plot of these estimates. A forest plot shows the effect estimate and confidence intervals of each study around a risk ratio of 1 (no effect), along with a summary estimate of the overall effect size of all studies in the analysis. The position of the boxes for each study represents the effect found, with the size

Table 7.1 Reported effect estimates and corresponding 95% confidence intervals (left hand columns) and posterior mean log odds ratios and 95% credible intervals (right hand columns).

	Observed			Posterior		
Study reference	Max. odds ratio	2.5%	97.5%	Mean	2.5%	97.5%
Kallianpur	1.042	0.964	1.124	0.056	−0.017	0.126
Shannon	1.360	1.113	1.662	0.200	0.058	0.372
Ferrucci	1.094	1.000	1.196	0.096	0.016	0.175
Taylor	1.161	1.046	1.294	0.139	0.049	0.233
De Stefani	1.425	1.098	1.849	0.195	0.042	0.397
Cho	1.109	0.982	1.249	0.108	0.009	0.210
Kabat	0.987	0.932	1.042	0.002	−0.054	0.059
van der Hel	1.128	0.341	1.374	0.122	−0.075	0.339
Lee	1.507	1.143	1.989	0.207	0.047	0.421
Delfino	0.783	0.601	1.017	0.159	0.010	0.342
Mills	1.021	0.883	1.182	0.062	−0.058	0.175
Zheng	1.189	0.846	1.673	0.132	−0.035	0.322
Ambrosone	1.042	0.856	1.259	0.083	−0.057	0.217
Vatten	1.082	0.908	1.259	0.097	−0.026	0.220
Toniolo	1.312	1.038	1.659	0.175	0.030	0.354
μ_0	—	—	—	0.122	0.055	0.205
tau_0	—	—	—	298.200	29.690	858.200

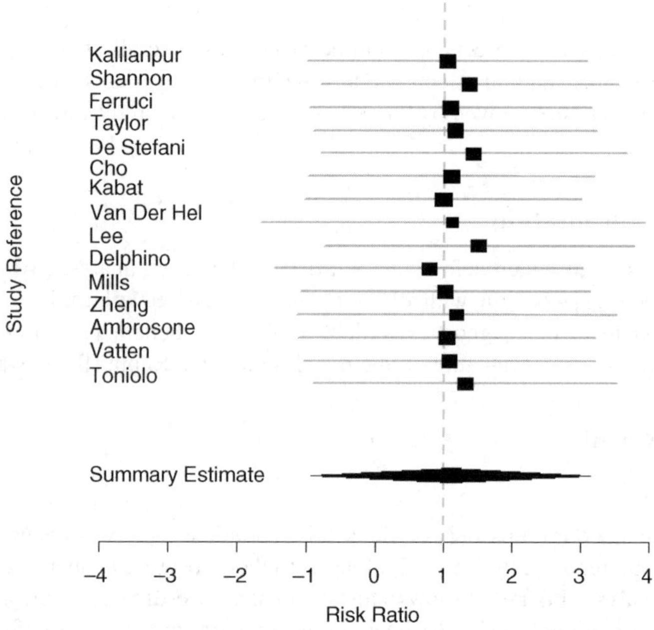

Figure 7.1 Forest plot of reported maximum odds ratios and 95% confidence intervals for all studies, and overall posterior mean odds ratio (summary estimate) with 95% credible interval.

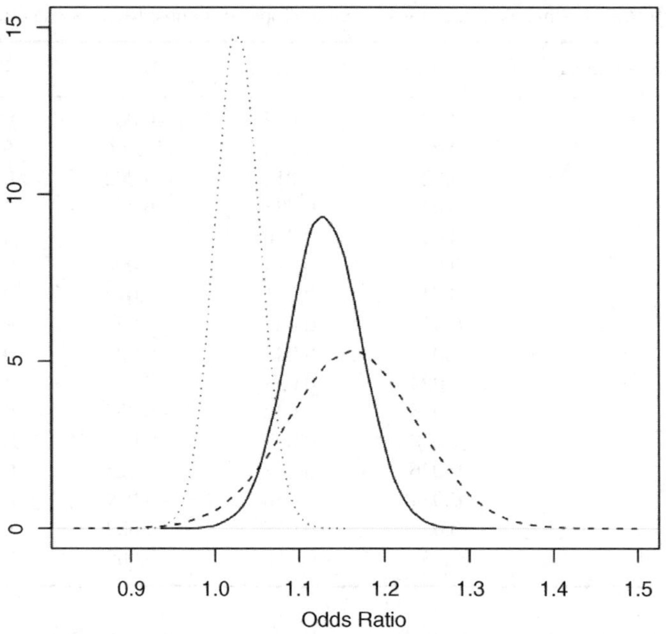

Figure 7.2 Posterior distributions of the odds ratios (horizontal axis) reflecting the association between red meat consumption and breast cancer for post-menopausal (left hand dotted line) and pre-menopausal (right hand dashed line) women, and all women combined (solid line).

of the box denoting the weight of the study. The summary estimate denotes the overall effect estimate, with the edges of the diamond denoting the overall confidence interval.

Analysis includes: Ambrosone *et al.* (1998), Cho *et al.* (2006), Delfino *et al.* (2000), Ferrucci *et al.* (2009), Kabat *et al.* (2007), Kallianpur *et al.* (2008), Lee *et al.* (1991), Mills *et al.* (1989), Shannon *et al.* (2003), Stefani *et al.* (1997), Taylor *et al.* (2009), Toniolo *et al.* (1994), van der Hel *et al.* (2004), Vatten and Kvinnsland (1990), Zheng *et al.* (1998).

Aim 2

The subgroup analyses revealed a distinct difference in the posterior distributions of the overall odds ratio for pre-menopausal and post-menopausal women (Figure 7.2). The overall odds ratio for the former group is 1.162 (95% CI = 1.041, 1.344), compared with 1.027 (95% CI = 0.916, 1.204) for the latter group. The posterior probability that the association (odds ratio) is stronger in the pre-menopausal group compared with the post-menopausal group (i.e. $\Pr(\mu_1 > \mu_2)$) is 0.953.

Aim 3

The exposure–response analysis indicated an overall positive trend in log odds ratios of 0.033 with 95% credible interval (0.005–0.068). Table 7.2 shows the posterior estimates of the study-specific slopes. Figure 7.3 depicts the slopes of all studies

Table 7.2 Estimates of the study-specific slopes of exposure–response effects.

Study reference	β_i	SD	2.5%	97.5%
1	0.022	0.017	−0.008	0.058
2	0.050	0.032	−0.006	0.114
3	0.026	0.015	−0.002	0.057
4	0.046	0.023	6.2E-5	0.089
5	0.053	0.036	−0.008	0.130
6	0.017	0.014	−0.008	0.047
7	−1.7E-4	0.011	−0.016	0.028
8	0.035	0.038	−0.034	0.118
9	0.041	0.042	−0.030	0.141
10	0.024	0.038	−0.054	0.103
11	0.026	0.032	−0.035	0.095
12	0.038	0.038	−0.030	0.123
13	0.026	0.027	−0.025	0.082
14	0.034	0.038	−0.039	0.118
15	0.052	0.027	0.002	0.107
Overall	0.033	0.016	0.005	0.068

along with a summary slope of the overall exposure–response effect. The overall slopes show that there is in fact a dose–response effect across all studies, despite one study in particular finding a weak effect in the opposite direction, and all others finding variable dose–response effects.

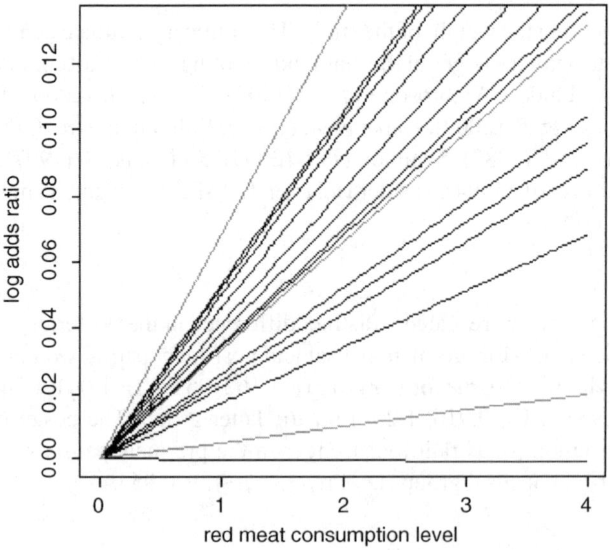

Figure 7.3 Linear exposure—response relationships based on the posterior means of the slope estimate for each study (solid lines) and the overall relationship based on the posterior mean and 95% upper and lower credible bounds (light grey lines).

7.2.5 Discussion

This case study has illustrated a number of Bayesian meta-analysis models, including a simple univariate random effects model for estimation of the overall association between red meat consumption and breast cancer, subset models and meta-regression for evaluation of different subgroups, and an exposure–response model for assessment of whether the risk of breast cancer increases with increasing levels of red meat consumption.

There is a large multi-disciplinary literature on issues related to systematic reviews and meta-analysis. See, for example, Vamkas (1997) for meta-analysis targeted to medicine, Koricheva *et al.* (2011) for meta-analysis targeted at ecology and evolution, and Davis *et al.* (2011) for issues arising in meta-analysis in social science research.

As illustration of these larger issues, we considered the impact of two possible biases arising from non-publication of borderline or negative results, and temporal changes in the conduct and reporting of studies on a topic. For the first issue, a funnel plot of the estimates from the 15 studies (Figure 7.4) did not indicate systematic publication bias. The second issue is typically addressed by undertaking a cumulative meta-analysis, whereby a series of meta-analyses are undertaken sequentially on the chronologically ordered set of studies. This is akin to 'Bayesian learning', whereby knowledge accumulated on the basis of existing data can be formulated as a prior in the analysis of subsequent data.

We illustrate this by running a meta-analysis of the first eight studies (1989–2000) and using the resultant posterior distributions of μ_0 and τ_0^2 to formulate the priors for a meta-analysis of the last seven studies (2003–2009). In the first stage (analysis of the first eight studies), we used the uninformative priors described above. In the second stage we used a normal prior $N(\mu_\alpha, \sigma_\alpha^2)$ for μ_0, where μ_α, σ_α^2 were the posterior mean and variance of μ_0 obtained from the first stage. Using the information that was now

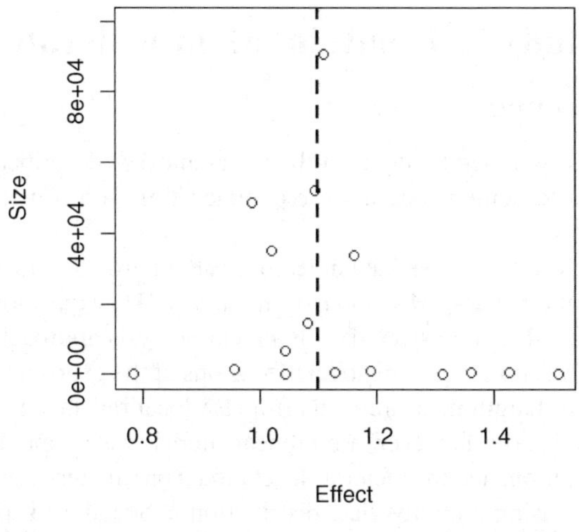

Figure 7.4 Funnel plot of the estimates of the 15 studies against their sample size.

available about r_0^2, in the second stage we replaced the uniform prior by a gamma distribution $Ga(\alpha, \beta)$) with parameters $\alpha = \mu_\beta^2/\sigma_\beta^2$, $\beta = \mu_\beta/\sigma_\beta^2$, where μ_β and σ_β^2 were the posterior means and variances of τ_0^2 from the first stage. The resultant priors were $\mu_0 \sim N(0.113, 304)$ and $\tau_0^2 \sim Ga(2.6\text{E-}4, 1.8\text{E-}7)$.

The posterior mean of the overall log odds ratio based on the studies published in 1989–2000 was 1.12 with 95% credible interval (1.01–1.14). The analogous values accumulated from both stages of analysis were 1.13 (1.05, 1.22). This may be compared with the analogous estimate of 1.13 (95% credible interval 1.04, 1.23) based on the analysis of all the data combined. (Note that this equivalence is expected mathematically.) These results indicate evidence of a small change in the observed association over time, but they do illustrate the increased certainty of a raised risk based on the accumulated evidence.

The issue of study quality must also be addressed. Wolpert and Mengersen (2004) present a Bayesian approach to this issue, in which the data inform about 'apparent' model parameters (which are subject to biases and misclassification), which are themselves considered to be estimates of corresponding 'true' model parameters. The potential biases and misclassification are thus represented by prior distributions on the apparent model parameters.

Finally, it is important to note that while this analysis has confirmed both a positive association and an exposure-dependent relationship between red meat consumption and breast cancer, this does not yet establish a causal relationship. For example, the criteria presented by Hill (1977) include nine conditions for assessing causation, three of which have not been addressed in the current study. These include the confirmation of time order (that the exposure preceded the outcome in a biologically credible manner) and the rigorous dismissal of confounders or other alternative explanations of the results.

7.3 Case study 2: Trends in fish growth rate and size

7.3.1 Background

This case study is designed to augment the meta-analyses described in the first case study. The aim is to demonstrate a more complex time series meta-regression and multivariate analysis.

In this example we are interested in learning about growth rate and size of North American largemouth bass, *Micropterus salmoides*. This question was posed by Helser and Lai (2004) and answered with a meta-analysis approach akin to that described in the above subsection on model extensions of the exposure–response model using annual mean length measurements from 25 locations taken over periods of 6 to 10 years. In this case, however, a nonlinear model was adopted to describe fish growth at each location, and the location-level model parameters were then combined at an overall level using a multivariate distribution to account for parameters in the nonlinear model being correlated.

The original data used by Helser and Lai (2004) were not published, so for the purposes of exposition we generated a synthetic data set (see the supplementary material on the book's website) based on the nonlinear growth model they adopted, which can be described as follows.

Let T_{ij} be the mean length at year Y_j in the ith study. Then

$$T_{ij} = L_{\omega i}\{1 - \exp[-K_j(Y_j - t_{0i})]\} + \epsilon_{ij}; \epsilon_{ij} \sim N(0, \sigma_z^2) \tag{7.1}$$

where $L_{\omega i}$ is the expected asymptotic length of fish in the ith study, K_i is a growth parameter, and t_{0i} is the expected length of fish at birth in the ith study.

Equation (7.1) contains two sources of variation: sampling variance around the reported annual mean lengths (due to taking a sample of the population of fish at each site and time period); and regression variance around the nonlinear model (since the values will typically not sit exactly on the regression line). Thus the equation can be rewritten as

$$\begin{aligned} T_{ij} &\sim N(\mu_{ij}, \sigma_{ij}^2) \\ \mu_{ij} &\sim N(\mu'_{ij}, \gamma_i^2) \\ \mu'_{ij} &= L_{\omega i}\{1 - \exp[-K_i(Y_j - t_{0i})]\}. \end{aligned}$$

The value μ'_{ij} is the expected average length of the fish at site i at time j, derived from the nonlinear growth model. The value μ_{ij} is the true average length at this site and time, which (as in any regression model) will deviate from the modelled value μ'_{ij} according to the regression variation γ_i^2. Finally, the value L_{ij} is the corresponding observed average length, which is an estimate of μ_{ij} and is reported with some sampling variance σ_{ij}^2 that depends on the population variance and the sample size. The three site-specific parameters of interest can be described by the vector $\theta_i = (L_{\omega i}, \log K_i, t_{0i})$.

7.3.2 Meta-analysis models

Case study model

The vector of parameters that describe the regression equations in this chapter can be combined in a random effects meta-analysis using a multivariate normal (MVN) distribution as follows:

$$\theta_i \sim \text{MVN}(\theta', \Gamma)$$

where θ' represents the vector of overall population parameters for these largemouth bass. This mean vector comprises three terms, the overall asymptotic length, log growth rate and length at birth, and the associated variance–covariance matrix Γ. Note that since K_i is a growth parameter, for reasons given below it is more convenient to transform it to the log scale so that it is more normally distributed.

For this case study, uninformative priors were employed, comprising a vague multivariate normal distribution for θ' and a vague inverse Wishart distribution for Γ.

This model accommodates a number of sources of uncertainty, arising from variation around reported mean length-at-age estimates, variation around the nonlinear growth model within each study, non-independence of model parameters, variation between studies, and uncertainty about overall estimates. The model can be extended to accommodate further uncertainty due to missing data, measurement error, covariates, study quality, and so on. This is easily achieved in a Bayesian model. For example, missing data can be treated as unknown variables to be estimated as part of the analysis, and the uncertainty of these estimates is propagated naturally through to the uncertainty in estimation of the other model parameters.

Model extensions

In other meta-analysis situations, the analyst may not have access to the primary data as in this case study, but may only have the reported regression estimates from each study, with associated variances. The above models can be easily adapted to this situation. For exposition, and because it is a typical situation in practice, we describe such a model for a linear time series (linear regression) meta-analysis. Thus we assume that each study has provided estimates of the intercept and slope parameters β_{0i} and β_{li}, denoted by b_{i0} and b_{il}. Taking into account that these estimates may be correlated, then

$$\begin{bmatrix} b_{i0} \\ b_{i1} \end{bmatrix} \sim N\left(\begin{bmatrix} \beta_{i0} \\ \beta_{i1} \end{bmatrix}, \Sigma_i \right)$$

where

$$\Sigma_i = \begin{bmatrix} \sigma_{i0}^2 & \sigma_{i01} \\ \sigma_{i10} & \sigma_{i1}^2 \end{bmatrix}$$

and σ_{i01} is the covariance between the intercept and the slope in the ith study. The corresponding matrix of estimated variance and covariance terms is represented by $\mathbf{C}_i$, so that

$$C = \begin{bmatrix} S_{i0}^2 & S_{i01} \\ S_{i10} & S_{i1}^2 \end{bmatrix} \tag{7.2}$$

Using the same notation, Equation (7.2) can also be written as

$$\begin{bmatrix} b_{i0} \\ b_{i1} \end{bmatrix} \sim N\left(\begin{bmatrix} \beta_{i0} \\ \beta_{i1} \end{bmatrix}, \Sigma_i\right)$$
$$\begin{bmatrix} \beta_{i0} \\ \beta_{i1} \end{bmatrix} \sim N\left(\begin{bmatrix} \mu_{0} \\ \mu_{1} \end{bmatrix}, \Gamma\right)$$
$$\Gamma = \begin{bmatrix} \gamma_0^2 & \gamma_{01} \\ \gamma_{10} & \gamma_1^2 \end{bmatrix}.$$

Combining the levels of the model, we find that

$$\begin{bmatrix} b_{i0} \\ b_{i1} \end{bmatrix} \sim N\left(\begin{bmatrix} \mu_{i0} \\ \mu_{i1} \end{bmatrix}, \Gamma + \Sigma_i\right).$$

Ideally, the estimated effects (b_{0i}, b_{1i}) and corresponding variance–covariance matrix $\mathbf{C}_i$ are reported for each study, or may be calculated from reported statistics, or can be derived by the meta-analyst from the primary data if these are available. Typically, however, even if the variance estimates are available, the covariance term is not. In the case of missing variance estimates, 'best guesses' can be made on the basis of existing data or other information, and a sensitivity analysis can be conducted in which a range of values are assumed and the effect on the estimates of interest is assessed. Alternatives are also being devised. For example, Riley *et al.* (2008) suggests a slight reparameterization of the random effects meta-analysis model that does not involve the covariance terms, at the expense of modified inferences. Although the synthesis of regression parameters was not explicitly discussed, it is conceptually possible to transfer these results to this context.

Note that if we ignore the covariances between parameter estimates within and between studies, then this model can be written as

$$T_{ij} = \beta_{i0} + \beta_{il}\mathbf{X}_{ij} + e_{ij}; \qquad e_{ij} \sim N(0, \sigma_T^2)$$
$$\beta_{i0} = \mu_0 + \epsilon_{i0}; \qquad \epsilon_{i0} \sim N(0, \gamma_0^2)$$
$$\beta_{i1} = \mu_1 + \epsilon_{i1}; \qquad \epsilon_{i1} \sim N(0, \gamma_1^2)$$

or alternatively

$$T_{ij} \sim N(\theta_{ij}, \sigma_T^2)$$
$$\theta_{ij} = \beta_{i0} + \beta_{i1}X_{ij}$$
$$\beta_{i0} \sim N(\mu_0, \gamma_0^2); \qquad \beta_{i1} \sim N(\mu_1, \gamma_1^2)$$

where μ_1 is the overall trend, γ_0^2 describes the variation between the study-specific intercepts β_{i0} and γ_1^2 describes the variation between the study-specific trends β_{i1}. Thus the first level of the hierarchical model (that describes the distribution of the T_{ij}) remains the same, but the second level is reduced to two separate univariate models. Similarly, the above model (based on the regression estimates from each study) becomes

$$
\begin{aligned}
b_{i0} &= \beta_{i0} + e_{i0}; & e_{i0} &\sim N(0, \sigma_{i0}^2)\\
b_{i1} &= \beta_{i1} + e_{i1}; & e_{i1} &\sim N(0, \sigma_{i1}^2)\\
\beta_{i0} &= \mu_0 + \epsilon_{i0}; & \epsilon_{i0} &\sim N(0, \gamma_0^2)\\
\beta_{i1} &= \mu_1 + \epsilon_{i1}; & \epsilon_{i1} &\sim N(0, \gamma_1^2).
\end{aligned}
$$

Alternatively,

$$
\begin{aligned}
b_{i0} &\sim N(\beta_{i0}, \sigma_{i0}^2); & \beta_{i0} &\sim N(\mu_0, \gamma_0^2)\\
b_{i1} &\sim N(\beta_{i1}, \sigma_{i1}^2); & \beta_{i1} &\sim N(\mu_1, \gamma_1^2)
\end{aligned}
$$

where σ_{i0}^2 is the study-specific variance of the estimate b_{i0}, and σ_{i1}^2 is the study-specific variance of the trend estimate b_{i1}. Finally, note that if the focus of the meta-analysis is on a single regression parameter, such as the linear trend, then the above model reduces to a simple univariate model involving only b_{i1}, β_{i1} and μ_1.

7.3.3 Computation

The model described above for the case study was analysed using WinBUGS; the code is provided on the book's website. The MCMC run comprised three chains (to demonstrate an alternative method from the previous example), 100 000 iterations for burn-in and a further 100 000 iterations for parameter estimation. Convergence of the MCMC chains was assessed using trace and density plots and the BGR statistic, and model fit was assessed using the deviance information criterion (DIC) and posterior probability checks.

7.3.4 Results

The output of this analysis comprised the following: posterior densities and corresponding summary statistics (mean, mode, variance, 95% credible interval) for model parameters ($L_{\omega i}$, $\log K_i$, t_{0i}), correlation between parameters, probability that a parameter exceeds a biologically important threshold, and ranking of studies with respect to posterior estimates. For example, the posterior mean log growth rate was estimated as -1.45 ($\pm$ 0.102) and, assuming for illustration that a growth rate of 0.2 cm per annum is biologically important, there is a 7% chance of this event (i.e. $\log K < -1.6$) based on this data set and model (see Figure 7.5).

Comparison of studies on the basis of estimated growth parameters is achieved by examining the posterior distribution of ranks for each study. For example, Study 17 had a median rank of 3 (indicating the third smallest growth rate among the 25 studies) but a 95% credible interval of $[1, 9]$, indicating that with 95% probability the study could have been ranked smallest or ninth smallest. This is much more informative than simple ranking based only on the mean or median values.

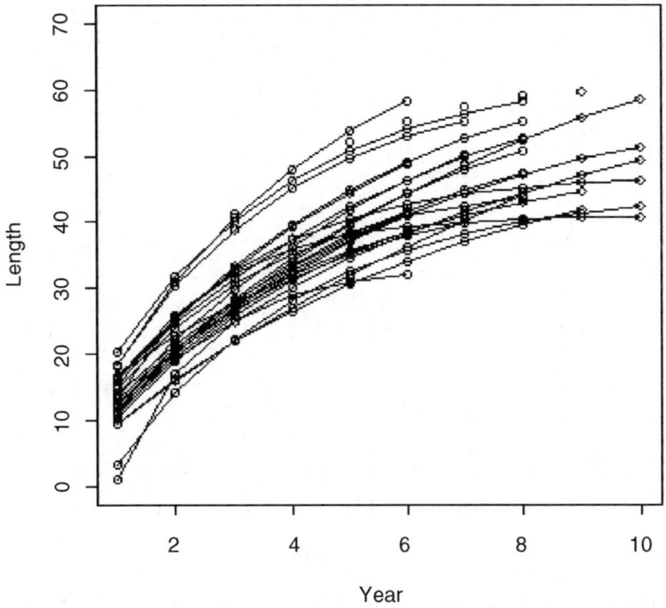

Figure 7.5 Data for Case study 2. Each point is an observed mean length with an associated sampling variance (not shown).

Finally, it is interesting to compare the relative magnitude of the different sources of variation in the meta-analysis model. Variation between studies was by far the greatest contributor in this particular analysis (see Figure 7.6). This motivates further investigation of covariates that might explain this heterogeneity. Helser and Lai (2004) also examined the impact of latitude. The comparison also shows that within-study variation is considerably larger than within-estimate variation, indicating that more precise overall estimates might be obtained by refining the nonlinear model in Equation (7.1) rather than refining estimates within studies (e.g. by increased sample sizes).

Helser and Lai (2004) compare their modelling approach with the traditional alternative of fitting the growth model separately to back-calculated length-at-age measurements of individuals within a given population and obtaining least squares estimates of growth parameters summarized by means and variances. The traditional model treats all growth parameters as fixed, assumes common variance for all residuals across all populations, and assumes no relationship among populations. Although the two models gave similar overall estimates, Helser and Lai (2004) obtained greater precision of estimation, a biologically important result of substantially smaller correlation between growth rate and initial length, and larger residual variation between studies (due to fitting study-specific growth models).

7.3.5 Discussion

The models described above can be extended to describe a variety of problems. For example, a dose–response relationship can be considered in the same manner

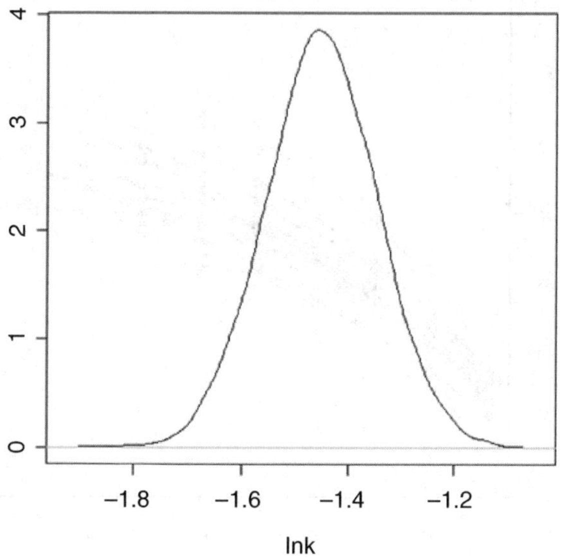

Figure 7.6 Case study 2: plot of the posterior distribution for log K.

as a time series analysis, with time replaced by 'dose' as the independent variable in the regression equation. The objectives of the meta-analysis may therefore be to use all of the information available from the multiple studies to estimate an overall relationship, better estimate study-specific relationships and/or identify sources of variation in relationships across studies.

A general mixed model was developed by DuMouchel (1995) for meta-analysis of dose–response estimates. This paper and later ones by DuMouchel allow combination of studies representing multiple outcomes or multiple treatments, accommodating both binary and continuous outcomes in the model (although a common scale is required for all the outcomes in any particular model). This method does not require correlations between outcomes within studies to be known since these are estimated by the model. This model extends the standard random effects model of DerSimonian and Laird (1986) allowing heterogeneous study designs to be accounted for and combined. No common treatment or control is required in all studies, unlike in the methods of Gleser and Olkin (1994) for multiple treatment designs. Different treatment groups can consist of results from separate (sub)groups of subjects, or from groups that cross over and are subject to multiple treatments (e.g. repeated measures of same individuals). A key difference from the univariate random effects models is that each group of subjects is modelled separately, so the model does not treat them as truly multivariate.

Houwelingen *et al.* (2002) also adopted a mixed model extension to their 1993 approach to investigate if covariates at the study level are associated with baseline risk via multivariate regression. In a similar vein, Ende (2001) discusses how to analyse a repeated measures response over levels of a treatment; Becker and Wu (2007) describe a polynomial regression meta-analysis, and Arends (2006) describes these models in the context of a wide range of methods for multivariate meta-analysis.

A number of Bayesian approaches to the full multivariate model have been proposed. General Bayesian mixed model approaches for meta-analysis have been promoted and illustrated by DuMouchel and co-authors, with particular focus on combining studies of interest to the pharmaceutical sciences (DuMouchel 1990). Dominici *et al.* (1999) proposed a hierarchical Bayesian group random effects model and presented a complex meta-analysis of both multiple treatments and multiple outcomes. This model assumes different random effects from different levels of study variation and evaluates 18 different treatments for the relief of migraine headaches, which were grouped within three classes. Multiple heterogeneous reported outcomes (including continuous effect sizes, differences between pairs of continuous and dichotomous outcomes) and multiple treatments were incorporated via relationships between different classes of treatments.

Three Bayesian multivariate models were proposed and compared by Nam *et al.* (2003). Their preferred model was the full hierarchical representation as described above. Nam *et al.* (2003) detail a Bayesian approach to the same problem considered by Raudenbush *et al.* (1988) and Berkey *et al.* (1998), focusing on multiple outcomes after adjusting for study-level covariates. Turner *et al.* (2006) provide details of a Bayesian hierarchical model for multivariate outcome data applicable to randomized cluster trials. These are studies in which entire clusters of subjects are randomized across treatments, but outcomes are measured on individuals (e.g. fields are given different fertilizer treatments and then growth of individual plants is measured). Here, correlation between observations in the same cluster, as well as correlation between the outcomes, creates dependence between the data that must be taken into account in the model. The authors also consider a parameterization of the model that allows for the description of a common intervention effect based on data measured on different scales. The models proposed by the authors are applicable to a multivariate meta-analysis with clustering among studies.

Finally, Becker (1995) describes methods of combining correlations. Interrelationships between variables in the correlation matrix can be analysed using fixed or random effects models. A set of linear models is derived from the correlation matrix that describes both direct and indirect relationships between outcomes of interest and explanatory variables. Inferences are made on the coefficients of the resultant standardized regression equations, including tests for partial relationships.

For examples of meta-analyses of repeated measures, see Marsh (2001), Barrowman *et al.* (2009), Harley *et al.* (2001), Harley and Myers (2001) and Blenckner *et al.* (2007).

Acknowledgements

The second case study described in this chapter was originally developed by the second author as part of a book chapter co-authored with Christopher Schmid and Michael Jennions, and then not included in that chapter. The first case study was developed with assistance from Aleysha Thomas and Jacqueline Davis.

References

Ambrosone C, Freudenheim J, Sinha R, Graham S, Marshall J, Vena J, Laughlin R, Nemoto T and Shields P 1998 Breast cancer risk, meat consumption and n-acetyltransferase (nat2) genetic polymorphisms. *International Journal of Cancer* **75**(6), 825–830.

Arends L 2006 Multivariate meta-analysis: modelling the heterogeneity. Technical report, Erasmus University, Rotterdam.

Bae Y, Yeon J, Sung C, Kim H and Sung M 2009 Dietary intake and serum levels of iron in relation to oxidative stress in breast cancer patients. *Journal of Clinical Biochemistry and Nutrition* **45**, 355–360.

Barrowman N, Myers R, Hilborn R, Kehler D and Field C 2009 The variability among populations of coho salmon in the maximum reproductive rate and depensation. *Ecological Applications* **13**, 784–793.

Becker B 1995 Corrections to using results from replicated studies to estimate linear model. *Journal of Educational Statistics* **20**, 100–102.

Becker B and Wu M 2007 The synthesis of regression slopes in meta-analysis. *Statistical Science* **22**, 414–429.

Beral V 2003 Breast cancer and hormone-replacement therapy in the million women study. *Lancet* **362**(9390), 419–427.

Berkey C, Hoaglin D, Antczak-Bouckoms A, Mosteller F and Colditz G 1998 Meta-analysis of multiple outcomes by regression with random effects. *Statistics in Medicine* **17**, 2537–2550.

Blenckner T, Adrian R and Livingstone D 2007 Large-scale climatic signatures in lakes across Europe: a meta-analysis. *Global Change Biology* **13**, 1314–1326.

Boyd N, Stone J, Voght K, Connelly B, Martin L and Minkin S 2003 Dietary fat and breast cancer: a meta-analysis of the published literature. *British Journal of Cancer* **89**(9), 1672–1685.

Byrne C, Schairer C, Wolfe J, Parekh N, Salane M, Brinton L, Hoover R and Haile R 1995 Mammographic features and breast cancer risk: effects with time, age, and menopause status. *Journal of the National Cancer Institute* **87**(21), 1622–1629.

Cho E, Chen W, Hunter D, Stampfer M, Colditz G, Hankinson S and Willett W 2006 Red meat intake and risk of breast cancer among premenopausal women. *Archives of Internal Medicine* **166**, 2253–2259.

Davis J, Mazerolle L and Bennett S 2011 Methods for criminological meta-analysis. In Review.]

Delfino R, Sinha R, Smith C, West J, White E, Lin H, Liao S, Gim J, Ma H, Butler J and Anton-Culver H 2000 Breast cancer, heterocyclic amines from meat and n-acetyltransferase 2 genotype. *Carcinogenesis* **21**(4), 607–615.

DerSimonian R and Laird N 1986 Meta-analysis in clinical trials. *Controlled Clinical Trials* **7**, 177–188.

Dominici F, Parmigiani G, Wolpert R and Hasselblad V 1999 Meta-analysis of migraine headache treatments: combining information from heterogeneous designs. *Journal of the American Statistical Association* **94**, 16–28.

DuMouchel W 1990 *Bayesian Meta-Analysis*, pp. 509–529. Marcel Dekker, New York.

DuMouchel W 1995 Meta-analysis for dose-response models. *Statistics in Medicine* **14**, 679–685.

Koricheva J, Gurevitch J and Mengersen K (eds) 2011 *Handbook of Meta-analysis in Ecology*. Princeton University Press, Princeton, NJ.

Ende NV 2001 Repeated-measures analysis: growth and other time dependent measures. In *Design and Analysis of Ecological Experiments*. Oxford University Press, Oxford.

Ferrucci L, Cross A, Graubard B, Brinton L, McCarthy C, Ziegler R, Ma X, Mayne S and Sinha R 2009 Intake of meat, meat mutagens, and iron and the risk of breast cancer in the prostate, lung, colorectal, and ovarian cancer screening trial. *British Journal of Cancer* **101**, 178–184.

Gleser L and Olkin I 1994 Stochastically dependent effect sizes. In *The Handbook of Research Synthesis*, pp. 16–28. Sage, New York.

Harley S and Myers R 2001 Hierarchical Bayesian models of length-specific catchability of research trawl surveys. *Canadian Journal of Fisheries and Aquatic Sciences* **58**, 1569–1584.

Harley S, Myers R, Barrowman N, Bowen K and Amiro R 2001 Estimation of research trawl survey catchability for biomass reconstruction of the eastern Scotian Shelf. Technical report, Canadian Science Advisory Secretariat, Ottawa.

Helser T and Lai H 2004 A Bayesian hierarchical meta-analysis of fish growth: with an example for the North American largemouth bass, Micropterus salmoides. *Ecological Modelling* **178**, 399–416.

Hill A 1977 *A Short Textbook of Medical Statistics*. Hodder and Stoughton, London.

Houwelingen HV, Arends L and Stijnen T 2002 Advanced methods in meta-analysis: multivariate approach and meta-regression. *Statistics in Medicine* **21**, 589–624.

Kabat G, Miller A, Jain M and Rohan T 2007 Dietary iron and heme iron intake and risk of breast cancer: a prospective cohort study. *Cancer Epidemiology, Biomarkers and Prevention* **16**(6), 1306–1308.

Kallianpur A, Asha R, Lee S, Lu YG, Zheng Y, Ruan Z, Dai Q, Gu K, Shu X and Zheng W 2008 Dietary animal-derived iron and fat intake and breast cancer risk in the Shanghai breast cancer study. *Breast Cancer Research and Treatment* **107**, 123–132.

Kingsmore D, Ssemwogerere A, Gillis C and George W 2003 Increased mortality from breast cancer and inadequate axillary treatment. *The Breast* **12**(1), 36–41.

Lee H, Gourley L, Duffy S, Esteve J, Lee J and Day N 1991 Dietary effects on breast cancer risk in Singapore. *Lancet* **337**(8751), 1197–1201.

Marsh D 2001 Fluctuations in amphibian populations: a meta-analysis. *Biological Conservation* **101**, 327–335.

Mills P, Phillips R and Fraser G 1989 Dietary habits and breast cancer incidence among 7th day adventists. *Cancer Causes and Control* **64**(3), 582–590.

Missmer S, Smith-Warner S, Spiegelman D, Yaun S, Adami H, Beeson W, van den Brandtand P, Fraser G, Freudenheim J, Goldbohm RA, Graham S, Kushi L, Miller A, Potter J, Rohan T, Speizer F, Toniolo P, Willett W, Wolk A, Zeleniuch-Jacquotte A and Hunter D 2001 Meat and dairy food consumption and breast cancer: a pooled analysis of cohort studies. *International Journal of Epidemiology* **31**(1), 78–85.

Nam I, Mengersen K and Garthwaite P 2003 Multivariate meta-analysis. *Statistics in Medicine* **22**, 2309–2333.

Raudenbush S, Becker B and Kalaian H 1988 Modeling multivariate effect sizes. *Psychology Bulletin* **103**, 111–120.

Riley R, Thompson J and Abrams K 2008 An alternative model for bivariate random-effects meta-analysis when the within-study correlations are unknown. *Biostatistics* **9**, 172–186.

Shannon J, Cook L and Stanford J 2003 Dietary intake and risk of premenopausal breast cancer (United States). *Cancer Causes and Control* **14**, 19–27.

Stefani ED, Ronco A, Mendilaharsu M, Guidobono M and Deneo-Pellegrini H 1997 Meat intake, heterocyclic amines, and risk of breast cancer: a case-control study in Uruguay. *Cancer Epidemiology, Biomarkers & Prevention* **6**(8), 537–581.

Stevens R, Jones D, Micozzi M and Taylor P 1988 Body iron stores and the risk of breast cancer. *New England Journal of Medicine* **319**, 1047–1052.

Sutton A and Abrams K 2001 Bayesian methods in meta-analysis and evidence synthesis. *Statistical Methods in Medical Research* **10**(4), 277–303.

Taylor V, Misra M and Mukherjee S 2009 Is red meat intake a risk factor for breast cancer among premenopausal women? *Breast Cancer Research and Treatment* **117**, 1–8.

Terry P, Miller A and Rohan T 2002 Cigarette smoking and breast cancer risk: a long latency period? *International Journal of Cancer* **100**(6), 723–728.

Toniolo P, Riboli E, Shore R and Pasternack B 1994 Consumption of meat, animal products, protein, and fat and risk of breast cancer – a prospective cohort study in New York. *Epidemiology* **5**(4), 391–397.

Turner R, Omar R and Thompson S 2006 Modelling multivariate outcomes in hierarchical data, with application to cluster randomized trials. *Biometrical Journal* **48**, 333–345.

Vamkas E 1997 Meta-analysis in transfusion medicine. *Transfusion* **37**(3), 329–345.

van der Hel OL, Peeters P, Wein D, Doll M, Grobbee D, Ocke M and de Mesquita H 2004 Stm1 null genotype, red meat consumption and breast cancer risk (The Netherlands). *Cancer Causes and Control* **15**, 295–303.

Vatten L and Kvinnsland S 1990 Body mass index and risk of breast cancer: a prospective study of 23,826 Norwegian women. *International Journal of Cancer* **45**, 440–444.

Wolpert R and Mengersen K 2004 Adjusted likelihoods for synthesizing empirical evidence from studies that differ in quality and design: effects of environmental tobacco smoke. *Statistical Science* **19**(3), 450–471.

Zheng W, Gustafson D, Moore D, Hong C, Anderson K, Kushi L, Sellers T, Folsom A, Sinha R and Cerhan J 1998 Well-done meat intake and the risk of breast cancer. *Journal of the National Cancer Institute* **90**(22), 1724–1729.

8

Bayesian mixed effects models

Clair L. Alston[1], Christopher M. Strickland[1], Kerrie L. Mengersen[1] and Graham E. Gardner[2]

[1]*Queensland University of Technology, Brisbane, Australia*
[2]*Murdoch University, Perth, Australia*

8.1 Introduction

In this chapter we introduce a Bayesian approach to linear mixed models. This model is an extension to the linear regression model in Chapter 4, when data are complex and hierarchical.

Mixed models, often described as a mixture of *fixed* and *random* terms, are used in cases were the data are clustered due to subpopulations, such as *sires* in genetics trials, *years* in trials that are conducted annually, *assessors* in experiments where the person obtaining the measurements may be subjective, *individual* in experiments that contain measurements in time on traits such as growth (longitudinal studies) and where higher level terms are considered random. For mixed models, the different sources of error in data are captured by treating the variation as within and between clusters.

The terms fixed and random are used somewhat loosely, throughout this chapter, in reference to their usage under the classical paradigm. In the Bayesian framework, as all parameters are treated as random, the real difference between these terms is their prior specification. This differentiation will become apparent in Section 8.3.

Demidenko (2004), Wu (2010) and Verbeke and Molenberghs (2009) provide substantial references for these models in the frequentist framework, and provide much of the mathematical foundations required for the Bayesian framework. Sorensen and Gianola (2002) provide a comprehensive likelihood and Bayesian outline of these

Case Studies in Bayesian Statistical Modelling and Analysis, First Edition. Edited by Clair L. Alston, Kerrie L. Mengersen and Anthony N. Pettitt.

models. Introductory-level Bayesian methods are outlined by Carlin and Louis (2009) and Hoff (2009). Efficient Bayesian algorithms have been developed for the mixed model by Chib and Carlin (1999); however, in this chapter we use a less efficient approach in order to convey the essence of the Bayesian mixed model.

Mixed models can be estimated in several Bayesian software packages: for example, WinBUGS (Lunn *et al.* 2000), library MCMCglmm (Hadfield 2010) in the R package and PyMCMC (see Chapter 25). We use PyMCMC to analyse the case studies in this chapter.

8.2 Case studies

8.2.1 Case study 1: Hot carcase weight of sheep carcases

Hot carcase weight (HCWT) is the weight of a carcase when it has had its non-usable meat products, such as head, intestinal tract and internal organs, removed, as well as the hide. This weight is taken on room temperature carcases.

In this case study, the HCWT of 3282 animals has been recorded. Researchers are interested in the effect of age at observation, sex, sire breed and body fat (and possible interactions) on the hot carcase weight. However, as an observational data set, these factors are highly unbalanced. For example, there are 1040 females and 2242 males and the six different breeds have samples sizes ranging from 164 to 1134.

Figure 8.1 (top) is a plot of hot carcase weight against age. Generally, these animals reach maturity by around 150 days, and as such we would not be expecting a large response in terms of growth in this data set. However, it can be observed that there may be an increase in weight with age. However, the bottom plot indicates that other factors, such as breed, may also account for this behaviour.

At first glance, there does not appear to be too much 'activity' in this data set; however, we will investigate the contributions of the random effects, as removing these may make the signal of the data clearer. There are several factors that are likely to contribute to the overall variance, but are not of interest in themselves. These factors, known as random effects, are sire (149), flock (15) and the group in which the animals were slaughtered (kill group, 59). Figure 8.2 illustrates the mean and range of HCWTs in each of these factors.

Apart from the obvious mean differences of HCWT between the 149 sires, it is also possible that there may be a sire by age interaction, as shown in Figure 8.3. These are data taken for a sub-sample of sires, and indicate that these sires may have a different response in terms of weight gain with increasing age. For example, sires 1, 2 and 6 seem to have a clear increase with age, whereas sires 3, 4 and 5 seem to have reached full maturity and are not increasing with additional age. Therefore, the fixed effect of age may be 'blurred' by the different sires. In our analysis, we will also test the importance of this random term.

8.2.2 Case study 2: Growth of primary school girls

Another use of mixed models is identifying clusters associated with individuals' growth curves. In this widely available study, 20 girls with a starting age of 6 years

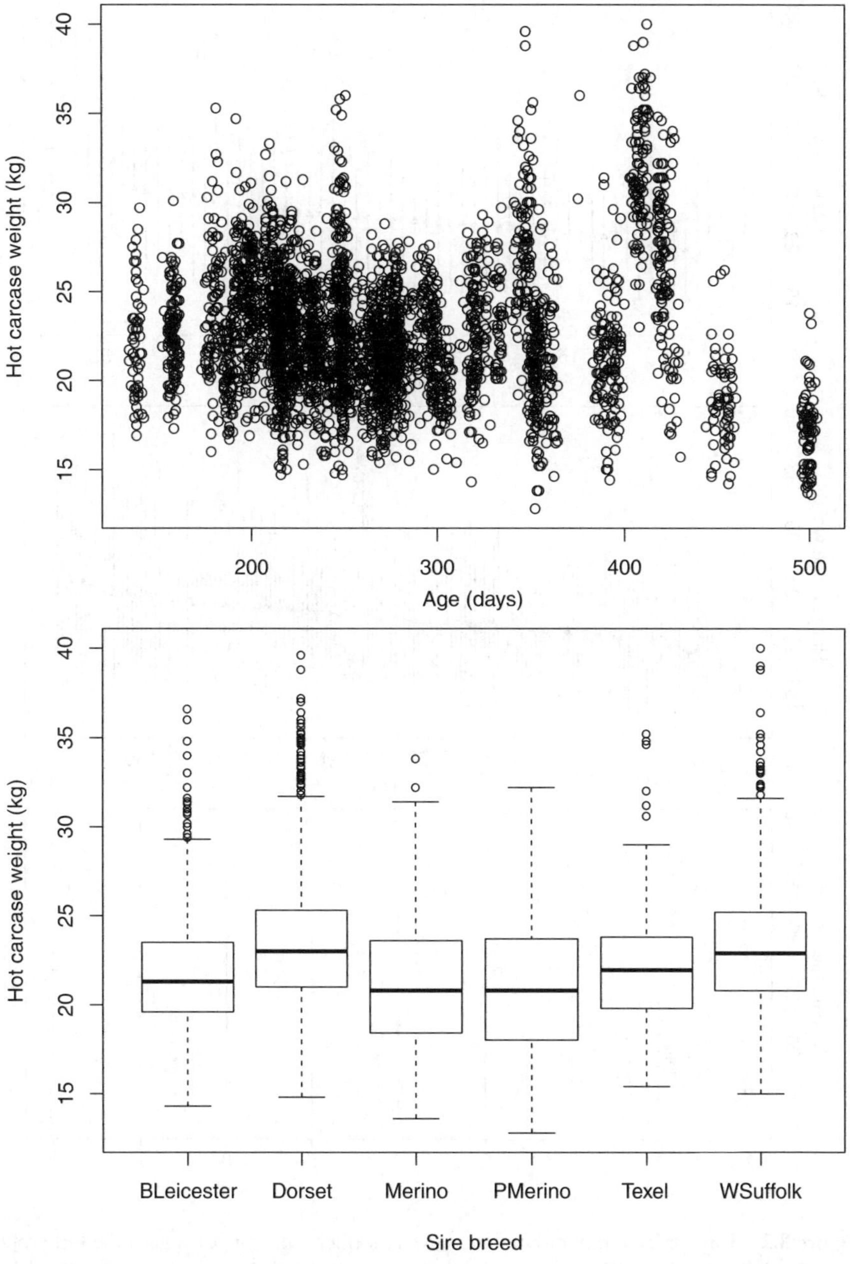

Figure 8.1 Top: Plot of hot carcase weight against the age of the animal at slaughter. Researchers would like to know if additional age (after maturity) increases the carcase weight. Bottom: Boxplot of hot carcase weight against sire breed, which is one of several factors that may influence final weight.

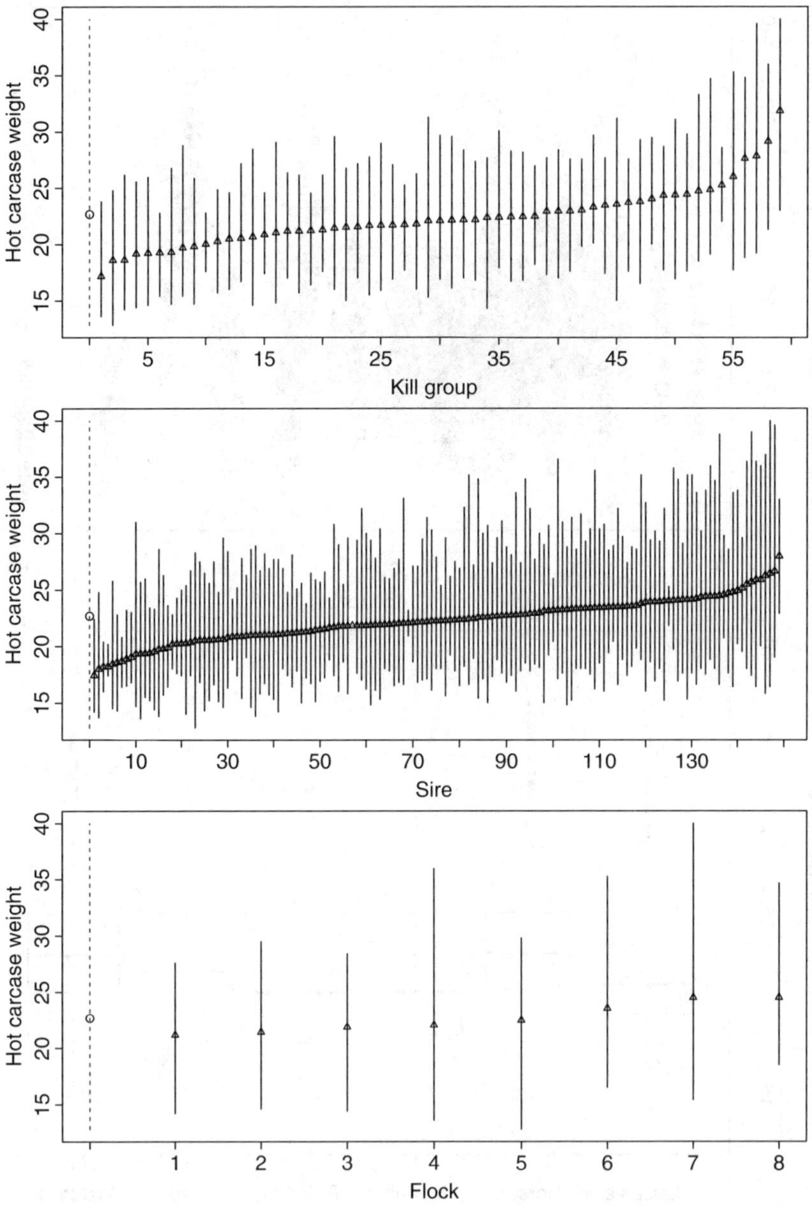

Figure 8.2 Top: Ordered plot of hot carcase weight (mean and range) against the kill group for animals at slaughter. These groups contain unbalanced data in terms of the fixed effects (sire breed, age, sex) and can be seen to vary widely in mean and range. The grey dashed line represents mean and range for whole data set. Middle: Ordered plot of hot carcase weight (mean and range) against the individual sires for animals at slaughter. Bottom: Ordered plot of hot carcase weight (mean and range) against the flock from which the animals originated at slaughter.

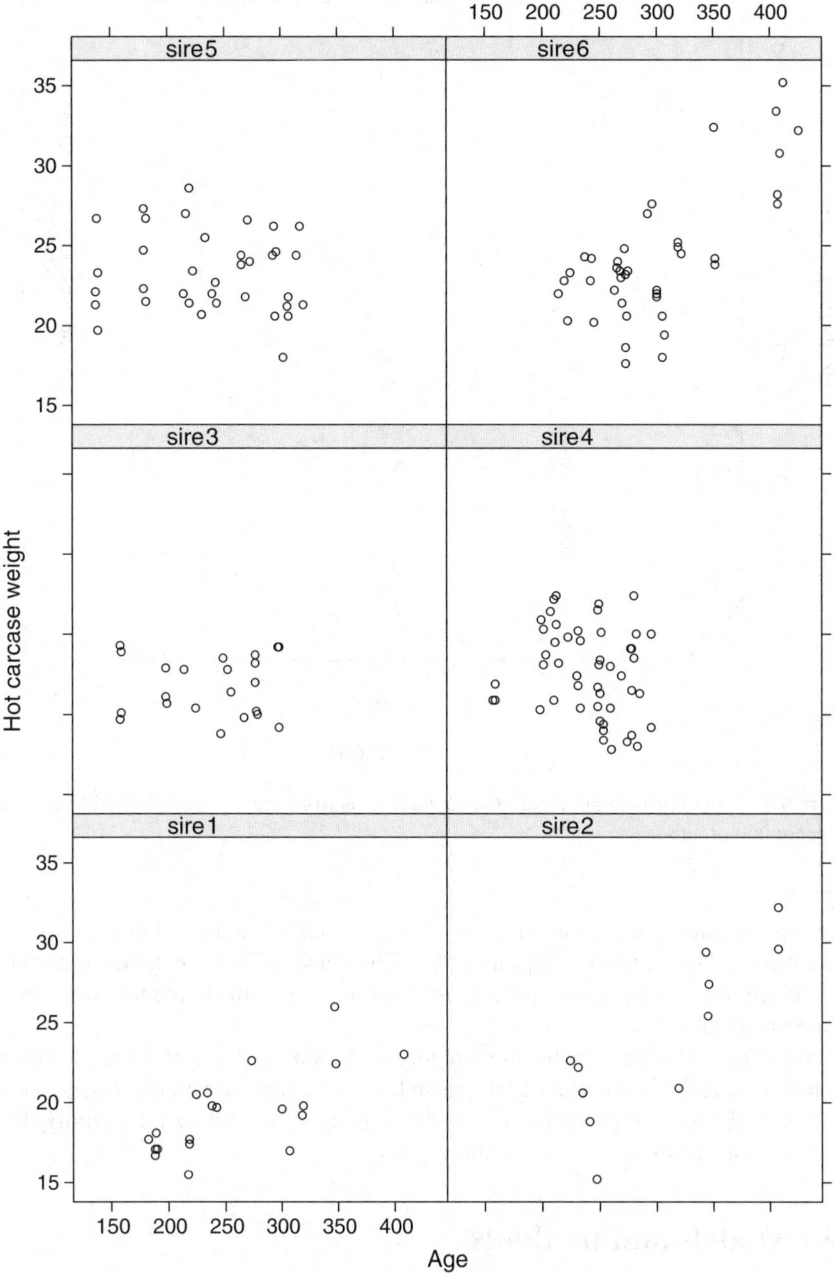

Figure 8.3 A selection of the sire response against age against hot carcase weight.

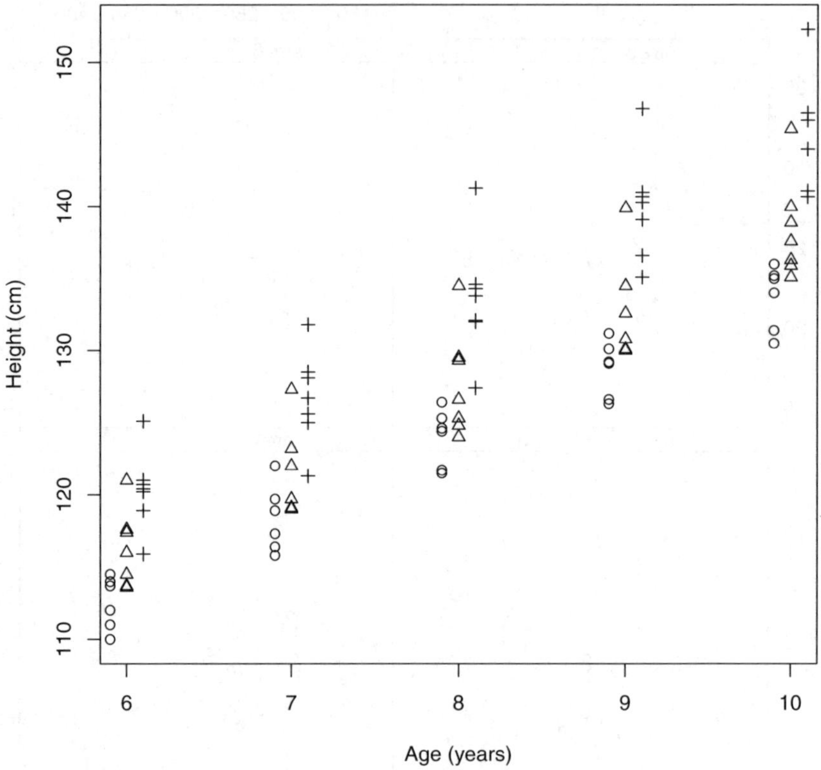

Figure 8.4 Plot of growth in height of school-age girls from ages 6 to 10 (jittered to display mother height groups).

have their heights measured annually for 5 years. As this is a limited range in age, a linear mixed effects model is appropriate, as illustrated in Figure 8.4. Had the study taken place over a longer time period, a nonlinear mixed effects model might have been more suitable.

From Figure 8.4 we note that the category of the mothers' height (small, medium and tall) may also be important in the model, as there seem to be differences in the mean heights for these group clusters at different ages, as well as a possibility that the growth rate is different between these groups.

8.3 Models and methods

Denoting the $(N \times 1)$ vector of observations as $\mathbf{y}$, then the linear mixed model may be represented as

$$\mathbf{y} = \mathbf{X}\boldsymbol{\beta} + \mathbf{Z}\mathbf{u} + \boldsymbol{\varepsilon}; \quad \boldsymbol{\varepsilon} \sim \mathbf{N}\left(\mathbf{0}, \sigma^2\mathbf{I}_N\right) \tag{8.1}$$

where $\mathbf{X}$ is an $(N \times k)$ matrix of regressors, $\boldsymbol{\beta}$ is a $(k \times 1)$ vector of fixed effects, $\mathbf{Z}$ is an $(N \times m)$ matrix of regressors for the random effects, $\mathbf{u}$ is an $(m \times 1)$ vector of random effects and $\boldsymbol{\varepsilon}$ is an $(N \times 1)$ normally distributed random vector, with a mean vector $\mathbf{0}$ and a covariance matrix $\sigma^2\mathbf{I}_N$. Here $\mathbf{I}_N$ refers to an N-dimensional identity matrix. The random effects are defined such that

$$\mathbf{u} \sim \mathbf{N}\left(\mathbf{0}, \mathbf{D}^{-1}\right), \tag{8.2}$$

where $\mathbf{D}$ is an $(m \times m)$ precision matrix. To complete the model specification, we need to define priors for $\boldsymbol{\beta}$, $\mathbf{D}$ and σ. We assume that

$$\boldsymbol{\beta} \sim \mathbf{N}\left(\underline{\boldsymbol{\beta}}, \underline{\mathbf{V}}^{-1}\right)$$

and

$$\sigma \sim \mathbf{IG}\left(\frac{\underline{\nu}}{2}, \frac{\underline{s}}{2}\right), \tag{8.3}$$

where $\underline{\nu}$ is the prior degrees of freedom and $\underline{s}$ is the prior scale term for the inverted-gamma distribution. We assume that $\mathbf{u}$ can be separated into K independent parts, $\mathbf{u}_i$, where

$$\mathbf{u}_i \sim \mathbf{N}\left(\mathbf{0}, \mathbf{D}_i^{-1} \otimes \mathbf{I}_{m_i}\right),$$

and where the variance of $\mathbf{u}_i$, $\mathbf{D}_i^{-1} \otimes \mathbf{I}_{m_i}$, implies that $\mathbf{D} = \oplus_i \left(\mathbf{D}_i \otimes \mathbf{I}_{m_i}\right)$; here $\oplus$ refers to the matrix direct sum and $\otimes$ is the Kronecker product. It is further assumed that the $(l_i \times l_i)$ precision matrix $\mathbf{D}_i$, for $i = 1, 2, \ldots, K$, is to be distributed following

$$\mathbf{D}_i \sim \mathbf{Wishart}\left(\underline{w}_i, \underline{\mathbf{S}}_i\right), \tag{8.4}$$

where $\underline{w}_i$ is the degrees of freedom parameter and $\underline{\mathbf{S}}_i$ is the inverse of the scale matrix for the Wishart distribution. The precision matrix $\mathbf{D}_i$ models the correlation between the l_i components in $\mathbf{u}_i = (\mathbf{u}_{i,1}^T, \mathbf{u}_{u,2}^T, \ldots, \mathbf{u}_{i,l_i}^T)^T$, where each component $u_{i,j}$, for $j = 1, 2, \ldots, l_i$, is a $(J_i \times 1)$ vector and $J_i = m_i/l_i$.

It is interesting to note that the only difference between the random effects, $\mathbf{u}$, and the fixed effects, $\boldsymbol{\beta}$, is the extra layer of hierarchy in the prior specification of the random effects.

The generic form, for the linear mixed model in Equation (8.1), is a compact representation that nests many useful forms, including the specific models used in Case studies 1 and 2, which we now describe.

8.3.1 Model for Case study 1

The observed variable, **HCWT**, is modelled as

$$\begin{aligned}\mathbf{HCWT} = {} & \beta_0 + \mathbf{Age}\beta_1 + \mathbf{S}\beta_2 + \mathbf{AS}\beta_3 + \mathbf{B}\boldsymbol{\beta}_4 + \mathbf{AB}\boldsymbol{\beta}_5 \\ & + \mathbf{PF}\beta_6 + \mathbf{Z}_s\boldsymbol{u}_s + \mathbf{Z}_{fl}\boldsymbol{u}_{fl} + \mathbf{Z}_{kg}\boldsymbol{u}_{kg} + \boldsymbol{\varepsilon},\end{aligned} \tag{8.5}$$

where **Age** is the age of the sheep, **S** is the sex of the sheep, **AS** is an interaction term between the age and sex of the sheep, **B** is the breed of the sheep, **AB** is an interaction term between the age and the breed of the sheep, **PF** is the percentage of fat in the sheep, $\mathbf{Z}_s$ is an $(N \times m_s)$ design matrix for the random effects that defines the sire of the sheep, $\mathbf{Z}_{fl}$ is an $\left(N \times m_{fl}\right)$ design matrix for the random effects that specifies the flock, $\mathbf{Z}_{kg}$ is an $\left(N \times m_{kg}\right)$ design matrix for the random effects for the kill group, and $\boldsymbol{\varepsilon}$ is the residual vector. The random effects $\mathbf{u}_s$, $\mathbf{u}_{fl}$, $\mathbf{u}_{kg}$ are further defined, such that

$$\begin{aligned} \mathbf{u}_s &\sim \mathbf{N}\left(0, d_s^{-1}\right), \\ \mathbf{u}_{fl} &\sim \mathbf{N}\left(0, d_{fl}^{-1}\right), \\ \mathbf{u}_{kg} &\sim \mathbf{N}\left(0, d_{kg}^{-1}\right). \end{aligned} \tag{8.6}$$

The design matrices $\mathbf{Z}_s$, $\mathbf{Z}_{fl}$ and $\mathbf{Z}_{kg}$ are designed in a similar fashion. As an example, if we define the ith row of $\mathbf{Z}_s$ as $\mathbf{z}_{s,i}^T$, then if the ith carcase belongs to group j, where $j \in \{1, 2, \ldots, m_s\}$, then set the jth element in $z_{s,i}^T$ equal to 1 and the remaining elements equal to 0. For instance, if $m_s = 3$ and the ith carcase is from the second sire, then $z_{s,i}^T = [0\ 1\ 0]$.

The model in Equations (8.5) and (8.7) can be expressed in terms of the general model in Equations (8.1) and (8.2), by defining $\mathbf{X} = (\mathbf{1}_N; \mathbf{Age}; \mathbf{S}; \mathbf{AS}; \mathbf{B}; \mathbf{AB}; \mathbf{PF})$, where the generic notation $(\mathbf{A}; \mathbf{B})$ is used to denote that the columns of $\mathbf{A}$ are concatenated with the columns of $\mathbf{B}$, and $\mathbf{1}_N$ is a vector of ones of order N. Further, we define $\boldsymbol{\beta} = (\beta_0, \beta_1, \beta_2, \boldsymbol{\beta}_3^T, \boldsymbol{\beta}_4^T, \beta_5, \beta_6)^T$,

$$\mathbf{Z} = \begin{bmatrix} \mathbf{Z}_s & \mathbf{0} & \mathbf{0} \\ \mathbf{0} & \mathbf{Z}_{fl} & \mathbf{0} \\ \mathbf{0} & \mathbf{0} & \mathbf{Z}_{kg} \end{bmatrix}, \quad \mathbf{u} = \left(\mathbf{u}_s^T, \mathbf{u}_{fl}^T, \mathbf{u}_{kb}\right)^T$$

and

$$\mathbf{D} = \begin{bmatrix} d_s\mathbf{I}_{m_s} & \mathbf{0} & \mathbf{0} \\ \mathbf{0} & d_{fl}\mathbf{I}_{m_{fl}} & \mathbf{0} \\ \mathbf{0} & \mathbf{0} & d_{kg}\mathbf{I}_{m_{kg}} \end{bmatrix}.$$

The prior hyperparameters for this case study are defined as $\underline{w} = 10$ and $\underline{S} = 0.01$, for each random effect, $\underline{\boldsymbol{\beta}} = \mathbf{0}$, $\underline{\mathbf{V}} = 0.01\mathbf{I}_{15}$.

8.3.2 Model for Case study 2

The observed variable, **Height**, can be expressed as

$$\mathbf{Height} = \beta_0 + \mathbf{Age}\beta_1 + \mathbf{Mother}\beta_2 + \mathbf{AM}\beta_3 + \mathbf{Z}_c\mathbf{u}_c + \mathbf{Z}_{ca}\mathbf{u}_{ca} + \boldsymbol{\varepsilon}, \tag{8.7}$$

where **Age** is the age of the child, **Mother** is a categorical variable, referencing the height of the child's mother, **AM** is an interaction term between **Age** and **Mother**, the design matrix $\mathbf{Z}_c$ defines the individual intercepts for each child's growth, $\mathbf{Z}_{ca}$ is a design matrix for an interaction term for each child's intercept and age, and $\boldsymbol{\varepsilon}$ is the residual vector. Further, it is assumed that the $(m_c \times 1)$ vectors, $\mathbf{u}_c$ and $\mathbf{u}_{ca}$ are jointly defined, such that

$$\begin{bmatrix} \mathbf{u}_c \\ \mathbf{u}_{ca} \end{bmatrix} \sim \mathbf{N}\left(\begin{bmatrix} \mathbf{0} \\ \mathbf{0} \end{bmatrix}, \mathbf{D}_{ca}^{-1} \right). \tag{8.8}$$

To express the model for heights in Equations (8.7) and (8.8) in terms of the general model in Equations (8.1) and (8.2), we define the vector of observations as $\mathbf{y} = (\mathbf{Height}_1^T, \mathbf{Height}_2^T, \ldots, \mathbf{Height}_n^T)^T$, where the $(n_i \times 1)$ vector $\mathbf{Height}_i$ contains the n_i repeated measures for the ith child. If we define **Age**, **Mother** and **AM** in a similar fashion, then the set of regressors, $\mathbf{X} = (\mathbf{Age}; \mathbf{Mother}; \mathbf{AM})$, and $\boldsymbol{\beta} = (\beta_0, \beta_1, \beta_2, \beta_3)^T$. The design matrix $\mathbf{Z} = (\mathbf{A}; \mathbf{AC})$, with $\mathbf{AC} = \oplus_i \mathbf{Age}_i$, where $\mathbf{Age}_i$ is the $(n_i \times 1)$ vector that contains the n_i repeated measures of age for the ith child. Here,

$$\mathbf{u} = \begin{bmatrix} \mathbf{u}_c \\ \mathbf{u}_{ca} \end{bmatrix} \quad \text{and} \quad \mathbf{D} = \mathbf{D}_{ca} \otimes \mathbf{I}_{mc}.$$

The prior hyperparameters for this case study are defined as $\underline{w} = 0$ and $\underline{\mathbf{S}} = \mathbf{0}$, for each random effect, $\underline{\boldsymbol{\beta}} = \mathbf{0}$, $\underline{\mathbf{V}} = \mathbf{0}$.

8.3.3 MCMC estimation

The aim of MCMC estimation is to sample from the joint posterior distribution for the unknown parameters, $\boldsymbol{\beta}$, $\mathbf{u}$, $\mathbf{D}$ and σ. An MCMC scheme at iteration j is as given in Algorithm 5.

Algorithm 5:

---

1. Jointly sample $\boldsymbol{\beta}^{(j)}$ and $\mathbf{u}^{(j)}$ from $p\left(\boldsymbol{\beta}, \mathbf{u} | \mathbf{y}, \mathbf{X}, \mathbf{Z}, \mathbf{D}^{(j-1)}, \sigma^{(j-1)}\right)$.
2. Sample $\mathbf{D}^{(j)}$ from $p\left(\mathbf{D} | \mathbf{y}, \mathbf{X}, \mathbf{Z}, \boldsymbol{\beta}^{(j)}, \mathbf{u}^{(j)}, \sigma^{(j-1)}\right)$.
3. Sample $\sigma^{(j)}$ from $p\left(\sigma | \mathbf{y}, \mathbf{X}, \mathbf{Z}, \boldsymbol{\beta}^{(j)}, \mathbf{u}^{(j)}, \mathbf{D}^{(j)}\right)$.

Step 1 of the algorithm is undertaken using standard Bayesian linear regression theory. Observe that Equation (8.1), once conditioned on **D**, is simply a standard Bayesian linear regression model; see Chapter 4 for specific details. If we define $\mathbf{W} = (\mathbf{X}; \mathbf{Z})$ and $\boldsymbol{\delta} = (\boldsymbol{\beta}^T, \mathbf{u}^T)^T$, then we can express Equation (8.1) as

$$\mathbf{y} = \mathbf{W}\boldsymbol{\delta} + \boldsymbol{\varepsilon}$$

and it follows that

$$\boldsymbol{\delta}|\mathbf{W},\sigma \sim \mathbf{N}\left(\overline{\boldsymbol{\delta}},\overline{\mathbf{H}}\right)^{-1}, \tag{8.9}$$

where

$$\overline{\boldsymbol{\delta}} = \left(\frac{\mathbf{W}^T\mathbf{W}}{\sigma^2}+\underline{\mathbf{H}}\right)^{-1}\left(\frac{\mathbf{W}^T\mathbf{y}}{\sigma^2}+\underline{\mathbf{H}\boldsymbol{\delta}}\right),$$

with

$$\underline{\boldsymbol{\delta}} = \left(\boldsymbol{\beta}^T,\mathbf{0}^T\right)^T \quad \text{and} \quad \underline{\mathbf{H}} = \begin{bmatrix} \underline{\mathbf{V}} & \mathbf{0} \\ \mathbf{0} & \mathbf{D} \end{bmatrix}.$$

It is important to note that step 2 of Algorithm 5 takes advantage of the conjugacy of the prior specification for $\mathbf{D}$, in Equation (8.3). To sample $\mathbf{D}$, we make use of the fact that the blocks, $\mathbf{D}_i$, described in Equation (8.4) are independent of each other. We therefore sample each precision matrix, $\mathbf{D}_i$, for $i = 1, 2, \ldots, K$, separately, using

$$\mathbf{D}_i|\mathbf{u}_i \sim \textbf{Wishart}\left(\overline{w}_i,\overline{\mathbf{S}}_i\right),$$

where $\overline{w}_i = \underline{w}_i + J_i$ and $\overline{\mathbf{S}}_i = \underline{\mathbf{S}}_i + \sum_{j=1}^{J_i}\mathbf{u}_{i,j}\mathbf{u}_{i,j}^T$.

Step 3 of Algorithm 5 also takes advantage of the conjugacy of the prior specification of σ, in Equation (8.3). In this case the posterior for σ is given as

$$\sigma|\mathbf{y},\mathbf{X},\mathbf{Z},\boldsymbol{\beta},\mathbf{u} \sim \mathbf{IG}\left(\frac{\overline{\nu}}{2},\frac{\overline{s}}{2}\right),$$

where $\overline{\nu} = \underline{\nu} + N$ and $\overline{s} = \underline{s} + (\mathbf{y}-\mathbf{W}\boldsymbol{\delta})^T(\mathbf{y}-\mathbf{W}\boldsymbol{\delta})$.

8.4 Data analysis and results

HCWT analysis

The HCWT data are initially fitted using a Bayesian linear regression (fixed effects only), with a Jeffreys prior (Chapter 4, Section 4.4.2). The model included the fixed effects of intercept, age, sex, breed, percentage body fat, and the appropriate interaction of age with both sex and breed. The model is then fitted including the random effects of sire, flock and kill group in a linear mixed model frameworks using Equation (8.5).

The comparison of the estimates for both models is given in Table 8.1. It is evident that inclusion of the random effects improves the fit of the fixed effects in that the credible intervals around the point estimates are generally smaller than in the standard regression, which ignores these terms. In addition, several fixed effects (signified with *) are significant, in that zero is not contained in their credible intervals, whereas these effects were not significant in the linear regression. These include terms involving the birth weight and response to age of animal with Texel breed, as well as the response to age of Dorset and White Suffolk breeds.

Table 8.1 Parameter estimates for fixed effects in the hot carcase weight model. Comparison is made between estimates for regular Bayesian regression and the mixed effects model.

Parameter	Linear regression estimate (95% CI)	Mixed model estimate (95% CI)	
Intercept	13.280 (11.430, 15.120)	16.596 (14.090, 19.077)	
Age	0.030 (0.022, 0.037)	0.016 (0.007, 0.025)	
Sex (male)	3.158 (1.703, 4.412)	2.159 (1.249, 3.072)	
Dorset	4.216 (2.545, 5.887)	2.620 (1.269, 3.974)	
Merino	14.220 (11.470,16.970)	9.125 (3.504, 14.768)	
Poll Merino	13.960 (10.390, 17.530)	10.740 (5.228, 16.228)	
Texel	1.734 (−1.446, 4.913)	3.163 (0.558, 5.684)	*
White Suffolk	3.366 (1.545, 5.187)	3.530 (2.036, 5.027)	
Percentage fat	0.087 (−0.080, 0.254)	0.134 (−0.130, 0.400)	
Sex (male): age	−0.007 (−0.013, −0.002)	−0.005 (−0.008, −0.001)	
Dorset: age	−0.008 (−0.014, −0.001)	−0.002 (−0.007, 0.003)	*
Merino: age	−0.046 (−0.054, −0.038)	−0.031 (−0.046, −0.017)	
Poll Merino: age	−0.046 (−0.055, −0.036)	−0.027 (−0.042, −0.012)	
Texel: age	−0.004 (−0.016, 0.007)	−0.009 (−0.017, 0.000)	*
White Suffolk: age	−0.005 (−0.012, 0.002)	−0.006 (−0.012, −0.001)	*

The magnitude of the random sources can be seen in Table 8.2. The overall estimate of σ^2 when the random effects are ignored is 12.20; however, this residual error is reduced to 5.52 when the random sources of variation are accounted for in the model. Kill group is the biggest source of variation, with an estimated variance of 4.86, followed by sire (0.87) and flock (0.01). The deviance information criterion (DIC) (Sorensen and Gianola 2002) improved with each addition of these terms, with a final DIC value of 15 083. We then tested the effects of sire:age, based on Figure 8.3; however, the DIC for this model increased to 15 096, and is therefore an unnecessary addition.

Results from the MCMC draws for sex and body fat coefficients are given in histogram form in Figure 8.5. It can be seen that the simulations are roughly symmetrical around their mean point estimates of 2.136 for males, with zero not included in the

Table 8.2 Estimated effects for variance components in the hot carcase weight model. Comparison is made between estimates for regular Bayesian regression and the mixed effects model.

Component	Regression model parameter estimate	Mixed effects model parameter estimate
σ_{error}	3.50	2.349 (2.291, 2.410)
σ_{sire}		0.933 (0.780, 1.100)
σ_{flock}		0.034 (0.022, 0.056)
$\sigma_{\text{kill group}}$		2.205 (1.849, 2.647)

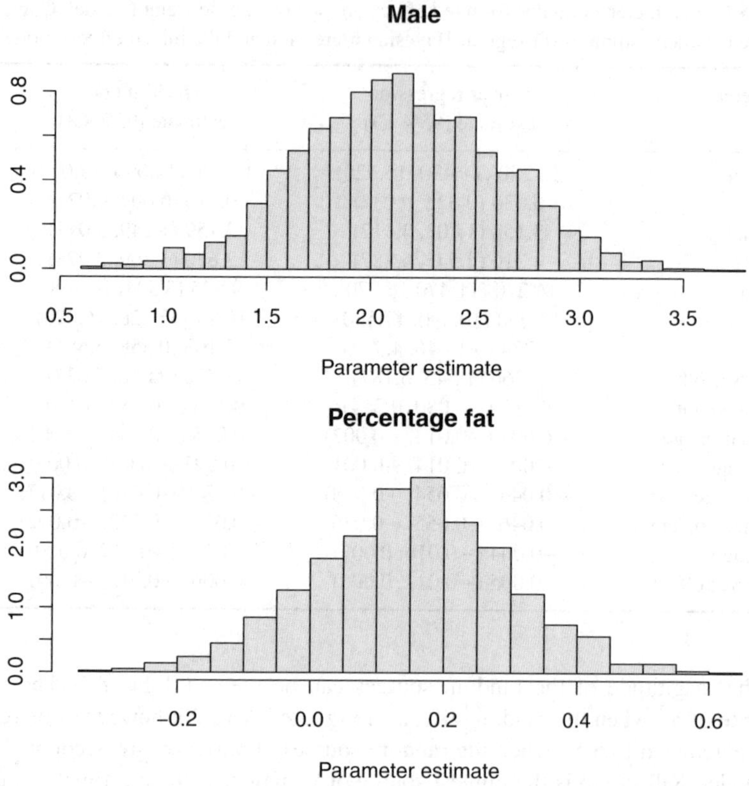

Figure 8.5 Density of the MCMC draws for the parameters of sex (top) and percentage body fat (bottom). The variable sex is significant, as zero is not contained in the 95% credible intervals, whereas the variable involving the percentage of body fat is non-significant as Prob($\beta_{\text{body fat}}$) $< 0 = 0.1515$ (i.e. 15.15% of values drawn in the MCMC chain are less than zero).

credible interval, and 0.150 for body fat, which is not significant as zero is included in the credible interval, with 15.15% of estimates being negative.

We can also plot the results for other variables, such as Border Leicester, and the additional effects of the other breeds. We show a selection of these in Figure 8.6, where the effect of a given breed needs to be added to the base breed. From this graph we observe that at the base level, the intercept (birth HCWT) and age are both significant, with point estimates of 16.596 and 0.016 respectively, and zero not being included in their credible intervals (Table 8.1). The animals which are Dorset breed are significantly heavier at birth (2.620); however, they do not have a significantly different response to age than the base Border Leicester breed (with 21.2% of values drawn being greater than zero, and zero being included in the parameter's credible interval Table 8.1). The Merino and Texel breeds have significantly different intercepts and slopes to the base Border Leicester breed.

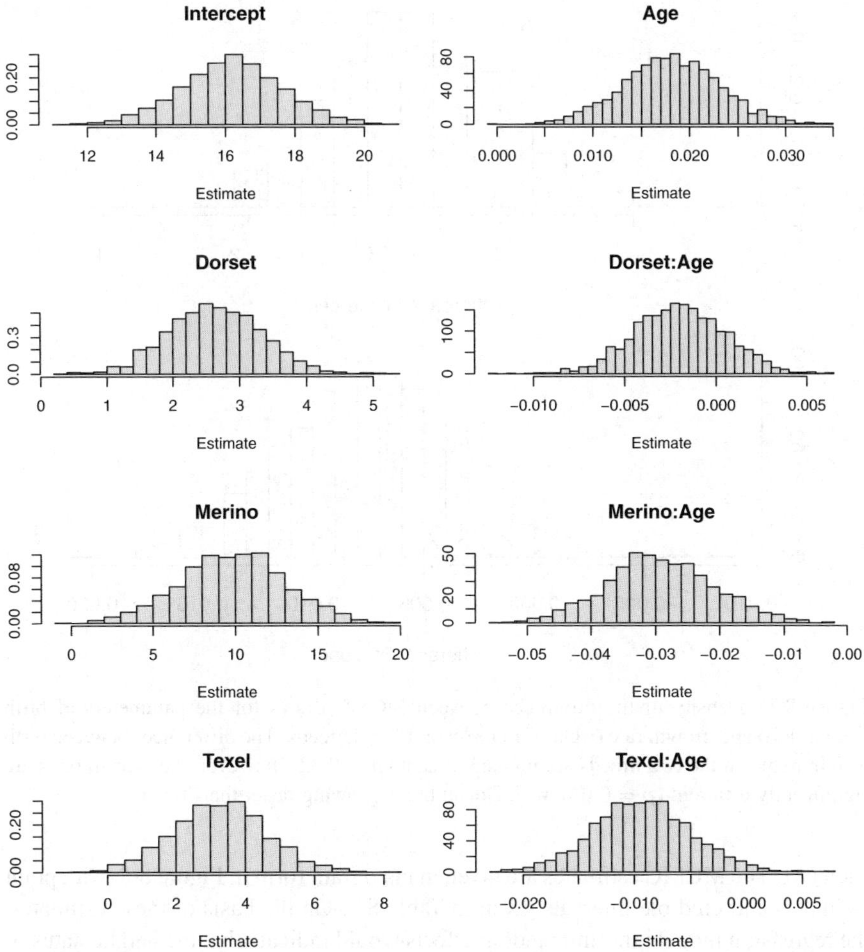

Figure 8.6 Density of the MCMC draws for the parameters of birth weight (top) and growth rate (bottom) for base breed Border Leicester and a selection of other breeds (Dorset, Merino and Texel).

It is relatively easy to test for differences between breeds by comparing the values at each iteration of the MCMC chain. An example of this is given in Figure 8.7 where we test for differences in birth weight (HCWT) and growth rate (response to age) between the Texel and Dorset breeds. We note from the top graph that Texel breeds are not significantly heavier than Dorset at birth (intercept); however, there is evidence of Texel being less responsive to age, with 4% of values falling below zero.

Girls' growth rate

The linear Bayesian regression model, with a Jeffreys prior, was first applied to the girls' height data. This model ignores the probable clustering of individual growth for

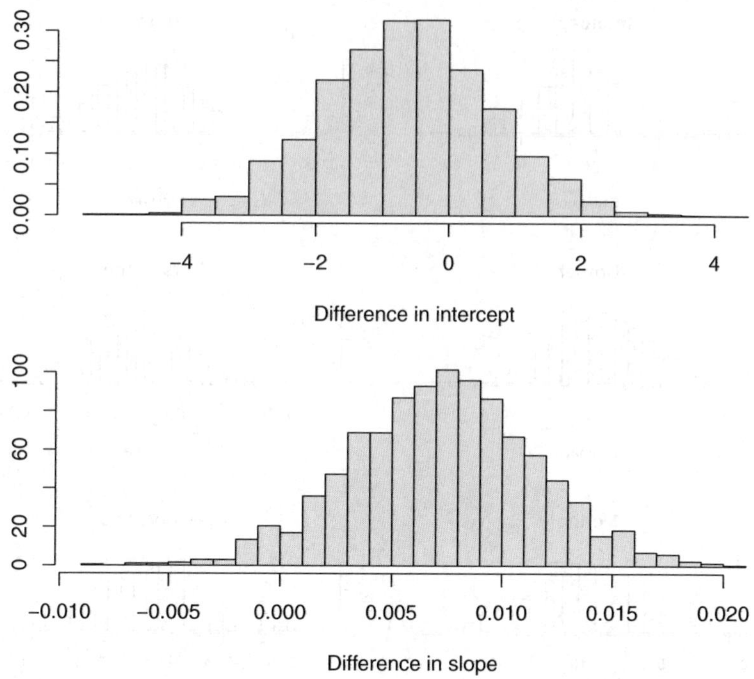

Figure 8.7 Density of the difference between MCMC draws for the parameters of birth weight (left) and growth rate (right) for Dorset and Texel breeds. The differences between birth weight between the two breeds are non-significant ($p = 0.32$); however, the growth rates are significantly different ($p = 0.04$), with Dorset breed growing faster than Texel.

each girl. The MCMC estimates are given in histogram form in Figure 8.8, with point estimates and credible intervals given in Table 8.3. On the basis of these estimates, the regression model ignoring random effects would indicate that the height status of the child's mother is not a significant factor in growth.

Table 8.3 Parameter estimates for fixed effects in the girls' height growth model. Comparison is made between estimates for regular Bayesian regression and the mixed effects model.

Parameter	Standard regression estimate (95% CI)	Mixed model estimate (95% CI)
Intercept	81.300 (75.100, 87.500)	81.298 (78.617, 83.971)
Age	5.270 (4.506, 6.034)	5.270 (4.901, 5.637)
Mother (medium)	1.674 (−6.779, 10.130)	1.679 (−2.008, 5.328)
Mother (tall)	1.823 (−6.630, 10.280)	1.811 (−1.804, 5.509)
Mother (medium): age	0.297 (−0.743, 1.338)	0.296 (−0.210, 0.801)
Mother (tall): age	0.979 (−0.062, 2.019)	0.980 (0.471, 1.487)

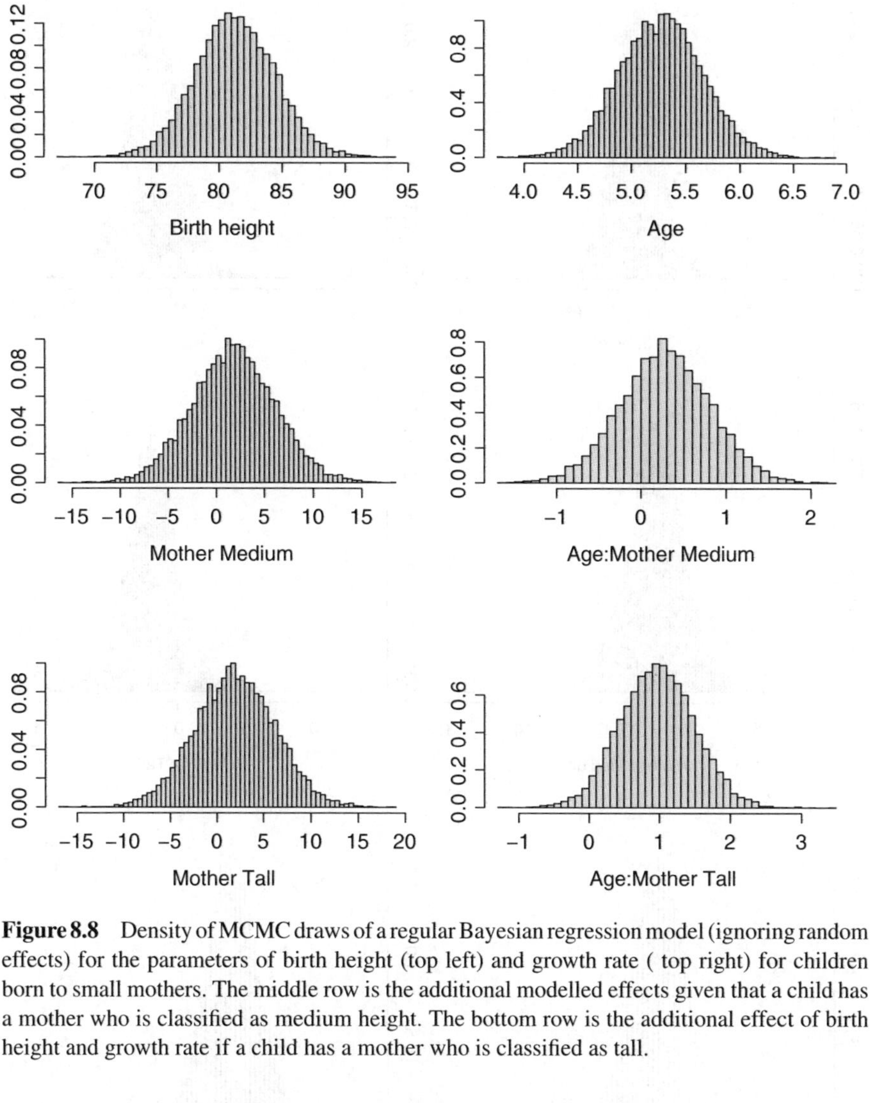

Figure 8.8 Density of MCMC draws of a regular Bayesian regression model (ignoring random effects) for the parameters of birth height (top left) and growth rate (top right) for children born to small mothers. The middle row is the additional modelled effects given that a child has a mother who is classified as medium height. The bottom row is the additional effect of birth height and growth rate if a child has a mother who is classified as tall.

When we fit the model using random effects for individual child and individual growth rates, with an unstructured correlation structure (see Section 8.3.2), we obtain the estimates for fixed effects given in Table 8.3 and illustrated in the histograms given in Figure 8.9. We note that the credible intervals and range of the MCMC chain of estimates are noticeably smaller around the point estimates than in the straight regression model. In addition the previously non-significant factor of tall mothers by age is now deemed significant. That is, children with tall mothers have a higher growth rate.

Table 8.4 shows the estimated variance components for both the regular regression and the mixed model. It can be seen that the largest amount of variation in the mixed

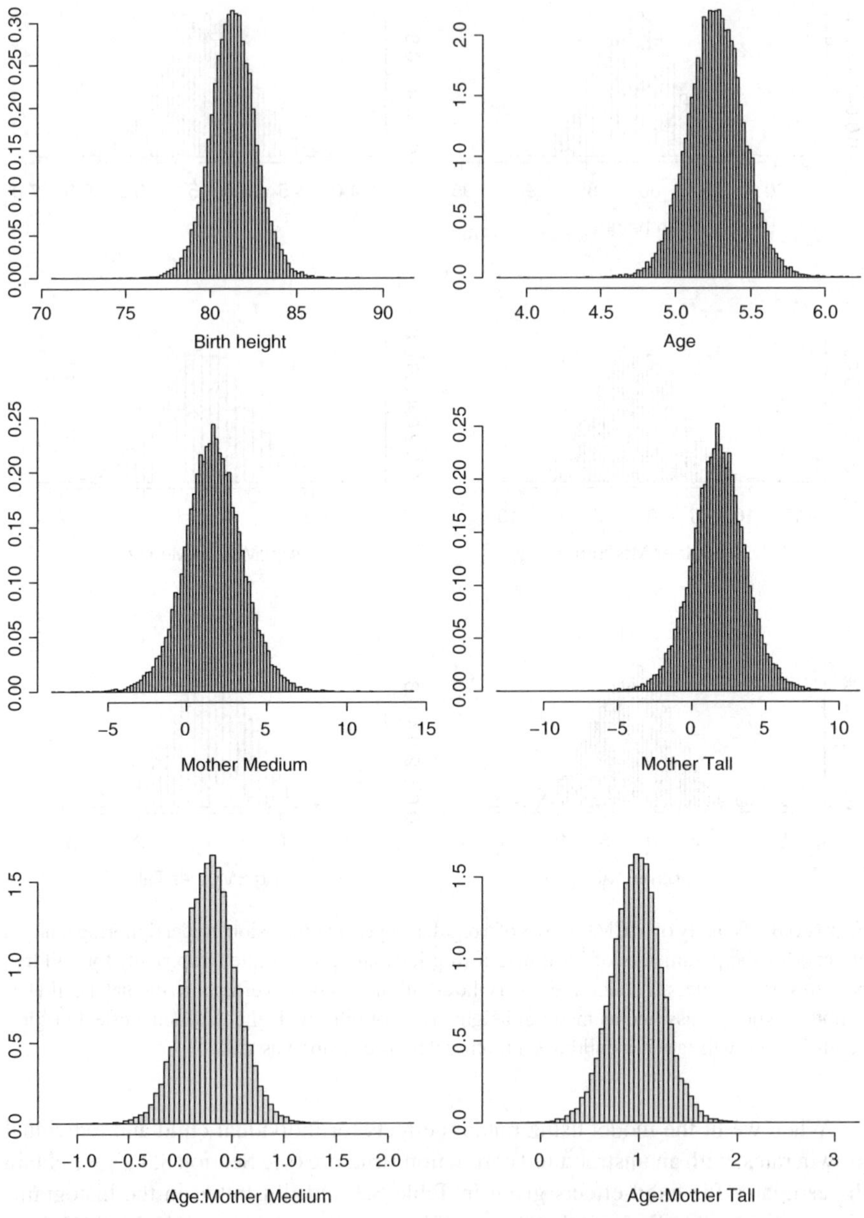

Figure 8.9 Density of MCMC draws of a Bayesian random effects regression model for the parameters of birth height (top left) and growth rate (top right) for children born to small mothers. The middle row is the additional modelled effects given that a child has a mother who is classified as medium height. The bottom row is the additional effect of birth height and growth rate if a child has a mother who is classified as tall.

Table 8.4 Estimated effects for variance components in the girls' height growth model. Comparison is made between estimates for regular Bayesian regression and the mixed effects model.

Component	Regression model parameter estimate	Mixed effects model parameter estimate
σ_{error}	3.003	0.843 (0.764, 0.949)
σ_{child}		2.712 (0.169, 4.373)
$\sigma_{\text{child:age}}$		0.395 (0.212, 0.607)

model is attributable to the random intercept (birth height) of the individual children. The estimated error variance (σ_{error}) has shrunk from 3.003 in the regression model to 0.843 in the mixed model, which accounts for the smaller credible intervals seen around the fixed effects estimates.

In some cases, the data analyst may be interested in the posterior distribution of some of the random effects. This information is readily obtained from the MCMC chains. The individual mean and credible intervals for the random intercept and slope

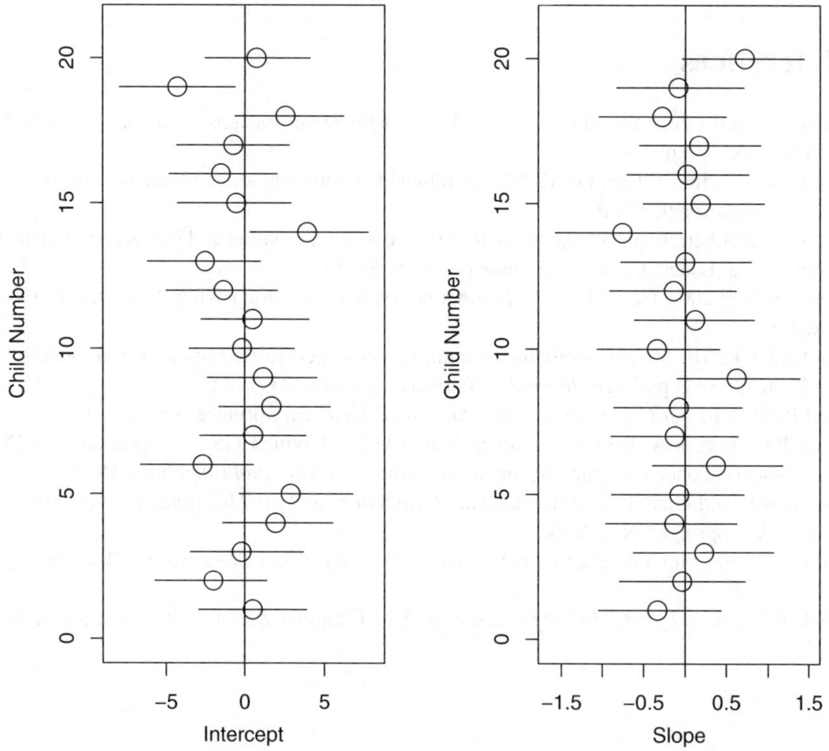

Figure 8.10 MCMC estimated mean and credible intervals for the random effect of child (left) and child growth rate (right).

of each child are given in Figure 8.10. From this graph we note that the model seems justified, as the mean intercepts for each individual seem to be different, as do the slopes. This graph could also be used to see if there are any individuals that are problematic; for instance, is it reasonable that child Id 14 has a large intercept and small slope relative to the other children?

As the effect of the individual slopes seems to be lesser in magnitude then the individual means (Table 8.4 and Figure 8.10), we can test its inclusion in the model by dropping it from the random effects and comparing DIC values. The DIC increased from 249 to 283; hence, the random slope is required.

8.5 Discussion

Linear mixed models are a standard and useful technique in the analysis of data which have clustering and other forms of random effects. The resulting improved modelling of the variance allows for more accurate estimation of the fixed effects and their associated credible intervals. Furthermore, in cases where data are either non-Gaussian or nonlinear, these models can be adapted, see for example Chib and Carlin (1999) and Chib *et al.* (1998).

References

Carlin BP and Louis TA 2009 *Bayesian Methods for Data Analysis*. Chapman & Hall/CRC Press, Boca Raton, FL.

Chib S and Carlin BP 1999 On MCMC sampling in hierarchical longitudinal models. *Statistics and Computing* **9**, 17–26.

Chib S, Greenberg E and Winkelmann R 1998 Posterior simulation and Bayes factors in panel count data model. *Journal of Econometrics* **86**, 33–54.

Demidenko E 2004 *Mixed Models: Theory and Applications*. John Wiley & Sons, Inc., Hoboken, NJ.

Hadfield JD 2010 MCMC methods for multi-response generalised linear mixed models: the MCMCglmm R package. *Journal of Statistical Software* **33**, 1–22.

Hoff P 2009 *A First Course in Bayesian Statistical Methods*. Springer, Seattle.

Lunn JD, Thomas A, Best N and Spiegelhalter D 2000 WinBUGS – a Bayesian modelling framework: concepts, structure, and extensibility. *Statistics and Computing* **10**, 325–337.

Sorensen D and Gianola D 2002 *Likelihood, Bayesian, and MCMC Methods in Quantitative Genetics*. Springer, New York.

Verbeke G and Molenberghs G 2009 *Linear Mixed Models for Longitudinal Data*. Springer, New York.

Wu L 2010 *Mixed Effects Models for Complex Data*. Chapman & Hall/CRC Press, Boca Raton, FL.

9

Ordering of hierarchies in hierarchical models: Bone mineral density estimation

Cathal D. Walsh[1] and Kerrie L. Mengersen[2]

[1] *Trinity College Dublin, Ireland*
[2] *Queensland University of Technology, Brisbane, Australia*

9.1 Introduction

In combining evidence from multiple sources, the use of hierarchical Bayesian models has become commonplace. There has been a strong growth in the use of such models in meta-analysis and in mixed treatment comparisons.

Where there are study-level properties that change (e.g. the exact outcome measure used or the patient type examined), it is natural to take account of this in the nesting of the hierarchies used in the analysis.

In this chapter we examine the impact that the ordering of hierarchies within such analyses has. In particular the overall conclusions that are drawn depend on the nesting of levels within the analysis, thus emphasizing the importance of careful consideration of model structure when fitting models in this framework.

Case Studies in Bayesian Statistical Modelling and Analysis, First Edition. Edited by Clair L. Alston, Kerrie L. Mengersen and Anthony N. Pettitt.

9.2 Case study

9.2.1 Measurement of bone mineral density

Fractures are a major contributor to failing health in an ageing population. When an older person fractures a bone it causes an immediate reduction in mobility and this in turn has knock-on effects on their ability to maintain social interactions, to care for themselves and to carry out normal activities of daily living. One major risk factor for fractures is a reduction in bone mineral density (BMD). This is a measure of the quality of the bone, and it is recognized that some people will suffer from a substantial reduction in BMD as they age. It is estimated that the mean cost associated with fractures due to this reduction is in excess of $30 000 per case. Fortunately, when a reduction in BMD is identified it can be treated using pharmaceutical agents and supplements.

In order to assess the amount of impairment in the population of patients attending hospital, bone density screening was carried out for a number of cohorts at a university teaching hospital, namely St James's Hospital in Dublin, Ireland. The studies were carried out as part of routine assessments of attendees at clinics and of inpatients. A question that arose is to what extent the general population seen by physicians has depleted bone density and therefore may require treatment or clinical intervention.

BMD is the average amount of bone material in grams per cubic centimetre. This can then be standardized by reference to a population of the same gender and ethnicity. The result of this is a 'T-score' which takes a value of 0 for the mean of the reference group and then each point on the scale is one standard deviation from the mean. Thus an individual with a T-score of −2.0 has a BMD that is two standard deviations lower than the mean of the reference population.

There are a number of different methods used to obtain BMD and related T-scores. Three types of measurement technique are of interest here.

The first technique is the estimation of bone density from DXA (Dual-energy X-ray Absorptiometry) scans of the hip. This is a standard technique and requires a patient to undergo a low-dose X-ray of their hips, the scan being carried out by an experienced operator.

The second technique is the use of a portable ultrasound machine to estimate the density of bones by scanning the heel of the individual. These devices are faster to use and scans may be carried out on patients in the ward without needing to transfer them to where a DXA machine is located.

The third method of interest is the estimate of bone density that may be obtained from an individual using a DXA scan of their spine. This scan involves capturing an image from a different angle and the estimates are sensitive to features that may be apparent in the individual, such as vertebral fractures or other anomalies.

In the data that have been recorded in this study, the information was obtained from a mixture of patients who attended the hospital as day visitors and those who are inpatients, with the reason for attendance or admission unrelated to their bone density. Thus, it may be considered that there are two distinct patient types here: namely, Group 1, the day hospital attendees; and Group 2, the inpatients.

Table 9.1 Summary statistics: mean T-score, standard error of the mean (in parentheses) and sample size for each cohort used in the study.

Measurement	Group	Value	Sample Size (n)
DXA hip	Outpatient	−2.58 (0.14)	101
DXA hip	Inpatient	−1.87 (0.15)	95
US heel	Outpatient	−3.07 (0.14)	91
US heel	Inpatient	−3.38 (0.45)	6
DXA spine	Outpatient	−2.88 (0.17)	105
DXA spine	Inpatient	−1.81 (0.20)	99

In the summary data shown in Table 9.1, there are six collections of data, each of the two groups and each of the three modes of measurement. Since the method of measurement and standardization already adjusts for ethnic background and gender, it is not necessary to include these covariates in the models that we examine. Summary statistics in the form of mean T-Score and standard error are available for each of the measures for each of the studies. However, within a given study, the measurements were not carried out on the same group of individuals.

9.3 Models

In describing the models the following notation is used.

Data that are observed are denoted by y. The overall mean value for the population of interest is denoted μ.

For the purposes of clarity, specific notation is used for each of two hierarchical structures examined in this chapter.

The hierarchical structures investigated are illustrated in Figure 9.1. At what is termed level 2, the lower level in the hierarchy, the parameters are denoted by ξ or λ. At what is referred to as level 1, η or ν.

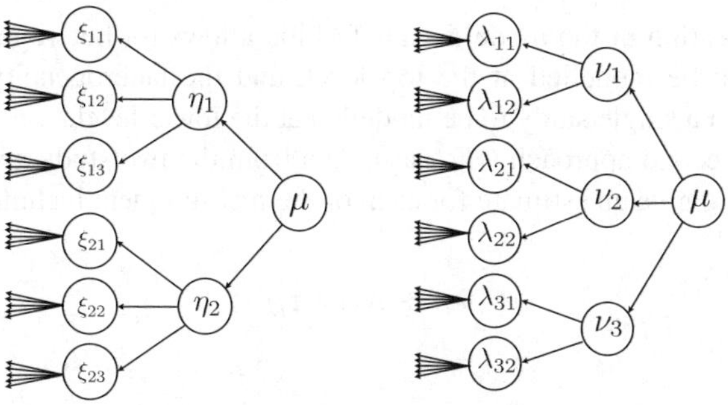

Figure 9.1 Schematic representation of the models, H1 on the left and H2 on the right. In H1 the three different measurement techniques are first combined to give estimates for groups and then these are combined to estimate the overall mean. In H2 we chose to combine information from our groups to estimate the mean response under each of the three techniques. The estimates from these techniques are then combined to give an estimate of the overall mean.

In this instance, an individual may belong to either of the two study groups of interest, and was examined using one of the three different measurement techniques.

9.3.1 Hierarchical model

Where there is a nested structure in the collection of data, this may be represented with multiple levels in the statistical model. Such models can be fitted in a classical or Bayesian framework, and there is a substantial literature dealing with the development of these methods in both frameworks. For an overview of multi-level models, including commentary on fitting them in a Bayesian framework, Goldstein (2011) provides an accessible text. An overview of how they apply to applications such as this is provided by Greenland (2000).

The case examined here involves aggregate data from six cohorts. The analytical challenge is to combine information from these groups while retaining the structure relating to measurement technique and patient source. This is similar to the challenge posed when combining studies in a meta-analysis (Sutton and Abrams 2001), and it is natural that the hierarchical structure of the chosen model matches that used for these.

Consider, then, a hierarchical model representing the type of data described above. Two approaches are possible. In the following, for specificity of illustration, Gaussian distributions are assumed for the various quantities, although the arguments follow for any distributions with a central tendency.

Under the first approach (H1, say), the hierarchical model might first combine across the measurement techniques, $j = 1, \ldots, J$, within a given study group, $i = 1, \ldots, I$, to give an overall estimate for the mean size of the property of interest for a given study η_i, of the property of interest for the ith study. Here this refers to mean T-score within inpatients/outpatients. Hence

$$\xi_{ij} \sim N(\eta_i, \tau_i).$$

The study means may then be combined in order to give an estimate of the overall mean T-score across studies. Thus

$$\eta_i \sim N(\mu, \tau_0).$$

The specification of the model in this fashion allows for heterogeneity between *study groups* to be modelled at the top level, and the heterogeneity between the *measures* within a single study to be modelled at the lower level.

Under the second approach (H2, say), data from the two study groups are combined together to give an estimate for each of the measurement techniques.

Thus

$$\lambda_{ij} \sim N(\nu_j, \tau_j)$$

and then

$$\nu_j \sim N(\mu, \tau_0).$$

The specification of the model in this fashion allows for heterogeneity between *measures* to be modelled at the highest level, and the heterogeneity between the *study groups* for each of the measures to be modelled at a lower level.

Note that the models are different: there are different total numbers of parameters. In particular it may readily be shown (see the appendix to the chapter) that the likelihoods for μ are, in general, different for each model.

Note also that a fully hierarchical model, rather than a multivariate one, is chosen in each case. The impact of this in a practical sense is to assume exchangeability of the measures within each of the subject groupings. In the case examined, this is a reasonable assumption. But it is something that requires careful consideration for each situation in which similar models are used.

9.3.2 Model H1

Firstly, a two-level hierarchical random effects model is considered. In this case, it is posited that within each of the two groups, the three measures provide an estimate of average bone density within the group. Between-measure variability is allowed for with a precision term at this level.

At the top level, the estimates for each group are combined to give an estimate of overall mean bone density for the hospital population. Variability between groups is allowed for at this level.

A schematic representation of the model is shown on the left in Figure 9.1. Notice that the three measures for a given group are combined. Then the estimates for each of the groups come together to give an estimate of overall mean T-score.

Thus the model may be written in terms of the following conditional densities:

$$\begin{aligned} y_{ijk} &\sim N(\xi_{ij}, \tau_{ij}), \\ \xi_{ij} &\sim N(\eta_i, \tau_i), \\ \eta_i &\sim N(\mu, \tau). \end{aligned}$$

A normal prior with mean 0 and standard deviation 1000 was used for μ. Inverse Gamma(0.1,0.1) priors were used for the standard deviations. Mindful of the work of Gelman (2006) and other authors in this context, we also examined the impact of uniform priors for standard deviation, but this made little difference to the results.

9.3.3 Model H2

For the second analysis, the information on bone density measured using a particular technique was combined at the lowest level. Thus, for example, an overall estimate of US heel across the two groups is obtained. This is then combined with the other measures to give an overall estimate of bone density.

The model is shown on the right in Figure 9.1. Notice that the groups are combined pairwise under each measure and then the three measures are combined to give an overall estimate of the mean BMD.

Here, the model may be written in terms of the conditional densities in a fashion similar to the previous section, but with the hierarchy specified in the other direction:

$$
\begin{aligned}
y_{ijk} &\sim N(\lambda_{ij}, \tau_{ij}) \\
\lambda_{ij} &\sim N(\nu_j, \tau_j) \\
\nu_j &\sim N(\mu, \tau)
\end{aligned}
$$

Similar priors were chosen as in Model H1.

9.4 Data analysis and results

The models were fitted using WinBUGS 1.4 (Spiegelhalter *et al.* 2004). Processing was done using R 2.9.0 (Team 2009) and the R2WinBUGS package (Sturtz *et al.* 2005). In fitting these models, three chains were used, with a burn-in of 2000 iterations and thinning of every second iteration. The convergence of chains was assessed using CODA (Plummer et al. 2004).

9.4.1 Model H1

Summary statistics for the posterior distributions at the top and second level are provided in Table 9.2.

As is the case in many such analyses, the key statistics of interest are those at the top level – in this case the mean T-score for the population.

Based on the quantiles from the posterior sample, a 95% confidence interval (CI) for the mean T-score is (−4.9,−0.3). This is a rather wide interval but is to be expected since the uncertainty in the overall estimate depends somewhat on the number of groups at the next level down.

The values of the estimates for the next level down have been included. We therefore have the mean of each of the measures for each of the groups in this case.

Table 9.2 Point estimates and quantiles for sampled values of mean T-score for model H1.

Variable	Mean posterior	2.5%	50.0%	97.5%
μ	−2.6	−4.9	−2.6	−0.3
η_1	−2.7	−3.5	−2.7	−1.9
η_2	−2.4	−3.2	−2.7	−1.7
ξ_{11}	−2.6	−3.1	−2.6	−2.0
ξ_{12}	−2.8	−3.3	−2.9	−2.1
ξ_{13}	−3.0	−3.8	−3.2	−1.9
ξ_{21}	−2.9	−3.3	−3.0	−2.1
ξ_{22}	−2.0	−2.9	−1.9	−1.6
ξ_{23}	−2.0	−2.9	−1.9	−1.5

It is possible to obtain p_D and the DIC (Spiegelhalter *et al.* 2002). These are 168.8 and 161.9 respectively.

9.4.2 Model H2

Summaries from the posterior distributions are provided in Table 9.3.

In this instance a 95% confidence interval for the mean T-score is (−3.5,−1.6). This interval is noticeably narrower, although centred at the same location as H1.

In this case, p_D was 162.1 with a corresponding DIC of 158.8. Although the DIC is a little lower here, the difference between the fits is small. One possible approach that could be taken is to weight the models according to the DIC. This is a standard approach that one might take in the context of model averaging (BUGS team 2012), and is a practical approach for the purposes of prediction.

In practice, clinicians are interested in the 'cutoff' value of −1.0, since this is the level that represents osteopenia, a state of abnormally low bone density. Thus, depending on whether H1 or H2 is fitted, their qualitative interpretation of the findings will differ.

The impact of ordering is clear from Figure 9.2. The y-axis is the T-score scale. The plotted points are the estimated values a posteriori for each of the parameters highlighted in Figure 9.1. The box on the right of each plot is proportional to the confidence interval width. In Model H1, this is influenced by the fact that, at level 1 in the hierarchy, there are two parameters being estimated, and there is a modest difference between them. In H2, however, there is less variability between parameters at level 1 and there are more of them.

The example used involves multiple measurements in a single hospital. However, it could just as easily be the case that these 'Groups' are presented in different studies, or some measurements are presented in one paper and other measurements in another. Thus the situation is analogous to any situation where we wish to combine information contained in multiple measures over multiple studies.

Table 9.3 Point estimates and quantiles for sampled values of mean T-score for model H2.

Variable	Mean posterior	2.5%	50.0%	97.5%
μ	−2.5	−3.5	−2.5	−1.6
ν_1	−2.6	−3.4	−2.6	−1.8
ν_2	−2.5	−3.3	−2.5	−1.7
ν_3	−2.5	−3.4	−2.5	−1.6
λ_{11}	−2.6	−3.2	−2.6	−2.0
λ_{12}	−2.9	−3.4	−3.0	−2.1
λ_{21}	−2.7	−3.3	−2.8	−1.9
λ_{22}	−2.1	−3.0	−1.9	−1.6
λ_{31}	−2.9	−3.7	−3.0	−1.6
λ_{32}	−2.1	−3.1	−1.9	−1.4

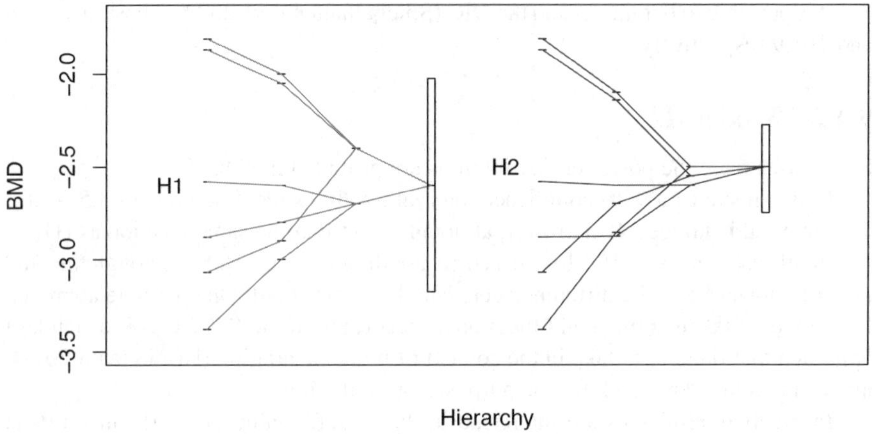

Figure 9.2 Plot showing the shrinkage from data, through each level, to the estimate of overall mean. The vertical box on the right hand side of each of the plots represents a confidence interval for the overall mean. Note how much larger this is in H1 than H2.

9.4.3 Implication of ordering

The results for H1 and H2 indicate that in the first case there is substantial uncertainty in the overall mean T-score. This uncertainty is largely due to the fact that the 'population' of patient groups is being estimated by just two groups in the sample. When a mean is estimated by a sample of two, it should be appreciated that there is substantial residual uncertainty in the point estimate.

In the second case, the estimate of mean T-score suffers only from the fact that there is between-measure variability; the fact that there are three measures provides a better estimate of the 'population' of measures.

Again, it is emphasized that these are different models; however, when estimating overall mean T-score, the summaries from either model may plausibly be presented.

The key point that we draw attention to is the fact that the choice of ordering in a hierarchical model has a material effect on what is concluded from the analysis. That is, given the same data, and the same 'overall' parameter being estimated, two analysts may end up with different findings because of a seemingly arbitrary choice in the formulation of the hierarchical model.

9.4.4 Simulation study

In order to examine the impact of model specification in situations where a larger number of studies or groups are included, a simulation study was presented.

9.4.5 Study design

In the simulation, a similar setup was used, where there are three measures which can be used to estimate the property of interest. The number of studies examined in this instance was 10. This size was chosen since it is similar to the number of studies used in evidence syntheses such as this, for example in the case of meta-analyses.

The overall mean was set at −2.5. The mean effect for each of the studies was then simulated to be normally distributed about this, with a precision of 1.0. The mean size for each of three measures was then simulated as normally distributed about these values with a precision of 1.0.

After simulating a complete set of observations, the model was fit using WinBUGS 1.4, and the mean and standard error (based on the standard deviation of the sampled values) were recorded.

Since simulation involves generating data, having conditioned on particular values of the parameters, the results are only interpretable up to sampling variability. In order to quantify the impact of this, the entire exercise was repeated 10 times.

9.4.6 Simulation study results

The width of the confidence interval under the different structures is very sensitive to the ordering of the hierarchy in the analysis. Examples of the confidence intervals for the overall mean are illustrated in Figure 9.3 for 10 realizations of this process under each of the hierarchical structures.

It is clearly the case here that the CI width (although not its location) is affected by the ordering in the hierarchy and the number of groupings at each level. And, as

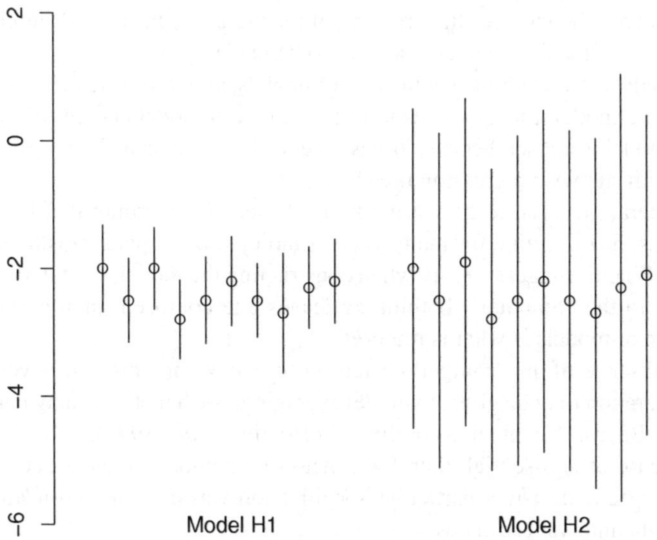

Figure 9.3 Plot showing the estimate of the overall mean in the simulation study. Results for 10 realizations are shown. The vertical lines correspond to the width of the confidence interval. Note how much larger this is in H2 than in H1. For this *simulation* section the number of groups has been increased to 10 at the top level of H1, with the number of measures being fixed at three for H2.

mentioned above, the qualitative interpretation in a clinical context may be altered, depending on which of the two models is fitted.

9.5 Discussion

This work demonstrates that in the case of a real example, and an associated simulation study, the way in which the model is specified is of material importance. Thus, if there is any confusion about which way the hierarchy should be arranged in these models, then care should be taken in the interpretation of the results.

Overall, the feature of note is that the hierarchical models that are fitted are not the same. The differences between the specification of the models are the assumptions about which parameters are exchangeable. The estimate of the overall mean within the group of interest is the same, independent of the model that has been fitted. However, uncertainty in this estimate is different for the two models. The reason for this difference is simple – the two models are quite different. The choice of a particular structure makes assumptions about which parameters are exchangeable in the model. This impacts on the precision of our estimate of overall effect.

This fact is mirrored in the simulation study, where the point estimates are the same (up to Monte Carlo error and rounding), but the uncertainty in the estimates differs.

In practice, the model chosen arises for a mixture of philosophical, physical or biological reasons. It may be chosen because of the way in which the study has been designed. It may be that there is a natural ordering for the hierarchy in a hierarchical model. However, the fact that the ordering of the hierarchy makes a difference to the results highlights the fact that a choice has to be made.

If the analyst is examining a collection of models, with a view to allowing the data to decide on a model, this is a question of data-driven model choice. In this setting the 'best' model is chosen because it has a better fit to the data. This is the case, for example, with stepwise regression models.

The general question of how a model is chosen is something that has engaged minds across the discipline for many years. Lindley (2000), paraphrasing and exaggerating de Finetti, suggests '*think* when constructing the model; with it, do not think but leave it to the computer'. If thinking leaves one confused, then perhaps some combination of models is what is required.

In the absence of any compelling reason to choose one ordering over another, then consideration may be given to model averaging. Such methods may naturally be applied in a Bayesian context, as outlined in Hoeting *et al.* (1999).

In any case, it is essential, therefore, that other authors are clear about *how* the model was specified, *why* a particular specification was chosen, and *what* is being reported in the interval estimates.

References

BUGS team 2012 *The BUGS Project – DIC.* http://www.mrc-bsu.cam.ac.uk/bugs/winbugs/dicpage.shtml (accessed 11 May 2012).

Gelman A 2006 Prior distributions for variance parameters in hierarchical models. *Bayesian Analysis* **1**(3), 515–534.

Goldstein H 2011 *Multilevel Statistical Models*, 4th edn. John Wiley & Sons, Ltd, Chichester.

Greenland S 2000 Principles of multilevel modelling. *International Journal of Epidemiology* **29**, 158–167.

Hoeting JA, Madigan D, Raftery AE and Volinsky CT 1999 Bayesian model averaging: a tutorial. *Statistical Science* **14**(4), 382–417.

Lindley DV 2000 The philosophy of statistics. *Journal of the Royal Statistical Society, Series D (The Statistician)* **49**(3), 293–337.

Plummer M, Best N, Cowles K and Vines K 2004 *CODA: Output Analysis and Diagnostics for MCMC. R package, version 0.9-1.*

Spiegelhalter DJ, Best NG, Carlin BP and van der Linde A 2002 Bayesian measures of model complexity and fit. *Journal of the Royal Statistical Society, Series B (Statistical Methodology)* **64**(4), 583–639.

Spiegelhalter D, Thomas A, Best N and Lunn D 2004 *WinBUGS version 1.4.1: User Manual.* Medical Research Council Biostatistics Unit, Cambridge.

Sturtz S, Ligges U and Gelman A 2005 R2WinBUGS: a package for running WinBUGS from R. *Journal of Statistical Software* **12**(3), 1–16.

Sutton A and Abrams KR 2001 Bayesian methods in meta-analysis and evidence synthesis. *Statistical Methods in Medical Research* **10**(4), 277–303.

Team RDC 2009 *R: A Language and Environment for Statistical Computing*. R Foundation for Statistical Computing, Vienna.

9.A Appendix: Likelihoods

The likelihood for H1 may be written as

$$f(\mathbf{y}|\mu, \tau) = \int_V f(\mathbf{y}|\xi) f(\xi|\eta) f(\eta|\mu) d\xi d\eta.$$

The likelihood for H2 may be written as

$$f(\mathbf{y}|\mu, \tau) = \int_V f(\mathbf{y}|\lambda) f(\lambda|\nu) f(\nu|\mu, \tau) d\lambda d\nu.$$

These likelihoods are convolutions of different dimension.

In the simple case of the fixed effects model, it is noted, for example, that

$$f(\xi_{ij}|\eta_i) = \delta(\xi_{ij}, \eta_i)$$
$$f(\eta_i|\mu) = \delta(\eta_i, \mu)$$

and so the convolution simplifies the likelihood for a model of the form of H1 and H2 to the same thing, namely

$$\prod_{i=1}^{2} \left[\prod_{j=1}^{3} \{ f(y_{ij}|\mu, \tau) \} \right],$$

a straightforward product of the same dimension.

10

Bayesian Weibull survival model for gene expression data

Sri Astuti Thamrin[1,2], James M. McGree[1] and Kerrie L. Mengersen[1]

[1]*Queensland University of Technology, Brisbane, Australia*
[2]*Hasanuddin University, Indonesia*

10.1 Introduction

In many fields of applied studies, there has been increasing interest in developing and implementing Bayesian statistical methods for modelling and data analysis. In medical studies, one problem of current interest is the analysis of survival times of patients, with the main aim of modelling the distribution of failure times and the relationship with variables of interest.

Gene expression data can be used to explain some variability in patient survival times. Making the best use of current DNA microarray technology in biomedical research allows us to study patterns of gene expression in given cell types, at given times, and under given set of conditions (Segal 2006). Estimation of patient survival times can be based on a number of statistical models. For example, a straightforward approach is a proportional hazards (PH) regression model (Nguyen and Rocke 2002). This model can be used to study the relationship between the time to event and a set of covariates (gene expressions) in the presence of censoring (Park and Kohane 2002).

In a microarray setting, there may be other information available about the regression parameters that represent the gene expressions. The traditional (frequentist) PH regression model uses present data as a basis to estimate patient survival times

Case Studies in Bayesian Statistical Modelling and Analysis, First Edition. Edited by Clair L. Alston, Kerrie L. Mengersen and Anthony N. Pettitt.

and does not takes into account available prior information, other unknown features of the model or the model structure itself. In a Bayesian framework, such information can be informed through prior distributions and uncertainty in the model structure can be accommodated (Kaderali *et al.* 2006; Tachmazidou *et al.* 2008).

The Weibull distribution is a popular parametric distribution for describing survival times (Dodson 1994). In this chapter, we implement Bayesian methodology to fit a Weibull distribution to predict patient survival given gene expression data. The issue of model choice through covariate selection is also considered. A sensitivity analysis is conducted to assess the influence of the prior for each parameter.

This methodology was motivated by, and applied to, a data set comprising survival times of patients with diffuse large-B-cell lymphoma (DLBCL) (Rosenwald *et al.* 2002). DLBCL (Lenz *et al.* 2008), which is a type of cancer of the lymphatic system in adults, can be cured by anthracycline-based chemotherapy in only 35 to 40% of patients (Rosenwald *et al.* 2002). In general, types of this disease are very diverse and their biological properties are largely unknown, meaning that this is a relative difficult cancer to cure and prevent. Rosenwald *et al.* (2002) proposed that there are three subgroups of patients of DLBCL: activated B-like DLBCL, germinal centre (GC)-B like and type III DLBCL. The GC B-like DLBCL is less dangerous than the others in the progression of the tumour; the activated B-like DLBCL is more active than the others; and the type III DLBCL is the most dangerous in the progression of tumour (Alizadeh *et al.* 2000). These groups were defined using microarray experiments and hierarchical clustering. The authors showed that these subgroups were differentiated from each other by distinct gene expressions of hundreds of different genes and had different survival time patterns.

The data set published by Rosenwald *et al.* (2002) can be downloaded at http://llmpp.nih.gov/DLBCL/. This data set contains 7399 gene expression measurements from $N = 220$ patients with DLBCL, including 138 patient deaths during follow-up. Patients with missing values for a particular microarray element were excluded from all analyses involving that element. In all, 220 patients were randomly assigned to the preliminary and validation groups. The first group consisting of 147 patients is used to estimate the model, while the second group consisting of 73 patients is used to explore how useful the model might be for predicting patient survival from such gene expression data. In the training group there are 71 censored patients (48%) and in the validation group there are 25 censored patients (34%).

10.2 Survival analysis

Survival analysis in a nutshell aims to estimate the three survival (survivorship, density and hazard) functions, denoted by $S(t)$, $f(t)$ and $h(t)$, respectively (Collet 1994). There exist parametric as well as non-parametric methods for this purpose (Kleinbaum and Klein 2005).

The survival function $S(t)$ gives the probability of surviving beyond time t, and is the complement of the cumulative distribution function, $F(t)$. The hazard function $h(t)$ gives the instantaneous potential per unit time for the event to occur, given that the individual has survived up to time t.

The Weibull model is one of the most widely used parametric models in the context of survival and reliability (Kim and Ibrahim 2000; Marin *et al.* 2005) to characterize the probabilistic behaviour of a large number of real-world phenomena (Kaminskiy and Krivtsov 2005). Johnson *et al.* (1995), Collet (1994) and Kundu (2008) have discussed the model in detail.

Suppose we have independent and identically distributed (iid) survival times $\mathbf{t} = (t_1, t_2, \ldots, t_n)'$, which follow a Weibull distribution, denoted by $W(\alpha, \gamma)$ with a shape parameter α and a scale parameter γ. The parameter α represents the failure rate behaviour. If $\alpha < 1$, the failure rate decreases with time; if $\alpha > 1$, the failure rate increases with time; when $\alpha = 1$, the failure rate is constant over time, which indicates an exponential distribution. The parameter γ has the same units as t, such as years, hours, etc. A change in γ has the same effect on the distribution as a change of the abscissa scale. Increasing the value of γ while holding $\alpha = 1$ constant has the effect of stretching out the pdf. If γ is increased, the distribution is stretched to the right and its height decreases, while maintaining its shape. If γ is decreased, the distribution is pushed in towards the left and its height increases.

The density function for Weibull-distributed survival times is as follows:

$$f(t_i \mid \alpha, \gamma) = \begin{cases} \alpha\gamma t_i^{\alpha-1} \exp(-\gamma t_i^{\alpha}), & t_i > 0, \alpha > 0, \gamma > 0 \\ 0, & \text{otherwise.} \end{cases}$$

Since the logarithm of the Weibull hazard is a linear function of the logarithm of time, it is more convenient to write the model in terms of the parameterization $\lambda = \log(\gamma)$ (Ibrahim *et al.* 2001), so that

$$f(t_i \mid \alpha, \lambda) = \alpha t_i^{\alpha-1} \exp(\lambda - \exp(\lambda) t_i^{\alpha}).$$

The corresponding survival function and the hazard function, using the λ parameterization, are as follows:

$$S(t_i \mid \alpha, \lambda) = \exp(-\exp(\lambda) t_i^{\alpha}),$$
$$h(t_i \mid \alpha, \lambda) = f(t_i \mid \alpha, \lambda)/S(t_i \mid \alpha, \lambda) = \alpha \exp(\lambda) t_i^{\alpha-1}.$$

The likelihood function of (α, λ) is as follows:

$$L(\alpha, \lambda \mid D) = \prod_{i=1}^{n} f(t_i \mid \alpha, \lambda)^{\delta_i} S(t_i \mid \alpha, \lambda)^{(1-\delta_i)}$$
$$= \alpha^d \exp\left\{ d\lambda + \sum_{i=1}^{n} [\delta_i(\alpha - 1)\log(t_i) - \exp(\lambda) t_i^{\alpha}] \right\},$$

where $D = (t, \delta)$ and $d = \sum_i^n \delta_i$.

Let x_{ij} be the jth covariate associated with t_i for $j = 1, 2, \ldots, p + 1$. In our case study, x_{ij} indicates the p gene expressions from DNA microarray data, and x_{i0}

indicates the multi-category phenotype covariate. The data structure is as follows:

$$\begin{bmatrix} \text{Survival time} \\ t_1 \\ t_2 \\ \vdots \\ t_n \end{bmatrix} \begin{bmatrix} \text{Category} & \text{Gene 1} & \dots & \text{Gene } p \\ x_{10} & x_{11} & \dots & x_{1p} \\ x_{20} & x_{21} & \dots & x_{2p} \\ \vdots & \vdots & \vdots & \vdots \\ x_{n0} & x_{n1} & \dots & x_{np} \end{bmatrix}.$$

The gene expression data can be included in the model through λ. Given that λ must be positive, one option is to include the covariates as follows:

$$\gamma_i = \exp(\mathbf{x}_i'\beta), \text{ so that}$$

$$\lambda_i = \log(\gamma_i) = \mathbf{x}_i'\beta. \tag{10.1}$$

Thus, the log-likelihood function thus becomes

$$\begin{aligned} \log L(\alpha, \beta \mid D) = & \sum_{i=1}^{n} \delta_i \left[\log(\alpha) + (\alpha - 1)\log(t_i) + \mathbf{x}_i'\beta\right] \\ & - \exp(\mathbf{x}_i'\beta)t_i^{\alpha}. \end{aligned}$$

We can also extend Equation (10.1) to include additional variation, ϵ_i, perhaps due to explanatory variables that are not included in the model. In this case, we obtain

$$\lambda_i = \mathbf{x}_i'\beta + \epsilon_i, \tag{10.2}$$

where $\epsilon_i \sim \text{N}(0, \sigma^2)$.

10.3 Bayesian inference for the Weibull survival model

10.3.1 Weibull model without covariates

Under the assumptions stated in Section 10.2, a Bayesian formulation of the Weibull model takes the form

$$\begin{aligned} t_i &\sim W(t_i \mid \alpha, \lambda), \quad i = 1, 2, \dots, n, \\ \alpha &\sim p(\alpha \mid \theta_\alpha), \\ \lambda &\sim p(\lambda \mid \theta_\lambda), \end{aligned}$$

where θ_α and θ_λ are the hyperparameters of the prior distribution $(p \mid .)$ of α and λ, respectively. In this model, inference may initially focus on the posterior distribution of the shape parameters α and the scale parameter λ. We assume that (α, λ) are independent (Marin *et al.* 2005), and assign gamma distributions:

$$\alpha \sim \text{Gamma}(u_\alpha, v_\alpha)$$
$$\lambda \sim \text{Gamma}(u_\lambda, v_\lambda),$$

where typically we might choose small positive values for $u_\alpha, v_\alpha, u_\lambda, v_\lambda$ to express diffuse prior knowledge (Marin *et al.* 2005). In the case study in the next section, we choose a value of 0.0001 for all four hyperparameters. With these parameter values, the gamma distribution has mean 10^{-8} and variance 10^{-12}.

Combining the likelihood function of (α, λ) with the prior distributions, the joint posterior distribution of (α, λ) is given by

$$\begin{aligned} p(\alpha, \lambda \mid D) &\propto L(\alpha, \lambda \mid D) p(\alpha) p(\lambda) \\ &\propto \alpha^{u_\alpha + d - 1} \exp\left[-\alpha\left(v_\alpha + \sum_{i=1}^{n} t_i^\alpha\right)\right] \lambda^{u_\lambda + d - 1} \\ &\quad \times \exp\left[-\lambda\left(v_\lambda + \sum_{i=1}^{n} t_i^\alpha\right)\right]. \end{aligned}$$

Since the joint posterior distribution of $(\alpha, \lambda \mid D)$ does not have a closed form, we use Markov chain Monte Carlo (MCMC) methods for computation (Gilks *et al.* 1996). Given the conditional distributions defined in Algorithm 6, the conditional distribution of λ does not have an explicit form. There are several algorithms that can be used to simulate from $p(\lambda \mid \alpha, \text{D})$, including Metropolis–Hastings (MH) and slice sampling.

Algorithm 6: Weibull survival model without covariates

1. $k = 0$. Set initial values $[\alpha^{(0)}, \lambda^{(0)}]$.
   For $k = 1 : T$, where T is large,
2. Sample $\alpha^{(k+1)} \sim \text{Gamma}(u_\alpha + d, v_\alpha + \sum_{i=1}^{n} t_i^\lambda)$.
3. Sample $\lambda^{(k+1)} \sim \lambda^{d+u_\alpha-1} \exp\left[-\lambda\left(v_\alpha - \sum_{i=1}^{n} \delta_i \log t_i\right) - \alpha \sum_{i=1}^{n} t_i^\lambda\right]$.

10.3.2 Weibull model with covariates

In order to build up a Weibull model with covariates, we develop a hierarchical model, introducing the covariates $\mathbf{x}_i$ through λ using Equation (10.2). As described in Section 10.2, the priors are now given by

$$\lambda_i \sim \text{N}(\mathbf{x}_i'\beta, \sigma^2)$$

$$\beta \sim \mathrm{N}(\mathbf{0}, \Sigma), \text{ where } \Sigma = \begin{bmatrix} \sigma_{10}^2 & 0 & \dots & 0 \\ 0 & \sigma_{20}^2 & \dots & \vdots \\ \vdots & \vdots & \ddots & 0 \\ 0 & \dots & 0 & \sigma_{i0}^2 \end{bmatrix}$$

$$\alpha \sim \text{Gamma}(u_\alpha, v_\alpha)$$

$$\tau^2 = \frac{1}{\sigma^2} \sim \text{Gamma}(a_0, b_0).$$

Diffuse priors are represented by large positive values for σ^2 and σ_0^2, and small positive values for u_α, v_α, a_0 and b_0.

The joint posterior distribution of (α, β) is given by

$$\begin{aligned} p(\beta, \alpha \mid \mathrm{D}) \propto & L(\alpha, \beta \mid \mathrm{D}) p(\alpha) p(\beta) \\ \propto & \alpha^{\alpha_0+d-1} \exp\left\{ \sum_{i=1}^{n} \left[\delta_i x_i' \beta + \delta_i (\alpha - 1) \log(t_i) - t_i^\alpha \exp\left(x_i'\beta\right) \right] \right. \\ & \left. - b_0 \alpha - \frac{1}{2} (\beta - \mu_0) \Sigma_0^{-1} (\beta - \mu_0) \right\}. \end{aligned}$$

MCMC analysis is done using the conditional distributions of the parameters, as described in Algorithm 7. As discussed earlier, the conditional distribution of α does not have an explicit form and as such can be estimated using algorithms such as MH or slice sampling.

Algorithm 7: Weibull survival model with covariates

1. $k = 0$, Set initial values $[\alpha^{(0)}, \beta^{(0)}, \tau^{2(0)}]$.
   For $k = 1 : T$, where T is large,
2. Sample $\alpha^{(k+1)} \sim \exp\left\{ d \log(\alpha) + \sum_{i=1}^{n} [\delta_i \lambda_i + \delta_i(\alpha - 1)\log(t_i)] - \sum_{i=1}^{n} (\exp(\lambda_i) t_i^\alpha) \right\} \alpha(u_\alpha - 1) \exp(-v_\alpha \alpha) \alpha$.
4. Sample $\beta^{(k+1)} \sim N(0, \sigma^2 (\mathbf{x}'\mathbf{x})^{-1} \mathbf{x}'\lambda)$.
5. Sample $\tau^{2(k+1)} \sim \text{Gamma}((n/2) + a_0, 2\left[(\lambda - \mathbf{x}\beta)' (\lambda - \mathbf{x}\beta + b_0)\right])$.

10.3.3 Model evaluation and comparison

The appropriateness of the Weibull model can be checked by applying goodness-of-fit measures which summarize the discrepancy between observed values and the values expected under the model in question (Gupta *et al.* 2008). The most commonly used assessments of model fit are in the form of information criteria, such as the Bayesian information criterion (BIC) (Schwarz 1978),

$$\text{BIC} = -2 \log L(t \mid \theta) + k \log(n),$$

and the deviance information criterion (DIC) (Spiegelhalter *et al.* 2002),

$$\text{DIC} = -2E(\log(L(t \mid \theta) \mid t)) + pD.$$

For the BIC, θ are unknown parameters of the model and k is the number of free parameters in the model. The term $k\ \log(n)$ in the BIC is also a complexity measure. The DIC penalizes complexity slightly differently through the term pD. The term estimates the effective number of parameters as follows:

$$pD = \overline{D(\theta)} - D(\bar{\theta}),$$

where $\overline{D(\theta)}$ represents the mean deviance and $D(\bar{\theta})$ denotes the deviance at the posterior means of the parameters. Both the BIC and DIC can be calculated from the simulated values based on MCMC results; smaller values indicate a more suitable model in terms of goodness of fit and short-term predictions (McGrory and Titterington 2007; Spiegelhalter *et al.* 2002).

A major goal in applying survival analysis to gene expression data is to determine a highly predictive model of patients' time to event (such as death, relapse, metastasis) using a small number of selected genes. The selection of variables in regression problems has occupied the minds of many statisticians. Several Bayesian variable selection methods have been applied to gene expression and survival studies. For example, these approaches have been used to identify relevant markers by jointly assessing sets of genes (Sha *et al.* 2006). Volinsky and Raftery (2000) investigated the BIC for variable selection in models for censored survival data and Ibrahim *et al.* (2008) developed Bayesian methodology and computational algorithms for variable subset selection in Cox PH models with missing covariate data. Other papers that deal with related aspects are Cai and Meyer (2011) and Gu *et al.* (2011). Cai and Meyer (2011) used conditional predictive ordinates and the DIC to compare the fit of hierarchical PH regression models based on mixtures of B-spline distributions of various degrees. Gu *et al.* (2011) presented a novel Bayesian method for model comparison and assessment using survival data with a cured fraction.

Comparing the predictive distribution with the observed data is generally termed a posterior predictive check (Gelman *et al.* 2004). If a model fits the data well, the observed data should be relatively likely under the posterior predictive distribution.

In our analysis, for demonstration we compute the DIC and BIC for all possible subsets of variables and select these models with smallest DIC and BIC values (Burnham and Anderson 2002). We also evaluate the model by applying posterior predictive checks based on the validation data set.

We also conduct a sensitivity analysis to understand how changes in the model inputs influence the outputs (Oakley and O'Hagan 2004). Like any method for statistical inference, the modelling approach and results will depend on various assumptions. A sometimes controversial aspect of the Bayesian approach is the need to specify prior distributions for the unknown parameters. In certain situations these priors may be very well defined. However, for complex models with many parameters, the choice of prior distributions and/or hyperparameters and conclusions of the subsequent Bayesian analysis can be validated through a sensitivity analysis

(Nur *et al.* 2009). In our analysis, described in the following section, we illustrate this by changing the Gamma prior distribution to exponential and uniform distributions and assessing the impact on the posterior estimates of interest.

10.4 Case study

10.4.1 Weibull model without covariates

Here, we used the model described in Section 10.3.1 to analyse the DLBCL survival times, ignoring gene expression data (and any other covariates). The model was fitted to the training data set and then evaluated based on the validation data set as described in Section 10.1. All analyses were undertaken using WinBUGS; the code is provided on the book's website. We ran the corresponding MCMC algorithm for 100 000 iterations (using only the first 10 000 iterations as burn-in). Two parallel chains were run from different sets of starting values. Convergence was assessed by using the two MCMC chains with a variety of different starting values. Posterior distributions and iteration history plots of the MCMC output for these parameters are presented in Figure 10.1 and show well-behaved, fast mixing processes. The

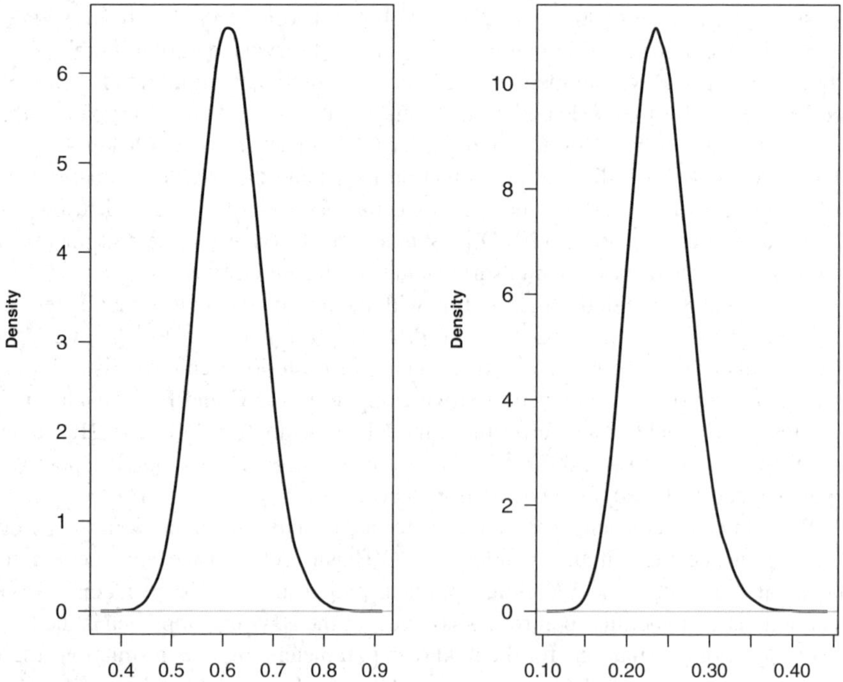

Figure 10.1 Posterior densities of α (left) and λ (right) using Weibull survival model without covariates for DLBCL data.

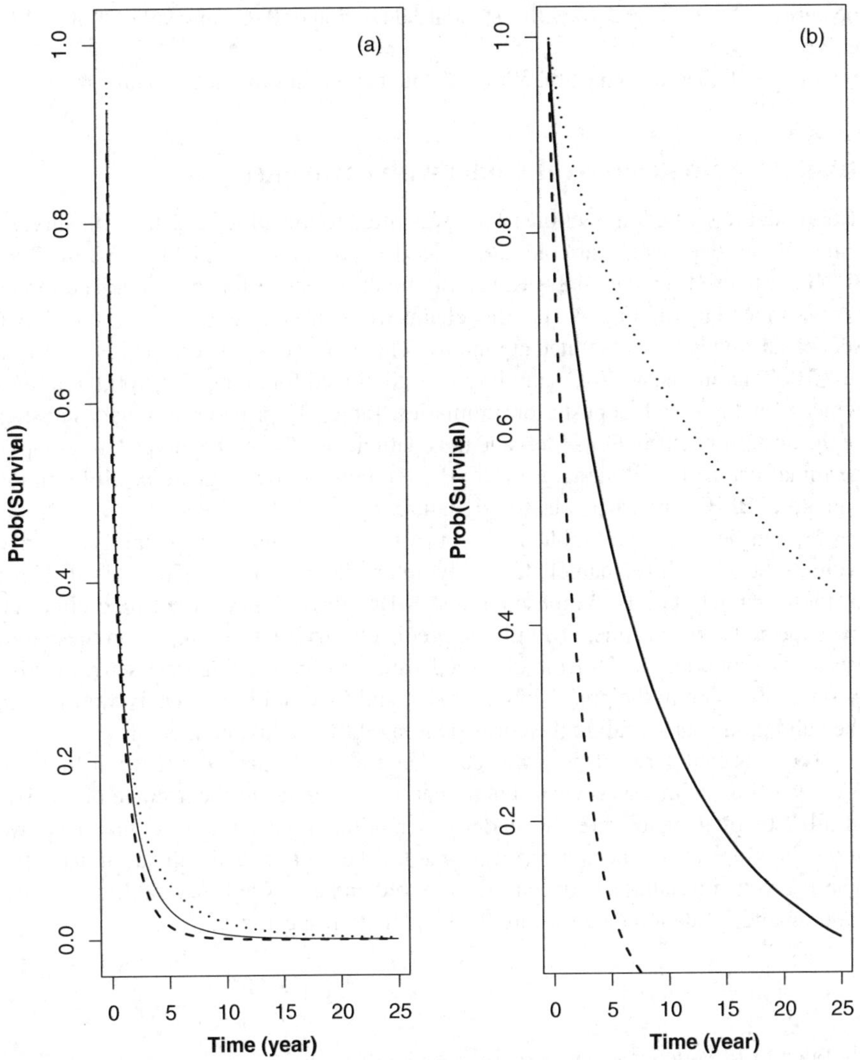

Figure 10.2 Posterior estimated survival function $S(t)$ for DLBCL data: graph (a) without covariates and (b) with covariates.

Gelman–Rubin 50% and 97.5% shrink factors (based on two dispersed starting values) were found to be 1.00 and 0.99 respectively for α, which confirm good mixing.

The estimated posterior mean of α is 0.614 with a standard deviation of 0.059, and 95% CI of (0.501, 0.735). With these data, we can see that the evidence is in favour of $\alpha < 1$; the posterior probability of this was estimated as 1. This is a strong indication that the hazard function is not constant over time. Moreover, the estimated posterior mean of λ is 0.240 with a standard deviation of 0.036 and 95% CI of (0.175, 0.316).

Figure 10.2a graphs the 5th, 50th and 95th percentiles for the survival curves $S(t)$.

10.4.2 Weibull survival model with covariates

The model described in Section 10.3.2 was fitted to the DLBCL data, with survival times as the dependent variable and $\mathbf{x}_i$ as the independent variables. Again, WinBUGS was used for the analyses; see the book's website for the code. The model was estimated by running two parallel chains from different starting values, with two sets of starting values generated at random. The two chains converged after a burn-in of 10 000 iterations. MCMC sampling was continued for another 100 000 iterations which were then used for posterior summaries. Table 10.1 provides summary statistics of the posterior distributions for α and the coefficients ($\boldsymbol{\beta}$) for phenotype (subgroups), germinal centre B-cell signature (GC-B), lymphoma node signature, proliferation signature, BMP6 and MHC class II signature.

As can be seen from Table 10.1, four of these variables substantially describe patients' survival times, namely GC-B, lymph node, BMP6 and MHC class II. Three of these, namely GC-B, lymph node and MHC class II, have a negative effect on the expected survival time. Overall, the predicted survival times based on these four variables show that 50% of the observed values in the training data set respective posterior 95% lie in the prediction intervals, and 51% of observed survival times in the validation data set fall in the corresponding 95% prediction intervals.

The expected survival time was calculated by integrating the survival function $S(t)$ over time, with the survival function at time t defined as the average of the $S(t)$ of all individuals who were still under observation at that time. In Figure 10.3, we show the graphs for expected survival time for the training and validation data sets. The Pearson correlations between the observed and estimated survival times for the training and validation data sets are 0.53 and 0.49, respectively.

Table 10.1 Posterior summary statistics for DLBCL data.

Parameters	Variables	Mean	Std dev.	95% CI
α		0.784	0.075	(0.643, 0.939)
β_0	Intercept	−1.742	0.235	(−2.218, −1.296)
β_1	Subgroups	−0.033	0.234	(−0.503, 0.416)
β_2	GC-B	−0.349	0.156	(−0.653, −0.048)
β_3	Lymph	−0.369	0.115	(−0.597, −0.145)
β_4	Proliferation	0.253	0.180	(−0.099, 0.608)
β_5	BMP6	0.377	0.148	(0.086, 0.668)
β_6	MHC	−0.489	0.127	(−0.738, −0.238)
σ	Variation	0.346	0.064	(0.248, 0.499)

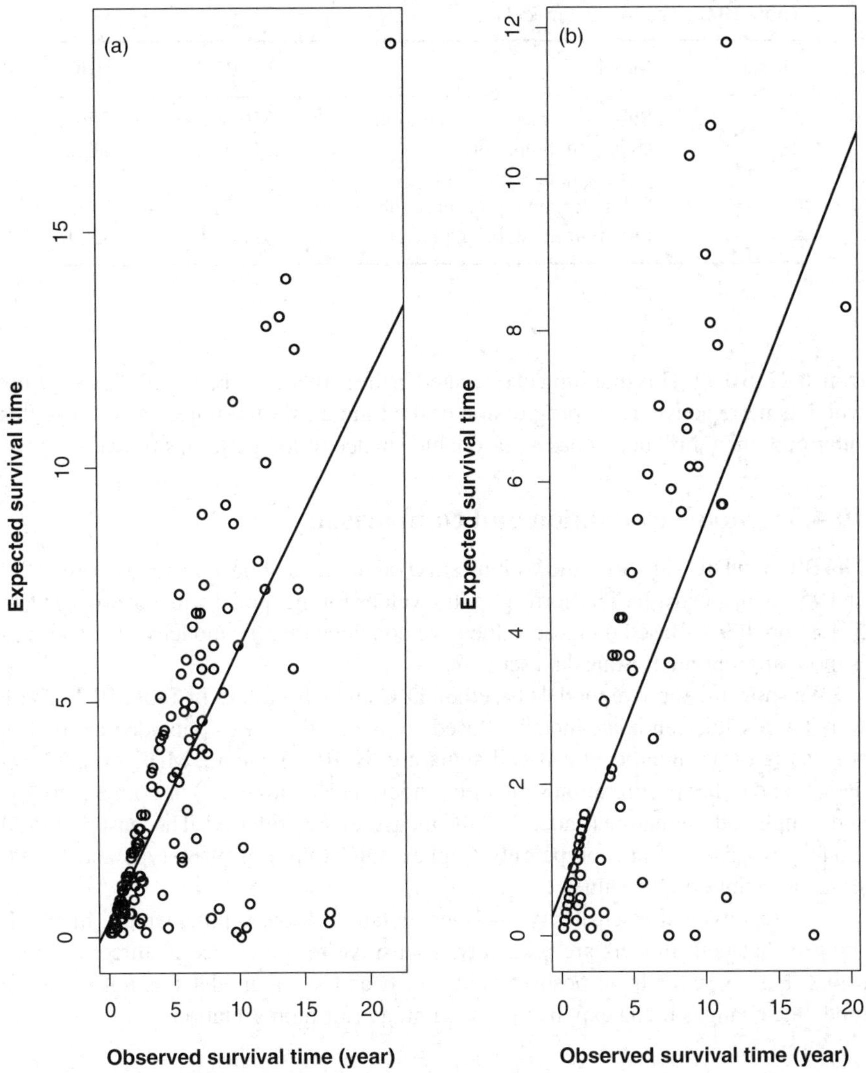

Figure 10.3 The graph for observed survival time versus the expected survival time for (a) training data set and (b) validation dataset.

In Figure 10.2b, we illustrate the predicted population survival function for a selected set of covariates based on the first subject in the validation dataset. The predicted survival falls quite rapidly towards 0.10 when the survival time is about 10 years but then decays very slowly. The posterior 95% CI for survival time of this individual, with the observed survival time 5.8 years, GC-B signature 2.24, lymph node signature 0.43, BMP6 signature −0.22 and MHC class II signature 1.35, ranges

Table 10.2 The top five models.

Model	Variables	DIC	BIC
1	Subgroup, gcb, lym, bmp, mhc	310.367	286.2
2	Gcb, lym, bmp, mhc	311.375	282.3
3	Gcb, lym, pro, bmp, mhc	311.495	285
4	Subgroup, lym, pro, bmp, mhc	324.316	311.5
5	Subgroup, lym, bmp, mhc	327.544	309

from 0.22 to 0.92. This first subject is located in the activated B-like DLBCL subgroup, which is more active in the progression of the tumour. If this subject survives in this subgroup, then this subject has a reasonable chance of longer term survival.

10.4.3 Model evaluation and comparison

The BIC and DIC values for the Weibull survival model without covariates were 459.7 and 453.6, respectively. The corresponding values for the model with covariates were 371.8 and 409.2. Based on these values, we conclude that the model with covariates is more appropriate for this data set.

We show the top five models based on DIC and BIC values in Table 10.2. There were 63 possible candidate models. Based on these values, we conclude that models consisting of germinal centre B-cell signature, BMP6 signature, MHC class II signature, and either proliferation signature (under the BIC measure) or both phenotype and lymph node signature (under the DIC measure) are preferred. The least preferred models consisted of phenotype only (highest DIC value) or phenotype and BMP6 signature (highest BIC value).

The results of the sensitivity analyses, detailed in the supplementary material, showed that both models are relatively insensitive to the choice of three different priors. However, for large positive values of α and λ, the model was not robust to moderate changes in the exponential or gamma prior representations.

10.5 Discussion

This chapter has presented the Bayesian Weibull model with MCMC computational methods. The case study that we considered involved lymphoma cancer survival, with covariates given by phenotype and gene expressions.

Based on two goodness-of-fit criteria, we showed that selected genes are substantially associated with survival in this study. Computation of DIC and BIC estimates for all possible combinations of the covariates facilitated full consideration of competing models. For example, models that are not best in the sense of goodness of fit (based on these two criteria) may be interpretable with respect to their biological and medical implications.

A popular alternative approach to post-analysis model comparison via measures such as the Bayes factor and their variants (including the DIC and BIC) is reversible jump Markov chain Monte Carlo (RJMCMC), proposed by Green (1995) and described in a general context by Lunn *et al.* (2009). At the time of writing this chapter, WinBUGS only supports normal, lognormal and Bernoulli distributions for RJMCMC, and while the Weibull model may be transformed to allow this, we did not pursue this option here.

There is a substantial literature on constructing Bayesian survival models using gene expression levels as covariates. For example, Lee and Mallick (2004) proposed an MCMC-based stochastic search technique with a Gibbs algorithm for variable selection in a survival model. As an alternative to the Weibull model, Kaderali *et al.* (2006) used a piecewise constant hazard rate in the Cox PH model to predict survival times of patients based on gene expression. Tadesse *et al.* (2005) used an alternative Bayesian error-in-variables model. Van *et al.* (2009) reviewed and compared the various methods with model survival using gene expression but not in a Bayesian context. After applying the methods to data sets for a quantitative comparison, Van *et al.* (2009) concluded that an L_2-penalized Cox regression or a random forest ensemble method is the best survival prediction method, depending on the evaluation measure. Furthermore, they also advised that to produce the best survival prediction method, characteristics such as tissue type, microarray platform, sample size, etc., should be considered.

To check the sensitivity of the model, we changed the prior distributions of selected parameters in the survival model. The results showed that the model was reasonably robust to moderate changes in the prior distribution.

Apart from accuracy and precision criteria used for the comparison study, the Bayesian approach coupled with MCMC enable us to estimate the parameters of Weibull survival models and probabilistic inferences about the prediction of survival times. This is a significant advantage of the proposed Bayesian approach. Furthermore, flexibility of Bayesian models, ease of extension to more complicated scenarios such as a mixture model, relief of analytic calculation of likelihood function, particularly for non-tractable likelihood functions, and ease of coding with available packages should be considered as additional benefits of the proposed Bayesian approach to predict survival times.

References

Alizadeh AA, Eisen MB and Davis RE 2000 Distinct types of diffuse large B-cell lymphoma identified by gene expression profiling. *Nature* **403**, 503–511.

Burnham KP and Anderson DR 2002 *Model Selection and Multimodel Inference: A Practical Information Theoretic Approach*. Springer, New York.

Cai B and Meyer R 2011 Bayesian semiparametric modelling of survival data based on mixtures of b-spline distributions. *Computational Statistics and Data Analysis* **55**, 1260–1272.

Collet D 1994 *Modelling Survival Data in Medical Research*. Chapman & Hall, London.

Dodson B 1994 *Weibull Analysis*. American Society for Quality, Milwaukee.

Gelman A, Carlin JB, Stern HS and Rubin DB 2004 *Bayesian Data Analysis*. Chapman & Hall, Boca Raton, FL.

Gilks WR, Richardson S and Spiegelhalter DJ 1996 *Markov Chain Monte Carlo in Practice*. Chapman & Hall, London.

Green PJ 1995 Reversible jump Markov chain Monte Carlo computation and Bayesian model determination. *Biometrika* **82**(4), 711–732.

Gu Y, Sinha D and Banerjee S 2011 Analysis of cure rate survival data under proportional odds model. *Life Time Data Analysis* **17**(1), 123–134.

Gupta A, Mukherjee B and Upadhyay SK 2008 Weibull extension model: a Bayes study using Markov chain Monte Carlo simulation. *Reliability Engineering and System Safety* **93**(10), 1434–1443.

Ibrahim JG, Chen MH and Kim S 2008 Bayesian variable selection for the Cox regression model with missing covariates. *Life Time Data Analysis* **14**, 496–520.

Ibrahim JG, Chen MH and Sinha D 2001 *Bayesian Survival Analysis*. Springer, Berlin.

Johnson NL, Kotz S and Balakrishnan N 1995 *Continuous Univariate Distributions*. John Wiley & Sons, Inc., New York.

Kaderali L, Zander T, Faigle U, Wolf J, Schultze JL and Schrader R 2006 CASPAR: a hierarchical Bayesian approach to predict survival times in cancer from gene expression data. *Bioinformatics* **22**(12), 1495–1502.

Kaminskiy MP and Krivtsov VV 2005 A simple procedure for Bayesian estimation of the Weibull distribution. *IEEE Transactions on Reliability* **54**(4), 612–616.

Kim SW and Ibrahim JG 2000 On Bayesian inference for proportional hazards models using noninformative priors. *Life Time Data Analysis* **6**(4), 331–341.

Kleinbaum DG and Klein M 2005 *Survival Analysis: A Self-learning Text*. Springer, Berlin.

Kundu D 2008 Bayesian inference and life testing plan for the Weibull distribution in presence of progressive censoring. *Technometrics* **50**(2), 144–154.

Lee KE and Mallick BK 2004 Bayesian methods for variable selection in survival models with application to DNA microarray data. *Indian Journal of Statistics* **66**(4), 756–778.

Lenz G, Wright GW, Emre NT, Kohlhammer H, Dave SS, Davis RE, Carty S, Lam LT, Shaffer AL, Xiao W, Powell J, Rosenwald A, Ott G, Muller-Hermelink HK, Gascoyne RD, Connors JM, Campo E, Jaffe ES, Delabie J, Smeland EB, Rimsza LM, Fisher RI, Weisenburger DD, Chan WC and Staudt LM 2008 Molecular subtypes of diffuse large B-cell lymphoma arise by distinct genetic pathways. *Proceedings of the National Academy of Sciences of the United States of America* **105**(36), 13520–13525.

Lunn DJ, Best N and Whittaker J 2009 Generic reversible jump MCMC using graphical models. *Statistics and Computing* **19**, 395–408.

Marin JM, Bernal MR and Wiper MP 2005 Using Weibull mixture distributions to model heterogeneous survival. *Communications in Statistics – Simulation and Computation* **34**(3), 673–684.

McGrory CA and Titterington DM 2007 Variational approximations in Bayesian model selection for finite mixture distributions. *Computational Statistics and Data Analysis* **51**(11), 5352–5367.

Nguyen DV and Rocke DM 2002 Partial least squares proportional hazard regression for application to DNA microarray survival data. *Bioinformatics* **18**(12), 1625–32.

Nur D, Allingham D, Rousseau J, Mengersen KL and McVinish R 2009 Bayesian hidden Markov model for DNA sequence segmentation: a prior sensitivity analysis. *Computational Statistics and Data Analysis* **53**(5), 1873–1882.

Oakley JE and O'Hagan A 2004 Probabilistic sensitivity analysis of complex models: a Bayesian approach. *Journal of the Royal Statistical Society, Series B* **66**(3), 751–769.

Park LT and Kohane IS 2002 Linking gene expression data with patients survival times using partial least squares. *Bioinformatics* **18**(1), 120–7.

Rosenwald A, Wright G, Wiestner A, Chan WC, Connors JM, Campo E, Gascoyne RD, Grogan T, Muller HK, Smeland EB, Chiorazzi M, Giltnane JM, Hurt EM, Zhao H, Averett L, Henrickson S, Yang L, Powell J, Wilson WH, Jaffe ES, Simon R, Klausner RD, Montserrat E, Bosch F, Greiner TC, Weisenburger DD, Sanger WG, Dave BJ, Lynch JC, Vose J, Armitage JO, Fisher RI, Miller TP, LeBlanc M, Ott G, Kvaloy S, Holte H, Delabie J and Staudt LM 2002 The use of molecular profiling to predict survival after chemotherapy for diffuse large B-cell lymphoma. *The New England Journal of Medicine* **346**(25), 1937–1947.

Schwarz GE 1978 Estimating the dimension of a model. *Annals of Statistics* **6**(2), 461–464.

Segal ML 2006 Microarray gene expression data with linked survival phenotypes: diffuse large B-cell lymphoma revisited. *Biostatistics* **7**(2), 268–85.

Sha N, Tadesse MG and Vannucci M 2006 Bayesian variable selection for the analysis of microarray data with censored outcome. *Bioinformatics* **22**(18), 2262–2268.

Spiegelhalter D, Best N, Carlin B and van der Linde A 2002 Bayesian measures of model complexity and fit. *Journal of the Royal Statistical Society, Series B* **64**(4), 583–639.

Tachmazidou I, Andrew T, Verzilli C, Johnson M and De IM 2008 Bayesian survival analysis in genetic association studies. *Bioinformatics* **24**(18), 2030–2036.

Tadesse MG, Ibrahim JG, Gentleman R, Chiaretti S, Ritz J and Foa R 2005 Bayesian error-in-variable survival model for the analysis of GeneChip arrays. *Biometrics* **61**, 488–497.

Van WN, Kun D, Hampel R and Boulesteix AL 2009 Survival prediction using gene expression data: a review and comparison. *Computational Statistics and Data Analysis* **53**, 1590–1603.

Volinsky CT and Raftery AE 2000 Bayesian information criterion for censored survival models. *Biometrics* **56**, 256–262.

11

Bayesian change point detection in monitoring clinical outcomes

Hassan Assareh[1], Ian Smith[2] and Kerrie L. Mengersen[1]

[1]*Queensland University of Technology, Brisbane, Australia*
[2]*St. Andrew's War Memorial Hospital and Medical Institute, Brisbane, Australia*

11.1 Introduction

A control chart is typically used to monitor the behaviour of a process over time, and to signal if a significant change in either the stability or dispersion of the process is detected. The signal can then be investigated to identify potential causes of the change and implement corrective or preventive actions (Montgomery 2008). This approach is well established in industry, and has more recently been taken up in other fields such as health care. In the clinical context, control charts can be used to monitor processes such as mortality after surgery. Here, patients comprise the units being monitored, and the underlying process parameter of interest is the mortality rate. One of the difficulties with this approach in a health-care context is that the process parameter is not constant: a patient's probability of death depends on their personal risk profile. In light of this, if is often preferable to construct a risk-adjusted control chart, that is one that takes into account patient mix (Cook *et al.* 2008; Grigg and Farewell 2004). For example, Steiner and Cook (2000) developed a risk-adjusted version of the cumulative sum control chart (CUSUM) to monitor surgical outcomes,

Case Studies in Bayesian Statistical Modelling and Analysis, First Edition. Edited by Clair L. Alston, Kerrie L. Mengersen and Anthony N. Pettitt.

death and survival, accounting for patients' health, age and other factors. This has been extended to exponentially weighted moving average (EWMA) control charts (Cook 2004; Grigg and Spiegelhalter 2007).

A common issue of concern is the detection of a step change in the process of interest. There is typically a lag between the occurrence of a change point and the control chart's signal. Thus the inferential aim is to accurately and precisely estimate the magnitude and time of the change, given a signal from the chart. Built-in change point estimators have been proposed by Page (1954, 1961) for CUSUM charts and by Nishina (1992) for EWMA charts. More recently, maximum likelihood methods have been used for the estimation of change points in several industrial processes (Perry and Pignatiello 2005; Samuel and Pignatiello 2001). In this chapter we describe a Bayesian approach proposed by Assareh *et al.* (2011).

11.2 Case study: Monitoring intensive care unit outcomes

This approach was motivated by the desire to monitor mortality of patients admitted to the intensive care unit (ICU) in a local hospital in Brisbane, Australia. Covariates such as age, gender and comorbidities along with the rate of mortality of the ICU within the hospital affect the probability of death of a patient who has been admitted to ICU. The Acute Physiology and Chronic Health Evaluation II (APACHE II), an ICU scoring system (Knaus *et al.* 1985), is used to quantify the risk of death, after adjusting for patient mix. APACHE II is based on a logistic regression using 12 physiological measurements taken in the first 24 hours after admission to the ICU, as well as chronic health status and age. The outcomes of patients entering the ICU are then monitored using risk-adjusted control charts.

The chapter proceeds as follows. After an introduction to control charts, we describe the model for a step change in the odds ratio of death and the associated risk-adjusted control charts. The setup and data from the case study are then used as a basis for a simulation experiment that investigates and analyses the performance of the Bayesian change point model. The Bayesian estimator is then compared with the alternative built-in estimators described above.

11.3 Risk-adjusted control charts

A risk-adjusted CUSUM (RACUSUM) control chart is a sequential monitoring scheme that accumulates evidence of the performance of the process and signals when either a deterioration or an improvement is detected, where the weight of evidence has been adjusted according to patient's prior risk (Steiner and Cook 2000).

For the ith patient, we observe a binary outcome y_i from a Bernoulli distribution. It is common to base the control charts on the odds ratio instead of the Bernoulli probability, if the underlying risks are not too small. The RACUSUM continuously evaluates a hypothesis of an unchanged risk-adjusted odds ratio, OR_0, against an alternative hypothesis of changed odds ratio, OR_1, in the Bernoulli process

(Cook *et al.* 2008). A weight W_i, termed the CUSUM score, is given to each patient considering the observed outcomes $y_i \in (0, 1)$ and their prior risks p_i,

$$W_i^{\pm} = \begin{cases} \log\left[\frac{(1-p_i+OR_0\times p_i)\times OR_1}{1-p_i+OR_1\times p_i}\right] & \text{if } y_i = 0 \\ \log\left[\frac{1-p_i+OR_0\times p_i}{1-p_i+OR_1\times p_i}\right] & \text{if } y_i = 1. \end{cases} \tag{11.1}$$

Upper and lower CUSUM statistics are obtained through

$$X_i^+ = \max\{0, X_{i-1}^+ + W_i^+\}\ ;\ X_i^- = \min\{0, X_{i-1}^- - W_i^-\}$$

respectively, and then plotted over patients, i, $i = 1, \ldots, T$, where T is the time (number of patients) at which the chart signals that there has been a change. The usual null hypothesis is of no change in risk, so that OR_0 is set to 1 and CUSUM statistics, X_0^+ and X_0^-, are initialized at 0. Therefore an increase in the odds ratio, $OR_1 > 1$, is asserted if a plotted X_i^+ exceeds a specified decision threshold h^+; similarly, a decrease in the odds ratio is asserted if X_i^- exceeds h^- (Steiner and Cook 2000).

A risk-adjusted EWMA (RAEWMA) control chart compares an exponentially weighted estimate of the observed process mean with the corresponding predicted process mean obtained through the underlying risk model. The EWMA statistic is given by $Z_{oi} = \lambda \times y_i + (1-\lambda) \times Z_{oi-1}$. The control chart has a centre line of $Z_{pi} = \lambda \times p_i + (1-\lambda) \times Z_{pi-1}$ and control limits of $Z_{pi} \pm L \times \sigma_{Z_{pi}}$, where $\sigma^2{}_{Z_{pi}} = \lambda^2 \times p_i(1-p_i) + (1-\lambda)^2 \times \sigma^2{}_{Z_{pi-1}}$. It is typical to let $\sigma_{Z_{p0}} = 0$ and $Z_{o0} = Z_{p0} = p_0$ in the calibration stage of the risk model and control chart; see Cook (2004) and Cook *et al.* (2008) for more details.

The magnitude of the decision thresholds in RACUSUM, h^+ and h^-, and the coefficient of the control limits in RAEWMA control charts, L, are set to ensure a specified performance of the charts in terms of false alarm and detection of shifts in odds ratio. See Grigg *et al.* (2003), Montgomery (2008) and Steiner and Cook (2000) for more details.

11.4 Change point model

Denoting the odds ratio by δ, $\delta_0 = 1$ indicates no change and $\delta \neq \delta_0$ indicates a step change in the odds ratio of the process. Let $y_i, i = 1, \ldots, T$, be independent in-control observations coming from a Bernoulli distribution with known variable rates p_{0i} that can be explained by an underlying risk model $p_{0i} \mid x_i \sim f(x_i)$, where $f(.)$ is a link function and x is a vector of covariates. At an unknown point in time, τ, the Bernoulli rate parameter changes from p_{0i} to p_{1i} obtained through

$$\delta = \frac{p_{1i}/1 - p_{1i}}{p_{0i}/1 - p_{0i}} \quad \text{and} \quad p_{1i} = \frac{\delta \times p_{0i}/(1-p_{0i})}{1 + (\delta \times p_{0i}/(1-p_{0i}))},$$

where $\delta \neq 1$ and > 0 so that $p_{1i} \neq p_{0i}$, $i = \tau, \ldots, T$. The Bernoulli process step change model in the presence of covariates can thus be parameterized as follows:

$$pr(y_i \mid p_i) = \begin{cases} p_{0i}^{y_i}(1 - p_{0i})^{1-y_i} & \text{if } i = 1, 2, \ldots, \tau \\ p_{1i}^{y_i}(1 - p_{1i})^{1-y_i} & \text{if } i = \tau + 1, \ldots, T \end{cases} \tag{11.2}$$

where the time, τ, and the magnitude, δ, of a step change in odds ratio are the unknown parameters of interest, and the posterior distributions of these parameters will be investigated in the change point analysis.

A prior distribution for δ is given by a zero left-truncated normal distribution $(\mu = 1, \sigma^2 = k)I(0, \infty)$, where k is study specific. A prior for τ is given by a uniform distribution on the range of $(1, T - 1)$.

11.5 Evaluation

The performance of the Bayesian model described above can be evaluated through a Monte Carlo experiment, as detailed in Assareh *et al.* (2011). The baseline risks in the control charts were based on a data set of available APACHE II scores for 4644 patients who were admitted to the ICU between 2000 and 2009 in a local hospital. The scores led to a distribution of logit values with a mean of −2.53 and a variance of 1.05, and an overall risk of death of 0.082 with a variance of 0.012. A value of $k = 25$ was set; see Assareh *et al.* (2011) for details.

Realizations of the process in the in-control state, y_i, $i = 1, \ldots, \tau$, were obtained by generating associated risks, $p_{0i} \sim \mathrm{N}(\mu = -2.53, \sigma^2 = 1.05), i = 1, \ldots, \tau$, and then drawing $y_i \sim \text{Bernoulli}(p_{0i})$. A step change was induced at $\tau = 500$ and observations $y_i \sim \text{Bernoulli}(p_{1i})$ were obtained until the control chart signalled; the time of signal was denoted by T. The experiment was repeated for each of the following step changes: increases of $\delta = \{1.25, 1.5, 2, 3, 5\}$ and decreases of $\delta = \{0.2, 0.33, 0.5, 0.66, 0.8\}$.

The RACUSUM charts were constructed to detect a doubling and a halving of the odds ratio in the in-control rate, $p_0 = 0.082$, and have an in-control average run length $(A\hat{R}L_0)$ of approximately 3000 observations. Monte Carlo simulation was used to determine decision intervals, $h^+ = 5.85$ and $h^- = 5.33$; see Assareh *et al.* (2011) for details. The associated CUSUM scores were also obtained through Equation (11.1) where y_i is 0 and 1, respectively.

The smoothing constant of RAEWMA was set at $\lambda = 0.01$ since the in-control rate was low and detection of small changes was desired; see Somerville *et al.* (2002), Cook (2004) and Grigg and Spiegelhalter (2007) for more details. The value of $L = 2.83$ was chosen to give the same in-control average run length $(A\hat{R}L_0)$ as the RACUSUM. A negative lower control limit in the RAEWMA was replaced by zero.

The step change and control charts were simulated in the R package (http://www.r-project.org). Posterior distributions of the time and the magnitude of the changes were obtained using the R2WinBUGS interface (Sturtz *et al.* 2005) to generate 100 000 samples through MCMC iterations in WinBUGS (Spielgelhalter *et al.* 2003) for all change point scenarios, with the first 20 000 samples treated as burn-in. The results were analysed using the CODA package in R (Plummer *et al.* 2010). If the posterior was asymmetric and skewed, the mode of the posteriors was used as an estimator for the change point model parameter (τ and δ). WinBUGS pseudo-code for the change point model is shown in Algorithm 8.

Algorithm 8: Pseudo-code for the change point model in WinBUGS

model {

for(i in 1:T){
y[i] $\sim$ Bernoulli (p[i])
p[i]=p_0[i]+step(i-τ) $\times(-p_0[i] + (\delta \times p_0[i]))/(p_0[i] \times (\delta - 1) + 1)$
p_0[i]=f_{apache}(u[i]) }
Priors
$\delta \sim$ Normal (1,0.04)I(0,)
$\tau \sim$ Uniform (1,T-1) }

11.6 Performance analysis

Table 11.1 summarizes the posterior estimates for step changes of size $\delta = 2.0$ and $\delta = 3.0$.

With a step change of size $\delta = 3.0$, the RACUSUM and RAEWMA charts signalled a positive jump in the odds ratio at $\tau = 594$ and 562 respectively, corresponding to respective delays of 94 and 62 observations (Figure 11.1(a1) and (b1)). The posterior distribution of the time of the change, τ, concentrated on the 500th observation, approximately, for both control charts (Figure 11.1(a2) and (b2)) and the magnitude of the change, δ, was also well estimated to be 3 (Figure 11.1(a3) and (b3)). The slight differences between the distributions obtained following RACUSUM and RAEWMA signals are expected since non-identical series of binary values were used for the two procedures.

When the odds ratio of death was doubled in the step change ($\delta = 2.0$), the RAEWMA and RACUSUM signalled after 157 and 297 observations, respectively.

Table 11.1 Posterior estimates (mode, SD) of step change point model parameters (τ and δ) following signals (RL) from RACUSUM ($(h^+, h^-) = (5.85, 5.33)$) and RAEWMA charts ($\lambda = 0.01$ and $L = 2.83$) where $E(p_0) = 0.082$ and $\tau = 500$. Standard deviations are shown in parentheses.

	RACUSUM			RAEWMA		
δ	RL	$\hat{\tau}$	$\hat{\delta}$	RL	$\hat{\tau}$	$\hat{\delta}$
2.0	797	508 (91)	1.8 (0.9)	657	506 (66)	2.3 (1.5)
3.0	594	502 (38)	3.3 (1.2)	562	509 (41)	3.6 (1.9)

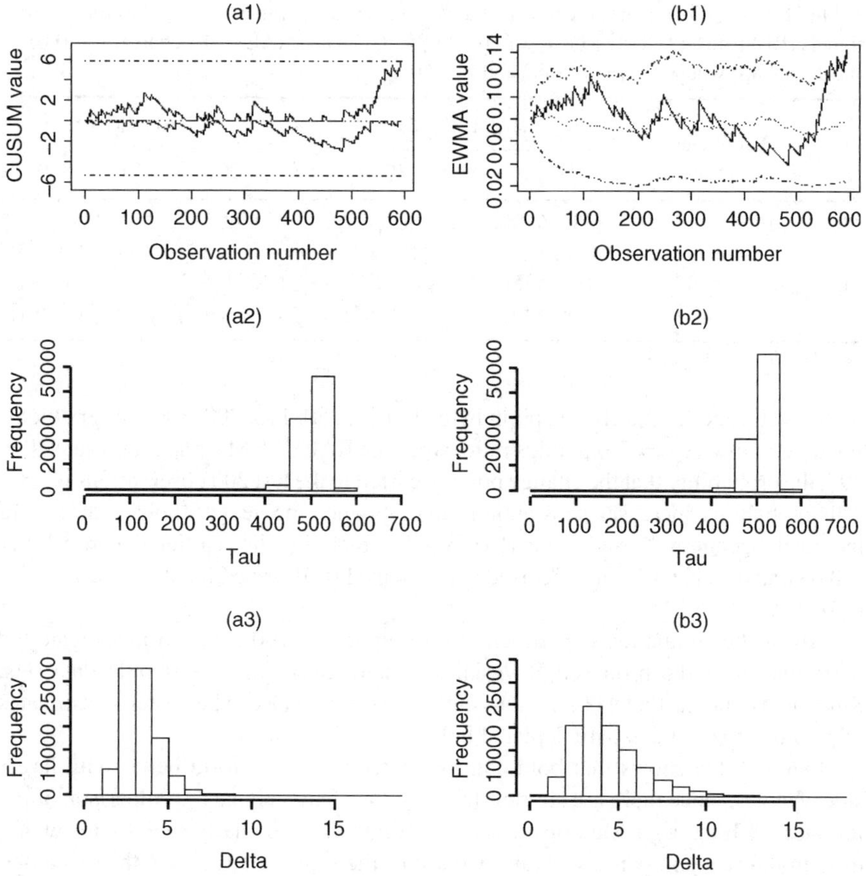

Figure 11.1 Risk-adjusted (a1) CUSUM ($(h^+, h^-) = (5.85, 5.33)$) and (b1) EWMA ($\lambda = 0.01$ and $L = 2.83$) control charts and obtained posterior distributions of (a2, b2) time τ and (a3, b3) magnitude δ of an induced step change of size $\delta = 3.0$ in odds ratio where $E(p_0) = 0.082$ and $\tau = 500$.

Despite the substantial delay in the RACUSUM signal, the posterior estimates of the time of the change were similar, at $\tau = 506$ and 508, respectively (Table 11.1). The magnitude of the change was also estimated relatively accurately (Figure 11.1(a3) and (b3)), especially in light of their corresponding standard deviations.

The 50% and 80% credible intervals (CIs) for the estimated time and the magnitude of changes in odds ratio for RACUSUM and RAEWMA control charts are presented in Table 11.2. For the RAEWMA, the 80% CI for the time of the change of size $\delta = 3.0$ covers 61 samples around the 500th observation whereas this increases to 127 observations for $\delta = 2.0$. This is due to the larger standard deviation; see Table 11.1.

The Bayesian model also enables us to make probabilistic inferences about the location of the change point prior to the chart's signal. Table 11.3 shows the probability

Table 11.2 Credible intervals for step change point model parameters (τ and δ) following signals (RL) from RACUSUM ($(h^+, h^-) = (5.85, 5.33)$) and RAEWMA charts ($\lambda = 0.01$ and $L = 2.83$) where $E(p_0) = 0.082$ and $\tau = 500$.

		RACUSUM		RAEWMA	
δ	Parameter estimate	50%	80%	50%	80%
2.0	$\hat{\tau}$	[482, 528]	[462, 593]	[491, 538]	[473, 594]
	$\hat{\delta}$	[1.5, 2.1]	[1.3, 2.4]	[1.7, 2.7]	[1.3, 3.5]
3.0	$\hat{\tau}$	[500, 535]	[477, 542]	[504, 535]	[480, 541]
	$\hat{\delta}$	[2.3, 3.6]	[1.9, 4.4]	[2.4, 4.5]	[1.8, 6.1]

of the occurrence of the change point in the last {25, 50, 100, 200, 300} observations. For a jump of size $\delta = 2.0$ in odds ratio, since the RACUSUM signals very late, $T = 797$, the probability that the change point occurred in the last 200 observations is 0.12. This probability increases substantially to 0.76 when the next 100 observations are included. In contrast, based on the RAEWMA, which signalled earlier, the probability is 0.68 that the change point occurred between the last 100 and 200 observations prior to signal.

All of the simulation experiments were replicated 100 times to investigate the performance of the proposed Bayesian estimator over different sample data sets. Simulated data sets that were obvious outliers were excluded. The posterior estimates of parameters of interest are depicted in Figure 11.2.

Figure 11.2a shows that both control charts tend to perform better with larger step changes, since there are shorter delays in the charts' signals. Comparison of the left side with the right side of this graph reveals that the charts signal later with a drop in the risk compared with an increase of the same relative size; this is caused by the dependency of the mean and the variance of the Bernoulli distribution. The RACUSUM and the RAEWMA performed differently over jumps and drops in the odds ratio: the RAEWMA was outperformed by the RACUSUM for drops, whereas it was superior for jumps. Similar behaviour was shown for the posterior precision of the charts' signals in Figure 11.2c.

As expected, both control charts tended to signal faster for medium step changes than for small steps (Figure 11.2a). However, even over large drops in the odds ratio,

Table 11.3 Probability of the occurrence of the change point in the last 25, 50, 100, 200 and 300 observed samples prior to signalling for RACUSUM ($(h^+, h^-) = (5.85, 5.33)$) and RAEWMA charts ($\lambda = 0.01$ and $L = 2.83$) where $E(p_0) = 0.082$ and $\tau = 500$.

	RACUSUM					RAEWMA				
δ	25	50	100	200	300	25	50	100	200	300
2.0	0.03	0.05	0.07	0.12	0.76	0.11	0.13	0.28	0.96	0.98
3.0	0.00	0.03	0.67	0.99	0.99	0.13	0.53	0.95	0.99	0.99

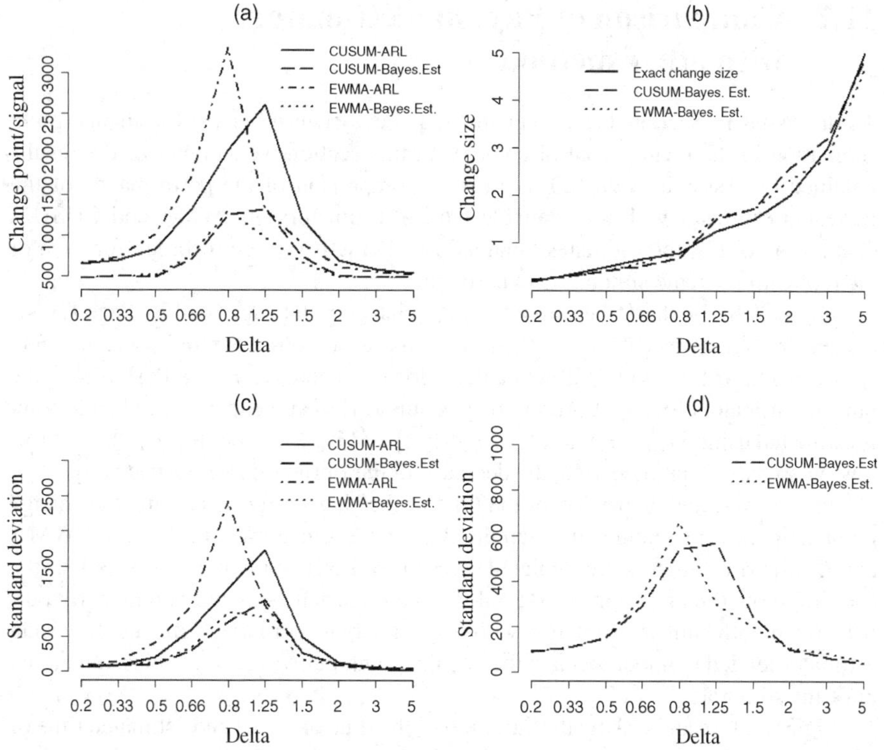

Figure 11.2 Charts' signals and posterior estimates of step change point model parameters, change in odds ratio, following signals from RACUSUM ($(h^+, h^-) = (5.85, 5.33)$) and RAEWMA control charts ($\lambda = 0.01$ and $L = 2.83$) where $E(p_0) = 0.082$ and $\tau = 500$: (a) average of charts' signals and posterior modes of time, τ; (b) posterior modes of change size, δ; (c) standard deviation of charts' signals and posterior modes of time, τ; (d) average of standard deviation of posterior distributions of time, τ.

there were still large delays associated with the charts' signals, although the posterior estimates of the time of the change were still acceptable. In the best scenarios, the RACUSUM and RAEWMA signalled at the 545th and 533rd observations for a very large increase in the odds ratio, $\delta = 5.0$, respectively. The time of the shift, τ, still tended to be underestimated for large step changes.

For all scenarios, the magnitudes of the change were estimated reasonably well (Figure 11.2b). The posterior estimates tended to be biased downwards for medium to large shifts and biased upwards for small shifts. The bias was less for moderate to large shifts. The precision of the posterior estimates of δ was better for small to medium shift sizes than larger shift sizes (Figure 11.2c). Less precise estimates were seen for drops compared with jumps of the same size. The standard deviation of the posterior distributions of time also decreased as the shift sizes increased (Figure 11.2d).

11.7 Comparison of Bayesian estimator with other methods

As discussed in Section 11.1, other methods have been proposed for change point estimation in the quality control context. In this section we summarize the results obtained by Assareh *et al.* (2011) in their comparison of the performance of the Bayesian estimator with the available built-in estimators of EWMA and CUSUM charts. To do this, the estimates obtained by alternatives were recorded for each of the replications in the simulation experiment.

For a CUSUM chart, the estimate of the change point suggested by Page (1954) is used: $\hat{\tau}_{cusum} = \max\{i : X_i^+ = 0\}$ if an increase in a process rate is detected, and $\hat{\tau}_{cusum} = \max\{i : X_i^- = 0\}$ following detection of a decrease. For an RAEWMA, the built-in estimator of EWMA proposed by Nishina (1992) is adapted. The change point is estimated using $\hat{\tau}_{ewma} = \max\{i : Z_{oi} \leq Z_{pi}\}$ and $\hat{\tau}_{ewma} = \max\{i : Z_{oi} \geq Z_{pi}\}$ following signals of an increase and a decrease in the Bernoulli rate, respectively.

The average and dispersion of the time of the charts' signals (T) and the change point estimates (τ) obtained through the Bayesian estimator and the built-in EWMA and CUSUM estimators are depicted in Figure 11.3. For moderate to small shifts, $\delta = 0.5, 0.66, 0.8$ and their inverse values, the Bayesian estimator significantly outperformed both built-in estimators with respect to bias. For large shifts, the Bayesian estimator tended to underestimate the change point, but was still less biased than the built-in estimators.

Figures 11.3c and d show the variation of the time of signal and estimated time of step change in the replication study. When the magnitude of the change increased, less dispersed estimates were observed. The variation in the time of signal was equivalent for all three methods, but more precise estimates of the time and magnitude of the step change were obtained by the Bayesian estimators across the various changes in the size and direction of the odds ratio.

11.8 Conclusion

Knowing the time when a process has changed is an important component of quality improvement programmes in a wide range of fields. In this chapter, we described a Bayesian approach to estimating the magnitude and time of a step change in a process, based on a post-change signal in the corresponding control chart. We considered a process with dichotomous outcomes, where the underlying probability for each observation was influenced by covariates. The case study that motivated the development of this approach involved monitoring the mortality rate in the intensive care unit of a local hospital, using risk-adjusted CUSUM and EWMA control charts in which the risk of death was evaluated by APACHE II, a logistic prediction model.

The performance of the proposed approach was evaluated and confirmed through a simulation study covering a range of scenarios of magnitude and direction of change in the underlying process. We also illustrated the advantage of the approach in deriving other measures of interest, for example the location of the change point prior to the

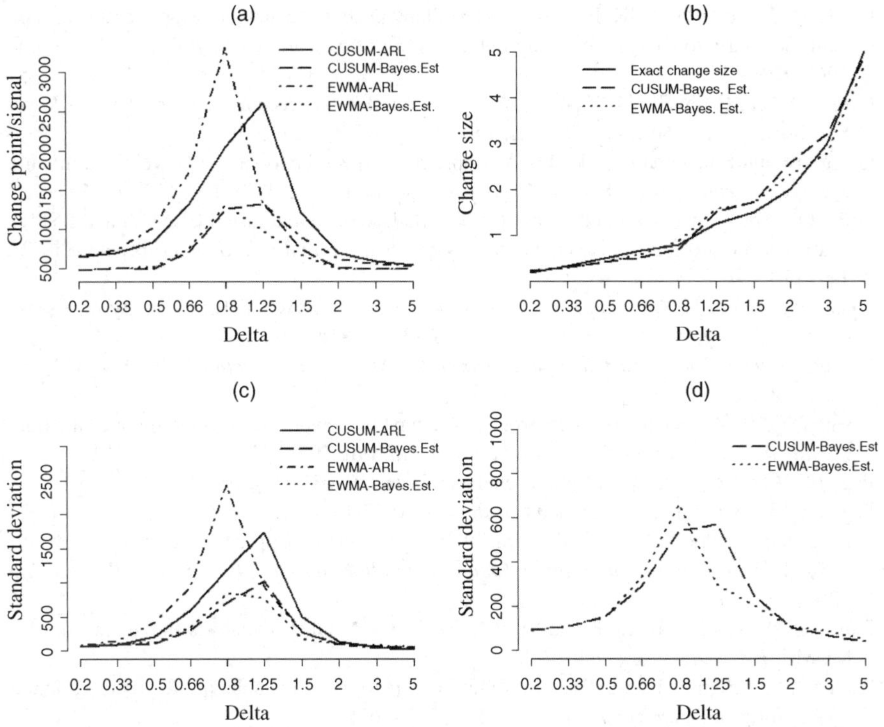

Figure 11.3 Charts' signals and estimated time of change in odds ratio obtained by the Bayesian estimator, CUSUM and EWMA built-in estimators following signals from RACUSUM ($(h^+, h^-) = (5.85, 5.33)$) and RAEWMA control charts ($\lambda = 0.01$ and $L = 2.83$) where $E(p_0) = 0.082$ and $\tau = 500$: (a) average of RACUSUM signals and estimates of time, τ; (b) average of RAEWMA signals and estimates of time, τ; (c) standard deviation of RACUSUM signals and estimates of time, τ; (d) standard deviation of RAEWMA signals and estimates of time, τ.

chart's signal. The Bayesian estimator was also favourably compared with built-in estimators of EWMA and CUSUM charts, particularly when the precision of the estimators was taken into account. Finally, the flexibility of Bayesian models, the ease of extension to more complicated change scenarios such as linear and nonlinear trends in odds ratios, and the ease of coding with available packages are additional benefits of the proposed Bayesian approach to monitoring for quality improvement.

References

Assareh H, Smith I and Mengersen K 2011 Change point detection in risk adjusted control charts. *Statistical Methods in Medical Research*. doi: 10.1177/0962280211426356.

Cook D 2004 *The development of risk adjusted control charts and machine learning models to monitor the mortality of intensive care unit patients*. PhD thesis. University of Queensland.

Cook DA, Duke G, Hart GK, Pilcher D and Mullany D 2008 Review of the application of risk-adjusted charts to analyse mortality outcomes in critical care. *Critical Care Resuscitation* **10**(3), 239–251.

Grigg OV and Farewell VT 2004 An overview of risk-adjusted charts. *Journal of the Royal Statistical Society, Series A (Statistics in Society)* **167**(3), 523–539.

Grigg OV and Spiegelhalter DJ 2007 A simple risk-adjusted exponentially weighted moving average. *Journal of the American Statistical Association* **102**(477), 140–152.

Grigg OV, Spiegelhalter DJ and Farewell VT 2003 Use of risk-adjusted CUSUM and RSPRT charts for monitoring in medical contexts. *Statistical Methods in Medical Research* **12**(2), 147–170.

Knaus W, Draper E, Wagner D and Zimmerman J 1985 APACHE II: a severity of disease classification system. *Critical Care Medicine* **13**(10), 818–829.

Montgomery DC 2008 *Statistical Quality Control: A Modern Introduction*. John Wiley & Sons, Inc., Hoboken, NJ.

Nishina K 1992 A comparison of control charts from the viewpoint of change-point estimation. *Quality and Reliability Engineering International* **8**(6), 537–541.

Page ES 1954 Continuous inspection schemes. *Biometrika* **41**(1/2), 100–115.

Page ES 1961 Cumulative sum charts. *Technometrics* **3**(1), 1–9.

Perry M and Pignatiello J 2005 Estimating the change point of the process fraction nonconforming in SPC applications. *International Journal of Reliability, Quality and Safety Engineering* **12**(2), 95–110.

Plummer M, Best N, Cowles K and Vines K 2010 CODA: Output analysis and diagnostics for MCMC. *R Package, version 0.14-2*.

Samuel T and Pignatiello J 2001 Identifying the time of a step change in the process fraction nonconforming. *Quality Engineering* **13**(3), 357–365.

Somerville SE, Montgomery DC and Runger GC 2002 Filtering and smoothing methods for mixed particle count distributions. *International Journal of Production Research* **40**(13), 2991–3013.

Spielgelhalter D, Thomas A and Best N 2003 *WinBUGS version 1.4. Bayesian inference using Gibbs sampling*. MRC Biostatistics Unit, Institute for Public Health, Cambridge.

Steiner SH and Cook RJ 2000 Monitoring surgical performance using risk adjusted cumulative sum charts. *Biostatistics* **1**(4), 441–452.

Sturtz S, Ligges U and Gelman A 2005 R2WinBUGS: a package for running WinBUGS from R. *Journal of Statistical Software* **12**(3), 1–16.

12

Bayesian splines

Samuel Clifford[1] and Samantha Low Choy[1,2]

[1] *Queensland University of Technology, Brisbane, Australia*
[2] *Cooperative Research Centre for National Plant Biosecurity, Australia*

12.1 Introduction

Smoothing splines are an example of the class of functions called 'scatterplot smoothers' which can be used in non-parametric regression. Smoothing splines are functions which are able to model the effect of a covariate on a response variable without assuming the functional form of the relationship or even that the effect of the covariate remains the same across the domain of the covariate. For this reason, smoothing splines have found application in generalized additive models (GAMs), an extension to the generalized linear model (GLM) (Hastie 1993; Wood 2006).

In section 12.2 we will give a brief introduction to some Bayesian statistical theory and the Metropolis–Hastings sampler, describe various spline models and discuss the interpretation of posterior summary statistics. In Section 12.3 we will give a number of examples of the use of smoothing splines within generalized additive models and use different data sets which call for different approaches. We will summarize the key points in Section 12.4 and discuss various ways in which the approach here can be extended.

12.2 Models and methods

12.2.1 Splines and linear models

Splines are piecewise-defined functions constructed from simpler functions (usually polynomials) between control points (or 'knots'). Splines have found use in regression

Case Studies in Bayesian Statistical Modelling and Analysis, First Edition. Edited by Clair L. Alston, Kerrie L. Mengersen and Anthony N. Pettitt.

due to their ability to model the nonlinear behaviour of a covariate in a flexible and robust manner without having to specify the functional form of the relationship between the response and covariate.

Splines used for interpolation or regression (or 'smoothing' splines) can be constructed from any spline basis and this is what makes them so flexible. The resulting spline basis is equivalent to the design matrix for a linear regression and as such we can fit spline models using any linear regression method. The general form for a linear regression model is

$$y = X\boldsymbol{\beta} + \boldsymbol{\varepsilon}$$

where X contains the values of the covariates corresponding to each observation y_i, $\boldsymbol{\beta}$ are the regression coefficients relating the covariates to the expected value of the observations and $\boldsymbol{\varepsilon}$ are the residuals, the leftover unexplained variation.

The inclusion of spline basis functions in a parametric regression model allows the expression of the model as a linear mixed effects model (LMM),

$$y = X\boldsymbol{\beta} + Z\boldsymbol{\gamma} + \boldsymbol{\varepsilon}$$

where the spline terms are equivalent to random effects, $Z\boldsymbol{\gamma}$ (Ruppert *et al.* 2003). As a result, the GAM is a special case of the generalized linear mixed model (GLMM) and we can use any GLMM technique to fit a GAM.

12.2.2 Link functions

The extension of a spline model to a GAM requires the inclusion of a link function, g, which expresses the relationship between the linear predictor, $\mu = X\boldsymbol{\theta}$, and the mean response, $E(\boldsymbol{y}|\boldsymbol{x}, \boldsymbol{\theta})$ (Hastie and Tibshirani 1990).

That is, the general form of a GAM is

$$\begin{aligned} E(\boldsymbol{y}|\boldsymbol{x}, \boldsymbol{\theta}) &= g(\mu) \\ g(\mu) &= \sum_i f_i(\boldsymbol{x}) \\ &= X\boldsymbol{\theta} \end{aligned} \tag{12.1}$$

for non-parametric functions f_i.

Because GAMs are highly flexible it is tempting to use them without thinking particularly hard about what kinds of data are being fitted. For example, count data require that the fitted values be strictly non-negative and a GAM fitted to count data using the normal likelihood with an identity link may, if the mean is almost zero, have a symmetric credible interval with negative values. These values are not physically possible and will compromise the ability to correctly infer from the fitted model. As such, the link function is important for ensuring fitted models are valid.

12.2.3 Bayesian splines

Splines fitted in a Bayesian framework can be treated as any other regression problem through specification of the design matrix (see Chapter 4). For a linear regression the columns of the design matrix comprise the covariate values corresponding to each observation. The columns of the Bayesian spline design matrix are the spline basis functions of the covariates. Although each column is not so directly interpretable, the role of the design matrix is the same and the coefficients of the model are the coefficients of the linear combination of spline basis functions which provide the optimal fit.

The interpretation of a spline basis function is not as obvious as it is for a polynomial basis function, primarily due to the placement of knots and the relationship that these knots have with the data (through the spline model). As such, inference tends to focus on the entire fitted spline function rather than the effect size of each individual spline basis function.

Thin-plate splines

Thin-plate splines are so called because they are equivalent to the solution of the equation giving the deflection of a thin plate of metal under load (Duchon 1977).

A 'low rank' thin-plate spline of order p, for a single covariate, x, and response variable y, can be written as

$$E(y|x, \boldsymbol{\theta}) = \beta_0 + \beta_1 x + \sum_{k=1}^{K} u_k |x - \kappa_k|^p \tag{12.2}$$

where the κ are fixed knots, either spaced at the quantiles of x (Lunn *et al.* 2000) or uniformly spaced throughout the covariate space. The coefficients β_0 and β_1 are the familiar linear regression coefficients and the u_k are the coefficients of the spline terms. Together, we will denote the parameters β_0, β_1, u_k as the vector $\boldsymbol{\theta}$.

We assume that each data point is independently normally distributed with a mean given by Equation (12.2) and a common standard deviation, σ.

We can express (12.2) as a linear model with design matrix $X \in \mathbf{R}^{n\times(K+2)}$ and vector of coefficients $\boldsymbol{\theta} \in \mathbf{R}^{(K+2)\times 1}$,

$$X = \begin{bmatrix} 1 & x_1 & |x_1 - \kappa_1|^3 & |x_1 - \kappa_2|^3 & \cdots & |x_1 - \kappa_K|^p \\ 1 & x_2 & |x_2 - \kappa_1|^3 & |x_2 - \kappa_2|^3 & \cdots & |x_2 - \kappa_K|^p \\ \vdots & \vdots & \vdots & \vdots & \ddots & \vdots \\ 1 & x_n & |x_n - \kappa_1|^3 & |x_n - \kappa_2|^3 & \cdots & |x_n - \kappa_K|^p \end{bmatrix}, \quad \boldsymbol{\theta} = \begin{bmatrix} \beta_0 \\ \beta_1 \\ u_1 \\ u_2 \\ \vdots \\ u_K \end{bmatrix}.$$

The corresponding likelihood for a normally distributed response is

$$p(y_i|\boldsymbol{\theta}) = \mathrm{N}\left(y_i; X_{i,*}\boldsymbol{\theta}, \sigma^2\right) = \mathrm{N}\left(y_i; \beta_0 + \beta_1 x_i + \sum_{k=1}^{K} u_k |x_i - \kappa_k|^p, \sigma^2\right).$$

Hence the thin-plate spline model shown in (12.2) can be interpreted as a mixed effects model where the intercept and slope are fixed effects and the spline terms are

Table 12.1 Fitting a thin-plate spline model with normal likelihood.

Get knots	$\boldsymbol{\kappa}$, the K quantiles of x
Generate basis splines	$b_{ik} = \lvert x_i - \kappa_k \rvert^p$
Likelihood	$p(y\lvert\boldsymbol{\theta}) = \prod_{i=1}^{n} \mathrm{N}\left(y_i; \mu_i, \sigma^2\right)$
	$\mu_i = \beta_0 + \beta_1 x_i + \sum_{k=1}^{K} u_k b_{ik}$
Prior for splines	$\boldsymbol{u} \sim \mathrm{MVN}(\mathbf{0}, \sigma_u I_K)$
Prior for linear terms	$\beta_0 \sim \mathrm{N}(0, \sigma_{\beta_0})$
	$\beta_1 \sim \mathrm{N}(0, \sigma_{\beta_1})$
Prior for model variance	$\sigma \sim \sqrt{\text{Inv-}\chi_1^2}$
Sample from posterior	$p(\boldsymbol{\theta}\lvert y) \propto p(y\lvert\boldsymbol{\theta})\, p(\boldsymbol{\theta})$

random effects. The random effects structure for the spline terms assumes that all u_k have the same standard deviation and mean, that is $\boldsymbol{u} \sim \mathrm{MVN}(\mathbf{0}, \sigma_u^2 I_K)$, which we can use as our prior for the spline coefficients.

We will use the non-informative multivariate normal priors from above with a large value of σ_u, say, 100, for the splines and weakly informative normals for the linear terms, $\beta_0 \sim \mathrm{N}(0, \sigma_{\beta_0})$, $\beta_1 \sim \mathrm{N}(0, \sigma_{\beta_1})$ with $\sigma_{\beta_1}, \sigma_{\beta_2}$ large as well.

We will use a conjugate inverse-gamma prior for the variance of the modelled values (Gelman et al. 2004),

$$\sigma^2 \sim \Gamma^{-1}\left(\sigma^2; 1/2, 1/2\right) = \frac{1}{\sqrt{2\pi}}\sigma^{-3} e^{-1/2\sigma^2} = \text{Inv-}\chi_{\nu=1}^2$$

$$\sigma \sim \sqrt{\text{Inv-}\chi_1^2}.$$

This prior has a mode at zero encouraging the standard deviation to be as small as possible. Updating the prior with data causes the mode to move away from zero according to the strength of the data relative to the prior (Gelman et al. 2004).

Combining the likelihood and priors allows us to fit the thin-plate spline model according to Table 12.1.

B-splines

It is not necessary that each spline basis function has a global effect. For example, a cubic regression spline consists of piecewise smooth third-degree polynomials. Spline basis functions may have compact support, equivalent to allowing a spline model to be treated as a structured random effect. Consider the data rather than the splines. It may be a reasonable assumption that two data points separated by a large distance should not affect the fitted function in each other's neighbourhood; this is one of the assumptions of loess regression (Cleveland *et al.* 1993) and other simple smoothers such as local linear regression or a moving mean or median.

B-splines are an example of splines whose basis functions have compact support (Eilers and Marx 1996). B-spline bases are defined as non-zero between two knots, where they are at least piecewise continuous (and piecewise smooth for splines of degree greater than one), and zero outside those knots. As such, the B-spline is very flexible at modelling drastic changes in response across the covariate space.

B-splines are defined on a mesh of knots and are a piecewise sum of low-order polynomials. A first-degree B-spline basis element is defined as two linear terms piecewise defined to be zero at κ_k and κ_{k+2} with a maximum at κ_{k+1}. A second-degree B-spline basis element is defined as the piecewise sum of three quadratic polynomials such that the spline basis element is zero at κ_k and κ_{k+3}, piecewise smooth at κ_{k+1} and κ_{k+2} and has a maximum halfway between κ_{k+1} and κ_{k+2}. Calculation of the B-spline basis is performed according to a recursive formulation given in Equation (12.3). Here, we will use only B-spline bases with equally spaced knots, for which code can be found in the appendix of Eilers and Marx (1996).

The recursion relationship for calculating B-spline bases (de Boor 1978; Eilers and Marx 1996) is

$$\frac{B_{j,k}(x)}{t_{j+k} - t_j} = \frac{x - t_j}{t_{j+k-1} - t_j} \frac{B_{j,k-1}(x)}{t_{j+k-1} - t_j} + \frac{t_{j+k} - x}{t_{j+k} - t_j} \frac{B_{j+1,k-1}(x)}{t_{j+k} - t_{j+1}} \tag{12.3}$$

on a grid of knots t_i, with $B_{j,0}(x) = 1$ between knots t_{j-1} and t_j. That is, the zero-order B-spline basis is constant between successive knots and higher order bases are created according to the recurrence relation.

While B-splines of arbitrary order (up to the number of knots minus one) can be constructed, the second-order B-spline is sufficient for the construction of smooth fits as it is a sum of piecewise C^2 smooth functions. A first-order B-spline will produce a fitted function which is piecewise continuous but not piecewise smooth. A zero-order B-spline can be fit which is neither smooth nor continuous and is equivalent to calculating the mean between successive knots. Examples of first- and second-order B-splines are given in Figure 12.1.

B-splines have been specified in a frequentist framework for some time and have been extended to include a smoothness penalty (converting a B-spline to a 'P'-spline) via a penalty matrix, outlined in Equation (12.4) (Eilers and Marx 1996). The smoothness penalty controls the fitting of a model so that the resulting spline is not too 'wiggly', indicating that the data are being overfit (Silverman 1985).

The penalty matrix to convert a frequentist B-spline to a P-spline with an order k penalty is based on the matrix form of the kth difference operator which has dimension equal to the width of the spline basis matrix. Let $Q = D^T D$, where D is the discretized kth difference operator; then the first-order penalty for a basis of n B-splines is

$$Q_{ij} = \begin{cases} 1, & i = j \in \{1, n\} \\ 2, & 1 < i = j < n \\ -1, & |i - j| = 1 \\ 0, & \text{otherwise.} \end{cases} \tag{12.4}$$

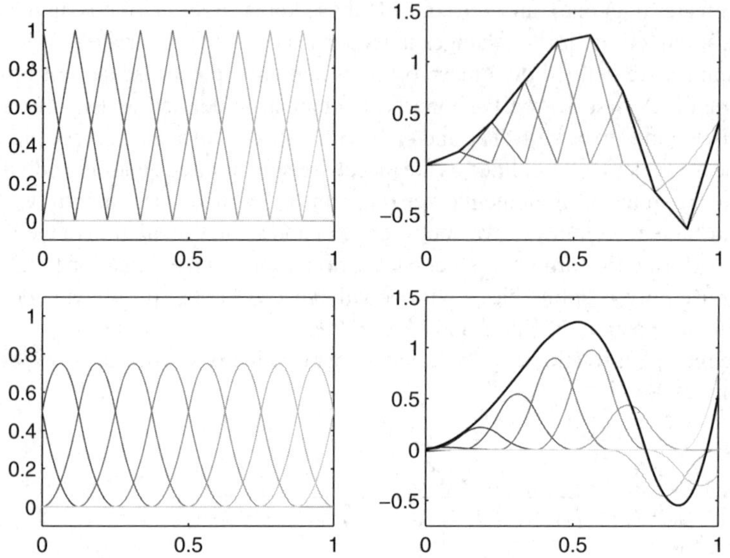

Figure 12.1 Ten first- and second-order B-spline basis splines defined on the domain (0, 1) and a linear combination of the bases to approximate $y = \sin(2\pi x^2) + 0.5x$.

For example, let $n = 5$; then

$$Q = \begin{bmatrix} 1 & -1 & 0 & 0 & 0 \\ -1 & 2 & -1 & 0 & 0 \\ 0 & -1 & 2 & -1 & 0 \\ 0 & 0 & -1 & 2 & -1 \\ 0 & 0 & 0 & -1 & 1 \end{bmatrix}. \tag{12.5}$$

The frequentist use of the penalty matrix is to determine the value of the smoothing parameter, λ, which solves the normal equations (where B is the B-spline basis matrix and Q the penalty matrix in Equation 12.4)

$$\left(B^T B + \lambda Q\right) \boldsymbol{a} = B^T \boldsymbol{y}$$

for a given value of λ, such that the coefficients of the P-spline model are $\boldsymbol{a}$. The choice of λ is determined by generalized cross-validation (Wahba 1990).

It is possible to include the smoothing in the likelihood and to treat λ as one of the model parameters and therefore derive its posterior. In a Bayesian context, a roughness penalty based on the difference matrix may be incorporated as a prior for the spline parameters rather than by smoothing the spline in the likelihood (Brezger 2004). This can be justified by viewing the smoothing as a belief regarding the spline parameters. In essence, a Bayesian P-spline is a B-spline with a prior belief that the resulting spline should not be too 'wiggly', that is the second difference of the spline parameters (analogous to the second derivative of the resulting smooth) should be as small as possible (Table 12.2).

The form of the penalty prior is proportional to a multivariate normal

$$\boldsymbol{\theta}|\sigma \propto \exp\left(-\frac{\boldsymbol{\theta}^T Q\boldsymbol{\theta}}{2\sigma^2}\right) \tag{12.6}$$

where $\boldsymbol{\theta}$ are the spline coefficients, σ is the standard deviation of the multivariate normal density and Q is the penalty matrix from Equation (12.4), imposing the structure of the penalty. Q is rank deficient and so is not symmetric, positive definite and its inverse is undefined; these problems are avoided by adding a small value to the diagonal of D^2, such as 0.000 01. Smoothing is controlled via the variance, σ^2, making it analogous to the frequentist smoothing parameter, λ (Brezger 2004).

We can reparameterize the variance of a normal density as a precision, that is $\lambda = 1/\sigma^2$, so that the Bayesian smoothing prior for spline coefficients is (Speckman *et al*. 2003)

$$\boldsymbol{\theta}|\lambda \propto \lambda^{(n-k)/2} \exp\left(-\frac{\lambda}{2}\boldsymbol{\theta}^T Q\boldsymbol{\theta}\right). \tag{12.7}$$

For the prior for the smoothing parameter, λ, we will use a gamma prior with $a = 1$ so that the prior has its mode at zetro (no smoothing). The scale parameter of the gamma, b, is chosen so that the prior is not too diffuse (Lang and Brezger 2004). The $\Gamma(1, b)$ prior is mathematically equivalent to an Exp(b) prior. Many other sensible priors exist for the smoothing parameter as the precision of a multivariate normal and these need not be conjugate (see Gelman et al. 2004, appendix A). An overview of the steps required to generate and fit a Bayesian B- or S-spline regression model with normal likelihood is given in Table 12.2.

Periodic basis

For a covariate which has some sort of physical interpretation as being on a circle (e.g. wind direction, hour of the day, season) we may decide to fit the effect of that covariate with a periodic spline; that is, the fitted function is (at least) continuous at the upper and lower values of the covariate space. To convert a B-spline basis to a

Table 12.2 Fitting a B- or P-spline model with normal likelihood.

Get knots	t, K evenly spaced values of x
Generate basis splines	Equation (12.3)
Likelihood	$p(y\|\boldsymbol{\theta}) = \prod_{i=1}^{n} \mathrm{N}\left(y_i; \mu_i, \sigma^2\right)$
	$\mu_i = B\boldsymbol{\theta}$
Prior for splines	Equation (12.7)
Prior for smoothing parameter	$\lambda \sim \Gamma(1, b)$
Prior for model variance	$\sigma \sim \sqrt{\text{Inv-}\chi_1^2}$
Sample from posterior	$p(\boldsymbol{\theta}\|y) \propto p(y\|\boldsymbol{\theta})\, p(\boldsymbol{\theta})$

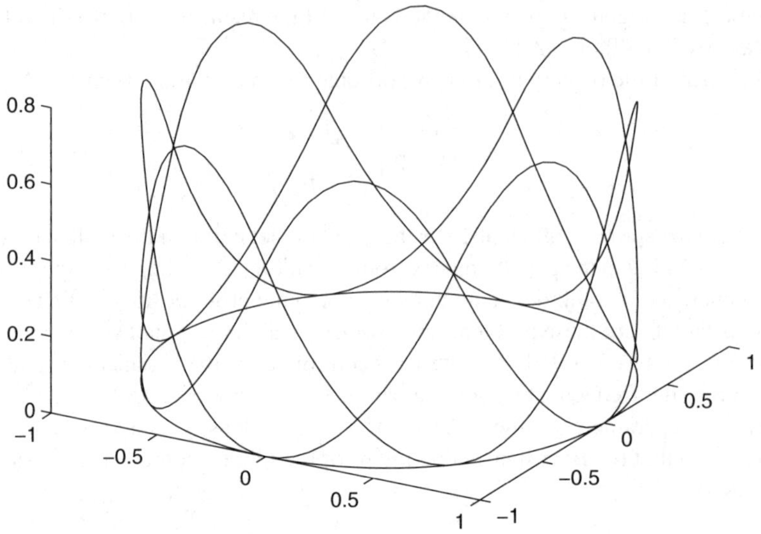

Figure 12.2 Eight periodic, second-order B-spline basis functions plotted on a circle.

periodic B-spline basis, the basis columns corresponding to the B-splines which are non-zero at the boundaries of the covariate space are added together such that any order-preserving permutation still represents a piecewise continuous B-spline.

The algorithm for the creation of the periodic basis is given in Algorithm 9. The *i*th column of the B-spline design matrix, B, is denoted as $B_{*,i}$. Figure 12.2 shows eight periodic, second-order B-spline bases plotted on a circle. Each spline is piecewise smooth around the entire circle.

Algorithm 9: Periodic B-spline basis

bdeg = degree of spline
for $i = 1$ **to** bdeg **do**
  $B_{*,i} = B_{*,i} + B_{*,n-\text{bdeg}+i}$
end for
delete $B_{*,n-\text{bdeg}+1} \dots B_{*,n}$

For a periodic B-spline, the penalty matrix would be a Toeplitz block circulant matrix, similar in structure to the penalty matrix in Equation (12.5) but with the first row being $(2, -1, 0, 0, -1)$ and the last row $(-1, 0, 0, -1, 2)$.

12.2.4 Markov chain Monte Carlo

The Bayesian statistical framework provides for multi-level models, which include mixed effects models (Gelman 2006). As such, any mixed model, such as a GLMM, can be fitted in a Bayesian framework. Software for fitting Bayesian GLMMs includes

WinBUGS (Lunn *et al.* 2000). Crainiceanu *et al.* (2007) give examples (including code) for fitting Bayesian GLMMs and spline models in WinBUGS. The R package spikeSlabGAM (Scheipl 2011) provides functions for fitting generalized additive mixed models, including surface fitting via bivariate splines.

Here we will demonstrate a method for fitting Bayesian splines using the Metropolis–Hasting (MH) algorithm (Hastings 1970; Metropolis *et al.* 1953). This algorithm is one of the family of Markov chain Monte Carlo (MCMC) processes. MCMC simulation is used to explore the properties of the 'target distribution' by constructing a continuous parameter space Markov chain whose stationary distribution is the Bayesian posterior distribution.

A special case of the MH algorithm proposes a random walk in the multidimensional parameter space and accepts the proposed value with a probability that is the minimum of the ratio of the posterior densities at the current and proposed values. All examples shown in this chapter are solved using the Metropolis algorithm with an adaptive random walk proposal. For further information about other popular MCMC samplers such as MH and the Gibbs sampler see Gelman et al. (2004,section 11.4).

The pseudo-code for the MH sampler is given in Algorithm 10 for a symmetric jumping rule, J_t, which simplifies the MH algorithm to the Metropolis algorithm, and a random walk, δ, for the model parameters, $\boldsymbol{\theta}$ (Gelman et al. 2004). The algorithm involves proposing a new set of parameter values and then accepting that proposal based on the posterior density of the target distribution at the current values and proposed values.

Algorithm 10: Metropolis sampler

initialize $\boldsymbol{\theta}^{(0)}$
for $t = 1$ **to** nsamples **do**
  $\boldsymbol{\theta}^{\star} = \boldsymbol{\theta}^{(t-1)} + N(0, \delta^{(t)})$
  $r = p\left(\boldsymbol{\theta}^{\star}|y\right) / p\left(\boldsymbol{\theta}^{(t-1)}|y\right)$
  generate u from U(0, 1):
  if $u < r$ **then**
    $\boldsymbol{\theta}^{(t)} = \boldsymbol{\theta}^{\star}$
  else
    $\boldsymbol{\theta}^{(t)} = \boldsymbol{\theta}^{(t-1)}$
  end if
end for

We will use an adaptive random walk procedure and provide for thinning and burn-in to obtain good convergence and autocorrelation properties for our chains. The adaptive random walk procedure allows for a target acceptance rate, $\overline{\tau}$, and adjusts the standard deviation terms of the random walk in order that the cumulative acceptance rate, r, converges to the target acceptance rate (Atchadé and Rosenthal 2005). The adaptive random walk procedure is outlined in Algorithm 11, where the proposed new standard deviation for the random walk for parameter s is denoted $\delta_s^{\star}$.

The series $\gamma_n \sim n^{-\lambda_1}$, for λ_1 a constant between $1/2$ and 1, controls the convergence of the standard deviation of the random walk.

Algorithm 11: Adaptive MCMC

$\epsilon_1 = 0.001$, $A_1 = 5$, $\overline{\tau} = 0.234$
$\delta_s^\star = \delta_s^{(t-1)} + \gamma_t(r - \overline{\tau})$
if $\delta_s^\star \leq \epsilon_1$ **then**
  $\delta_s^{(t)} \leftarrow \epsilon_1$
else if $\delta_s^\star \geq A_1$ **then**
  $\delta_s^{(t)} \leftarrow A_1$
else
  $\delta_s^{(t)} \leftarrow \delta_s^\star$
end if

12.2.5 Model choice

When choosing which Bayesian spline model is most appropriate (e.g. the number of knots required) it is important to choose a model which provides a sensible fit to the data and high-precision parameter estimates but does not overfit the model by using a large number of spline terms. The worst case of non-parametric model over-smoothing is to use a model which joins every data point so that the smoothing is interpolation.

The DIC (Spiegelhalter *et al.* 2002) is a Bayesian alternative to frequentist measures of goodness of fit based on the likelihood, such as the Akaike information criterion (AIC) (Akaike 1974) and Schwarz's Bayesian criterion (BIC) (Schwarz 1978). In this case, the DIC is useful as we cannot apply Bayes' factors to models with improper priors (such as the rank-deficient penalty matrices used for P-splines) as they are undefined.

The DIC measures the goodness of fit by the expected deviance and penalizes by the effective number of parameters, similar to the AIC. This takes into account the contribution of random effects, such as splines, when treated as parameters and their contribution can depend heavily on the data. The deviance is defined as

$$D(\boldsymbol{\theta}) = -2\log\{p(y|\boldsymbol{\theta})\} + C \tag{12.8}$$

where C is a constant dependent on the data. The DIC, then, can be calculated from the MCMC samples of the posterior, such that

$$p_D = \overline{D}(\boldsymbol{\theta}) - D(\overline{\boldsymbol{\theta}})$$
$$\text{DIC} = \overline{D}(\boldsymbol{\theta}) + p_D$$

where $\overline{\boldsymbol{\theta}} = E(\boldsymbol{\theta})$ and $\overline{D}(\boldsymbol{\theta}) = E(D(\boldsymbol{\theta}))$ from the MCMC samples.

12.2.6 Posterior diagnostics

To make inferences for a fitted model we report several summary statistics:

1. The 95% credible interval for the fitted values.
2. Each variable in the model's
   (a) kernel density estimate
   (b) mean, median and 95% credible interval
   (c) MCMC chain trace plot
   (d) MCMC chain autocorrelation.

We also expect that posterior densities of our parameters will be unimodal and approximately normal (Gelman et al. 2004).

These diagnostics allow us to determine the properties of the Markov chains and the variables they represent. To assess whether the MCMC chains have converged in distribution to their target posterior distributions, we assess the cross-correlations, autocorrelations, trace plots over MCMC iterations, and the cumulative acceptance rate of the random walk. The fitted model will be plotted with the data and the resulting effective number of parameters, p_D, will be provided.

It is usual to assume that the parameters are distributed normally when they have support over the reals, so a good indicator of convergence is the approximate normality of the kernel density estimate of the values of the chain.

12.3 Case studies

12.3.1 Data

For this chapter we have chosen case studies which represent the flexibility of the GAM and illustrate a range of different spline types in a Bayesian setting: thin-plate smoothing splines, B-splines and P-splines. Each example will showcase a type of spline and a link function for the GAM.

Temperature This data set is the temperature, recorded every 10 minutes, in Brisbane, Australia, for the entirety of 1 January 2009 (Commonwealth of Australia 2010). We have chosen to use this data set to illustrate spline regression with a periodic basis. That is, we assume that the temperature is the same at 12:00 on 1 January 2009 and 12:00 on 2 January 2009 and fit a non-parametric model for the temperature throughout the day.

Simulated data We simulate some data from the function

$$y = \cos\left(2\pi x^2\right) + \mathrm{N}\left(-\int_0^1 \cos\left(2\pi x^2\right)\, dx, 0.1\right) \tag{12.9}$$

where the x values are 40 points drawn from a uniform distribution on (0,1). We have centred the y values around 0 by subtracting the mean of the simulation function.

This is an example of data which exhibit a known nonlinear behaviour which may be accurately modelled by fitting the nonlinearity locally with smooth functions.

Union membership This data set is part of a survey of Canadian workers in 1985 (Berndt 1991). The response is membership of a trade union (binary) and the covariate is the worker's hourly wage, for 534 workers. Ruppert *et al.* (2003) show that a logistic GLM is an inappropriate method (at least in the context of the data) for calculating the proportion of workers belonging to a trade union according to their wages. We will use this data set to compare the results of spline regression with two different bases.

Fuel efficiency This data set is taken from the Statistics Toolbox from MATLAB as it is a classic example of a logistic linear regression. The data in this case are the number of cars with 'poor' gas (petrol) mileage, y_i, out of n_i cars tested with engine weight x_i. These data are taken from the MATLAB documentation on logistic GLMs.

We use this data set to show that a spline with two basis elements (and hence 2 degrees of freedom) is equivalent to a linear regression.

Coal mining disasters A simple data set for examining Poisson spline regression is the number of British coal mine accidents in the years 1851–1962 (Jarrett 1985). These data represent the annual number of incidents at a mine that caused 10 or more fatalities.

Traditionally, this data set has been used to illustrate change point analysis (Akman and Raftery 1986; Boukai and Zhou 1997). Here we use a completely data-driven method for determining how the number of accidents changes continuously over time rather than assuming that there are two (or more) intervals of time with a constant rate.

12.3.2 Analysis

Gaussian regression

Link function The simplest link function (see Section 12.2.2) is the identity link, $g(\mu) = \mu$. This link function is used to fit the Gaussian model which has the likelihood

$$p(y_i|\boldsymbol{\theta}) \sim \mathrm{N}\left(y_i;\, \mu_i, \sigma^2\right) \propto \exp\left(-\frac{(y_i - \mu_i)^2}{2\sigma^2}\right)$$

$$\boldsymbol{\mu} = X\boldsymbol{\theta}.$$

Periodic B-spline We wish to fit a B-spline model for the time series of temperature in Brisbane, Australia. Here we will use a basis of 11 periodic B-splines, with the periodicity ensuring that the fitted function is smooth and continuous at midnight at each end of the day. While it may not be strictly valid to assume that the temperature will be the same 24 hours apart at midnight, we make this choice to illustrate the periodic spline basis. The periodic basis also encourages the addition of more days of data to estimate the smooth and continuous daily trend over multiple days.

The model for a periodic B-spline (with 11 basis elements, derived from 13 knots) fit to the temperature data is

$$p(\boldsymbol{y}|\boldsymbol{\theta}, \lambda, \sigma) \sim \text{MVN}\left(\boldsymbol{y};\ B\boldsymbol{\theta}, \sigma^2 I\right)$$

with priors

$$\begin{aligned} p(\boldsymbol{\theta}|\lambda) &\sim \text{MVN}(\mathbf{0}, \lambda I_{11}) \\ p(\lambda) &\sim \text{Exp}(0.1) \\ p(\sigma) &\sim \sqrt{\text{Inv}-\chi_1^2}. \end{aligned}$$

Figure 12.3 shows the results of the MCMC simulation with 5000 samples, discarding the first 1000. We see that the adaptive MCMC algorithm has converged to the target acceptance rate, 0.234, and that the chains have converged (trace plot). The posterior correlation of the spline coefficients exhibits correlation between immediate neighbours (about −0.5). This is due to the neighbouring basis splines both being able to explain variation in the region where they overlap, so the effect of an increase in the coefficient of one basis spline will be offset by a decrease in its neighbouring basis elements.

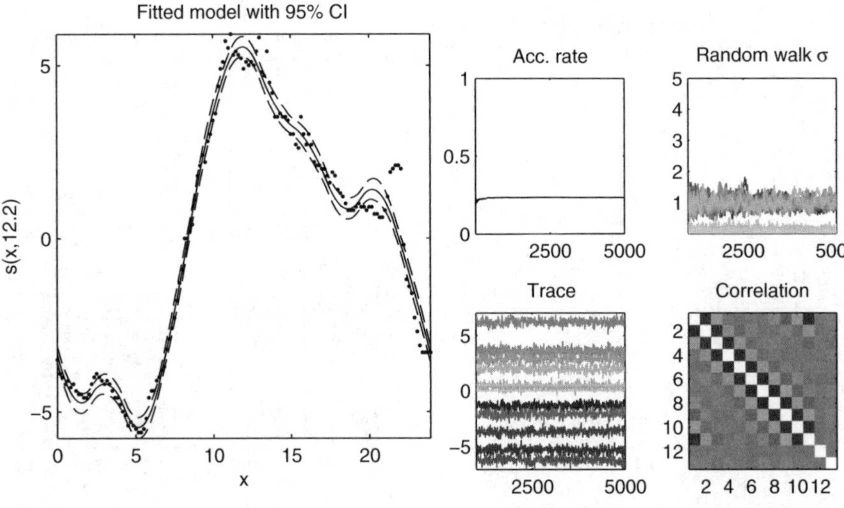

Figure 12.3 B-spline with 11 cyclic, second-order B-splines fit to the temperature data set: fitted function with credible intervals; cumulative acceptance rate of MH; standard deviations of random walks for each parameter; trace of MCMC chains; posterior correlation between MCMC chains for each variable.

We note that the B-spline fits the overall pattern in the data adequately but that there are times when local variation is not captured by the spline. We could increase the number of knots in order to minimize the deviance but we prefer a model which preserves parsimony. This can be achieved by fitting models with differing numbers of knots and choosing the one with the smallest DIC.

P-spline We now fit a P-spline model to the simulated data set.

The model for a non-periodic P-spline with 10 first-order B-spline basis elements and a first-order penalty fit to the simulated data is the likelihood

$$p(\boldsymbol{y}|\boldsymbol{\theta}, \lambda, \sigma) \sim \text{MVN}\left(\boldsymbol{y};\ B\boldsymbol{\theta}, \sigma^2 I\right)$$

and priors

$$\begin{aligned} p(\boldsymbol{\theta}|\lambda) &\sim \text{MVN}\left(\mathbf{0}, \lambda Q^{-1}\right) \\ p(\lambda) &\sim \text{Exp}(1) \\ p(\sigma) &\sim \sqrt{\text{Inv}-\chi_1^2}, \end{aligned}$$

where Q is the penalty matrix from Equation (12.4).

Figure 12.4 shows a first-order P-spline fit to the simulated data set. We have used prior smoothing with 10 knots and simulated 50 000 iterations with a burn-in of 1000

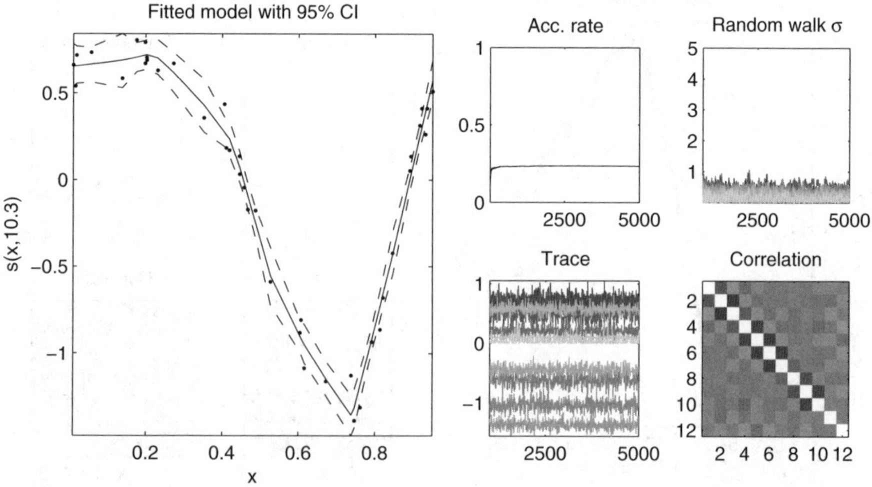

Figure 12.4 P-spline with 10 first-order B-splines penalized with a prior smoother, fit to the data simulated from Equation (12.9): fitted function with credible intervals; cumulative acceptance rate of MH; standard deviations of random walks for each parameter; trace of MCMC chains; posterior correlation between MCMC chains for each variable.

Table 12.3 Summary statistics for posterior distributions of variables in Figure 12.5, from fitting a model to simulated data in Equation (12.9).

Variable	2.5%	50%	97.5%	Mean	SE
θ_1	0.5390	0.6565	0.7772	0.6552	0.0612
θ_2	0.4341	0.6736	0.8869	0.6653	0.1098
θ_3	0.6178	0.7301	0.8472	0.7297	0.0569
θ_4	0.2572	0.4984	0.7170	0.4930	0.1148
θ_5	0.0939	0.2066	0.3083	0.2069	0.0535
θ_6	−0.8053	−0.6186	−0.4407	−0.6178	0.0920
θ_7	−1.1930	−1.0479	−0.8781	−1.0460	0.0782
θ_8	−1.4825	−1.3691	−1.2479	−1.3679	0.0613
θ_9	−0.5700	−0.4440	−0.3212	−0.4452	0.0624
θ_{10}	0.4729	0.5839	0.6831	0.5820	0.0535
σ	0.0811	0.1014	0.1353	0.1031	0.0130
λ	0.0003	0.0049	0.0270	0.0072	0.0071

and no thinning. The adaptive MCMC acceptance rate and standard deviation trace, the trace of the model parameters and corresponding autocorrelation are also given.

The posterior summaries for each parameter are also given in Table 12.3 and in Figure 12.5 we see that these are normally distributed except for the posterior estimates of σ and λ. These are still unimodal and because we have chosen conjugate priors of an inverse-χ^2 for σ^2 and a gamma for λ, we see that the posteriors are of the shape we would expect. While the mode for λ is not zero, the skewed shape of the posterior indicates that only a small degree of smoothing was required to obtain a model which is not overly wiggly.

Logistic regression

Link function The link function for logistic regression is the logit, $g(\mu) = \log(\mu/(1-\mu))$, which maps the linear predictor from the real line to the range (0, 1):

$$\begin{aligned} p(y_i|\boldsymbol{\theta}) &= \text{Bin}\,(y_i;\, n_i,\, p_i) \\ &\propto p_i^{y_i}\,(1-p_i)^{n_i-y_i} \\ \log\left(\frac{p_i}{1-p_i}\right) &= X_{i,*}\boldsymbol{\theta}. \end{aligned}$$

Bernoulli response We will now give an example of logistic regression with splines by analysing the relationship between wages and union membership using data from the 1985 Current Population survey (Berndt 1991). For the Bernoulli case of the binomial likelihood, we set $n_i = 1$ so that y_i is either 0 or 1.

We will fit two different spline models and use the DIC to choose one over the other.

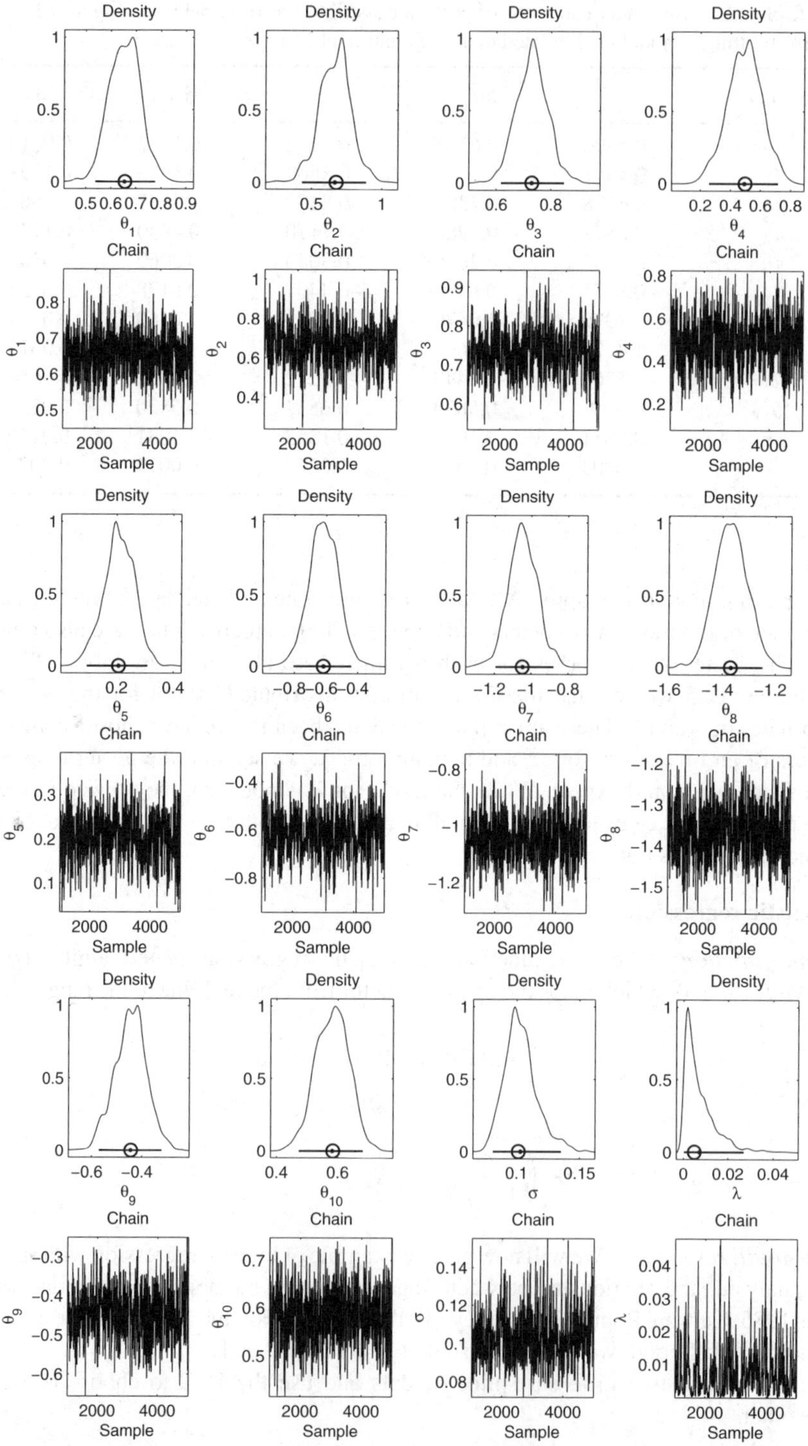

Figure 12.5 Posterior summaries for the variables from fitting the model in Figure 12.4.

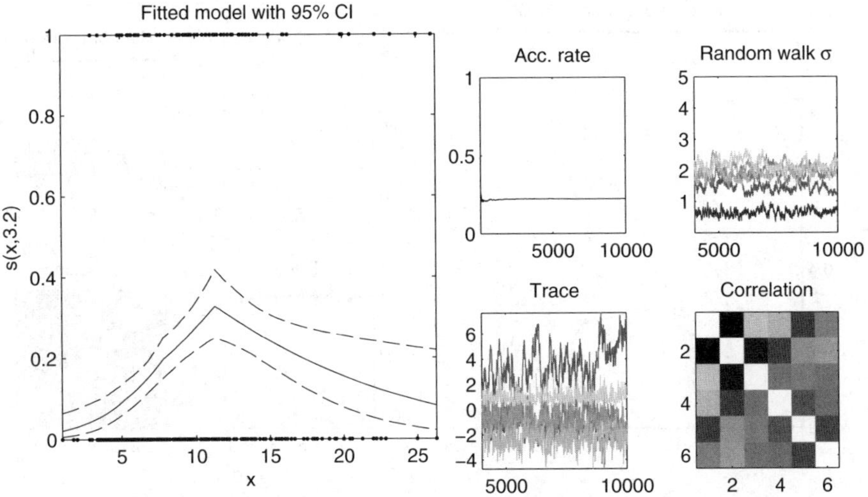

Figure 12.6 Low-rank thin-plate spline with three knots fitted to the Canadian wage–union membership data: fitted function with credible intervals; cumulative acceptance rate of MCMC; standard deviations of random walks for each parameter; history of values of MCMC chains; correlation between MCMC chains for each variable.

Figure 12.6 shows the fitted model and posterior diagnostics for a low-rank thin-plate spline of order 1 with three knots with uninformative normal priors on the spline coefficients. Figure 12.7 shows a P-spline model with six first-order splines fit to the same data with a second-order smoothing prior. In both cases, we see that the proportion of workers belonging to a trade union is almost zero for low-wage workers, has a global maximum around \$12 per hour and decreases as wages increase.

To choose between these two models we examine the DIC and number of parameters, given in Table 12.4. In this case, the DIC indicates that we should use the thin-plate spline model as it achieves a much better fit (in terms of the deviance) with a similar number of parameters.

The P-spline model is

$$\begin{aligned} p(y_i|\boldsymbol{\theta}) &= \text{Bin}\left(y_i;\, n_i,\, p_i\right) \\ \log\left(\frac{p_i}{1-p_i}\right) &= X_{i,*}\boldsymbol{\theta} \\ p\left(\boldsymbol{\theta}|\lambda\right) &\sim \text{MVN}\left(\mathbf{0}, \lambda Q^{-1}\right) \\ p(\lambda) &\sim \Gamma\left(1, 0.1\right). \end{aligned}$$

Binomial response We now give an example of binomial regression, using the car fuel efficiency data set. We fit a model with two first-order B-splines, equivalent in

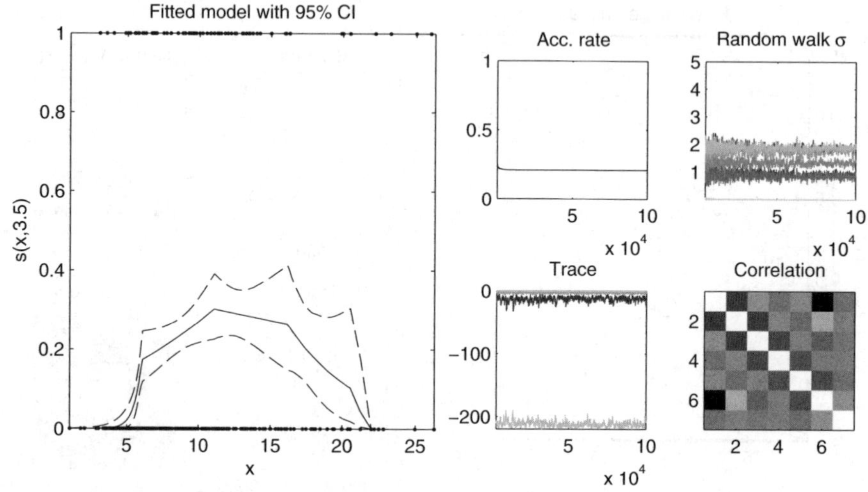

Figure 12.7 P-spline with six first-order B-splines fitted to the Canadian wage–union membership data: fitted function with credible intervals; cumulative acceptance rate of MCMC; standard deviations of random walks for each parameter; history of values of MCMC chains; correlation between MCMC chains for each variable.

Table 12.4 The DIC for the thin-plate spline (Figure 12.6) and P-spline models (Figure 12.7) fit to the Canadian wage–union membership data.

Model	DIC	p_D
Thin plate	475.72	3.22
P-spline	1453.74	3.5

the number of degrees of freedom to logistic linear regression. The model is

$$\begin{aligned} p(y_i|\boldsymbol{\theta}) &= \text{Bin}\,(y_i;\, n_i,\, p_i) \\ \log\left(\frac{p_i}{1-p_i}\right) &= X_{i,*}\boldsymbol{\theta} \\ p\,(\boldsymbol{\theta}) &\sim \text{U}\,(-10, 10)\,. \end{aligned}$$

The priors for the spline coefficients are weakly uninformative, uniform on $(-10, 10)$.

Figure 12.8 shows the fitted model with the data plotted at $(x_i, y_i/n_i)$, and posterior diagnostics.

We see that the fitted model in Figure 12.8 fits the data well and that there is no real reason to use a higher number of knots. With so few degrees of freedom in the model (there are only 12 engine weights) it would be inadvisable to use any more spline basis functions than have been used.

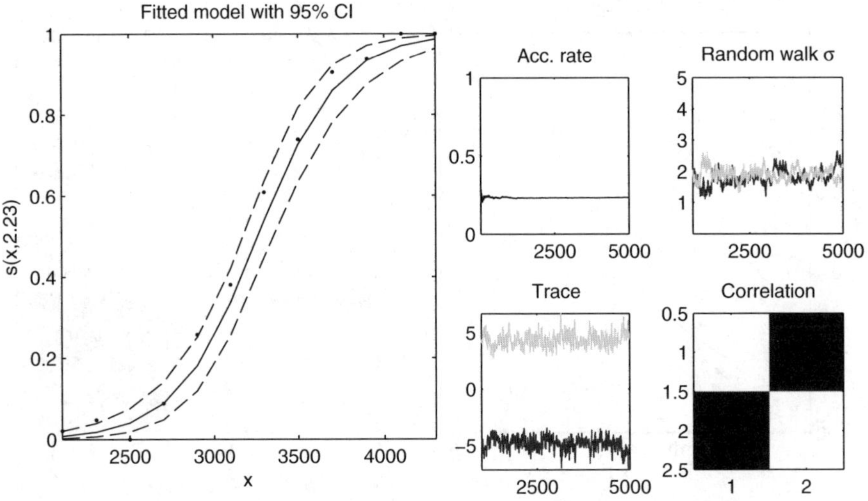

Figure 12.8 Logistic B-spline model with two first-order B-splines fit to the car data: fitted function with credible intervals; cumulative acceptance rate of MH; standard deviations of random walks for each parameter; trace of MCMC chains; posterior correlation between MCMC chains for each variable.

Poisson regression

Link function The log link function for Poisson regression, $g(\mu) = \log \mu$, maps the rate parameter, λ (which is non-negative), to the real line:

$$\begin{aligned} p(y_i|x_i, \boldsymbol{\theta}) &\sim \text{Po}(y_i; \lambda_i) \\ &= \frac{\lambda^{y_i}}{y_i!} e^{-\lambda_i} \\ \log \boldsymbol{\lambda} &= X\boldsymbol{\theta}. \end{aligned}$$

Fitting the coal mine data with 10 first-order B-splines with a second-order penalty (for 10 000 samples, discarding the first 5000 as burn-in) shows not only that there is a decrease in the rate of coal mine disasters but also that this decrease is nonlinear. Figure 12.9 shows that the rate is roughly constant until about 1880, and then a continual decline until the 1910s when the rate is again roughly constant until 1940 when it decreases again. The credible intervals are not symmetric about the mean of the spline because the data are not equally weighted; this is due to the link function.

The continued decrease in mining accidents can likely be attributed to the formation of a mining union in the 1880s, the National Union of Mineworkers (McGrayne 2011). Traditionally this data set has been used to illustrate change points in Poisson processes but here we have used a semi-parametric regression method, driven by data rather than parametric assumptions, to determine how the number of accidents changes continuously over time rather than assuming that there are two (or more) time periods with a constant rate.

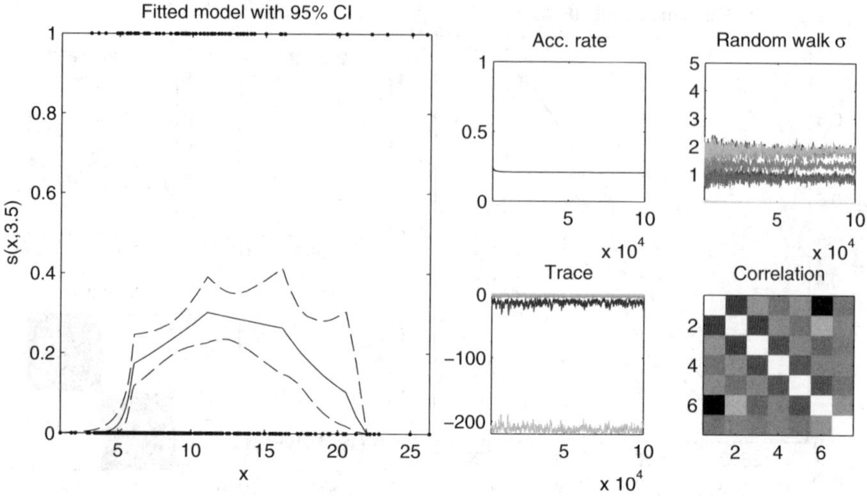

Figure 12.9 P-spline with 10 first-order B-splines fit to the coal mine disaster data set: fitted function with credible intervals; cumulative acceptance rate of MH; standard deviations of random walks for each parameter; trace of MCMC chains; posterior correlation between MCMC chains for each variable.

12.4 Conclusion

12.4.1 Discussion

In section 12.2 we introduced the Metropolis sampler and the adaptive MCMC algorithms and discussed the important summary statistics that allow us to analyse the fitting of a model to data (such as the DIC for model choice and the credible intervals for parameters). We introduced the thin plate and B-splines as a way to extend linear regression. We discussed choices of prior for the variables involved in those splines, including how to incorporate smoothing without affecting the likelihood.

In section 12.3 we extended our spline models by introducing the link function, converting our models into generalized additive models (GAMs). Examples were given of some common models which required slightly different approaches, including Bernoulli, Poisson and logistic regression. These examples required different likelihoods and link functions and were fitted with different bases and priors.

We saw that the fitted models were able to explain nonlinear behaviour without any special adaptation of the likelihood, gave asymmetric credible intervals where warranted and had chains with good convergence properties, particularly the normality of spline coefficients and inverse chi-squared distribution of the standard deviation in Figure 12.5.

In Figure 12.4 we showed the extension of a B-spline to a P-spline and that, while there may be a remaining trend in the residuals, the effective number of parameters was roughly half the number of specified parameters.

We saw in Figures 12.6 and 12.7 that the change of basis from thin-plate spline to P-spline maintained the same general behaviour (although the resulting fits were slightly different) even though the spline bases had very different formulations and degrees of freedom. The effect of the smoothing prior was evident in examining the values of the chains in Figure 12.7.

The plotting of the estimated smooth functions is a more informative way of describing the effect of a spline than reporting summaries of the individual coefficients, particularly when the choice of basis elements can change the properties of the fitted function. In Figure 12.1, the first-order B-spline approximation in 12.1 is only piecewise continuous, while the second-order B-spline approximation is piecewise smooth; the coefficients of the spline basis functions will no doubt be quite similar.

12.4.2 Extensions

This chapter has only dealt with smoothers of a single variable with the number of knots specified a priori. The models could be extended to multiple univariate splines by organizing a partitioning of the vector of parameters and design matrix according to the number of knots for each univariate smooth function. Summaries of model fits (e.g. plots, effective number of parameters) would need to be modified to summarize each covariate's fitted function separately.

Bivariate smoothers require a combination of the basis splines of the covariates to be jointly fit. This could be achieved by forming a tensor product of the appropriate basis vectors (Eilers and Marx 2003; Stone *et al.* 1997). A tensor product of two non-periodic spline bases will lead to a 2D spline on a plane. If one of the spline bases is periodic, the result is a 2D spline basis on a cylinder; if both are periodic, the result is a 2D spline basis on a torus. The bivariate smoothers could be generalized to multivariate smoothers, although the visualization of the resulting multivariate function may require some special treatment.

Any spline basis can be included as long as a function to create the appropriate design matrix exists. Such splines include B-splines with irregular knot spacing, O'Sullivan penalized splines, cubic regression splines, natural splines, Bézier splines, interpolating splines and Hermite splines (Ruppert *et al.* 2003; Wand and Ormerod 2008).

The link functions used in this chapter were examples of some of the more common links but they are by no means an exhaustive list of link functions. Any twice differentiable and monotonic function can be used as a link function (Hastie and Pregibon 1992). This allows the creation of link functions such as power laws (e.g. μ^{-2} for the inverse Gaussian).

The number of knots, specified here in advance, can also be modelled during the MCMC simulation by using the reversible jump MCMC algorithm (Green 1995) by allowing for the birth and death of knots throughout the MCMC sampling (Biller 1998).

It may also be desirable to model an autocorrelation structure in the residuals at the same time as fitting the model (Chib 1993). This may be of particular interest when the data are from a longitudinal study with periodic trends such as daily variation in air quality (Clifford *et al.* 2011).

The authors are currently pursuing some of these extensions in a forthcoming publication (Mølgaard *et al.* 2012).

12.4.3 Summary

Throughout this chapter we have discussed and shown what a spline is, how some particular splines are formulated, how they can be used in a Bayesian regression setting, how to check the models and infer from their posteriors. More specifically, we have described the Bayesian P-spline and thin-plate regression spline and given examples of how they are fit via MCMC methods. We have described how spline-based GAMs are an extension to GLMs and how to move from the frequentist B-spline penalty matrix to a Bayesian smoothing prior in a straightforward manner.

There are a number of benefits to the Bayesian approach for splines, particularly the treatment of smoothing parameters as unknown variables in an MCMC simulation, placing a prior on them and allowing the MCMC sampling to inform us of a range for the smoothing parameters by way of the posterior credible interval. This avoids the problem of iterating over a number of fixed values of the smoothing parameter(s) with generalized cross-validation, which may be quite cumbersome for a large number of unknown smooth functions.

Likewise, the issue of the number and location of knots can be addressed by treating them as unknown quantities, assigning a prior and allowing the MCMC simulation to inform us of appropriate values (e.g. reversible jump MCMC).

It is for these reasons that we recommend Bayesian GAMs with splines as a flexible alternative to the more familiar frequentist GLMs.

References

Akaike H 1974 A new look at the statistical model identification. *IEEE Transactions on Automatic Control* **19**, 716–723.

Akman VE and Raftery AE 1986 Asymptotic inference for a change-point Poisson process. *Annals of Statistics* **14**, 1583–1590.

Atchadé YF and Rosenthal JS 2005 On adaptive Markov chain Monte Carlo algorithms. *Bernoulli* **11**, 815–828.

Berndt ER 1991 *The Practice of Econometrics*. Addison-Wesley, Reading, MA.

Biller C 1998 Adaptive Bayesian regression splines in semiparametric generalized linear models. *Journal of Computational and Graphical Statistics* **9**, 122–140.

Boukai B and Zhou H 1997 Nonparametric estimation in a two change-point model. *Journal of Nonparametric Statistics* **8**, 275–292.

Brezger A 2004 *Bayesian P-splines in structured additive regression models*. PhD thesis. Ludwig-Maximilians University.

Chib S 1993 Bayes regression with autoregressive errors: a Gibbs sampling approach. *Journal of Econometrics* **58**(3), 275–294.

Cleveland WS, Grosse E and Shyu WM 1993 Local regression models. In *Statistical Models in S*, pp. 309–376. Chapman & Hall/CRC, Boca Raton, FL.

Clifford S, Low Choy S, Hussein T, Mengersen K and Morawska L 2011 Using the generalised additive model to model the particle number count of ultrafine particles. *Atmospheric Environment* **45**(32), 5934–5945.

Commonwealth of Australia 2010 Bureau of Meteorology. http://www.bom.gov.au/ (accessed 14 May 2012).

Crainiceanu C, Ruppert D and Wand MP 2007 Bayesian analysis for penalized spline regression using WinBUGS. Johns Hopkins University, Department of Biostatistics Working Paper.

de Boor C 1978 *A Practical Guide to Splines*. Springer, Berlin.

Duchon J 1977 Splines minimizing rotation-invariant seminorms in Sobolev spaces. In *Constructive Theory of Functions of Several Variables*. Lecture Notes in Mathematics 1, pp. 85–100. Springer, Berlin.

Eilers PHC and Marx BD 1996 Flexible smoothing with B-splines and penalties. *Statistical Science* **11**, 89–121.

Eilers PHC and Marx BD 2003 Multivariate calibration with temperature interaction using two-dimensional penalized signal regression. *Chemometrics and Intelligent Laboratory Systems* **66**(2), 159–174.

Gelman A 2006 Prior distributions for variance parameters in hierarchical models. *Bayesian Analysis* **1**, 515–533.

Gelman A, Carlin JB, Stern HS and Rubin DB 2004 *Bayesian Data Analysis*. Chapman & Hall, Boca Raton, FL.

Green PJ 1995 Reversible jump Markov chain Monte Carlo computation and Bayesian model determination. *Biometrika* **82**, 711–732.

Hastie TJ 1993 Generalized additive models. In *Statistical Models in S*, pp. 249–307. Chapman & Hall/CRC, Boca Raton, FL.

Hastie TJ and Pregibon D 1992 Generalized linear models. In *Statistical Models in S*, pp. 195–246. Chapman & Hall/CRC, Boca Raton, FL.

Hastie TJ and Tibshirani RJ 1990 *Generalized Additive Models*. Chapman & Hall/CRC, Boca Raton, FL.

Hastings WK 1970 Monte Carlo sampling methods using Markov chains and their applications. *Biometrika* **57**, 97–109.

Jarrett R 1985 Coal-mining disasters. In *Data, a Collection of Problems from Many Fields for the Student and Research Worker*. Springer, Berlin.

Lang S and Brezger A 2004 Bayesian P-splines. *Journal of Computational and Graphical Statistics* **13**(1), 183–212.

Lunn D, Thomas A, Best N and Spiegelhalter D 2000 WinBUGS – a Bayesian modelling framework: concepts, structure, and extensibility. *Statistics and Computing* **10**, 325–337.

McGrayne SB 2011 *The Theory That Would Not Die: How Bayes' Rule Cracked the Enigma Code, Hunted Down Russian Submarines, and Emerged Triumphant from Two Centuries of Controversy*. Yale University Press, New Haven, CT.

Metropolis N, Rosenbluth AW, Rosenbluth MN, Teller AH and Teller E 1953 Equation of state calculations by fast computing machines. *Journal of Chemical Physics* **21**, 1087–1092.

Mølgaard B, Clifford S, Corander J, Hämeri K, Low Choy S, Mengersen K and Hussein T 2012 Bayesian semi-parametric forecasting with autoregressive errors. *In preparation*.

Ruppert D, Wand MP and Carroll RJ 2003 *Semiparametric Regression*. Cambridge University Press, Cambridge.

Scheipl F 2011 *spikeSlabGAM: Bayesian variable selection and model choice for generalized additive mixed models*. Unpublished paper.

Schwarz G 1978 Estimating the dimension of a model. *Annals of Statistics* **6**, 461–464.

Silverman BW 1985 Some aspects of the spline smoothing approach to non-parametric regression curve fitting. *Journal of the Royal Statistical Society, Series B* **47**, 1–52.

Speckman P, Sun D, Speckman PL and Sun D 2003 Fully Bayesian spline smoothing and intrinsic autoregressive priors. *Biometrika* **90**(2), 289–302.

Spiegelhalter DJ, Best NG, Carlin BP and van der Linde Bayesian measures of model complexity and fit. *Journal of the Royal Statistical Society, Series B* **64**, 583–639.

Stone C, Hansen M, Kooperberg C and Truong Y 1997 Polynomial splines and their tensor products in extended linear modeling. *Annals of Statistics* **25**, 1371–1470.

Wahba G 1990 *Spline Models for Observation Data*. Society for Industrial and Applied Mathematics, Philadelphia, PA.

Wand MP and Ormerod JT 2008 On semiparametric regression with O'Sullivan penalized splines. *Australian and New Zealand Journal of Statistics* **50**, 179–198.

Wood SN 2006 *Generalized Additive Models: An Introduction with R*. Chapman & Hall/CRC, Boca Raton, FL.

13

Disease mapping using Bayesian hierarchical models

Arul Earnest[1,2], Susanna M. Cramb[3,4] and Nicole M. White[3,5]

[1]*Tan Tock Seng Hospital, Singapore*
[2]*Duke-NUS Graduate Medical School, Singapore*
[3]*Queensland University of Technology, Brisbane, Australia*
[4]*Viertel Centre for Research in Cancer Control, Cancer Council Queensland, Australia*
[5]*CRC for Spatial Information, Australia*

13.1 Introduction

Disease mapping falls under the category of spatial epidemiology, which is the description and analysis of geographically indexed health data with respect to demographic, environmental, behavioural, socioeconomic, genetic, and infectious risk factors (Elliott and Wartenberg 2004). Disease maps can be useful for estimating relative risk; ecological analyses, incorporating area and/or individual-level covariates; or cluster analyses (Lawson 2009). As aggregated data are often more readily available, one common method of mapping disease is to aggregate the counts of disease at some geographical areal level, and present them as choropleth maps (Devesa *et al.* 1999; Population Health Division 2006). Therefore, this chapter will focus exclusively on methods appropriate for areal data.

Some population data are often needed to make meaningful comparisons of disease counts across regions. For instance, an area may have a higher disease count

Case Studies in Bayesian Statistical Modelling and Analysis, First Edition. Edited by Clair L. Alston, Kerrie L. Mengersen and Anthony N. Pettitt.

simply because there are more people at risk living in that area. Expected counts of disease are usually calculated, and these expected counts are often standardized for main demographics such as sex and age group, which are collectively termed 'confounding variables'.

Standardization is performed to sieve out the effects of these main confounders. Two types of standardization are commonly available: external (or direct) and internal (indirect) standardization. The main difference between the two standardization methods is that, for direct standardization, the study sample provides the rates and the standard population the weights, and vice versa for indirect standardization. Therefore, for direct standardization, one would take the sex–age rates of the sample data and apply them to a standard population. Conversely, in indirect standardization, one would apply the sex–age rates of the overall region to the specific counts of population at risk of individual strata studied, to calculate the expected counts in each stratum. The sum of the expected counts provides the age–sex standardized expected counts for each area.

The ratio of the observed and expected count of disease in each area provides the indirectly standardized incidence ratio (SIR). The SIRs are also the maximum likelihood estimators of the area-specific relative risks. A relative risk of one would indicate that an area has a risk similar to the overall average of the entire region, while relative risks below one and above one would indicate a risk lower and higher than the overall region average, respectively.

However, the use of the crude SIR as described above has several disadvantages. Firstly, for areas which are sparsely populated, the SIR can be imprecise and unstable. This is due to the variance of the SIR estimate and its inverse relationship with the expected counts, as shown by the formula

$$\text{Var}(\theta_i) = \frac{O_i}{E_i^2}$$

where θ_i is the risk of being in the disease group rather than the background group for the ith area, O_i are the observed counts and E_i are the expected counts.

These sparsely populated areas often appear prominent in maps because of their large geographic areas. This heightens the problem of interpretation. The issue is further compounded when studying rare disease outcomes such as birth defects. Extremely sparse counts can be observed in many of these rural areas, with some including excessive numbers of zero counts.

Traditional regression models are not adequate for analysing spatial data because they fail to account for geographic correlation, if any, in the data. Correlation in counts of disease between neighbouring areas can arise due to several reasons. Firstly, the outcome measures themselves may be correlated or clustered. For instance, the count of cancer in one region can be high, and this region could also be surrounded by neighbours with a similarly high count of cancer. Secondly, there could be confounders or environmental factors, which themselves are unmeasured and spatially structured. Ignoring correlation when present may result in spurious conclusions (Choo and Walker 2008).

Bayesian approaches to disease mapping that aim to address these perceived issues and others have led to a substantial literature in recent decades. With their beginnings traced back to Clayton and Kaldor (1987), the vast majority of published approaches represent variants of the generalized linear mixed model (GLMM) (Clayton 1996), and form the focus of this chapter. Representing an extension to the well-recognized generalized linear model (GLM) (McCullagh and Nelder 1989), GLMMs accommodate both fixed and random effects, with the latter effect allowing for the modelling of excess variation attributed to different sources. In disease mapping, random effects may be used to capture excess heterogeneity in observed counts, after adjusting for covariates included in the model as fixed effects. Furthermore, they provide a means of accounting for latent geographic variation via the specification of spatially structured random effects. In the Bayesian framework, this spatial structure is encoded into the prior distribution for the random effects and involves the definition of relationships between spatially close areas.

The seminal paper by Besag *et al.* (1991) represents the first fully Bayesian GLMM for disease mapping that includes both spatial and non-spatial random effects. Variants of this model include different specifications of prior distribution for the random effects, for example the proper prior in Cressie (1993), and the mixture prior by Leroux *et al.* (1999), that aims to explore the relative dominance of spatial versus non-spatial random effects. In Lawson and Clark (2002), a mixture prior involving different forms of spatial variation was considered to better capture discontinuities in disease risk between areas, to lessen over-smoothing of risk estimates. The extension of GLMMs into the spatio-temporal domain began with Bernardinelli *et al.* (1995), with later contributions from Waller *et al.* (1997), Knorr-Held and Besag (1998) and Knorr-Held (2000). For the study of multiple diseases, a bivariate extension of the spatial prior featured in Besag *et al.* (1991) was proposed by Kim *et al.* (2001), and was later generalized to the multivariate case by Gelfand and Vounatsou (2003) and Carlin and Banerjee (2003).

Aside from GLMMs, alternative approaches to disease mapping include unsupervised spatial clustering (Green and Richardson 2002), dynamic linear models (Schmidt *et al.* 2010), generalized additive mixed models (MacNab 2007), spatio-temporal Dirichlet process mixtures (Kottas *et al.* 2007) and, for multiple diseases, spatial and spatio-temporal factor models (for a recent example see Tzala and Best (2007)). A more detailed review of spatial models in disease mapping is presented by Best *et al.* (2005).

The unifying feature of these varied approaches is their formulation as a Bayesian hierarchical model, allowing for the expression of complex relationships between parameters or effects of interest. In the case of GLMMs, the resulting hierarchy promotes the sharing of 'borrowing of information' across regions via the spatial random effects. In the literature, this feature is described as spatial smoothing. The natural incorporation of spatial smoothing in these models, as a level in the hierarchy, thus provides an appealing alternative to the estimation of relative risk, in place of the aforementioned SIR estimator. In particular, for sparsely populated regions, spatial smoothing allows for the borrowing of strength from neighbouring areas and shrinkage towards the overall relative risk. As discussed in Gelman *et al.* (2003), this also results in the robust, more precise estimation of point estimates, in this case relative risks, as it minimizes squared error loss.

Not only is disease mapping helpful for examining geographical differences in incidence or mortality via modelled SIRs or SMRs (Standardized Mortality Ratios), but it can also be used to examine differences in survival. Survival is defined as the percentage of patients who survive for a given time after diagnosis. Measuring survival is considered to be the single most important measure for monitoring and evaluating early diagnosis or treatment components of disease control (Dickman *et al.* 2011), and this is particularly important for diseases such as cancer.

When examining cancer survival using population-based data, relative survival is the most commonly employed method (Cramb *et al.* 2011). Relative survival estimates net survival for the condition of interest, correcting for estimated mortality from other causes (Dickman *et al.* 2004). The main advantage of using relative survival is that it avoids the dangers of misclassification of cause of death inherent in cause-specific analyses, including the difficulties in assigning death as primarily due to cancer. Although relative survival has mainly been used for cancer data, it has also been proposed as a useful measure for other chronic diseases such as cardiovascular disease (Nelson *et al.* 2008).

This chapter will give worked examples of two different disease mapping models using Bayesian hierarchical models: incidence and relative survival.

13.2 Case studies

13.2.1 Case study 1: Spatio-temporal model examining the incidence of birth defects

Across all states in Australia, there has been considerable variation in the reported rates of birth defects over the past 20 years. For instance, the rate of all malformations ranged from 159.4 per 10 000 births (1981–1995) to about 175.2 per 10 000 births (1997). There was also gross spatial variation in the reported rates between states in the period 1991–1997, with highest rates found in Victoria (229.2 per 10 000 births), followed by ACT (222.3 per 10 000 births) and Queensland (194.3 per 10 000 births) (AIHW National Perinatal Statistics Unit 2001). Within New South Wales (NSW), there was spatial variation in the reported birth defect rates for the eight different administrative health areas between 1999 and 2005. For example, the NSW Mothers and Babies Report 2005 found elevated rates of birth defects in the Hunter and New England area (Centre for Epidemiology and Research and NSW Department of Health 2007).

Using data that were probabilistically linked from two large birth and birth defect registries in NSW, standardized expected counts of total birth defects were calculated at the statistical local area (SLA) level, for which there were 198 SLAs defined within the NSW study area. These represent administrative districts that relate to local government jurisdictions. The median population size in an SLA (2006 data) was 22 717 (range: 364 to 139 476). Record linkage was performed in ChoiceMaker (http://oscmt.sourceforge.net/), the open source matching software based on machine learning techniques. A pre-trained model that contains weights for a set of clues was developed in NSW Health. Clues come in three varieties: *match-clues* which predict match, *differ-clues* which predicts differ (non-match), and *hold-clues* which predict hold (clerical review). A machine learning technique, maximum entropy, is used to weight all the possibly contradictory clues and produce a probabilistic decision.

SLA-specific relative risk estimates were calculated as the ratio of the observed and expected counts for each area, where the expected counts were defined as the age-specific birth defect rate for the overall region (i.e. NSW) applied to the population of the SLA for the specific age group, then summed over age groups to obtain an estimate for each SLA. The first two time periods 1995–1997 and 1998–2000 were used to build the model coefficients and then model fit was assessed using data from 2001–2003.

13.2.2 Case study 2: Relative survival model examining survival from breast cancer

In Queensland, breast cancer is the most commonly diagnosed cancer among females and the second highest cancer cause of death. Risk factors for developing breast cancer include increasing age, family history of breast cancer, reproductive and hormonal factors (i.e. delayed or no childbearing, not breastfeeding, early menstruation, late menopause), overweight/obesity, excessive alcohol consumption and insufficient physical activity (Youlden *et al.* 2009). Although breast cancer has comparatively high survival with a 5 year relative survival of 88.7% (2003–2007 data) (Viertel Centre for Research in Cancer Control 2010), there is some evidence to suggest there are geographical inequalities in survival throughout the state (Youlden *et al.* 2009). To examine this issue, de-identified individual-level cancer data covering 1996 to 2007 were obtained from the Queensland Cancer Registry (QCR).

The QCR is a population-based cancer registry which maintains records for all cancer cases diagnosed in Queensland since 1982, excluding basal and squamous cell carcinomas. Public and private hospitals, nursing homes and pathology laboratories are required under legislation to notify the QCR of any cancer diagnoses, while mortality data are obtained from the Registrar of Births, Deaths and Marriages and linked to hospital and pathology data (Queensland Cancer Registry 2010). Fields were provided for SLA at diagnosis (based on the 2006 SLA boundaries), cancer type, month and year of diagnosis, month and year of death, days survived from diagnosis to death or censoring, age at diagnosis and gender.

Queensland has a population of 4.1 million people (2006 data) dispersed over an area of 1.85 million km^2. There were 478 SLAs in the state in 2006. Also, Queensland has multiple islands, many of which are independent SLAs. Due to the small populations on these islands, there are generally few cases diagnosed. Therefore, it was important to use a method which was able to handle data sparseness.

13.3 Models and methods

13.3.1 Case study 1

In the context of disease mapping using hierarchical modelling, the conditional autoregressive (CAR) model is most commonly formulated as follows (Besag *et al.* 1991). Let O_i and E_i be the observed and expected counts of disease in an area i. Assuming that the disease is relatively rare and independently distributed in each area,

$$O_i \sim \text{Poisson}(E_i\theta_i), \quad i = 1, \ldots, n$$

where

$$\log(\theta_i) = x_i\beta_i + v_i.$$

The x_i are covariates measured at the area level, with the corresponding parameter coefficients given by the β_i. The θ_i represent the relative risk in each area. An intrinsic CAR prior is placed on the random effect terms $u_i = (u_1, \ldots, u_n)$ to capture spatially correlated unobserved heterogeneity. The joint probability density of the intrinsic CAR prior as described in Besag *et al.* (1991) is given by

$$p(u) \propto \exp\left\{-\sum_{i<j} w_{ij}\phi(u_i - u_j)\right\}, \qquad u \in R^n$$

based on pairwise differences among the u_i. The w_{ij} are given non-negative weights, with $w_{ij} = 0$ unless areas i and j are neighbours, commonly taken as spatially contiguous regions, and $\phi(z)$ is a specified monotonic function of z, increasing with $|z|$. The conditional density of u_i is therefore

$$p_i(u_i|\ldots) \propto \exp\left\{-\sum_{j\epsilon\delta_i} w_{ij}\phi(u_i - u_j)\right\}, \qquad u_i \in R$$

where $w_{ij} = w_{ji}$ defines w_{ij} for $i > j$ and δ_i denotes the SLAs that are neighbours to i. The various neighbourhood and weight matrices are discussed in a later section of this chapter.

As the distribution above is strictly improper due to only addressing differences in the u_i and not their overall level, Besag *et al.* (1991) proposed the following restriction of the v_i, where $\phi(z)$ is set to be equal to $z^2/2\kappa$, where κ is an unknown positive constant, giving rise to

$$p(u|\kappa) \propto \frac{1}{\kappa^{n/2}} \exp\left\{-\frac{1}{2\kappa}\sum_{i\sim j}(u_i - u_j)^2\right\}$$

where $i \sim j$ denotes that i and j are contiguous. A large value of κ is indicative of spatially structured variation, while a small κ value indicates the u_i values are more constant (Besag *et al.* 1991).

Tips and Tricks: There are a number of examples using the CAR model to analyse spatial data within WinBUGS (see Lawson (2009)). Further examples can be found in WinBUGS—Map—Manual.

The CAR model helps to address the issue of imprecise estimates of the SIR due to the sparse 'population at risk' issue. Essentially, it does this by smoothing the SIR to a value that is a compromise between the observed value and either a local mean (average of neighbouring values) or a global mean (often taken as one because of the standardization), depending on the amount of autocorrelation in the risk estimates.

Bernardinelli *et al.* (1995) developed a spatio-temporal model, where both the area-specific intercepts and temporal trends were modelled as random effects. They further extended the model to incorporate spatio-temporal effects and allow for correlation between the area and time random effects; however, these additional terms were not significant in their application. This model was restricted to a linear time trend, although inclusion of quadratic and higher order terms would be straightforward.

A modified form of the Bernardinelli model was used to analyse birth defects in NSW from 1995 to 2003 (Earnest *et al.* 2007). Provisions were also made within the spatio-temporal model to explore differences in the smoothing properties of the model between the contiguity (adjacency) and various distance-based methods of defining neighbourhood spatial weights.

The formulation of the CAR model used in the analyses is shown below:

$$O_{ik} \sim \text{Poisson}(\mu_{ik})$$
$$\log(\mu_{ik}) = \log(E_{ik}) + u_i + v_i + \beta_{1i}^* t_k + \beta_{1i}^* t_k^2$$

where O_{ik} and E_{ik} are the observed and expected birth defects for an SLA in the ith region and kth time period, u_i is a spatially structured random effect and v_i is a spatially unstructured random effect. A quadratic temporal random effect term was also added to capture time trends. The main difference from the Bernardinelli *et al.* (1995) model lies in the exclusion of the space–time interaction random effect term. Possible spatial correlation was accommodated in the model by introducing a CAR prior for the spatial random effects, as shown below:

$$[u_i | u_j, i \neq j, \tau_u^2] \sim \text{N}(\bar{\mu}_i, \tau_i^2)$$
$$\bar{\mu}_i = \frac{1}{\sum_j w_{ij}} \sum_j u_j w_{ij}$$
$$\tau_i^2 = \frac{\tau_u^2}{\sum_j w_{ij}}.$$

As can be seen from the above equations, estimation of the risk in any area is conditional on risks in neighbouring areas. Subscripts i and j refer to an SLA and its neighbour respectively, and $j \in N_i$ where N_i represents the set of neighbours of region i. Besides the identification of neighbours, the assigned weights also affect the risk estimation. The weights for the adjacency and distance models are given by $(w_{ij}) = 1$ if i,j are adjacent, and 0 otherwise. For the other distance-based models, various formulations of the weights (described in detail below) were used.

Neighbourhood matrices

Four different neighbourhood adjacent weight matrices were created. These are commonly used in spatial regression, namely Queen-1, Queen-2, Rook-1 and Rook-2. The numbers reflect the order of contiguity, and the main difference between the Queen and Rook method of assigning neighbours is that the latter uses only common boundaries to define neighbours, while the former includes all common points (boundaries and vertices). For instance, the Queen-1 neighbourhood matrix for an SLA would include all its immediate neighbours that share common points with that area, while a Queen-2 matrix would include the immediate neighbours of the neighbours as well. All neighbours for the adjacency weight matrix contribute equal weights.

Seven distance-based matrices were also created and are henceforth referred to as Weight-1 to Weight-7. The simple distance-based matrix included all SLAs as neighbours and assigned them equal weights. Distance between SLAs was calculated using the GeoDa software, with the distance measured between polygon centroids using Euclidean distance. The GeoDa software is freely available from http://geodacenter.asu.edu/software/downloads, and can be used to create basic neighbourhood weight matrices like Queen, Rook or simple distance-based matrices. Weight-1 to Weight-7 also include all SLAs as neighbours, but the weights were assigned differently. For Weights 1–3, the following formulations were used:

$$w_{ij} = \frac{1}{\text{dist}_{ij}},$$

$$w_{ij} = \frac{1}{\text{dist}_{ij}^2} \quad \text{and}$$

$$w_{ij} = \frac{1}{\text{dist}_{ij}^3}$$

respectively. The weight matrix for the 'Gravity' model was defined by

$$w_{ij} = \frac{e_i e_j}{\text{dist}_{ij}}.$$

The corresponding weight matrix for the 'Entropy' model was defined by

$$w_{ij} = \exp(-10 * \text{dist}_{ij}),$$

and for the 'Density' model by

$$w_{ij} = \frac{1}{\text{dist}_{ij}} \times \text{dens}_i \times \text{dens}_j,$$

with dist_{ij} being the distance (decimal degrees) between the two SLAs, e_i and e_j being the standardized counts of births for an SLA and its respective neighbour, and dens_i and dens_j being the respective standardized birth population densities. A distance decay parameter of 10 was chosen, based on a preliminary exploratory examination of the correlogram of the relative risks over distance. All weight matrices were computed in Stata (StataCorp LP, Texas).

The weights for the 'Gravity' and 'Density' models were standardized against their mean and standard deviation for the purpose of comparability. Finally, for the Weight-7 model, the weights were

$$w_{ij} = \frac{1}{\text{dist}_{ij} \times \text{absolute}(\text{ma}_i - \text{ma}_j + 0.0001)},$$

with ma_i and ma_j being the mean maternal ages in SLAs i and its neighbour j respectively. Maternal age was used as a variable as it was a strong determinant of birth defects. A small constant was needed to ensure that weights were defined for those pairs with identical values.

The priors for the uncorrelated random effects and betas were set to a normal distribution, with standard deviation set to cover a wide range of values. All priors for the standard deviations of the precision estimates were set to a uniform distribution with a wide yet plausible interval (i.e. range from 0.000 01 to 20). This range was selected from initial exploratory analysis of the data. The uniform prior was used for σ, instead of the conventional gamma distribution on τ as, dependent on the data scale, the latter may not be truly non-informative. This is discussed in Gelman (2006).

13.3.2 Case study 2

There are different approaches to calculating relative survival, but a recommended method is to model excess mortality (i.e. the mortality observed above that expected from population mortality rates) using a GLM based on collapsed data using exact survival times and a Poisson assumption (Dickman *et al.* 2004), as follows:

$$\begin{aligned} d_j &\sim \text{Poisson}(\mu_j) \\ \mu_j &= \lambda_j y_j \\ \log(\mu_j - d_j^*) &= \log(y_j) + x\beta \end{aligned}$$

where d_j is the number of observed deaths in each stratum j (for instance, age group), λ_j is the hazard rate, y_j is person-time at risk and is the offset, d_j^* is the expected number of deaths estimated from population mortality rates (assumed to be from causes other than the cancer of interest), and β represents the coefficients of the vector of covariates. Note that obviously $\mu_j > d_j^*$.

To incorporate spatial and random effects the model can be extended as in Fairley *et al.* (2008) to become

$$\log(\mu_j - d_j^*) = \log(y_j) + \alpha + x\beta + u_i + v_i$$

where α is the intercept, u_i is the spatial random effects in each area i modelled with the intrinsic CAR prior and v_i is the unstructured random effects (and has a normal distribution). The relative excess risk (RER) for each area is given by $\exp(u_i + v_i)$. Note that when the intrinsic CAR prior is used with a sum to zero constraint, it is essential to include an intercept term in the model. Also, there are likely to be population differences between SLAs, so unless age distributions are taken into account any observed differences in relative survival may simply result from different age structures. (Commonly, cancer patients diagnosed at an older age have poorer survival.) Broad age groups can be incorporated into the $x\beta$ matrix, or data can be age standardized.

13.4 Data analysis and results

13.4.1 Case study 1

Twelve different CAR models were run for the various adjacencies described above, using WinBUGS version 1.4.1 (Imperial College and Medical Research Council, UK). The models were run through WinBUGS interfaced with Stata (Thompson *et al.* 2006). The first 40 000 iterations were discarded, then a further 20 000 iterations were run and used in the calculation of the posterior estimates. Two different chains were used, starting from diverse initial values, and convergence was assessed using the Gelman–Rubin convergence statistic, as modified by Brooks and Gelman (1998). The Gelman and Rubin plots indicated convergence after about 20 000 iterations for the posterior estimates of the regression coefficients, three randomly selected relative risk estimates and three randomly selected posterior probability estimates. The 'fraction' parameters took a longer time to converge (around 40 000 iterations); thus for consistency, the first 40 000 iterations were discarded for all parameters.

Estimates for the smoothed relative risk, posterior probability of relative risk greater than one, spatially structured random effect, spatially unstructured random effect and their corresponding 95% credible intervals were derived from the posterior distribution. The fraction of total random variation explained by the model as a ratio of the empirical variance of the spatial component against the total variance was also calculated. This fraction provides us with a means to explore how much of the spatial variation in relative risk is explained by the model.

In order to determine the magnitude of smoothing in relation to epidemiologically meaningful cutoffs for the relative risks, we tabulated risk estimates into three groups, based on the 25th and 75th percentile (low, RR$<$ 0.65; neutral, 0.65 $<$RR$<$ 1.15; high, RR$>$ 1.15) and cross-tabulated the observed with the predicted relative risk estimates. Percentiles were used to obtain enough numbers in each group for reasonable sensitivity and specificity. The κ statistic (Cohen 1960) was calculated to quantify the extent of change and compare models:

$$\hat{\kappa} = \frac{(p_o - p_e)}{(1 - p_e)}$$

with p_o and p_e being the observed and expected proportion of agreement respectively.

A comparison of the characteristics of the neighbourhood types in terms of various measures is provided in Table 13.1. Generally, there was greater agreement between observed and predicted relative risks, categorized in quartiles, using the distance-based matrices compared with the adjacency-based neighbourhoods. The Queen-1 model had a low κ value of 0.05, indicating a larger amount of smoothing. Among the distance-based models, the 'Gravity' model (Weight-4) performed better with a κ of 0.15. The 'Covariate' model fell in between the adjacency and distance-based models.

When the DIC was examined as a basis of model selection, the 'Gravity' model had the lowest DIC of 2202, followed by Weight-2 (DIC = 2204) and 'Density' model

Table 13.1 Comparison of model fit and sensitivity of detecting areas with an elevated risk.

Neighbourhood type	Kappa	Fraction	DIC	AUC	Sensitivity	Specificity
Queen-1	0.05	22%	2283	0.61	77%	41%
Queen-2	0.05	51%	2282	0.62	78%	43%
Rook-1	0.05	14%	2282	0.62	76%	41%
Rook-2	0.05	41%	2282	0.62	78%	39%
Distance	0.07	41%	2274	0.62	80%	30%
Weight-1 (1/distance)	0.07	41%	2213	0.62	80%	30%
Weight-2 (1/distance2)	0.12	98%	2204	0.61	75%	44%
Weight-3 (1/distance3)	0.08	93%	2218	0.59	75%	46%
Weight-4 (Gravity)	0.15	99%	2202	0.60	74%	39%
Weight-5 (Entropy)	0.08	91%	2227	0.60	74%	47%
Weight-6 (Density)	0.14	97%	2208	0.60	74%	37%
Weight-7 (Covariate)	0.10	95%	2210	0.62	74%	47%

AUC: Area Under Curve from receiver operating characteristics (ROC) analysis.
DIC: Deviance Information Criteria.
Kappa: Measure of agreement between predicted and actual relative risks.
Fraction: Proportion of total variation in relative risks explained by spatial random effects.
Sensitivity: Ability of model to correctly classify SLAs with an increased risk of defects.
Specificity: Ability of model to correctly classify SLAs without an increased risk of defects.
Reproduced with permission from Earnest A., Morgan G., Mengersen K., Ryan L., Summerhayes R., Beard J. Evaluating the effect of neighbourhood weight matrices on smoothing properties of Conditional Autoregressive (CAR) models. *Int. Journal of Health Geography* 2007 Nov 29; **6**(1):54–58.

(DIC = 2208). Generally, the adjacency-based matrices had higher DICs, although the differences were marginal.

There were considerable differences demonstrated in the amount of smoothing performed by the CAR model, depending on the nature of weight matrix specified. In particular, specifying a weight matrix according to similarity in terms of a key confounder instead of a conventionally used Queen method of assignment helped improve the ability of the model to recover the underlying relative risk. If the aim were to remove the spatial relationship in the relative risk estimates, then distance-based models, such as 'Gravity' or Weight-2 models, might be useful. Conversely, if it is desired to preserve the spatial structure of the relative risks and examine the relationship between covariates that were spatial in nature, then one might prefer a model that has a low fraction of random effect due to spatially structured random effects, and to continue adding spatial covariates until there is no spatial correlation.

13.4.2 Case study 2

To conduct this analysis, the following steps were taken:

1. Neighbourhood matrices were generated in GeoDa (Anselin *et al.* 2006) using first-order Queen adjacencies. Many of the islands were grouped together, along with nearby mainland regions. The output .gal file was manipulated in Stata to convert it to a format suitable for WinBUGS. (Note that R has packages available which can read in .gal files.)

Tips and Tricks: When creating an adjacency matrix:
(a) Each SLA requires at least one neighbour (may need to manually adjust the GeoDa output if there are islands etc.).
(b) It is important to provide neighbours which have large enough populations to provide strength for SLAs with small numbers.
(c) Sensitivity analyses are beneficial, that is comparing distance with adjacency-based matrices.

2. General population mortality data were used to create lifetables (via the 'lifetabl' command in Stata, for each SLA), by gender and year group (1997–2002,2003–2007). The mortality rate was averaged for each individual age from 5 year rates (e.g. the rates for ages 0–4 were used for ages 0,1,2,3,4).
3. The expected deaths and person-time at risk for each SLA, gender, broad age group (0–49,50–69,70–89) and time period were calculated using the strs command in Stata. The expected number of deaths and person-time at risk were used from the output data set at the individual level (default name 'individ').
4. The model was run using WinBUGS, with 150 000 burn-in followed by 100 000 iterations. The output was 'thinned', that is every 10th iteration was kept.

   The long burn-in period was to allow for the arbitrarily selected initial values to move to the appropriate value, even if this was substantially different. Convergence of the RER variable was assessed using trace and density plots for the first 24 regions, and Geweke diagnostics (Geweke 1992) for all regions (results not shown). A Geweke p-value < 0.01 was considered to indicate non-convergence.

Tips and Tricks: Generally, using an improper, very wide uniform distribution for the intercept is recommended. However, this is not uniform over the whole set of real numbers, so it can be worthwhile to put more realistic boundaries on the intercept. This adjustment may also prevent the model crashing with messages such as 'undefined real result'.

5. Model goodness-of-fit tests: Model fit was assessed by removing the spatial and random model components and also comparing observed values with predicted data.

   It can also be helpful to delete the spatial and random components one at a time and rerun the model to check the effect on model fit, as determined by DIC values. From Table 13.2 it can be seen that excluding the random component has little impact on the fit of the model (the DIC is not sufficiently different to be able to conclude with certainty that any difference exists), but more shrinkage is occurring, as evidenced by the smaller p_D value. The p_D value represents the effective number of parameters, which due to 'borrowing of strength' in hierarchical models is often less than the total number of model parameters (Zhu and Carlin 2000). Excluding the spatial component reduces model fit (the DIC value is more than six above the full model). This is as would be expected.

   The observed versus predicted number of deaths showed the model was producing reasonable estimates (see Figure 13.1).

Table 13.2 Model comparisons of breast cancer survival.

Model	p_D	DIC
Full	43.7	9932.5
No random component (v excluded)	36.9	9933.2
No spatial component (u excluded)	67.8	9949.3

6. A basic Poisson model (excluding the spatial and random components) was run in Stata to generate unsmoothed, 'raw' estimates for comparative purposes.
7. Sensitivity analyses were conducted on the choice of prior for the hyperparameter. This is particularly important when conducting analyses involving a CAR prior.

   The model was run under six different, non-informative hyperparameter distributions:

   (1) $\tau_u \sim \Gamma(0.5, 1000)$, $\tau_v \sim \Gamma(0.5, 1000)$
   (2) $\tau_u \sim \Gamma(0.1, 10)$, $\tau_v \sim \Gamma(0.001, 1000)$
   (3) $\tau_u \sim \Gamma(0.1, 100)$, $\tau_v \sim \Gamma(0.1, 100)$
   (4) $\tau_u \sim \Gamma(0.5, 2000)$, $\tau_v \sim \Gamma(0.5, 2000)$
   (5) Reparameterized on σ: $\sigma_u \sim \text{Unif}(0, 1)$, $\sigma_v \sim \text{Unif}(0, 1)$
   (6) Reparameterized on σ: $\sigma_u \sim \text{Unif}(0, 1000)$, $\sigma_v \sim \text{Unif}(0, 1000)$.

   Note that gamma distributions are expressed above as (shape, scale), and uniform distributions as (minimum,maximum).

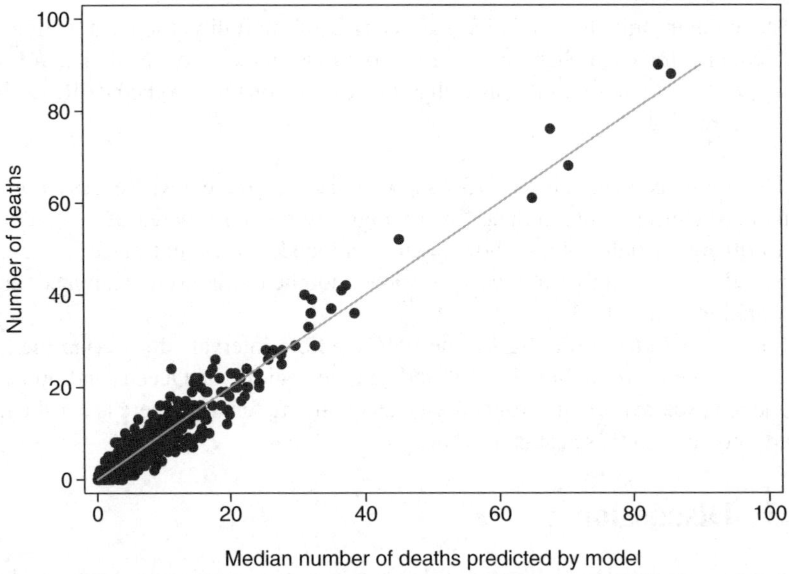

Figure 13.1 Comparison of the median predicted number of deaths for each SLA versus the observed number of deaths.

Table 13.3 Hyperprior distribution comparisons for breast cancer survival.

	Prior 1	Prior 2	Prior 3	Prior 4	Prior 5	Prior 6
Distribution of RER						
Maximum	1.44	1.56	1.53	1.43	1.54	1.57
75% quartile	1.19	1.20	1.18	1.18	1.20	1.19
Median	0.96	0.96	0.97	0.95	0.96	0.97
25% quartile	0.87	0.85	0.87	0.87	0.86	0.86
Minimum	0.71	0.65	0.70	0.73	0.69	0.69
90% ratio	1.7	1.9	1.6	1.6	1.7	1.7
p_D	43.7	54.1	66.9	45.3	62.6	66.4
DIC	9932.5	9930.6	9930.4	9932.7	9930.0	9931.4
Spatial fraction	0.96	0.97	0.58	0.90	0.70	0.68
(95% CI)	(0.60,1.00)	(0.64,1.00)	(0.28,0.87)	(0.40,0.99)	(0.30,0.97)	(0.24,0.98)

The spatial fraction checks whether the spatial or unstructured heterogeneity dominates, and is calculated as the marginal spatial variance divided by the sum of the marginal spatial variance and the unstructured random effect marginal variance (Best *et al.* 1999; Eberly and Carlin 2000). Spatial heterogeneity dominates if it is close to one, whereas random heterogeneity dominates if it is close to zero.

The comparison of priors in Table 13.3 does not suggest any one as clearly superior, and as the estimates generated are also fairly similar, we chose to use the prior 1 distribution.

8. The median smoothed relative excess risk of mortality for each region was calculated for each SLA, classified into categories and mapped (i.e. $RER_i = \exp(u_i + v_i)$). These results, as well as those based on unsmoothed RER, are given in Figure 13.2

The smoothed map in Figure 13.2 shows a marked gradient with excess mortality from breast cancer among females increasing as remoteness increases.

To obtain an indication of how much confidence we should place in these estimates, the 95% credible intervals for each smoothed RER were examined, and is illustrated in Figure 13.3.

For the majority of the SLAs, the 95% credible intervals do overlap the average RER. Therefore, although rural and remote regions of Queensland appear to experience excess mortality from breast cancer among females, there is considerable uncertainty around these estimates.

13.5 Discussion

Bayesian hierarchical models are useful to conduct disease mapping at the areal level. The examples in this chapter demonstrated the applicability for examining incidence in a spatio-temporal model, as well as excess mortality in a relative survival model.

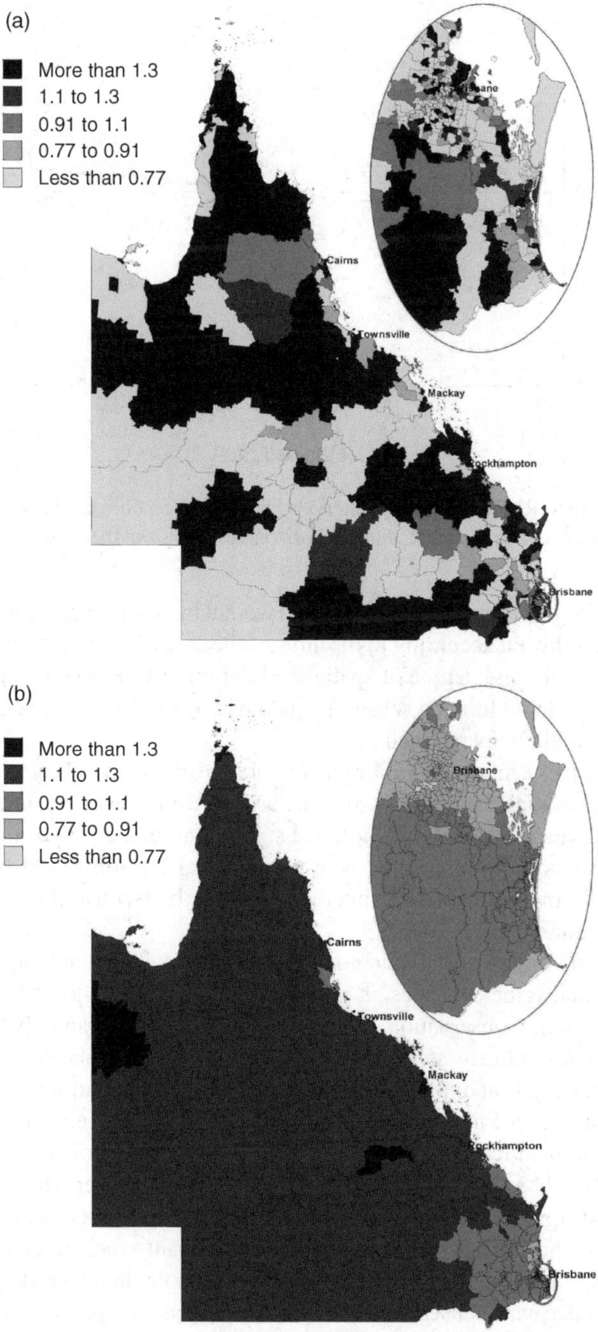

Figure 13.2 Median unsmoothed RER estimates (top) and smoothed RER estimates (bottom) by SLA.

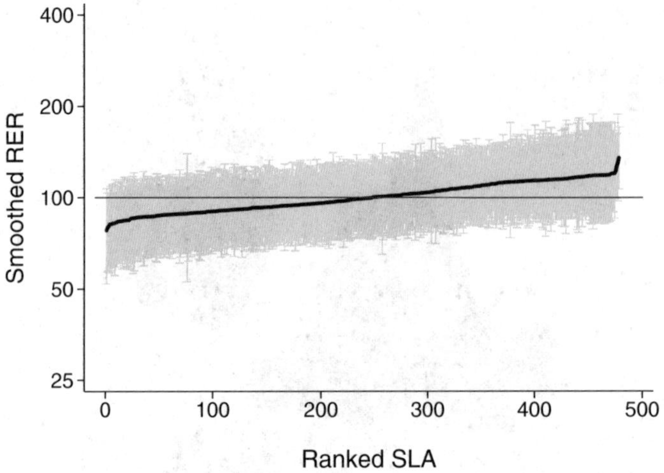

Figure 13.3 Median RER and credible intervals by SLA. Grey lines are 95% credible intervals. The thick black line is the RER, and the horizontal line shows the Queensland average.

Selection of an appropriate neighbourhood weight matrix is important. The weight matrix should be chosen according to the aims of the study or through an exploratory analysis of the nature and degree of spatial correlation prior to risk mapping or ecological modelling. In addition, a sensitivity analysis on the choice of neighbourhood weight matrix should be performed.

Modelling excess mortality in a relative survival model via Bayesian methods which allow for spatial correlation produces useful results, even when data sparseness is present. The model used for relative survival assumed proportional hazards during each year of follow-up time. Non-proportional hazards could be modelled by extending the model to include interaction terms between follow-up time and covariates (Dickman *et al.* 2004).

The additive nature of spatial and non-spatial random effects in both case studies raises the potential issue of Bayesian non-identifiability (Gelfand and Sahu 1999). In essence, for a single observation, this means that inferences can only be made for $(u_i + v_i)$, not on u_i and v_i as separate effects. This property can be shown through a simple reparameterization of the model that results in the full conditional distribution for either random effect not depending on the observed data (e.g. see Eberly and Carlin (2000)). In practice, except in cases when informative priors are used, estimation of these models is likely to result in slow or lack of convergence in u_i and v_i independently, despite observing convergence in their sum (Eberly and Carlin 2000).

Nonetheless, the impact of this issue is only relevant when inference of these separate effects is sought (Besag *et al.* 1995). For example, in the second case study presented, inference was focused on the RER for each area and, therefore, independent effect inferences were not of concern. That said, when inference on separate random effects forms the focus of a study, it may prove fruitful to consider alternative prior specifications that impose model identifiability. For instance, in Best *et al.* (1999), the proportion of total excess variation attributed to spatial clustering is explored via the ratio of posterior standard deviations for each random effect, estimated using draws from the stationary posterior distributions for both **u** and **v**. A suitable alternative prior specification for this purpose is the mixture prior proposed by Leroux *et al.* (1999).

References

AIHW National Perinatal Statistics Unit 2001 *Congenital malformations, 1981-1997*. AIHW National Perinatal Statistics Unit, Randwick, NSW.

Anselin L, Syabri I and Kho Y 2006 GeoDa: an introduction to spatial data analysis. *Geographical Analysis* **38**, 5–22.

Bernardinelli L, Clayton D, Pascutto C, Montomoli C, Ghislandi M and Songini M 1995 Bayesian analysis of space-time variation in disease risk. *Statistics in Medicine* **14**(21–22), 2433–2443.

Besag J, Green P, Higdon D and Mengersen K 1995 Bayesian computation and stochastic systems. *Statistical Science* **10**, 3–41.

Besag J, York J and Mollié A 1991 Bayesian image restoration with two applications in spatial statistics. *Annals of the Institute of Statistical Mathematics* **43**, 1–59.

Best N, Arnold R, Thomas A, Waller L and Conlon E 1999 Bayesian models for spatially correlated disease and exposure data. In *Bayesian Statistics 6*, pp. 131–156. Oxford University Press, Oxford.

Best N, Richardson S and Thomas A 2005 A comparison of Bayesian spatial models for disease mapping. *Statistical Methods in Medical Research* **14**, 35–59.

Brooks S and Gelman A 1998 Alternative methods for monitoring convergence of iterative simulations. *Journal of Computational and Graphical Statistics* **7**, 434–455.

Carlin BP and Banerjee S 2003 Hierarchical multivariate CAR models for spatio-temporally correlated survival data (with discussion). In *Bayesian Statistics 7*, pp. 45–63. Oxford University Press, Oxford.

Centre for Epidemiology and Research and NSW Department of Health 2007 New South Wales mothers and babies 2005. *New South Wales Public Health Bulletin* **18**(S1), 1–94.

Choo L and Walker S 2008 A new approach to investigating spatial variations of disease. *Journal of the Royal Statistical Society, Series A* **171**, 395–405.

Clayton D 1996 Generalized linear mixed models. In *Markov Chain Monte Carlo in Practice*, pp. 275–301. Chapman & Hall/CRC, Boca Raton, FL.

Clayton D and Kaldor J 1987 Empirical Bayes estimates of age-standardised relative risks for use in disease mapping. *Biometrics* **43**, 671–681.

Cohen J 1960 A coefficient of agreement for nominal scales. *Educational and Psychological Measurement* **20**, 37–46.

Cramb SM, Mengersen KL and Baade PD 2011 Developing the atlas of cancer in Queensland: methodological issues. *International Journal of Health Geographics* **10**, 9.

Cressie N 1993 *Statistics for Spatial Data*. John Wiley & Sons, Inc., New York.

Devesa S, Grauman D, Blot W, Pennello G and Hoover R 1999 *Atlas of Cancer Mortality in the United States, 1950-1994*. US Government Printing Office, Washington, DC.

Dickman PW, Lambert PC and Hakulinen T 2011 *Population-based Cancer Survival Analysis*. John Wiley & Sons, Ltd, Chichester.

Dickman PW, Sloggett A, Hills M and Hakulinen T 2004 Regression models for relative survival. *Statistics in Medicine* **23**(1), 51–64.

Earnest A, Morgan G, Mengersen K, Ryan L, Summerhayes R and Beard J 2007 Evaluating the effect of neighbourhood weight matrices on smoothing properties of Conditional Autoregressive (CAR) models. *International Journal of Health Geographics* **6**(1), 54–57.

Eberly LE and Carlin BP 2000 Identifiability and convergence issues for Markov chain Monte Carlo fitting of spatial models. *Statistics in Medicine* **19**, 2279–2294.

Elliott P and Wartenberg D 2004 Spatial epidemiology: current approaches and future challenges. *Environ Health Perspectives* **112**(9), 998–1006.

Fairley L, Forman D, West R and Manda S 2008 Spatial variation in prostate cancer survival in the Northern and Yorkshire region of England using Bayesian relative survival smoothing. *British Journal of Cancer* **99**(11), 1786–1793.

Gelfand A and Sahu S 1999 Identifiability, improper priors, and Gibbs sampling for generalized linear models. *Journal of the American Statistical Association* **94**, pp. 247–253.

Gelfand AE and Vounatsou P 2003 Proper multivariate conditional autoregressive models for spatial data analysis. *Biostatistics* **4**(1), 11–25.

Gelman A 2006 Prior distributions for variance parameters in hierarchical models. *Bayesian Analysis* **1**, 515–533.

Gelman A, Carlin J, Stren H and Rubin D 2003 *Bayesian Data Analysis*, 2nd edn. Chapman & Hall/CRC, Boca Raton, FL.

Geweke J 1992 Evaluating the accuracy of sampling-based approaches to calculating posterior moments. In *Bayesian Statistics 4*, pp. 169–194. Oxford University Press, Oxford.

Green P and Richardson S 2002 Hidden Markov models and disease mapping. *Journal of the American Statistical Association* **97**, 1055–1070.

Kim H, Sun D and Tsutakawa R 2001 A bivariate Bayes method for improving the estimates of mortality rates with a twofold conditional autoregressive model. *Journal of the American Statistical Association* **96**, 1506–1521.

Knorr-Held L 2000 Bayesian modelling of inseparable space-time variation in disease risk. *Statistics in Medicine* **19**, 2555–2567.

Knorr-Held L and Besag J 1998 Modelling risk from a disease in time and space. *Statistics in Medicine* **17**(18), 2045–2060.

Kottas A, Duan J and Gelfand A 2007 Modeling disease incidence data with spatial and spatio-temporal Dirichlet process mixtures. *Biometrical Journal* **5**, 1–14.

Lawson A 2009 *Bayesian Disease Mapping: Hierarchical Modeling in Spatial Epidemiology*. Chapman & Hall/CRC, Boca Raton, FL.

Lawson A and Clark A 2002 Spatial mixture relative risk models applied to disease mapping. *Statistics in Medicine* **21**, 359–370.

Leroux B, Lei X and Breslow N 1999 Estimation of disease rates in small areas: a new mixed model for spatial dependence. In *Statistical Models in Epidemiology, the Environment and Clinical Trials*, pp. 135–178. Springer, Berlin.

MacNab Y 2007 Spline smoothing in Bayesian disease mapping. *Environmetrics* **18**, 727–744.

McCullagh P and Nelder J 1989 *Generalized Linear Models*. Chapman & Hall/CRC, Boca Raton, FL.

Nelson CP, Lambert PC, Squire IB and Jones DR 2008 Relative survival: what can cardiovascular disease learn from cancer? *European Heart Journal* **29**, 941–947.

Population Health Division 2006 *The health of the people of New South Wales – Report of the Chief Health Officer*. NSW Department of Health, Sydney.

Queensland Cancer Registry 2010 *Cancer in Queensland: Incidence, Mortality, Survival and Prevalence 1982 to 2007*. QCR, Cancer Council Queensland and Queensland Health, Brisbane.

Schmidt A, Hoeting J, Pereira J and Vieira P 2010 Mapping malaria in the Amazon rain forest: a spatio-temporal mixture model. In *Oxford Handbook of Applied Statistics*, pp. 90–117. Oxford University Press, Oxford.

Thompson J, Palmer T and Moreno S 2006 Bayesian analysis in Stata using WinBUGS. *The Stata Journal* **6**(4), 530–549.

Tzala E and Best N 2007 Bayesian latent variable modelling of multivariate spatio-temporal variation in cancer mortality. *Statistical Methods in Medical Research* **17**, 97–118.

Viertel Centre for Research in Cancer Control 2010 *Queensland Cancer Statistics On-Line*, www.cancerqld.org.au/research/qcsol. Based on data released by the Queensland Cancer Registry (1982–2007; released June 2010).

Waller L, Carlin B, Xia H and Gelfand A 1997 Hierarchical spatio-temporal mapping of disease rates. *Journal of the American Statistical Association* **92**(438), 607–617.

Youlden DR, Cramb SM and Baade PD 2009 *Current status of female breast cancer in Queensland: 1982 to 2006*. Viertel Centre for Research in Cancer Control, Cancer Council Queensland, Brisbane.

Zhu L and Carlin BP 2000 Comparing hierarchical models for spatio-temporally misaligned data using the deviance information criterion. *Statistics in Medicine* **19**, 2265–2278.

14

Moisture, crops and salination: An analysis of a three-dimensional agricultural data set

Margaret Donald[1], Clair L. Alston[1], Rick Young[2] and Kerrie L. Mengersen[1]

[1]*Queensland University of Technology, Brisbane, Australia*
[2]*Tamworth Agricultural Institute, Department of Primary Industries, Tamworth, Australia*

14.1 Introduction

Large data sets over space and time are generated routinely by modern technology. Such data typically exhibit high autocorrelations over both spatial and time dimensions. We describe a WinBUGS application for a spatial three-dimensional data set in which conditional autoregressive (CAR) models are used for point-referenced data to account for spatial autocorrelations.

The CAR model was originally suggested by Besag (1974) in the context of image analysis and is also known as the intrinsic CAR model with a convolution prior, or the Besag, York and Mollie (BYM) model. It is used here, rather than the geostatistical models usually suggested for point-referenced data (Banerjee *et al.* 2004; Cressie 1991), because we found such models effectively impossible to fit in WinBUGS when

Case Studies in Bayesian Statistical Modelling and Analysis, First Edition. Edited by Clair L. Alston, Kerrie L. Mengersen and Anthony N. Pettitt.

there was a regression component with a large number of terms. This is not surprising since CAR models have a sparse precision matrix whereas geostatistical models do not. The CAR model is an example of a Gaussian Markov random field (GMRF).

There are two key elements to the analyses presented. Although the measurements are collected sufficiently closely across space to expect autocorrelation, there are considerable differences in scale across the three dimensions, with the two horizontal dimensions being of the same order, but the third, depth, being collected on a scale at least 50 times more dense. This poses problems for the use of a CAR model over three dimensions, where a weighting system based on distance would effectively reduce the neighbourhood to a depth neighbourhood only. However, there are further reasons why one would wish to treat the neighbourhoods as neighbourhoods within a depth layer. The treatments are applied across the horizontal surface and the questions of interest relate to how the treatments affect a response variable at each depth. Hence, it makes sense to define neighbours as neighbours within a horizontal layer. This separation by layers has the important advantage that each horizontal CAR model can have a different variance.

Even had we not wished to model the response as a function of depth, the differences in scale between the three dimensions mean that although data may be three dimensional, the data may well be better modelled with incommensurate dimensions modelled separately.

CAR models were used rather than kriging models, because of the complexity of the fixed part of the model, which varied from about 54 terms to 135 terms, which meant that a kriging model was essentially impossible to fit within the framework of WinBUGS (Lunn *et al.* 2000). Recent work (Besag and Mondal 2005; Lindgren *et al.* 2011) shows that CAR models may be used to approximate continuously indexed Gaussian fields.

14.2 Case study

The data used in this case study were collected to determine a cropping system for the Liverpool Plains, in New South Wales, eastern Australia. The aim was to maximize water use for grain production while minimizing leakage below the crop root zone. Leakage below the root zone leaches ground salts as it leaks to the underground aquifers, and leads to their salinification, with a consequent loss of their usefulness for agriculture. Hence, determining a cropping system under which salinification is less likely to occur is important for the continuing viability of agriculture in the region.

Soil moisture measurements at 15 depths (20 to 300 cm) below the surface were taken for nine different cropping systems on an area of about 26 ha, laid out in 6 rows and 18 columns (108 sites). Measurements were taken at these 108 sites over a period of roughly 5 years, and approximately a month apart, giving 61 days of measurements.

The current cropping practice in the Liverpool Plains is long-fallowing which operates on an 18 month cycle of sowing, growing, harvesting and lying fallow, and the two crops grown under this system are wheat and sorghum. The long cycle of

18 months in long-fallow cropping, together with the fact that each year is generally unlike the next year, meant that three treatments were allocated to long-fallowing and ran in differing phases. Treatment 4 was a continuous-cropping treatment where wheat and barley alternated as crops with roughly 6 months between harvest and resowing.

The treatments of most interest to the crop scientists were the response cropping treatments (5 and 6), where an appropriate crop (either a winter or summer crop) is planted when soil moisture exceeds a predetermined level. Treatments 7–9 had the sites sown to pasture, with lucerne, a mixture of lucerne and a winter perennial, and finally a mixture of native winter and summer perennial grasses.

We consider four models for the moisture profile by depth: the saturated model (9×15 fixed parameters), two linear spline models, and a linear spline model using an errors-in-variables model (Fuller 1987) for depth.

14.2.1 Data

We examine data from a single day from a randomized complete block experiment, which comprises soil moisture measurements taken at three dimensions in space, namely row, column and depth, with 3 rows $\times$ 18 columns $\times$ 15 depth measurements (20, 40, ..., 300 cm), or 810 measurements.

Nine experimental treatments consisting of three fully phased cropping systems and three types of perennial pasture were allocated as a randomized complete block design to the 18 plots, each containing 3 of the 54 measurement sites (Ringrose-Voase *et al.* 2003 p. 23). Treatments are as follows:

1. Treatments 1–3. Long-fallow wheat/sorghum rotation, where one wheat and one sorghum crop are grown in 3 years with an intervening 10–14 month fallow period. The first long-fallow cropping treatment started with wheat in the summer of 1995, the second started with wheat in the winter of 1995, and the last with wheat in the winter of 1996.
2. Treatment 4. Continuous cropping in winter with wheat and barley grown alternately.
3. Treatments 5 and 6. Response cropping, where an appropriate crop (either a winter crop or a summer crop) was planted when the depth of moist soil exceeded a predetermined level. The two response-cropping treatments were differentiated by the sequence of crop types.
4. Treatments 7–9. Perennial pastures. The three treatments were lucerne (a deep-rooted perennial forage legume with high water use potential), lucerne grown with a winter-growing perennial grass, and a mixture of winter- and summer-growing perennial grasses.

14.2.2 Aim of the analysis

The aim of the analysis was to model the cropping treatment effects on soil moisture, and to determine their differences at the various depths, taking spatial correlations into account.

14.3 Review

14.3.1 General methodology

When data are point referenced, the kriging models of Cressie (1991) and Banerjee *et al.* (2004) are usually the models of first choice. However, Besag and Mondal (2005) and Lindgren *et al.* (2011) have shown that CAR models may be represented as continuously indexed geostatistical models. Where the fixed part of the model is complex, it becomes difficult to fit a kriging model using MCMC, which is not surprising in that kriging models use the covariance structure, which is dense, while CAR models use the precision matrix defined by the neighbourhood matrix, which is typically sparse and, hence, computationally far more efficient. Here, several CAR models are fitted, one for each horizontal layer of the data.

14.3.2 Computations

The models discussed here were fitted within WinBUGS (Lunn *et al.* 2000) and took up to four and a half hours to run on a Dell PowerEdge 2950 with two cores of Xeon E5410 2.33 GHz and 2 GB RAM (with a burn-in of 20 000 and total iterations of 80 000).

14.4 Case study modelling

14.4.1 Modelling framework

The treatment effects are modelled as a function of depth, since we wish to determine the effect of a cropping regime for a site over the 15 depth measurements. There are a number of choices for these functions: the saturated model where 9 treatments $\times$ 15 depth means are fitted, or treatment curves with fewer parameters. In addition to the saturated model, we fit two linear spline models: one with four equally spaced knots, and another which uses nine knots. The nine-knot spline model fits mean treatment effects for each depth up to 200 cm, with a continuous line thereafter, for each treatment. This second spline addresses the possible lack of fit of the first spline model in the shallower depths and allows us to model the observed flatness of the curves beyond the root zone. The final model is an errors-in-variables model.

WinBUGS code for the saturated model is given on the book's website, and code showing the modifications necessary for the fixed knot linear spline model is also provided. A point to notice about the coding is that we update the estimates for the 9 $\times$ 15 treatment points, with indexing being used to assign them to the 810 measurement points. Additionally, the neighbourhood matrix may be given as a neighbourhood for 810 points with a total number of neighbours of 2610 (where the neighbours are first order), or, using the depth layering, as the neighbourhood of 54 points with just 174 neighbours. Here, again, indexing is used, creating more efficient code. We also provide code on the book's website for fitting a linear spline model. Additionally, we provide code for creating the errors-in-variables model.

The models discussed here fit the following framework, where the fixed part of the model represents a function of depth for each treatment, with the treatment being site dependent. Thus, if the sites are indexed by $i = 1, \ldots, 54$, at depths 20, 40, ..., 300 (indexed by $d = 1, \ldots, 15$), the moisture value, y_{id}, is modelled as nine depth functions, $f_{j(i)}(d)$, one for each treatment j, determined by the site index, i. These are functions of the depth (indexed by d). The residual from this fixed effects model is modelled as the sum of a spatial residual component, s_{id}, and a non-spatial residual component, ϵ_{id}, with $\epsilon_{id} \sim N(0, \tau^2)$. The conditional spatial residual component is an average of the neighbouring spatial residuals, (Besag and Kooperberg 1995; Besag *et al.* 1991). This local spatial smoothing specification ensures a global specification via Brooke's lemma (Banerjee *et al.* 2004), and allows us to account for spatial similarities. For site i at depth d, the full model is

$$\begin{aligned} y_{id} &= f_{j(i)}(d) + s_{id} + \epsilon_{id}, \\ \epsilon_{id} &\sim N(0, \tau^2), \end{aligned}$$

where $f_{j(i)}(d)$ is the treatment effect for treatment j at site i and depth index d, and is a function of depth.

The conditional probability of the spatial residual component, s_{id}, given its neighbours, $s_{kd} \in \partial_i$, is

$$s_{id}|s_{kd}, k \in \partial_i \sim N\left(\sum_{k\in\partial_i}\left(\frac{w_{ik}s_{kd}}{w_{i+}}\right), \frac{\sigma_d^2}{w_{i+}}\right),$$

where ∂_i is the set of indices for the neighbours of site i, w_{ik} is the weight of the kth neighbour of i, w_{i+} is the sum of the weights of the neighbours of i, and σ_d^2 is a variance component for the CAR model at depth d, and there is a common homogeneous variance component across all depths, τ^2.

In the models illustrated, the neighbourhood definitions are identical for each depth layer; first-order neighbours are used; and all neighbours are of equal weight. Earlier work by Donald *et al.* (2011), shown here as Table 14.1, compared various neighbourhood schemes for these data using the deviance information criterion (DIC) of Spiegelhalter *et al.* (2002) and showed that the first-order neighbourhood gave the lowest DIC of the many neighbourhoods and variance structures compared.

The fixed part may be modelled as various types of basis function of depth. Here we consider linear splines and a saturated model, both of which may be represented as

$$\begin{aligned} \hat{y}_{i,d} &= f_{j(i)}(d) \\ &= \mathbf{X}\beta, \end{aligned}$$

with $\mathbf{X}$ being an $n \times p$ design matrix based on the treatments $j(i)$ at site i, the basis functions of the depth index, d, and β a $p \times 1$ vector of coefficients.

Table 14.1 Comparing spatial neighbourhood modelling. Treatment effects model is identical for all models (orthogonal polynomial degree 8). Models have 15 spatial variance components (σ_d^2) and 1 homogeneous variance component (τ^2), except where otherwise stated.

Description	p_D	DIC	ΔDIC
Base model: no spatial component	81	−2690	—
Linear CAR (maximum 2 horiz. neighbours)	264	−2811	121
CAR (maximum 4 horiz. neighbours)	358	−2990	300†
CAR (maximum 8 horiz. neighbours)	320	−2930	240
AR(1), AR(1)	945	−2947	257
CAR (maximum 4 horiz. neighbours, 2 depth)*	109	−2752	62
CAR (maximum 4 horiz. neighbours)*	110	−2960	270
CAR (maximum 4 horiz. neighbours)**	121	−2766	76

ΔDIC = DIC(base model)– DIC.
†The favoured neighbourhood model.
*Models with 1 spatial variance and 15 non-spatial variance components.
**Model with 15 spatial and 15 non-spatial variance components.
Reproduced with permission from Elsevier from Donald M.R., Alston C.L., Young R.R. and Mengersen K.L. 2011. A Bayesian analysis of an agricultural field trial with three spatial dimensions. *Computational Statistics & Data Analysis* **12**, 3320–3332.

Splines

The penalized linear splines of Wand (2009) are fitted with $f_j(d)$, the treatment effect (j) at depth index d, given by

$$f_j(d) = \beta_{j0} + \beta_{j1} d + \sum_{1}^{K} u_{jk} x_k(d), \quad j = 1, \ldots, 9,$$

where $x_k(d) = (d - \kappa_k)_+$ for some knot sequence $\kappa_1, \ldots, \kappa_K$, the prior for $u_{jk} \sim \text{N}(0, \sigma_u^2)$ and K the number of knots.

Thus, in the linear spline model, the basis functions were truncated linear functions, with u_{jk} the coefficients of the knot functions. For cubic splines and cubic radial bases, see Ngo and Wand (2004). However, for these data, modelling via linear splines appears to give adequate descriptions of the data (Donald *et al.* 2011).

Errors-in-measurement model

Spline models allow us to consider the depth as being measured with error. This is a useful addition to the model for the reasons given below.

The depth is the measured depth through the soil. However, since soils shrink on drying and expand with moisture, the depth index, d, does not represent the depth within the soil profile (Ringrose-Voase *et al.* 2003). Hence, we fit a measurement error model, where the independent variable is assumed to be measured with error.

The errors-in-measurement model assumes that the true depth index z is interval censored and is related to the observed depth index, d, in the following way:

$$
\begin{aligned}
z_d | d &\sim N(d, \sigma_z^2) I(z_{d-1}, z_{d+1}) \quad \text{for} \quad d = 2, 3, \ldots, 14, \\
z_1 | d = 1 &\sim N(1, \sigma_z^2) I(0, z_2), \\
z_{15} | d = 15 &\sim N(15, \sigma_z^2) I(z_{14}, 16).
\end{aligned}
$$

Here, σ_z^2 is given by $1/\sigma_z^2 = \tau_z$, with $\tau_z \sim \text{Gamma}(0.5, 0.0005)$.

We experimented with half-Cauchy distributions but found they were too heavy tailed. Here we use the gamma prior suggested by Wakefield *et al.* (2000).

Where an errors-in-variable model is used for depth, d is replaced by z_d, the unobserved true depth index, and the knots adjusted accordingly. The treatment effects of the model, $f_{j(i)}(d)$, become $f_{j(i)}(z_d)$.

14.5 Model implementation: Coding considerations

14.5.1 Neighbourhood matrices and CAR models

Neighbourhood matrix definitions are based on the GeoBUGS manual (Lunn *et al.* 2000). With a layering scheme, this can be created either by hand if the number of sites is small, or by coding.

The data are entered in a particular order and the neighbourhood model is based on that order. If we wish to use a layered CAR model, where only horizontal neighbours are used, then the data must be ordered by depth. These data were ordered by depth, row and column. The same CAR model may be fitted as a single neighbourhood model, or, as done here, using indexing and a common neighbourhood model for each depth. With an arrangement of 3×18, there are 174 first-order neighbours in a layer, giving $15 \times 174 = 2610$ neighbours for a single neighbourhood model. We use the indexed model with 174 neighbours, to give the same statistical model.

Defining points as neighbours only within the same horizontal depth layer allows the variance components for each CAR model to be different for each depth, which is far more realistic for these data. Thus layering confers a considerable advantage in the modelling.

14.5.2 Design matrices vs indexing

Unlike many of the models in the GeoBUGS manual, where the fixed model is fitted in the main loop, indexing and a smaller loop have been used to minimize memory requirements and to speed convergence in all models except the errors-in-variables model, where the differing latent depths imply common treatment curves but differing estimates for the nominal depth and hence for the treatment at each site.

14.6 Case study results

Ideally, the errors-in-variables model for depth gives the true depth. More realistically, it may be seen as addressing the lack of fit of the fixed model in addition to picking up differences between the sites. To address possible lack of fit, the linear spline with nine knots was fitted. This model had interior knots at $d = 2, 3, \ldots, 10$. Thus, the line segments were fitted between adjacent depths up to a depth of 200 cm, giving a treatment by depth effect up to this depth and a linear segment from 200 to 300 cm ($d = 10, \ldots, 15$). See Figure 14.1.

The errors-in-variables model, shown in Figure 14.2, is clearly the best of the fitted models (Table 14.2). However, the model which fits the first nine depth by treatments using a final continuous line segment is the best of the models with no latent depth variable, and the saturated model is the worst of the three.

The soil scientists wished to determine the difference between long-fallow cropping and response cropping. Estimates for this may be set up in WinBUGS or calculated from the saved WinBUGS CODA output. Contrasts from the errors-in-variables model estimating the differences between cropping regimes and their associated 95% credible intervals are shown in Figure 14.3. These showed the hoped-for difference between long-fallow cropping and response cropping at the

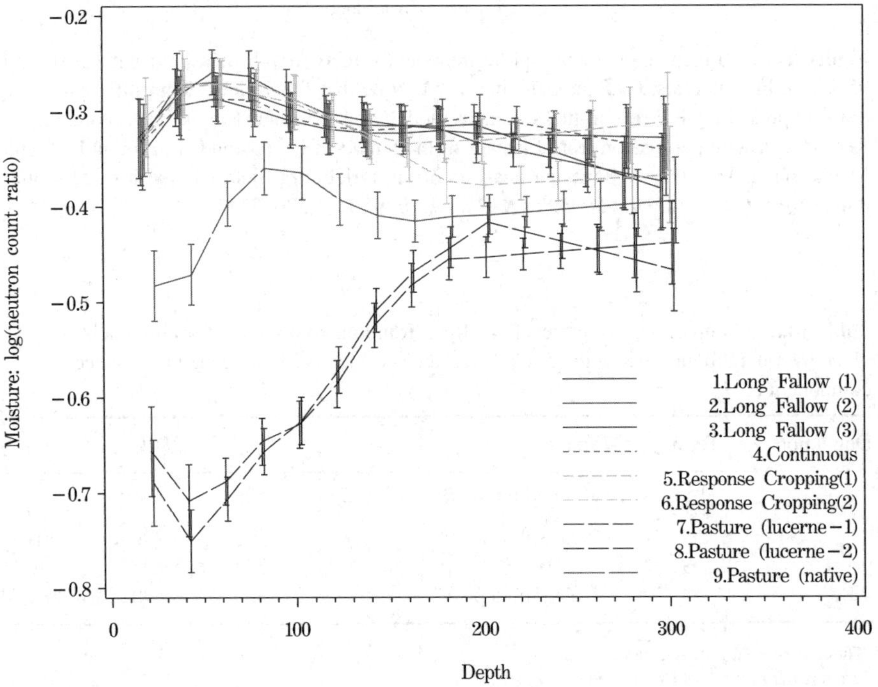

Figure 14.1 Linear spline which fits knots at all depths up to 200 cm and a continuous line over the remaining depths.

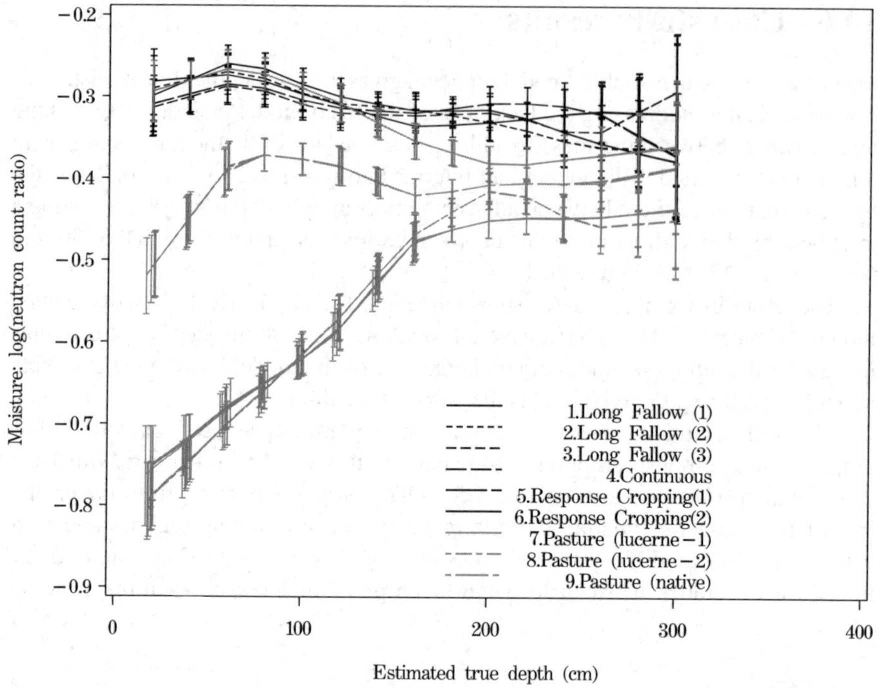

Figure 14.2 'Fixed' part: linear spline treatment effects, depth measured with error and 95% credible intervals, CAR model, sites 1–54. Note that the various depth differences are those implied by the errors-in-measurement model, and that there are 54 lines plotted here. Reproduced with permission from Elsevier from Donald M.R., Alston C.L., Young R.R. and Mengersen K.L. 2011. A Bayesian analysis of an agricultural field trial with three spatial dimensions. *Computational Statistics & Data Analysis* **12**, 3320–3332.

Table 14.2 Comparing fixed effects modelling. Random components for all models are given by 4 neighbour CAR with 15 depth variances (σ_d^2) and 1 homogeneous variance component (τ^2).

Deg./knots	Terms	Type	p_D	DIC	ΔDIC
	135	Saturated model (9 × 15 terms)	809	−2319	—
4	54	Linear spline	318	−2923	604
9	99		394	−2965	646
4	54	(+ error in depth)	369	−3002	683[†]

[†]The best model of those fitted.
Terms is the number of fitted fixed effects terms.
p_D, DIC given for the moisture value to allow comparison.
ΔDIC: DIC(saturated model) − DIC.

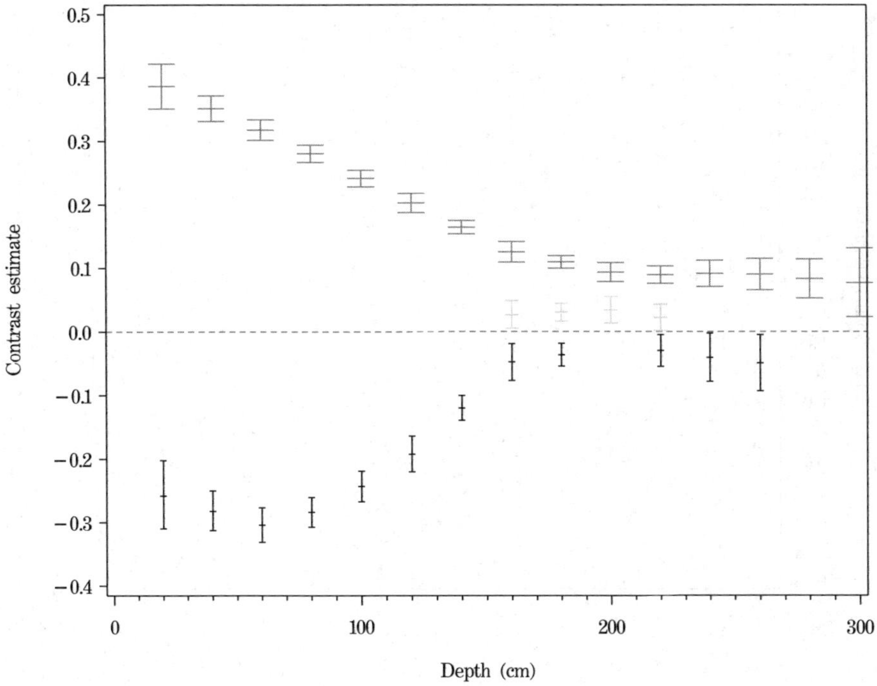

Figure 14.3 The 95% credible intervals for the contrast differences for the five-knot linear spline model with errors-in-measurement (graphed where the 95% CI did not include zero). The CIs with the widest bars show the contrast of cropping vs pastures; the narrowest bars, lucernes vs native pastures; and the medium-width bars, those for long-fallow cropping vs response cropping.

intermediate depths. This difference is sufficient in the critical part of the profile for response cropping to be recommended should such a difference be repeated in further data. The differences are in the mid-depth range where moisture uptake is needed to prevent salination.

Figure 14.4 shows that the CAR layered model captured an important property of the spatial variances. Donald *et al.* (2011) also showed that the CAR layered model, with a single homogeneous random component, modelled the data better than alternative CAR neighbourhood models and was markedly better than CAR models that included depth neighbours, which are unable to model CAR variances by depth.

14.7 Conclusions

We have shown how to model complex fixed models within three spatial dimensions, by using CAR models to capture spatial correlation. In particular, we have shown how to create a layered CAR model which builds CAR models in two dimensions for each of layer of the third dimension (depth). This separation into layered CAR

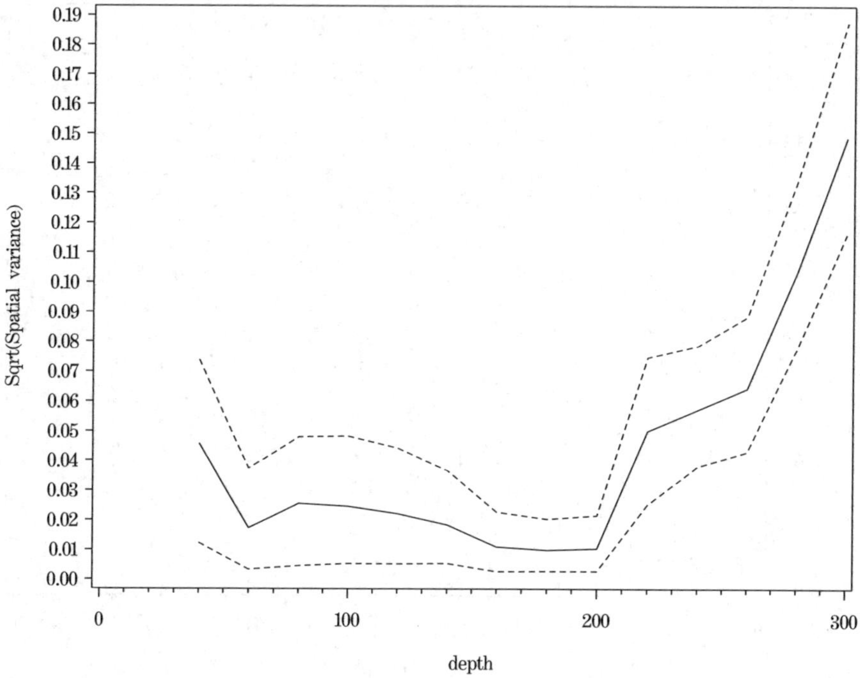

Figure 14.4 The 95% credible intervals for the square root of the spatial variance components at each depth.

models was shown to be useful, in that it permitted differing variances for each depth, but it also allowed a clear understanding of the treatment effect.

However, regardless of the fixed model component, when measurements are made on intervals which are not scaled commensurately, it makes little sense to fit a three-dimensional CAR model or a three-dimensional kriging model, because the closer measurements will dominate the smoothing almost to the exclusion of the measurements in the other dimensions, if weights dependent on distances between points are used.

The CAR layered model introduced here is applicable in many agricultural settings but may well also be useful in other three-dimensional settings.

References

Banerjee S, Carlin BP and Gelfand AE 2004 *Hierarchical Modeling and Analysis for Spatial Data*. Monographs on Statistics and Applied Probability. Chapman & Hall, Boca Raton, FL.

Besag JE 1974 Spatial interaction and the statistical analysis of lattice systems (with discussion). *Journal of the Royal Statistical Society, Series B* **36**(2), 192–236.

Besag JE and Kooperberg C 1995 On conditional and intrinsic autoregressions. *Biometrika* **82**(4), 733–746.

Besag JE and Mondal D 2005 First-order intrinsic autoregressions and the de Wijs process. *Biometrika* **92**(4), 909–920.

Besag JE, York J and Mollie A 1991 Bayesian image restoration with applications in spatial statistics (with discussion). *Annals of the Institute of Mathematical Statistics* **43**, 1–59.

Cressie NAC 1991 *Statistics for Spatial Data*. Wiley Series in Probability and Mathematical Statistics. John Wiley & Sons, Inc., New York.

Donald MR, Alston CL, Young RR and Mengersen KL 2011 A Bayesian analysis of an agricultural field trial with three spatial dimensions. *Computational Statistics & Data Analysis* **12**, 3320–3332.

Fuller WA 1987 *Measurement Error Models*. John Wiley & Sons, Inc., New York.

Lindgren F, Rue H and Lindstrom J 2011 An explicit link between Gaussian fields and Gaussian Markov random fields: the stochastic partial differential equation approach. *Journal of the Royal Statistical Society, Series B* **73**(4), 423–498.

Lunn DJ, Thomas A, Best N and Spiegelhalter D 2000 WinBUGS – a Bayesian modelling framework: concepts, structure, and extensibility. *Statistics and Computing* **10**(4), 325–337.

Ngo L and Wand M 2004 Smoothing with mixed model software. *Journal of Statistical Software* **9**, 1–56.

Ringrose-Voase A, Young RR, Payder Z, Huth N, Bernardi A, Cresswell H, Keating B, Scott J, Stauffacher M, Banks R, Holland J, Johnston R, Green T, Gregory L, Daniells I, Farquharson R, Drinkwater R, Heidenreich S and Donaldson S 2003 Deep drainage under different land uses in the Liverpool Plains Catchment. Technical Report 3, Agricultural Resource Management Report Series, NSW Agriculture, Orange, NSW.

Spiegelhalter DJ, Best NG, Carlin BP and van der Linde A 2002 Bayesian measures of model complexity and fit. *Journal of the Royal Statistical Society, Series B* **64**(4), 583–639.

Wakefield J, Best N and Waller L 2000 Bayesian approaches to disease mapping. In *Spatial Epidemiology: Methods and Applications* (eds P Elliott *et al.*), pp. 104–127. Oxford University Press, Oxford.

Wand MP 2009 Semiparametric and graphical models. *Australian and New Zealand Journal of Statistics* **51**(1), 9–41.

15

A Bayesian approach to multivariate state space modelling: A study of a Fama–French asset-pricing model with time-varying regressors

Christopher M. Strickland[1] and Philip Gharghori[2]

[1]*Queensland University of Technology, Brisbane, Australia*
[2]*Monash University, Melbourne, Australia*

15.1 Introduction

The state space framework provides a generic methodology for modelling time series. Algorithms such as the Kalman filter (Kalman 1960) provide generally applicable tools that facilitate estimation, see Anderson and Moore (1979) and Harvey (1989). Bayesian inference of state space models (SSMs) is currently most commonly undertaken using Markov chain Monte Carlo (MCMC). As MCMC methods typically require the state space algorithms to be used tens of thousands of times, particular care is required with their implementation. This is especially true when dealing with

Case Studies in Bayesian Statistical Modelling and Analysis, First Edition. Edited by Clair L. Alston, Kerrie L. Mengersen and Anthony N. Pettitt.

multivariate data. This chapter details a Bayesian approach to the estimation of linear Gaussian multivariate SSMs. The approach is applied to a system analysis of the Fama–French asset-pricing model, with time-varying regressors.

15.2 Case study: Asset pricing in financial markets

Modelling the behaviour of asset prices is at the core of finance. The first model that quantified the relationship between an asset's risk and return is the capital asset-pricing model (CAPM) of Sharpe (1964), Lintner (1965) and Black (1972). The model suggests that there is a positive relationship between the systematic risk of an asset and its expected return. The early empirical tests of the model were promising and, for a period, it was accepted as a good descriptor of stock returns. However, in the early 1980s, evidence started to emerge that is inconsistent with the model. In particular, Banz (1981) shows that after controlling for systematic risk, there is a negative relationship between the size of a firm and its returns. In a similar vein, Stattman (1980) shows that after risk adjustment, there is a positive relationship between the book-to-market equity ratio of a firm and its returns. These empirical observations are termed anomalies, as they cannot be explained by the CAPM.

The central intuition of the CAPM is that the systematic risk of an asset, which is the covariation between the return of the asset and the return of the entire market of assets, is the only determinant of its expected return. Although the CAPM is an elegant model, it is questionable whether systematic risk is the only determinant of asset prices. This contention coupled with the empirical evidence against the model has led to the creation of a number of alternative asset-pricing models. Perhaps the most prominent of these is the three-factor model of Fama and French (1993), hereafter the Fama–French model. The Fama–French model is an augmentation of the one-factor CAPM with a size and a book-to-market factor. Fama and French (1996) argue that the size and book-to-market factors could be capturing distress risk (the risk associated with the degree of financial distress of a firm) and that this risk is orthogonal to systematic risk and is thus separately priced in stock returns. The evidence on this conjecture is mixed. Regardless, the Fama–French model has emerged as the dominant asset-pricing model in the finance literature. The key reason for this is its performance in explaining variation in stock prices. The major criticism of the model is that it is atheoretical and derived from empirical observation.

In typical applications of the Fama–French model, the parameters on the factors are constant over time. However, it is reasonable to assume that the parameters are time varying. For example, when the stock market is more volatile, it is likely that the parameters on all three factors will be higher, as this implies that the return for bearing risk associated with these factors is expected to be higher when the market is more volatile. A large amount of literature, such as studies by Ferson *et al.* (1987), Avramov and Chordia (2006) and Lewellen and Nagel (2006) examine time variation in the parameter estimates of asset-pricing models. Further, Jostova and Philipov (2005) and Ang and Chen (2007) employ Bayesian techniques to model time variation in the parameters from asset-pricing models.

It is common in time series asset-pricing tests to examine the model on portfolios of stocks rather than on individual stocks. The typical test portfolios used are 25 size and book-to-market sorted portfolios, which together represent the entire stock market. One can examine the performance of the model in a univariate context by regressing the time series of returns of an individual test portfolio against the returns of the factors from the model. This allows inferences to be drawn on the performance of the model in explaining the returns of that single portfolio. However, a true test of the model must be conducted in a multivariate system context where the model is estimated concurrently on all 25 test portfolios. In doing so, the covariance between the test assets is taken into consideration when evaluating the performance of the model. There are a number of multivariate system-based tests in the finance and econometrics literature that have been developed to analyse asset-pricing models. Prominent examples are those of Gibbons *et al.* (1989), MacKinlay and Richardson (1991) and Hansen and Jagannathan (1997). A limitation of these tests is that they can only be applied when the parameters are not time varying. This study employs Bayesian techniques to assess the performance of the Fama–French model in a multivariate system context where the parameters are time varying.

15.2.1 Data

The data used are conventional for an asset-pricing study. The monthly returns of US stocks are analysed over the period 1963 to 2009, that is a time series of 564 months. The start date of 1963 is chosen for consistency with the seminal study of Fama and French (1993). The dependent variables or test portfolios are the returns on 25 size and book-to-market sorted portfolios. The independent variables or factors are the returns on the market, size and book-to-market factors. All the data used in this study are publicly available and can be obtained from Kenneth French's data library at the following URL: http://mba.tuck.dartmouth.edu/pages/faculty/ken.french/data_library.html. Specifically, the two files required are called 'Fama/French Factors' and '25 Portfolios Formed on Size and Book-to-Market (5 x 5)'.

15.3 Time-varying Fama–French model

The standard Fama–French model for the excess return on the ith portfolio, r_i, at time t, for $t = 1, 2, \ldots, n$, is

$$r_{i,t} = \beta_{i,0} + \text{MKT}_t\beta_{i,1} + \text{SMB}_t\beta_{i,2} + \text{HML}_t\beta_{i,3} + \varepsilon_{i,t}$$

where $\beta_{i,0}$ is a constant, MKT_t is the excess return on the market portfolio at time t, that is the return on the market portfolio less the return on a risk-free asset, SMB_t is the return on a zero-cost portfolio of small minus big stocks at time t, HML_t is the return on a zero-cost portfolio of high minus low book-to-market stocks at time t, $\beta_{i,1}$, $\beta_{i,2}$ and $\beta_{i,3}$ are the corresponding regression coefficients and $\varepsilon_{i,t}$ is a normally distributed random variable, with mean 0 and an unknown variance. If we define $\mathbf{x}_t^T = [1, \text{MKT}_t, \text{SMB}_t, \text{HML}_t]$ and denote the $(p \times 1)$ vector of portfolio returns at time t as $\mathbf{r}_t = \left(r_{1,t}, r_{2,t}, \ldots, r_{p,t}\right)^T$, then the system of equations that represent the

Fama–French model may be represented as follows:

$$\boldsymbol{r}_t = \left(\boldsymbol{x}_t^T \otimes \boldsymbol{I}_p\right) \boldsymbol{\beta} + \boldsymbol{\varepsilon}_t, \quad \boldsymbol{\varepsilon}_t \sim \mathrm{N}\left(\mathbf{0}, \boldsymbol{\Sigma}_\varepsilon^{-1}\right), \tag{15.1}$$

where $\boldsymbol{I}_p$ is a $(p \times p)$ identity matrix,

$$\boldsymbol{\beta} = \left(\beta_{1,0}, \beta_{2,0}, \ldots, \beta_{p,0}\beta_{1,1}, \beta_{2,1}, \ldots, \beta_{p,1}, \ldots, \beta_{p,3}\right)^T$$

and $\boldsymbol{\varepsilon}_t$ is an independent and identically distributed random variable, which follows a multivariate normal distribution with a mean vector $\mathbf{0}$ and a precision matrix $\boldsymbol{\Sigma}_\varepsilon$. The time-varying Fama–French model generalizes Equation (15.1), so that the regression coefficients vary across time, for $t = 1, 2, \ldots, n$, as follows:

$$\boldsymbol{r}_t = \left(\boldsymbol{x}_t^T \otimes \boldsymbol{I}_p\right) \boldsymbol{\beta}_t + \boldsymbol{\varepsilon}_t, \quad \boldsymbol{\varepsilon}_t \sim \mathrm{N}\left(\mathbf{0}, \boldsymbol{\Sigma}_\varepsilon^{-1}\right), \tag{15.2}$$

where the evolution of $\boldsymbol{\beta}_t$ is most easily described by defining $\boldsymbol{\beta}_t = \boldsymbol{R}\boldsymbol{\alpha}_t$, where $\boldsymbol{R}$ is a $(k \times m)$ selection matrix and for $t = 1, 2, \ldots, n-1$,

$$\boldsymbol{\alpha}_{t+1} = \boldsymbol{T}_t\boldsymbol{\alpha}_t + \boldsymbol{G}_t\boldsymbol{\eta}_t, \quad \boldsymbol{\eta}_t \sim \mathrm{N}\left(0, \boldsymbol{\Sigma}_\alpha^{-1}\right). \tag{15.3}$$

The $(m \times m)$ matrix $\boldsymbol{T}_t$ is typically referred to as the transition matrix, $\boldsymbol{G}_t$ is an $(m \times r)$ system matrix and the $(r \times 1)$ vector $\boldsymbol{\eta}_t$ is a normally distributed random variable with a mean vector $\mathbf{0}$ and an $(r \times r)$ precision matrix $\boldsymbol{\Sigma}_\alpha$. The matrices $\boldsymbol{T}_t$ and $\boldsymbol{G}_t$ can possibly depend on both known and unknown parameters. The model is completed through the specification of $\boldsymbol{\alpha}_1$, with

$$\boldsymbol{\alpha}_1 \sim \mathrm{N}(\boldsymbol{a}_1, \boldsymbol{P}_1) \tag{15.4}$$

where both the $(m \times 1)$ mean vector $\boldsymbol{a}_1$ and the $(m \times m)$ covariance matrix $\boldsymbol{P}_1$ are assumed known. For the following sections and for notational convenience, for $t > s$, we define $\boldsymbol{r}_{s:t} = \left(\boldsymbol{r}_s^T, \boldsymbol{r}_{s+1}^T, \ldots, \boldsymbol{r}_n^T\right)$. This notation extends to any other vector, for instance $\boldsymbol{\beta}_{s:t} = (\boldsymbol{\beta}_s, \boldsymbol{\beta}_{s+1}, \ldots, \boldsymbol{\beta}_t)$.

15.3.1 Specific models under consideration

The time-varying Fama–French model represented by Equation (15.2), Equation (15.3) and Equation (15.4) nests many specific variations of the model. For the analysis, it is assumed that $\boldsymbol{\beta}_t$ follows a multivariate random walk, such that $\boldsymbol{T}_t = \boldsymbol{I}_m$, $\boldsymbol{G}_t = \boldsymbol{I}_m$ and $\boldsymbol{R} = \boldsymbol{I}_m$. Three particular variations of the model are considered.

Model 1

Model 1 is specified such that the individual time series are contemporaneously uncorrelated. This is achieved by defining $\boldsymbol{\Sigma}_\varepsilon$ and $\boldsymbol{\Sigma}_\eta$ as diagonal matrices, such that $\boldsymbol{\Sigma}_\varepsilon = diag\left(\kappa_1, \kappa_2, \ldots, \kappa_p\right)$ and $\boldsymbol{\Sigma}_\alpha = diag\left(\tau_1, \tau_2, \ldots, \tau_m\right)$.

Model 2

Model 2 relaxes the assumption that the individual time series are contemporaneously uncorrelated; however, it retains the assumption that the time-varying regressors evolve independently of one another. That is, $\boldsymbol{\Sigma}_\varepsilon$ is assumed to be a dense precision matrix, while $\boldsymbol{\Sigma}_\alpha$ is still assumed to be diagonal.

Model 3

Model 3 allows for contemporaneous correlation across the individual time series and also allows for a degree of correlation in the evolution of the time-varying regressors. In particular, as for Model 2, $\boldsymbol{\Sigma}_\varepsilon$ is assumed to be a dense precision matrix; however, $\boldsymbol{\Sigma}_\beta$ is assumed to be block diagonal, such that $\boldsymbol{\Sigma}_\alpha = \text{block diagonal}\left(\boldsymbol{\Sigma}_{\alpha,0}, \boldsymbol{\Sigma}_{\alpha,1}, \boldsymbol{\Sigma}_{\alpha,2}, \boldsymbol{\Sigma}_{\alpha,3}\right)$, where $\boldsymbol{\Sigma}_{\alpha,i}$, for $i = 1, 2, 3, 4$, are dense $(p \times p)$ precision matrices.

15.4 Bayesian estimation

Bayesian inference aims to summarize uncertainty about the models of interest through the posterior distribution. For the time-varying Fama–French model, the posterior probability density function (pdf) of interest is

$$p\left(\boldsymbol{\beta}_{1:n}, \boldsymbol{\Sigma}_\varepsilon, \boldsymbol{\Sigma}_\alpha | \boldsymbol{r}_{1:n}, \boldsymbol{x}_{1:n}\right) \propto p\left(\boldsymbol{r}_{1:n} | \boldsymbol{\beta}_{1:n}, \boldsymbol{\Sigma}_\varepsilon, \boldsymbol{\Sigma}_\alpha, \boldsymbol{x}_{1:n}\right) \times p\left(\boldsymbol{\beta}_{1:n} | \boldsymbol{\Sigma}_\varepsilon, \boldsymbol{\Sigma}_\alpha, \boldsymbol{x}_{1:n}\right) \times p\left(\boldsymbol{\Sigma}_\varepsilon\right) \times p\left(\boldsymbol{\Sigma}_\alpha\right)$$

where $p\left(\boldsymbol{r}_{1:n} | \boldsymbol{\beta}_{1:n}, \boldsymbol{\Sigma}_\varepsilon, \boldsymbol{\Sigma}_\alpha, \boldsymbol{x}_{1:n}\right)$ is the joint pdf for $r_{1:n}$ conditional on $\boldsymbol{\beta}_{1:n}$, $\boldsymbol{\Sigma}_\varepsilon$, $\boldsymbol{\Sigma}_\beta$ and $\boldsymbol{x}_{1:n}$, $p\left(\boldsymbol{\beta}_{1:n} | \boldsymbol{\Sigma}_\varepsilon, \boldsymbol{\Sigma}_\alpha, \boldsymbol{x}_{1:n}\right)$ is the joint pdf for $\boldsymbol{\beta}_{1:n}$ conditional on $\boldsymbol{\Sigma}_\varepsilon$, $\boldsymbol{\Sigma}_\alpha$ and $\boldsymbol{x}_{1:n}$, $p\left(\boldsymbol{\Sigma}_\varepsilon\right)$ and $p\left(\boldsymbol{\Sigma}_\alpha\right)$ are the prior pdfs for $\boldsymbol{\Sigma}_\varepsilon$ and $\boldsymbol{\Sigma}_\alpha$, respectively. From Equation (15.3) it is clear that

$$p\left(\boldsymbol{r}_{1:n} | \boldsymbol{\beta}_{1:n}, \boldsymbol{\Sigma}_\varepsilon, \boldsymbol{\Sigma}_\alpha, \boldsymbol{x}_{1:n}\right) \propto \exp\left\{-\frac{1}{2}\sum_{t=1}^{n}\left(\boldsymbol{r}_t - \boldsymbol{X}_t\boldsymbol{\beta}_t\right)^T \boldsymbol{\Sigma}_\varepsilon^{-1}\left(\boldsymbol{r}_t - \boldsymbol{X}_t\boldsymbol{\beta}_t\right)\right\} \tag{15.5}$$

and

$$p\left(\boldsymbol{\beta}_{1:n} | \boldsymbol{\Sigma}, \boldsymbol{\Sigma}_\alpha, \boldsymbol{x}_{1:n}\right) \propto p\left(\boldsymbol{\beta}_1 | \boldsymbol{\Sigma}_\varepsilon, \boldsymbol{\Sigma}_\alpha, \boldsymbol{x}_{1:n}\right) \times \prod_{t=1}^{n-1} p\left(\boldsymbol{\beta}_{t+1} | \boldsymbol{\beta}_t, \boldsymbol{\Sigma}_\varepsilon, \boldsymbol{\Sigma}_\alpha, \boldsymbol{x}_{1:n}\right).$$

15.4.1 Gibbs sampler

The Gibbs sampler of Gelfand and Smith (1990) is implemented to facilitate estimation. At iteration j the Gibbs scheme is as follows in Algorithm 12.

Algorithm 12: Gibbs sampler at iteration j

1. Sample $\boldsymbol{\Sigma}_\varepsilon^{(j)}$ from $p\left(\boldsymbol{\Sigma}_\varepsilon | \boldsymbol{\beta}_{1:n}^{(j-1)}, \boldsymbol{\Sigma}_\alpha^{(j-1)}, \boldsymbol{r}_{1:n}, \boldsymbol{x}_{1:n}\right)$.
2. Sample $\boldsymbol{\beta}_{1:n}^{(j)}$ from $p\left(\boldsymbol{\beta}_{1:n} | \boldsymbol{\Sigma}_\varepsilon^{(j)}, \boldsymbol{\Sigma}_\alpha^{(j-1)}, \boldsymbol{r}_{1:n}, \boldsymbol{x}_{1:n}\right)$.
3. Sample $\boldsymbol{\Sigma}_\alpha^{(j)}$ from $p\left(\boldsymbol{\Sigma}_\beta | \boldsymbol{\beta}_{1:n}^{(j-1)}, \boldsymbol{\Sigma}_\varepsilon^{(j)}, \boldsymbol{r}_{1:n}, \boldsymbol{x}_{1:n}\right)$.

Sampling from the posteriors of $\boldsymbol{\Sigma}_\varepsilon$, $\boldsymbol{\beta}_{1:n}$ and $\boldsymbol{\Sigma}_\alpha$, for each of Models 1, 2 and 3, is described in the following subsections.

15.4.2 Sampling $\boldsymbol{\Sigma}_\varepsilon$

Model 1

Sampling from the posterior distribution of $\boldsymbol{\Sigma}_\varepsilon$ is achieved by taking advantage of its diagonal structure. Assuming that a priori κ_i, for $i = 1, 2, \ldots, p$, follows the gamma distribution, then

$$\kappa_i \sim \mathrm{G}\left(\frac{\underline{\nu}}{2}, \frac{\underline{s}}{2}\right), \qquad (15.6)$$

where $\underline{\nu}$ and $\underline{s}$ are prior hyperparameters. Given Equation (15.5), it is clear that the posterior distribution for κ_i is given by

$$\kappa_i | \boldsymbol{r}_{1:n}, \boldsymbol{\beta}_{1:n}, \boldsymbol{\Sigma}_\varepsilon, \boldsymbol{\Sigma}_\alpha, \boldsymbol{x}_{1:n} \sim \mathrm{G}\left(\frac{\overline{\nu}}{2}, \frac{\overline{s}}{2}\right),$$

where $\overline{\nu} = \underline{\nu} + n$ and $\overline{s} = \underline{s} + \sum_{i=1}^{n} e_{i,t}^2$, with $\boldsymbol{e}_t = \boldsymbol{r}_t - \left(\boldsymbol{x}_t^T \otimes \boldsymbol{I}_p\right) \boldsymbol{\beta}_t$, for $t = 1, 2, \ldots, n$, and $e_{i,t}$ is the ith element of $\boldsymbol{e}_t$.

Models 2 and 3

Assuming a priori that $\boldsymbol{\Sigma}_\varepsilon$ follows a Wishart distribution, then

$$\boldsymbol{\Sigma}_\varepsilon \sim \mathrm{Wishart}\left(\underline{\boldsymbol{W}}_\varepsilon^{-1}, \underline{\nu}_\varepsilon\right), \qquad (15.7)$$

where $\underline{\boldsymbol{W}}_\varepsilon$ and $\underline{\nu}_\varepsilon$ are the scale matrix and degrees of freedom parameter, respectively. Given Equation (15.7) and Equation (15.5), the posterior distribution for $\boldsymbol{\Sigma}_\varepsilon$ is given by

$$\boldsymbol{\Sigma}_\varepsilon | \boldsymbol{\beta}_{1:n}, \boldsymbol{\Sigma}_\alpha, \boldsymbol{r}_{1:n}, \boldsymbol{x}_{1:n} \sim \mathrm{Wishart}\left(\overline{\boldsymbol{W}}_\varepsilon^{-1}, \overline{\nu}_\varepsilon\right),$$

where

$$\overline{\boldsymbol{W}}_\varepsilon^{-1} = \left[\underline{\boldsymbol{W}}_\varepsilon^{-1} + \sum_{t=1}^{n} \boldsymbol{e}_t \boldsymbol{e}_t^T\right]$$

and

$$\overline{\nu}_\varepsilon = \underline{\nu}_\varepsilon + n.$$

15.4.3 Sampling $\boldsymbol{\beta}_{1:n}$

Sampling $\boldsymbol{\beta}_{1:n}$ from its full conditional posterior distribution,

$$p\left(\boldsymbol{\beta}_{1:n} | \boldsymbol{\Sigma}_\varepsilon, \boldsymbol{\Sigma}_\alpha, \boldsymbol{r}_{1:n}, \boldsymbol{x}_{1:n}\right),$$

is straightforward using tools that have been developed for SSMs. Specifically, any of the simulation filtering algorithms developed by Früwirth-Schnatter (1994), Carter and Kohn (1994), Koopman and Durbin (2002), de Jong and Shephard (1995) and

Strickland *et al.* (2009) can be applied. However, for multivariate state space models, the algorithm of Strickland *et al.* (2009) is in general the most efficient and as such is employed in this study. To use the algorithm of Strickland *et al.* (2009), the observation equation, Equation (15.2), needs to be transformed so that the errors are contemporaneously uncorrelated. A convenient method to achieve this is to pre-multiply the observation equation by $\boldsymbol{U}$, where $\boldsymbol{U}$ is the upper Cholesky triangle from the Cholesky factorization of $\boldsymbol{\Sigma}_\varepsilon$. Denoting $\boldsymbol{y}_t = \boldsymbol{U}\boldsymbol{r}_t$, $\boldsymbol{Z}_t = \boldsymbol{U}\left(\boldsymbol{x}_t^T \otimes \boldsymbol{I}_p\right)\boldsymbol{R}$ and $\boldsymbol{e}_t = \boldsymbol{U}\boldsymbol{\varepsilon}_t$, then the ith transformed observation equation, at time t, is given by

$$y_{t,i} = z_{t,i}^T \boldsymbol{\alpha}_t + e_{t,i}, \quad e_{t,i} \sim \mathrm{N}(0, 1), \tag{15.8}$$

where $y_{t,i}$ is the ith element of $\boldsymbol{y}_t$, $z_{t,i}^T$ is the ith row of $\boldsymbol{Z}_t$ and $e_{t,i}$ is the ith element of $\boldsymbol{e}_t$ and is independently normally distributed with a mean of 0 and a variance of 1. Using the transformed measurement in Equation (15.8) and the system matrices in the transition Equation (15.3), the simulation smoother of Strickland *et al.* (2009) can then be used to sample $\boldsymbol{\alpha}_{1:n}$ from its posterior distribution. Given $\boldsymbol{\alpha}_{1:n}$, $\boldsymbol{\beta}_{1:n}$ can be constructed using the identity $\boldsymbol{\beta}_t = \boldsymbol{G}\boldsymbol{\alpha}_t$, for $t = 1, 2, \ldots, n$. The approach proposed by Strickland *et al.* (2009) combines the filtering and smoothing algorithms for the univariate representation of the SSM of Anderson and Moore (1979) and Koopman and Durbin (2000) with the simulation smoother of Koopman and Durbin (2002) to efficiently sample $\boldsymbol{\alpha}_{1:n}$. The filtering equations can be summarized as in Algorithm 13.

Algorithm 13: Filtering equations

Initialize $a_{11} = a_1$ and $P_{1,1} = P_1$, then for $t = 1, 2, \ldots, n$ and for $i = 1, 2, \ldots, p$, compute

$$\begin{aligned}
v_{t,i} &= y_{t,i} - z_{t,i}^T \boldsymbol{a}_{t,i} \\
\boldsymbol{m}_{t,i} &= \boldsymbol{P}_{t,i} z_{t,i} \\
f_{t,i} &= z_{t,i}^T \boldsymbol{m}_{t,i} + 1 \\
\boldsymbol{k}_{t,i} &= \boldsymbol{m}_{t,i} / f_{t,i} \\
\boldsymbol{a}_{t,i+1} &= \boldsymbol{a}_{t,i} + \boldsymbol{k}_{t,i} v_{t,i} \\
\boldsymbol{P}_{t,i+1} &= \boldsymbol{P}_{t,i} - \boldsymbol{k}_{t,i} \boldsymbol{m}_{t,i}^T,
\end{aligned}$$

where $v_{t,i}$ and $f_{t,i}$ are scalars, $\boldsymbol{m}_{t,i}$, $\boldsymbol{k}_{t,i}$ and $\boldsymbol{a}_{t,i}$ are $(m \times 1)$ vectors and where $\boldsymbol{P}_{t,i}$ is an $(m \times m)$ matrix. At each time period t, for $t = 1, 2, \ldots, n$, calculate

$$\begin{aligned}
\boldsymbol{a}_{t+1} &= \boldsymbol{T}_t \boldsymbol{a}_{t,p+1} \\
\boldsymbol{P}_{t+1} &= \boldsymbol{T}_t \boldsymbol{P}_{t,p+1} \boldsymbol{T}_t^T + \boldsymbol{G}_t \boldsymbol{\Sigma}_\alpha^{-1} \boldsymbol{G}_t.
\end{aligned}$$

The smoothing algorithm requires $v_{t,i}$, $f_{t,i}$, $k_{t,i}$, $\boldsymbol{a}_{t,1}$ and $\boldsymbol{P}_{t,i}$ to be stored from the filtering pass described in Algorithm 13. The algorithms for smoothing are described as in Algorithm 14.

Algorithm 14: Smoothing equations

Initialize the $(m \times 1)$ vector $r_{t,i} = \mathbf{0}$, then for $t = n, n-1, \ldots, 1$ and for $i = p, p-1, \ldots, 1$, calculate

$$\boldsymbol{r}_{t,i} = z_{t,i}/f_{t,i}\nu_{t,i} + \boldsymbol{r}_{t,i} - z_{t,i}\left(\boldsymbol{k}_{t,i}^T \boldsymbol{r}_{t,i}\right).$$

At each time period, t, calculate

$$\hat{\boldsymbol{a}}_t = \boldsymbol{a}_{t,i} + \boldsymbol{P}_{t,i}\boldsymbol{r}_{t,0}$$
$$\boldsymbol{r}_{t-1,p} = \boldsymbol{T}_{t-1}^T \boldsymbol{r}_{t,0}.$$

The smoothing equations in Algorithm 14 are a rearrangement of the smoothing equations in Strickland *et al.* (2009). In particular, a small modification is made in calculating $\boldsymbol{r}_{t,i}$, which leads to a substantial improvement in computational efficiency when the dimension of $\boldsymbol{\alpha}$, m, is large. The following simulation smoothing algorithm uses both Algorithms 13 and 14 to sample $\boldsymbol{\alpha}_{1:n}$ and consequently $\boldsymbol{\beta}_{1:n}$ from its full conditional posterior distribution.

Algorithm 15: Simulation smoothing algorithm

1. Sample $\boldsymbol{e}_t^+$, for $t = 1, 2, \ldots, n$, from $\mathrm{N}\left(\mathbf{0}, \boldsymbol{I}_p\right)$, where $\boldsymbol{I}_p$ is an order p identity matrix.
2. Sample $\boldsymbol{\eta}_t^+$, for $t = 1, 2, \ldots, n-1$, from $p\left(\eta_t|\boldsymbol{\Sigma}_\alpha\right)$.
3. Sample $\boldsymbol{\alpha}_1^+$ from $p\left(\boldsymbol{\alpha}_1|\boldsymbol{\theta}, \boldsymbol{\Sigma}_\alpha\right)$ and then generate $\boldsymbol{\alpha}_{2:n}^+$ and $\boldsymbol{y}_{1:n}^+$.
4. Calculate $\tilde{\boldsymbol{y}}_{1:n} = \boldsymbol{y}_{1:n} - \boldsymbol{y}_{1:n}^+$.
5. Use the filtering and smoothing equations in Algorithms 13 and 14 to calculate $\hat{\boldsymbol{a}}_{1:n}$.
6. Take $\boldsymbol{\alpha}_{1:n} = \hat{\boldsymbol{\alpha}}_{1:n} - \boldsymbol{\alpha}_{1:n}^+$ as a draw from

$$p\left(\boldsymbol{\alpha}_{1:n}|\boldsymbol{\Sigma}_\varepsilon^{(j)}, \boldsymbol{\Sigma}_\alpha^{(j-1)}, \boldsymbol{\theta}^{(j-1)}, \boldsymbol{r}_{1:n}, \boldsymbol{x}_{1:n}\right).$$

15.4.4 Sampling $\boldsymbol{\Sigma}_\alpha$

Models 1 and 2

Assuming a gamma prior for τ_i, for $i = 1, 2, \ldots, p$, such that

$$\tau_i \sim \mathrm{G}\left(\frac{\vartheta}{2}, \frac{\omega}{2}\right),$$

then given Equation (15.5), it follows that the posterior distribution, for $i = 1, 2, \ldots, p$, is

$$\tau_i|\boldsymbol{\beta}_{1:n}, \boldsymbol{\Sigma}_\varepsilon, \boldsymbol{r}_{1:n}, \boldsymbol{x}_{1:n} \sim \mathrm{G}\left(\frac{\overline{\vartheta}}{2}, \frac{\overline{\omega}_i}{2}\right),$$

where $\overline{\vartheta} = \underline{\vartheta} + n$ and $\overline{\omega}_i = \underline{\omega} + \sum_{t=1}^{n} w_{i,t}^2$, with $\boldsymbol{w}_t = \boldsymbol{\beta}_{t+1} - \boldsymbol{\beta}_t$ and $w_{i,t}$ is the ith element of $\boldsymbol{w}_t$.

Model 3

Assuming a Wishart prior for $\boldsymbol{\Sigma}_{\alpha,i}$, for $i = 1, 2, \ldots, 4$, such that

$$\boldsymbol{\Sigma}_{\alpha,i} \sim \text{Wishart}\left(\underline{\boldsymbol{W}}_{\alpha}^{-1}, \underline{\nu}_{\alpha}\right)$$

it follows that

$$\boldsymbol{\Sigma}_{\alpha,i}|\boldsymbol{\beta}_{1:n}, \boldsymbol{\Sigma}_{\varepsilon}, \boldsymbol{r}_{1:n}, \boldsymbol{x}_{1:n} \sim \text{Wishart}\left(\overline{\boldsymbol{W}}_{\alpha,i}^{-1}, \overline{\nu}_{\alpha}\right),$$

with

$$\overline{\nu}_{\alpha} = \underline{\nu}_{\alpha} + n$$

and

$$\overline{\boldsymbol{W}}_{\alpha,i}^{-1} = \underline{\boldsymbol{W}}_{\alpha}^{-1} + \sum_{t=1}^{n} \tilde{\boldsymbol{w}}_{i,t}\tilde{\boldsymbol{w}}_{i,t}^{T},$$

where $\boldsymbol{w}_t$, for $t = 1, 2, \ldots, n$, is partitioned into equal blocks such that $\boldsymbol{w}_t = \left(\tilde{\boldsymbol{w}}_1^T, \tilde{\boldsymbol{w}}_2^T, \tilde{\boldsymbol{w}}_3^T, \tilde{\boldsymbol{w}}_4^T\right)^T$ and $\tilde{\boldsymbol{w}}_{i,t}$ is the ith block.

15.4.5 Likelihood calculation

The comparison of the three models is conducted using the Bayesian information criterion (BIC). As such, it is necessary to compute the likelihood. The calculation uses Algorithm 13 to compute $\boldsymbol{\nu}_{t,i}$ and $f_{t,i}$, for $i = 1, 2, \ldots, p$ and for $t = 1, 2, \ldots, n$, which can be used to evaluate the likelihood function for the time-varying Fama–French model in Equation (15.2), Equation (15.3) and Equation (15.4). The likelihood function is defined as

$$\begin{aligned} p\left(\boldsymbol{r}_{1:n}|\boldsymbol{\Sigma}_{\varepsilon}, \boldsymbol{\Sigma}_{\beta}\right) &= \int p\left(\boldsymbol{r}_{1:n}|\boldsymbol{\Sigma}_{\varepsilon}, \boldsymbol{\beta}_{1:n}, \boldsymbol{\Sigma}_{\alpha}\right) p\left(\beta_{1:n}|\boldsymbol{\Sigma}_{\alpha}\right) d\boldsymbol{\beta}_{1:n} \\ &= (2\pi)^{np/2}\,|\boldsymbol{U}|^n \prod_{t=1}^{n}\prod_{i=1}^{p} f_{t,i}^{-1/2} \exp\left(-\frac{1}{2f_{t,i}}\nu_{t,i}^2\right), \end{aligned}$$

where the quantity $|\boldsymbol{U}|^n$ is required as a result of the transformation to the observation equation in Equation (15.8). As $\boldsymbol{U}$ has a diagonal structure for Model 1 and a triangular structure for Models 2 and 3, the determinant is particularly easy to calculate, and in particular $|\boldsymbol{U}| = \prod_{i=1}^{p} \boldsymbol{U}_{ii}$, where $\boldsymbol{U}_{ii}$ is the ith diagonal element of $\boldsymbol{U}$. Note that, in practice, typically the log-likelihood is of interest, in which case the corresponding calculation will be $\log\left(|\boldsymbol{U}|^n\right) = n\sum_{i=1}^{p} \log\left(\boldsymbol{U}_{ii}\right)$.

15.5 Analysis

15.5.1 Prior elicitation

Finance theory can be used in eliciting the values for $\boldsymbol{a}_1$ and $\boldsymbol{P}_1$. In particular, if the asset-pricing model correctly prices the test portfolio, then the constant should be zero. Thus, it is expected that the constant will be around zero. A coefficient on the market factor of one indicates that the test portfolio has the same level of systematic risk as the market. As the thousands of stocks in the US market have been aggregated into 25 portfolios, then it is reasonable to assume that the coefficient on the market factor for these portfolios will be around one. For portfolios comprising small (high book-to-market) stocks, it is expected that the coefficients on SMB (HML) will be positive, whereas for the portfolios containing large (low book-to-market) stocks, the expectation is that the coefficients on SMB (HML) will be negative. These expectations are based on the findings of prior research. However, it is difficult to translate these predictions into a precise estimate or even an estimated range for the purposes of prior elicitation. A key reason for this is that the Fama–French model is atheoretical and was derived from empirical observation, so there is no theory to guide the estimates of the coefficients on SMB and HML. In the absence of any theory, it is assumed that the coefficients on SMB and HML are zero. In contrast, the theoretically derived CAPM, of which the Fama–French model is an augmentation, suggests that the constant should be around zero and that the coefficient on the market factor should be around one. Assuming a priori that each of the time-varying regressors is uncorrelated and has a prior variance of 4, it follows that

$$\boldsymbol{a}_1 = [0, 1, 0, 0]^T \otimes \boldsymbol{i}_p \quad textand \quad \boldsymbol{P}_1 = I_m \times 4,$$

where $\boldsymbol{i}_p$ is a $(p \times 1)$ vector of ones. A vague prior is assumed for $\boldsymbol{\Sigma}_\varepsilon$ and $\boldsymbol{\Sigma}_\alpha$, such that $\underline{\boldsymbol{W}_\varepsilon}^{-1} = \underline{\boldsymbol{W}_\alpha}^{-1} = I_p \times 0.01$, and for each of κ_i, for $i = 1, 2, \ldots, m$, and τ_j, for $j = 1, 2, \ldots, p$, a flat prior over the positive domain of the real-number line.

15.5.2 Estimation output

The analysis of Models 1, 2 and 3 is undertaken using 20 000 iterations of the MCMC sampler, described in Algorithm 12, where the first 5000 iterations are discarded as burn-in.

Table 15.1 contains the estimation output from the MCMC analysis, and in particular reports the Bayesian information criterion (BIC). It is clear that Model 2 is

Table 15.1 Estimation results from the MCMC analysis of the Fama–French model with time-varying regressors, based on 20 000 iterations, with the first 5000 discarded.

Model	1	2	3
BIC	−73 552	−76 608	−71 505

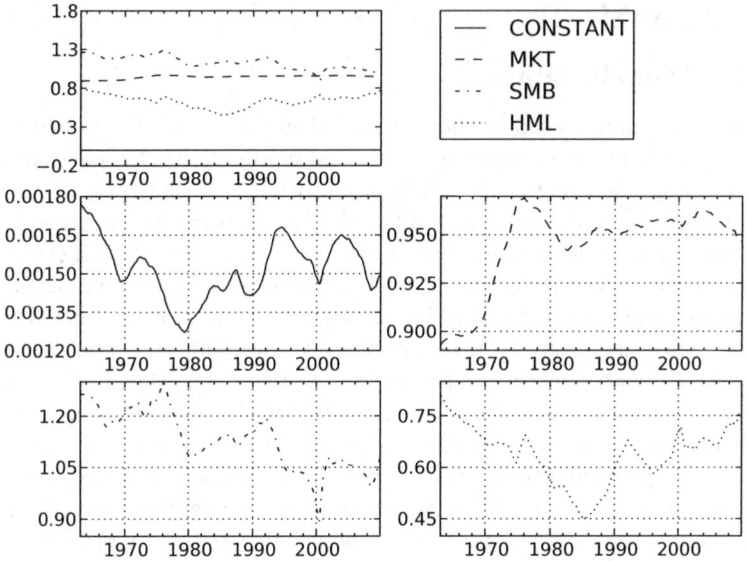

Figure 15.1 Plots of the marginal posterior mean estimates for the S1B5 portfolio, for the time-varying regressors for the intercept, MKT, SMB and HML.

the preferred model. The preference of Model 2 over Model 1 justifies the additional complexity brought about by analysing the time-varying Fama–French model as a system. The BIC indicates that the performance of Model 3 is relatively poor and the simpler specification in Model 2 should be preferred.

Figure 15.1 is produced from the MCMC analysis of Model 2 for the S1B5 portfolio. Specifically, it contains plots of the marginal posterior mean estimates for the time-varying regressors. The plot in the top left corner contains each of the four time-varying regressors, corresponding to the constant, MKT, SMB and HML, while they are plotted individually in the bottom four plots.

Plots for three of the 25 test portfolios are presented for discussion purposes and to demonstrate the time variation in the regressors. The first portfolio chosen, S1B5, contains the smallest stocks (S1) and the stocks with the highest book-to-market ratio (B5). Small stocks and high book-to-market stocks have historically been the best performers, so the S1B5 portfolio, which combines both small and high book-to-market stocks, has historically had a high return. The plots show that there is time variation in the parameters on all regressors, but particularly for the parameters on SMB and HML. The positive constant over the entire sample period indicates that the S1B5 portfolio has outperformed relative to the time-varying Fama–French model. This is expected because, as previously discussed, the S1B5 portfolio is known as a strong performer. The coefficient on MKT is always below one. This is surprising because, as well as being a strong performer, S1B5 is also one of the riskiest portfolios. This is because smaller stocks and high book-to-market stocks are typically riskier. Thus, one would expect this portfolio to have a systematic risk greater than one and

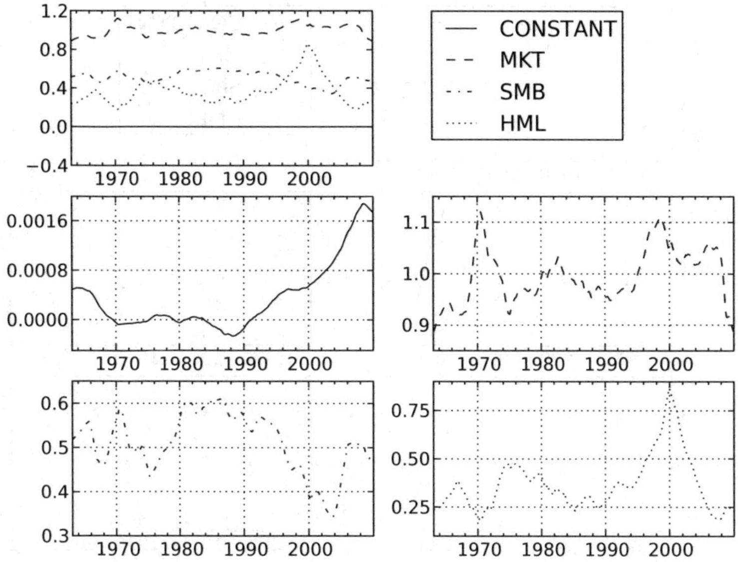

Figure 15.2 Plots of the marginal posterior mean estimates for the S3B3 portfolio, for the time-varying regressors for the intercept, MKT, SMB and HML.

thus a systematic risk greater than the overall market. The coefficient on SMB (HML) is positive, as expected, because smaller (higher book-to-market) stocks invariably have positive coefficients on SMB (HML).

Figure 15.2 is produced from the MCMC analysis of Model 2 for the S3B3 portfolio. Similar to Figure 15.1, it contains plots of the marginal posterior mean estimates for the time-varying regressors. All four regressors, corresponding to the constant, MKT, SMB and HML, are jointly plotted in the top left hand corner and individually plotted in the bottom four plots.

The S3B3 portfolio contains stocks of medium size (S3) and a medium level of book-to-market (B3). There is a fair amount of variation in the parameters on MKT, SMB and HML, although it appears that the HML coefficient has the greatest variability. The coefficient on the MKT factor fluctuates around one as expected. The coefficients on SMB and HML are consistently positive but are generally lower than the corresponding coefficients in Figure 15.1, which is expected considering they are relatively larger firms with a relatively lower book-to-market compared with the firms in S1B5. The constant fluctuates from positive to negative indicating that there are times when the S3B3 portfolio outperforms and underperforms relative to the Fama–French model. A positive (negative) coefficient indicates that the actual return on the portfolio is higher (lower) than the expected return from the Fama–French model and thus that the portfolio outperforms (underperforms) relative to the model.

Figure 15.3 is produced from the MCMC analysis of Model 2 for the S5B1 portfolio. It contains plots of the marginal posterior mean estimates for the

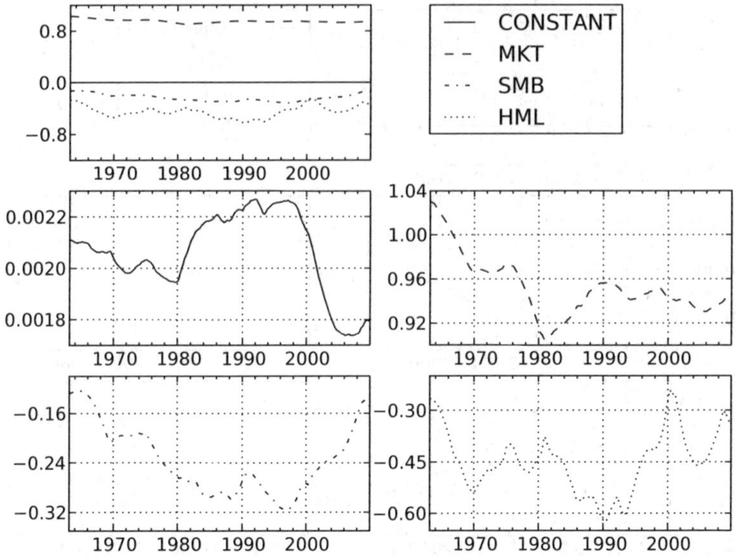

Figure 15.3 Plots of the marginal posterior mean estimates for the S5B1 portfolio, for the time-varying regressors for the intercept, MKT, SMB and HML.

time-varying regressors. The four regressors are jointly plotted in the top left hand corner plot and individually plotted in the bottom four plots.

The S5B1 portfolio contains the largest stocks (S5) and the stocks with the lowest book-to-market ratio (B1). The S5B1 portfolio is at the other end of the spectrum to S1B5. S5B1 (S1B5) firms are the largest (smallest) firms with the most (least) promising future growth prospects. S5B1 firms are known as large stable firms and this is reflected in their coefficient estimates, which are much less variable over time in comparison with the other two portfolios analysed. The constant is always positive indicating that the S5B1 portfolio has outperformed relative to the Fama–French model over the entire sample period. The MKT coefficient fluctuates around one whereas the SMB and HML coefficients are consistently negative. The negative coefficient on SMB (HML) is expected because large (low book-to-market) firms typically have negative coefficients on SMB (HML).

15.6 Conclusion

This chapter details an approach for the Bayesian analysis of multivariate SSMs. The approach is used to undertake a system analysis of the Fama–French model with time-varying regressors. The study considered three specific variants of the model. The preferred model did not ignore the possibility of contemporaneous correlation between the series and as such justified the use of a system-based method over an approach that models the time series individually. The multivariate Bayesian framework presented nests many more variants of the Fama–French model, which are possibly attractive alternatives to the ones examined in this chapter. In particular, it may be worth considering variants where the time variation in the regressors is derived from a stationary process.

References

Anderson BDO and Moore JB 1979 *Optimal Filtering*. Prentice Hall, Englewood Cliffs, NJ.

Ang A and Chen J 2007 CAPM over the long run: 1926-2001. *Journal of Empirical Finance* **14**, 1–40.

Avramov D and Chordia T 2006 Asset pricing models and financial market anomalies. *Review of Financial Studies* **19**, 1001–1040.

Banz RW 1981 The relationship between return and market value of common stocks. *Journal of Financial Economics* **9**, 3–18.

Black F 1972 Capital market equilibrium with restricted borrowing. *Journal of Business* **45**, 444–455.

Carter C and Kohn R 1994 On Gibbs sampling for state space models. *Biometrika* **81**, 541–553.

de Jong P and Shephard N 1995 The simulation smoother for time series models. *Biometrika* **82**, 339–350.

Fama EF and French KR 1993 Common risk factors in the returns on stocks and bonds. *Journal of Financial Economics* **33**, 3–56.

Fama EF and French KR 1996 Multifactor explanations of asset pricing anomalies. *Journal of Finance* **51**, 55–84.

Ferson WE, Kandel S and Stambaugh RF 1987 Tests of asset pricing with time-varying expected risk premiums and market betas. *Journal of Finance* **42**, 201–220.

Früwirth-Schnatter S 1994 Data augmentation and dynamic linear models. *Journal of Time Series Analysis* **15**, 183–202.

Gelfand AE and Smith AFM 1990 Sampling-based approaches to calculating marginal densities. *Journal of the American Statistical Association* **85**, 398–409.

Gibbons MR, Ross SA and Shanken J 1989 A test of the efficiency of a given portfolio. *Econometrica* **57**, 1121–1152.

Hansen LP and Jagannathan R 1997 Assessing specification errors in stochastic discount factor models. *Journal of Finance* **52**, 557–590.

Harvey AC 1989 *Forecasting Structural Time Series and the Kalman Filter*. Cambridge University Press, Cambridge.

Jostova G and Philipov A 2005 Bayesian analysis of stochastic betas. *Journal of Financial and Quantitative Analysis* **40**, 747–778.

Kalman RE 1960 A new approach to linear filtering and prediction problems. *Transactions of the ASME. Series D, Journal of Basic Engineering* **83**, 95–108.

Koopman SJ and Durbin J 2000 Fast filtering and smoothing for multivariate state space models. *Journal of Time Series Analysis* **21**, 281–296.

Koopman SJ and Durbin J 2002 Simple and efficient simulation smoother for times series models. *Biometrika* **81**, 341–353.

Lewellen J and Nagel S 2006 The conditional CAPM does not explain asset-pricing anomalies. *Journal of Financial Economics* **82**, 289–314.

Lintner J 1965 The valuation of risk assets and the selection of risky investments in stock portfolios and capital budgets. *Review of Economics and Statistics* **47**, 13–37.

MacKinlay AC and Richardson MP 1991 Using generalized method of moments to test mean-variance efficiency. *Journal of Finance* **46**, 511–527.

Sharpe WF 1964 Capital asset prices: a theory of market equilibrium under conditions of risk. *Journal of Finance* **19**, 425–442.

Stattman D 1980 Book values and stock returns. *The Chicago MBA: A Journal of Selected Papers* **4**, 25–45.

Strickland CM, Turner IW, Denham R and Mengersen KL 2009 Efficient Bayesian estimation for multivariate state space models. *Computational Statistics & Data Analysis* **53**, 4116–4125.

16

Bayesian mixture models: When the thing you need to know is the thing you cannot measure

Clair L. Alston[1], Kerrie L. Mengersen[1] and Graham E. Gardner[2]

[1]*Queensland University of Technology, Brisbane, Australia*
[2]*Murdoch University, Perth, Australia*

16.1 Introduction

There are many situations in which the data analyst may find finite mixture models helpful. In general, they are used when it is thought that there may be more than one population present in the data set, and either by oversight, or more usually as a result of being unable to identify subpopulations, the data may contain a mixture of measurements from several populations. Alternatively, they may be used in a semi-parametric density estimation setting.

This concept was first expounded in Pearson (1894), who fitted a mixture of two normal distributions to a data set which contained the forehead to body length ratio measurements of sampled crabs residing in the Bay of Naples. The skewness of the data prompted speculation of species evolution delivering two subpopulations, with mixtures being used as a confirmation (McLachlan and Peel 2000, pp. 2–3).

Case Studies in Bayesian Statistical Modelling and Analysis, First Edition. Edited by Clair L. Alston, Kerrie L. Mengersen and Anthony N. Pettitt.

The method of moments proposed by Pearson (1894) was computationally demanding and as a result mixture modelling was rarely used for data analysis until Dempster *et al.* (1977) produced their seminal paper using the EM algorithm, and illustrated its ability to simplify the application of maximum likelihood techniques to incomplete data settings. This approach had a direct application to the classic mixture model, the missing data being the unobserved component membership, which we discuss in Section 16.3.1. The likelihood could be simplified through using the expectation (E) step to estimate the unobserved components and then substituting these values into the maximization (M) step to obtain maximum likelihood estimates for the unknown mixture parameters. Bayesian mixture models began appearing in the literature in the early 1990s, with West (1992) and Diebolt and Robert (1994) being among the first published efforts which incorporated the missing data structure of mixture modelling with the use of Bayesian techniques. Bayesian mixture modelling is now routinely used in data analysis due to the natural setting for hierarchical models within the Bayesian framework, the availability of high-powered computing and developments in posterior simulation techniques, such as Markov chain Monte Carlo (MCMC).

The finite mixture model is now a staple in the modern data analyst's toolkit. It can be adapted to situations such as density determination (in which components themselves are not specifically of interest and instead the mixture represents a very flexible semi-parametric model useful for estimation, prediction and inference), clustering and other latent class analysis (in which the components are important and interpretable in their own right) and regression (e.g. variable selection).

In this chapter we introduce the finite mixture model as a method of density determination in computed tomography (CT) image analysis (Alston *et al.* 2004, 2005), then illustrate the use of spatial information using a Markov random field. Other work by the authors in this area include the use of the Bayesian mixture model as a clustering tool in CT scanning to allow volume calculation of whole carcases (Alston *et al.* 2009), which we will discuss briefly in Section 16.4.3.

16.2 Case study: CT scan images of sheep

The use of CT scan technology has been incorporated into animal studies in countries such as Australia, New Zealand, the UK and Norway since the early 1990s (Thompson and Kinghorn 1992). This technology allows the researcher to study the composition of tissue in a carcase without the need to slaughter live animals, hence allowing longitudinal observations, as well as being able to facilitate carcase assessment without the need for costly and wasteful dissections.

CT images are often reduced from their original measurement units to a 256 greyscale representation to be viewed on a screen. Often, these images may seem to clearly define tissue groupings. However, scientists do not wish to view and manually dissect these images. Instead, they need a mathematical model to estimate tissue proportions and process multitudes of images using the original Hounsfield unit values of the pixels (range $\pm$ 1000).

As a first step to analysing such data, researchers (Ball *et al.* 1998; Brenøe and Kolstad 2000; Jopson APMIS. 1997; Kolstad 2001; Kvame and Vangen 2007) have

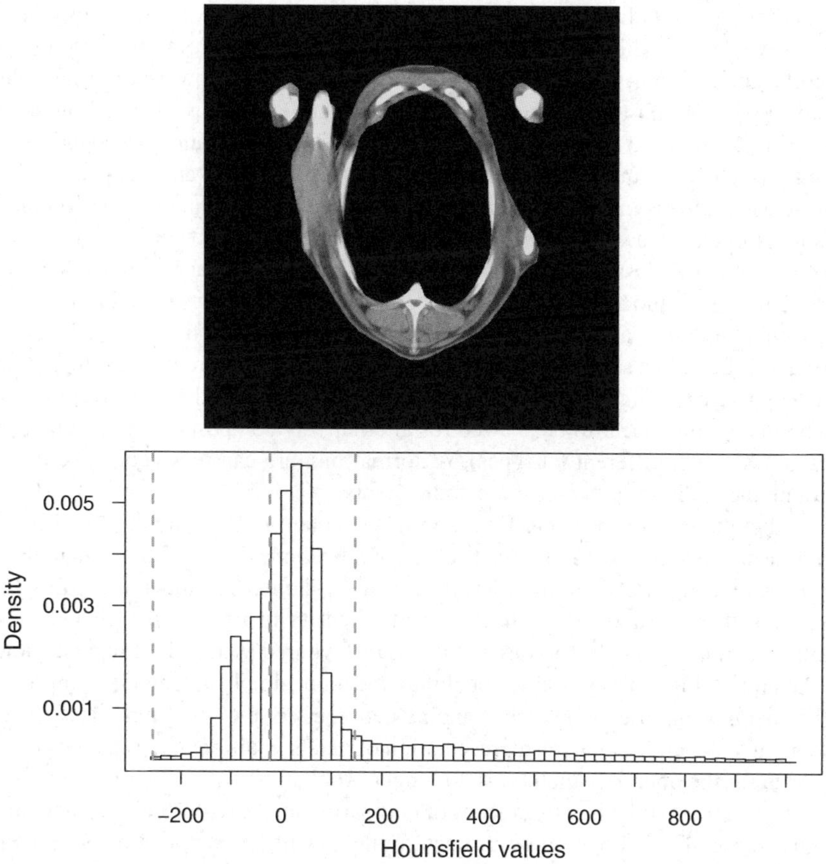

Figure 16.1 Top: CT scan from around the fourth to fifth thoracic vertebra region of a sheep. Dark grey regions are generally fat tissue, light grey regions are muscle, and white pixels represent bone. Bottom: Image represented by a histogram of pixel Hounsfield values. Dashed lines indicate boundaries commonly used in sheep studies (Kvame and Vangen 2007) to allocate pixels to fat ($y \leq -22$), muscle ($-22 < y \leq 146$) and bone ($y > 146$) components.

used a simple boundary technique, based on fat being less dense than muscle, which is considerably less dense than bone. This approach assumes that pixels within a specified range may be designated to one of these three tissues. If the tissue densities were clearly separated this would be a reasonable strategy; however, further research has shown that fat and muscle tissue are neither clearly separate nor consistent in their denseness between sheep or over a single carcase.

Figure 16.1 (top) illustrates a CT scan from around the fourth to fifth thoracic vertebra region of a sheep, with pixel values binned in order to obtain a 256 greyscale representation. The dark grey regions represent fat tissue, mid grey regions represent muscle, and white regions are general bone. It is natural at the first stage of analysis to view such data in histogram form (Figure 16.1, bottom). From this graph we can observe one minor peak in the region around -75, a distinct peak in the region around 50, and a tailing off of the data thereafter. From the known biology of the data it would seem likely that the first peak is associated with fat, the second with muscle, and the high-valued and spread-out data with bone. However, it is also clear from this histogram that there is not a distinct cutoff in terms of pixel values for any of these tissues. The histogram in Figure 16.1 also contains the boundary cut points which are frequently used by animal scientists (Kvame and Vangen 2007) to hard-classify pixels into tissue types for subsequent use in estimating the volume of tissue within a carcase. It can be observed that while this cutoff for bone may result in quite credible classifications, as large shifts in this boundary would affect very few actual pixel allocations, the boundary between fat and muscle is not clear and we would anticipate a high degree of mixing between the two tissues in this region; therefore, these data seem highly suitable to analysis by mixture models. In addition to the apparent mixing in the boundary regions, the densities of the tissues represented in the histogram appear to be skewed (or alternatively composed of subpopulations of the tissue, i.e. different fat types). A normal mixture can represent these tissues through the addition of two or more components.

Although we will focus our results in this chapter primarily on the analysis of a single scan, Alston *et al.* (2009) illustrate the application of this technique to multiple scans from a single carcase in order to calculate volume. As shown in Figures 16.2 and 16.3, these data clearly indicate that the density of tissue types changes over carcase regions, hence the mixture is even more favoured as a realistic representation of the underlying biology of the model than the more simplistic boundary approach. When using a mixture model, the changing denseness of the tissue types is captured, and as such the under- or over-representation of pixels in different tissue types that is an issue for the boundary method is no longer problematic.

The mixture model is also a means of comparing images, for example, in a longitudinal sense, it may be of interest to determine how tissue composition is changing through time, and the parameters of the mixture model could be used as a measure of increasing or decreasing denseness, which is related to meat quality.

16.3 Models and methods

16.3.1 Bayesian mixture models

The normal finite mixture model can be represented by the likelihood:

$$g(\mathbf{y}) \approx \hat{g}(\mathbf{y}) = \prod_{i=1}^{N} \sum_{j=1}^{k} \lambda_j \frac{1}{\sqrt{2\pi\sigma_j^2}} \exp\left[-\frac{1}{2}\left(\frac{y_i - \mu_j}{\sigma_j}\right)^2\right] \tag{16.1}$$

where N is the sample size, k is the number of components in the mixture, μ_j and σ_j^2 are the mean and variance of component j, and λ_j represents the weight of component j, that is the probability that a pixel is classified into the jth component.

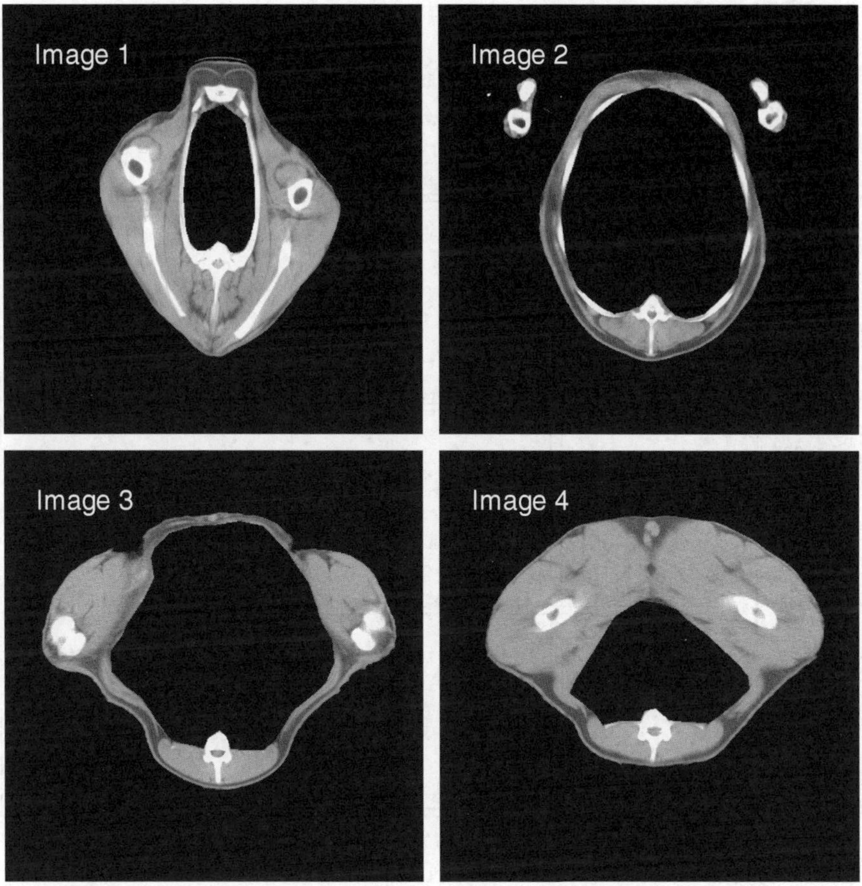

Figure 16.2 Images from four different regions of the same sheep carcase.

In Bayesian analysis it is usual to estimate the posterior distribution of the unknown parameters $(\mu_j, \sigma_j^2, \lambda_j)$ by allocating a prior distribution to these parameters and appealing to Bayes' rule:

$$p(\mu, \sigma^2, \lambda \mid y) \propto \prod_{i=1}^{N} \sum_{j=1}^{k} \lambda_j \frac{1}{\sqrt{2\pi\sigma_j^2}} \exp\left[-\frac{1}{2}\left(\frac{y_i - \mu_j}{\sigma_j}\right)^2\right] p(\mu, \sigma^2, \lambda) \quad (16.2)$$

where $p(\mu, \sigma^2, \lambda)$ represents the joint prior distribution. Note that, if an appropriate prior distribution is used in Equation (16.2), an explicit estimator is available for the unknown parameters. However, this posterior would involve the sum of k^N terms, making computations too intensive to contemplate in this case (Robert and Casella 2004, p. 433).

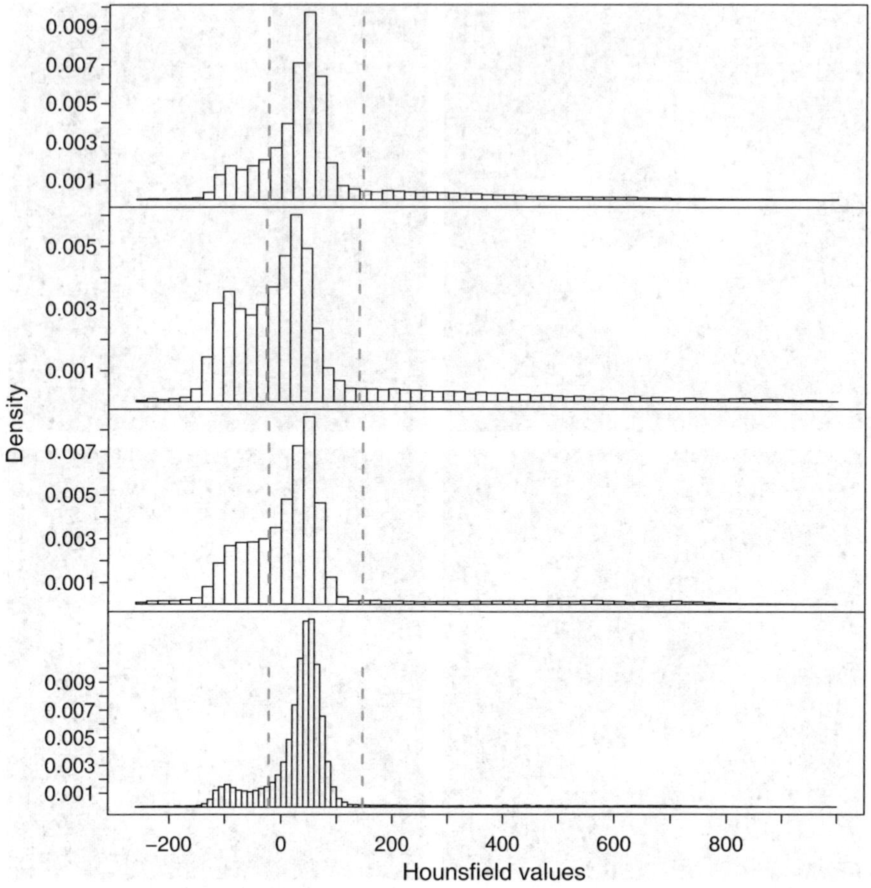

Figure 16.3 Histograms corresponding to Figure 16.2.

It is usual in the Bayesian setting to formulate the mixture model within a hierarchical framework. This is achieved by introducing a new variable, z_i, which is a $(1 \times k)$ indicator vector which represents the unobserved component membership of the data. In this analysis, z_i will represent the tissue type, or a subpopulation of this tissue, to which pixel i belongs. The variable z_i has a sampling distribution, given the weights, λ, which is usually taken to be a multinomial distribution. Then, the joint conditional distribution of the data (y_i) and the unobserved component membership (z_i) is given by

$$\begin{aligned} p(y, z \mid \mu, \sigma^2, \lambda) &\propto p(z \mid \lambda) p(y \mid \mu, \sigma^2, z) \\ &\propto \prod_{i=1}^{N} \prod_{j=1}^{k} \left\{ \lambda_j \left(\sqrt{2\pi\sigma_j^2} \right)^{-1} \exp\left[-\frac{1}{2} \left(\frac{y_i - \mu_j}{\sigma_j} \right)^2 \right] \right\}^{z_{ij}} \end{aligned} \tag{16.3}$$

where $z_{ij} = 1$ if y_i belongs to component j, otherwise $z_{ij} = 0$.

Interest usually lies in estimating the unknown model parameters, μ, σ^2 and λ, but, in this case, we are also interested in the allocation variable z_i as a soft clustering technique. In the case of mixtures we can use the property of independent priors between the proportions (λ) and the component parameters (μ and σ^2) and the joint conditional distribution of **y** and **z**, given by Equation (16.3), to obtain the joint posterior of the unknown parameters:

$$\begin{aligned} p(\boldsymbol{\mu}, \boldsymbol{\sigma}^2, \boldsymbol{\lambda} \mid \mathbf{y}, \mathbf{z}) &\propto p(\mathbf{z} \mid \boldsymbol{\lambda}) p(\mathbf{y} \mid \boldsymbol{\mu}, \boldsymbol{\sigma}^2, \mathbf{z}) p(\boldsymbol{\mu}, \boldsymbol{\sigma}^2, \boldsymbol{\lambda}) \\ &\propto p(\mathbf{z} \mid \boldsymbol{\lambda}) p(\boldsymbol{\lambda}) p(\mathbf{y} \mid \boldsymbol{\mu}, \boldsymbol{\sigma}^2, \mathbf{z}) p(\boldsymbol{\mu} \mid \boldsymbol{\sigma}^2) p(\boldsymbol{\sigma}^2). \end{aligned} \tag{16.4}$$

Simulating values from the joint posterior (Equation 16.4) would be challenging. MCMC techniques could be used with any prior (joint or otherwise), but the use of the Gibbs sampler, which we use to evaluate this model, is only possible with a judicious choice of conjugate priors.

16.3.2 Parameter estimation using the Gibbs sampler

The Gibbs sampler is an algorithm in which values of each parameter are simulated in turn from their respective full conditional distributions. It is frequently used when the joint posterior of the unknown parameters is difficult to simulate.

Using starting values $\boldsymbol{\theta}^0$, the Gibbs sampling algorithm simulates updated values as follows:

$$\begin{aligned} \theta_1^t &\sim p(\theta_1 \mid y, \theta_2^{t-1}, \ldots, \theta_p^{t-1}) \\ \theta_2^t &\sim p(\theta_2 \mid y, \theta_1^t, \theta_3^{t-1}, \ldots, \theta_p^{t-1}) \\ &\vdots \\ \theta_p^t &\sim p(\theta_p \mid y, \theta_1^t, \theta_2^t, \ldots, \theta_{p-1}^t). \end{aligned}$$

This process is iterated until a satisfactory convergence to the required stationary distributions is obtained. The resulting chains are then used to obtain the ergodic average for each of the parameters ($\boldsymbol{\theta}$).

Prior distributions

To analyse the mixture model using the Gibbs sampler we need to nominate suitable prior distributions in Equation (16.4). In this analysis we have used conjugate priors, that is priors which follow the same parametric form as their corresponding posterior distributions. These priors, which are given below, are both computationally convenient and seem to be biologically sensible:

$$\boldsymbol{\lambda} \sim \text{Dirichlet}(\alpha_1, \alpha_2, \ldots, \alpha_k) \tag{16.5}$$

$$\mu_j \mid \sigma_j^2 \sim \text{N}\left(\xi_j, \frac{\sigma_j^2}{m_j}\right) \tag{16.6}$$

$$\sigma_j^2 \sim \text{InverseGamma}\left(\frac{\nu_j}{2}, \frac{s_j^2}{2}\right) \tag{16.7}$$

where $\alpha_j, \xi_j, m_j, \nu_j$ and s_j are fixed hyperparameters. See Section 16.4.1 for a further discussion of these hyperparameters.

The two-stage Gibbs sampler for this model is outlined in Algorithm 16.

Algorithm 16: Gibbs sampler for estimation of parameters in normal mixture model

1. Update $z_i \sim \text{Multinomial}(1; \omega_{i1}, \omega_{i2}, \ldots, \omega_{ik})$,

$$\omega_{ij} = \frac{\lambda_j \left(\sqrt{2\pi\sigma_j^2}\right)^{-1} \exp\left[-\frac{1}{2}\left(\frac{y_i - \mu_j}{\sigma_j}\right)^2\right]}{\sum_{t=1}^{k} \lambda_t \left(\sqrt{2\pi\sigma_t^2}\right)^{-1} \exp\left[-\frac{1}{2}\left(\frac{y_i - \mu_t}{\sigma_t}\right)^2\right]}$$

.

2. Calculate $n_j = \sum_{i=1}^{N} z_{ij}$, $\bar{y}_j = \left(\sum_{i=1}^{N} z_{ij} y_i\right) / n_j$, $\hat{s}_j^2 = \sum_{i=1}^{N} z_{ij}(y_i - \bar{y}_j)^2$.
3. Update $\lambda \sim \text{Dirichlet}(n_1 + \alpha_1, n_2 + \alpha_2, \ldots, n_k + \alpha_k)$.
4. Update $\mu_j \sim \text{N}\left(\frac{n_j\bar{y}_j + m_j\xi_j}{n_j + m_j}, \frac{\sigma_j^2}{n_j + m_j}\right)$.
5. Update $\sigma_j^2 \sim \text{InverseGamma}\left(\frac{n_j + \nu_j + 1}{2}, \frac{1}{2}\left[s_j^2 + \hat{s}_j^2 + \frac{n_j m_j}{n_j + m_j}(\bar{y}_j - \xi_j)^2\right]\right)$.

16.3.3 Extending the model to incorporate spatial information

The independent mixture model is readily extended to incorporate some of the spatial information contained in the CT images. This technique, which was presented for CT data in Alston *et al.* (2005), incorporates spatial information through the use of a Markov random field (MRF) for the allocation vector z. The spatial information contained in the CT image could also be represented by considering z_i to be drawn from an MRF involving first-order neighbours. Using a hidden MRF in the mixture model allows neighbouring observations to be correlated through a common origin (pixel i). As a result, the independence assumption on the unobserved component membership (z_i) is removed, and information supplied by neighbouring pixels can influence the allocation of the origin pixel into a component group (McLachlan and Peel 2000, chapter 13).

An appropriate joint distribution for the unobserved component membership, z, is the Potts model (Potts 1952; Stanford 1999; Winkler 2003) which can be represented by

$$p(z \mid \beta) = C(\beta)^{-1} \exp\left(\beta \sum_{i \sim j} z_i z_j\right)$$

where $C(\beta)$ is a normalizing constant, $i \sim j$ indicates neighbouring pixels, then $z_i z_j = 1$ if $z_i = z_j$, otherwise $z_i z_j = 0$. The parameter β estimates the level of spatial homogeneity in component membership between neighbouring pixels in the image. Positive values for β would infer that neighbouring pixels have a similar component membership; a zero value for β would imply that the allocations of neighbouring pixels are independent of each other and negative values would imply dissimilarity between pixels which are neighbours.

Obtaining updated estimates of the spatial parameter β can be difficult as it involves an intractable normalizing constant, so, for estimation, we employ the suggestion of Rydén and Titterington (1998) and replace $p(z \mid \beta)$ with the pseudolikelihood:

$$\begin{aligned} p_{PL}(z \mid \beta) &= \prod_{i=1}^{N} p(z_i \mid z_{\partial i}, \beta) \\ &= \prod_{i=1}^{N} \frac{\exp[\beta \mathrm{U}(z_{\partial i}, z_i)]}{\sum_{j=1}^{k} \exp[\beta \mathrm{U}(z_{\partial i}, j)]}. \end{aligned}$$

The prior distribution of β in our analysis is taken as a Uniform(ζ, δ) and an updating scheme using the Metropolis–Hastings step described below is implemented to simulate new values at each iteration.

The Gibbs sampler for estimation of the parameters in this spatial mixture, based on conjugate priors for μ and σ^2, is detailed in Algorithm 17.

Algorithm 17: Gibbs sampler for estimation of parameters in spatial normal mixture model

1. Update $z_i \sim \text{Multinomial}(1; \omega_{i1}, \omega_{i2}, \ldots, \omega_{ik})$,

$$\omega_{ij} = \frac{\left(\sqrt{2\pi\sigma_j^2}\right)^{-1} \exp\left[-\frac{1}{2}\left(\frac{y_i - \mu_j}{\sigma_j}\right)^2 + \beta U(z_{\partial i}, j)\right]}{\sum_{t=1}^{k} \left(\sqrt{2\pi\sigma_t^2}\right)^{-1} \exp\left[-\frac{1}{2}\left(\frac{y_i - \mu_t}{\sigma_t}\right)^2 + \beta U(z_{\partial i}, t)\right]},$$

   where $U(z_{\partial i}, j)$ represents the number of neighbours of pixel i currently allocated to component j.

2. Calculate $n_j = \sum_{i=1}^{N} z_{ij}$, $\bar{y}_j = \left(\frac{\sum_{i=1}^{N} z_{ij} y_i}{n_j}\right)$, $\hat{s}_j^2 = \sum_{i=1}^{N} z_{ij}(y_i - \bar{y}_j)^2$.
3. Update $\mu_j \sim \mathrm{N}\left(\frac{n_j \bar{y}_j + m_j \xi_j}{n_j + m_j}, \frac{\sigma_j^2}{n_j + m_j}\right)$.
4. Update $\sigma_j^2 \sim \text{InverseGamma}\left(\frac{n_j + \nu_j + 1}{2}, \frac{1}{2}\left[s_j^2 + \hat{s}_j^2 + \frac{n_j m_j}{n_j + m_j}(\bar{y}_j - \xi_j)^2\right]\right)$.
5. Update $\beta_{\text{new}} \sim \text{Unif}(\gamma, \delta)$.
   Accept β_{new} with prob=$\min\left\{1, p(\beta_{\text{new}} \mid \mu, \sigma, z, y)/p(\beta_{\text{current}} \mid \mu, \sigma, z, y)\right\}$ using a Metropolis–Hastings step.

16.4 Data analysis and results

16.4.1 Normal Bayesian mixture model

Initial analysis was conducted using log transformed data and the standard Bayesian mixture model (Algorithm 16).

Tips and Tricks: In this case study, the data are binned due to the discrete output from a CT scan. Therefore, in stage 1 of the Gibbs sampler outline in Algorithm 16, the z_i can be drawn from the Multinomial$(N_u; \omega_{u1}, \omega_{u2}, \dots, \omega_{uk})$ for the unique values (u), of which there are 1572, as opposed to drawing the z individually over the 55 032 pixel values. This strategy will save substantial amounts of computation.

In other examples, data binning may be an issue which needs to be addressed. Alston and Mengersen (2010) propose a method of accounting for data binning within the Gibbs sampling framework for mixtures.

Hyperparameters for prior distributions

To initiate the modelling process, values need to be assigned to the hyperparameters of the priors in Equations (16.5), (16.6) and (16.7). As we had little prior information on the parameters or their expected behaviour, it was prudent to nominate values for the hyperparameters that reflected this uncertainty. It was also desirable to nominate priors that encapsulate all likely values of the unknown parameters. Alternatively, if the priors were allowed to be unreasonably vague, they would contribute little to the modelling process, possibly even hindering it, and as a result this technique would perform well below its capabilities. To better understand the influence of hyperparameters, readers are directed to the conditional distributions in Algorithm 16, by which they may compare the influence of the data and the hyperparameters.

In Equation (16.5), the prior distribution of $\boldsymbol{\lambda}$, the hyperparameters, $\boldsymbol{\alpha}$, need to be assigned a fixed value. In this analysis all of the α parameters have been given the value 1. This implies that from our prior state of knowledge we believe the data could be equally divided into the k components. In a Dirichlet distribution with these values, the mean value of each of the k proportions is $1/k$, and the distribution of each parameter is contained in the interval [0,1]. By allocating a value of unity for each of the α_k parameters we are also ensuring that the estimates of $\boldsymbol{\lambda}$ are largely influenced by the data, as the values of n_k will subsequently dominate the simulated Dirichlet values.

The hyperparameters in Equation (16.6) are ξ_j, the a priori mean for component j, and m_j, the number of observations we are prepared to commit to the jth component. In this analysis we have set $\boldsymbol{\xi}$ equal to the overall mean of the log transformed data (5.70) and $\mathbf{m}$ equal to 100. When compared with the likely number of pixels in each component, with a sample size in excess of 50 000, choosing a small value for m allows the variance of the prior distribution for the mean, $p(\mu_j \mid \sigma_j^2)$ to be quite broad, thereby taking our uncertainty with regards to this prior into account and allowing the data to largely drive the parameter estimates.

The prior distribution of the component variances (σ_j^2) is given by Equation (16.7) and involves the hyperparameters ν_j and s_j^2. In this study, we have only a vague idea of the likely values of the component variances, so we inflated our prior estimates to about four times the value we think likely. By doing so, we have covered the possibility that there is in fact only one large component which covers the full range of the data. As a consequence, the prior distribution we have used in Equation (16.7) has hyperparameters $\nu_j = 4.05$ and $s_j^2 = 0.5125$ for each of the k components.

Tips and Tricks: The hyperparameters for the prior on σ_j^2 are the most influential in terms of estimation of parameters via the Gibbs sampler. When viewing your data, make a realistic interpretation of all possible values this parameter could take and set your hyperparameters with this in mind. Unreasonably vague priors can hinder the successful estimation of parameters when drawing from their full conditional distributions.

Results

The Gibbs sampler was run for 110 000 iterations (10 000 burn-in), starting with $k = 3$ components and adding a component using a targeted addition scheme (Alston *et al.* 2007) until the BIC statistic failed to improve under a new addition. The final model contained five components: two associated with fat tissue, two associated with muscle tissue and one which comprises the bone component. The estimates for the parameters in the five-component mixture model, calculated using a thinning of 100 in the MCMC output, are given in Table 16.1.

Figure 16.4 (bottom) is a histogram of the data with an overlay of the density fit from the independent mixture model. From this graph we observe that the density appears to be a reasonable fit to the data, with the possible exception of a peak being missed at around 5.25. The final estimates of the proportion of each tissue present in

Table 16.1 Comparison of estimated mixture model parameters and credible intervals using independent and spatial mixture model for data given in Figure 1. The MCMC output was thinned to include every 100th iteration.

Mixture	Tissue	$\hat{\mu}$	95% CI	$\hat{\sigma}$	95% CI	$\hat{\lambda}$	95% CI
Independent	Fat	4.85	[4.71, 4.99]	1.38	[1.29, 1.47]	0.0164	[0.0140, 0.0191]
	Fat	5.36	[5.34, 5.38]	0.29	[0.28, 0.30]	0.3702	[0.3465, 0.3937]
	Muscle	5.62	[5.60, 5.63]	0.12	[0.11, 0.13]	0.2385	[0.2082, 0.2689]
	Muscle	5.72	[5.72, 5.73]	0.08	[0.07, 0.09]	0.1861	[0.1609, 0.2136]
	Bone	6.34	[6.30, 6.37]	0.48	[0.46, 0.50]	0.1888	[0.1789, 0.1993]
Spatial	Fat	4.55	[4.47, 4.63]	1.07	[1.01, 1.12]	0.0181	[0.0175, 0.0187]
	Fat	5.11	[5.09, 5.11]	0.17	[0.16, 0.18]	0.1419	[0.1394, 0.1450]
	Fat	5.40	[5.39, 5.41]	0.10	[0.09, 0.11]	0.1457	[0.1426, 0.1487]
	Muscle	5.56	[5.55, 5.57]	0.06	[0.05, 0.07]	0.1649	[0.1607, 0.1687]
	Muscle	5.67	[5.66, 5.68]	0.05	[0.04, 0.06]	0.1529	[0.1487, 0.1576]
	Muscle	5.76	[5.75 5.77]	0.05	[0.04, 0.06]	0.1803	[0.1776, 0.1829]
	Bone	5.98	[5.97, 5.99]	0.15	[0.14, 0.16]	0.0755	[0.0739, 0.0771]
	Bone	6.62	[6.60,6.63]	0.33	[0.32, 0.34]	0.1207	[0.1189, 0.1226]

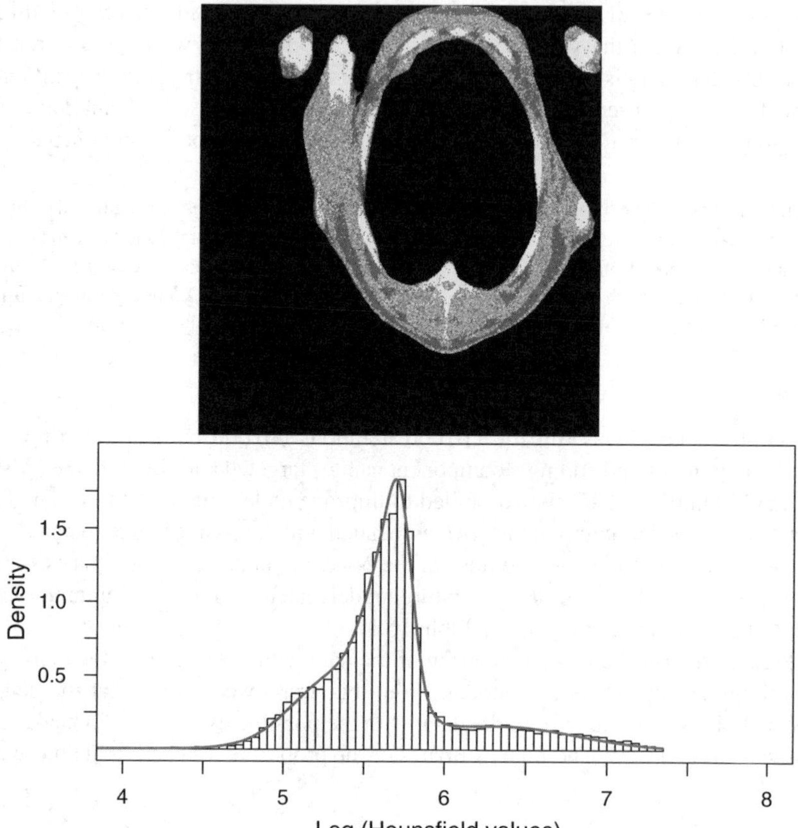

Figure 16.4 Top: Representation of CT scan using typical allocation in stage 1 of Algorithm 16. Darkest grey components are generally fat tissue, medium grey components are muscle, and whitest pixels represent bone. Bottom: Image represented by a histogram of pixel log Hounsfield values. The overlaid line is the estimated density (Equation 16.1) using parameter estimates from the independent mixture model (Table 16.1).

the scan are given in Table 16.2, from which we note that possibly the estimate for percentage fat is slightly high, but still may be biologically reasonable.

Figure 16.4 (top) represents a randomly chosen iteration of the allocation of the unknown component membership z to each of the five components. The salient features, in terms of location of fat deposits, muscles and bone, seem to be preserved; however, the appearance of the reproduced image is quite speckled. This phenomenon is due to the allocations being made without regard to the neighbouring pixels.

16.4.2 Spatial mixture model

As information on the spatial location of the pixels within a scan is available, it seems sensible to assess whether inclusion of this information can improve model fit.

The computational overhead of Algorithm 17 is much greater than the independent model, specifically in the allocation of the latent variable z_i, which can no longer take advantage of the binned nature of the data.

Hyperparameters for prior distributions

The hyperparameters for the prior distribution of μ and σ^2 are set to be the same as in Section 16.4.1. Then we set the prior distribution for the spatial parameter (β) to be

$$\beta \sim \text{Uniform}(0, 3)$$

where setting a minimum of 0 and maximum of 3 implies that there is either a positive or independent association between the pixels (negative association is not allowed). The maximum value of 3 allows the neighbouring pixels to have some influence on the allocation stage of Algorithm 17, but does not allow the neighbours to completely dominate the allocation.

Results

The Gibbs sampler was again run for 110 000 iterations (10 000 burn-in, thinning of 100), starting with $k = 3$ components and adding a component using a targeted addition scheme (Alston *et al.* 2007) until the BIC statistic failed to improve under a new addition. The final model contained eight components: three associated with fat tissue, three associated with muscle tissue and two which comprise the bone component.

The estimates for the parameters in the eight-component mixture model are given in Table 16.1. We observe that the credible intervals around parameter estimates are tighter, which is sensible when we note that the allocation vector from which component estimates are formed should also be less variable when neighbouring pixels are considered. The estimate for the spatial parameter was $\hat{\beta} = 1.417$ [1.389, 1.446], illustrating that there was a positive association between neighbouring pixels, and as such inclusion of spatial information in the model was beneficial.

Figure 16.5 (bottom) is the histogram of the data with an overlay of the density fit from the spatial mixture model. From this graph we observe that the density appears to be a reasonable fit to the data and the peak at around 5.25 appears to be better modelled using the spatial algorithm. The final estimates of the proportion of each tissue present in the scan are given in Table 16.2 which appears to have a more reasonable estimate for percentage fat using the spatial model, based on previous studies (Thompson and Atkins 1980)

Figure 16.5 (top) represents a typical allocation of the unknown component membership, z, to each of the eight components. The salient features of the image seem to be better preserved using the spatial model, and the speckled appearance under the independent model is no longer evident.

In terms of volume estimation, we are now able to apply the spatial mixture model to each of the CT images.

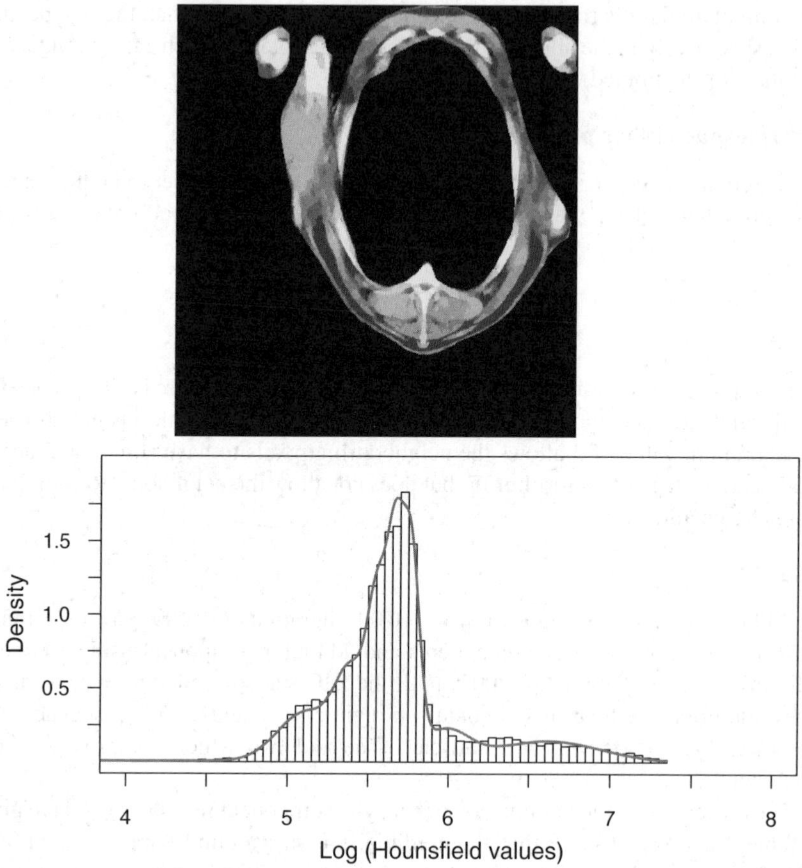

Figure 16.5 Top: Representation of CT scan using typical allocation in stage 1 of Algorithm 17. Darkest grey components are generally fat tissue, medium grey components are muscle, and whitest pixels represent bone. Bottom: Image represented by a histogram of pixel log Hounsfield values. The overlaid line is the estimated density (Equation 16.1) using parameter estimates from the spatial mixture model (Table 16.1).

Table 16.2 Comparison of estimated proportions and credible intervals using independent and spatial mixture model for data given in Figure 1.

Tissue	Independent mixture	95% CI	Spatial mixture	95% CI
Fat	0.3866	[0.3631, 0.4100]	0.3050	[0.2933, 0.3096]
Muscle	0.4245	[0.4068, 0.4431]	0.4984	[0.4937, 0.5071]
Bone	0.1887	[0.1790, 0.1993]	0.1965	[0.1949, 0.1976]

16.4.3 Carcase volume calculation

Alston *et al.* (2009) used the Cavalieri method to calculate tissue volumes in a whole carcase based on 15 equally spaced CT scans, taken 4 cm apart. Most animal science papers to date have used the Cavalieri method, and it has proven to be reliable in a wide range of situations (Gundersen and Jensen 1987; Gundersen et al. 1988). An estimate of volume using Cavalieri's method is calculated as

$$V_{\text{Cav}} = d \times \sum_{s=1}^{m} \text{area}_g - t \times \text{area}_{\text{max}}$$

where m is the number of CT scans taken (15 in this example), d is the distance the CT scans are apart, in this case 4 cm. The value of t is the thickness of each slice (s), which, in this example, is 1 cm, and area_{max} is the maximum area of any of the m scans. In our MCMC scheme, the area can be calculated at each iteration (or iterations after thinning is applied), and as such it is possible to obtain the posterior mean and credible intervals for the carcase volume of each tissue type.

In addition, calculating tissue weight is straightforward using the mixture model analysis. We use the established conversion to grams per cubic centimetre given by Fullerton (1980):

$$\text{Est. density}(\text{g/cm}^3) = 1.0062 + \text{Hu} \times 0.001\,06.$$

To estimate the weight of each tissue we use the equation

$$\widehat{\text{Tissue weight}} = \frac{\left(1.0062 + \overline{\text{Hu}}_{\text{tissue}} \times 0.001\,06\right) \times \text{NumPix}_{\text{tissue}} \times \left(\text{pixels/cm}^3\right)^{-1}}{1000}$$

where $\overline{\text{Hu}}$ is the mean Hounsfield number for pixels allocated to the tissue of interest, and $\text{NumPix}_{\text{tissue}}$ is the number of pixels allocated to the tissue type. The number of pixels per cubic centimetre is 113.78 in this data set.

Results

Using the allocated stage of our MCMC sampler, we are able to use each iteration of z_i to estimate the proportion of pixels allocated to each tissue group, along with their credible intervals. We have done this using the last 1000 iterations of the MCMC algorithm. The resulting estimates and 95% credible intervals for tissue proportions for each of the 15 scans in this animal can be seen in Figure 16.6, with the bold images being those depicted in Figure 16.2. As scans are taken at 4 cm spacing, the smooth transitions in tissue proportion are biologically sensible. The only scans displaying a sizable change were for the bone component between scans 13 and 14, this being explained by scan 14 passing through the pelvic region of the carcase where the bone component is proportionally large. The width of the 95% credible intervals, which were based on allocation of pixels to tissue types in the last 1000 iterations of the Gibbs sampler, indicates the model is stable around these estimates.

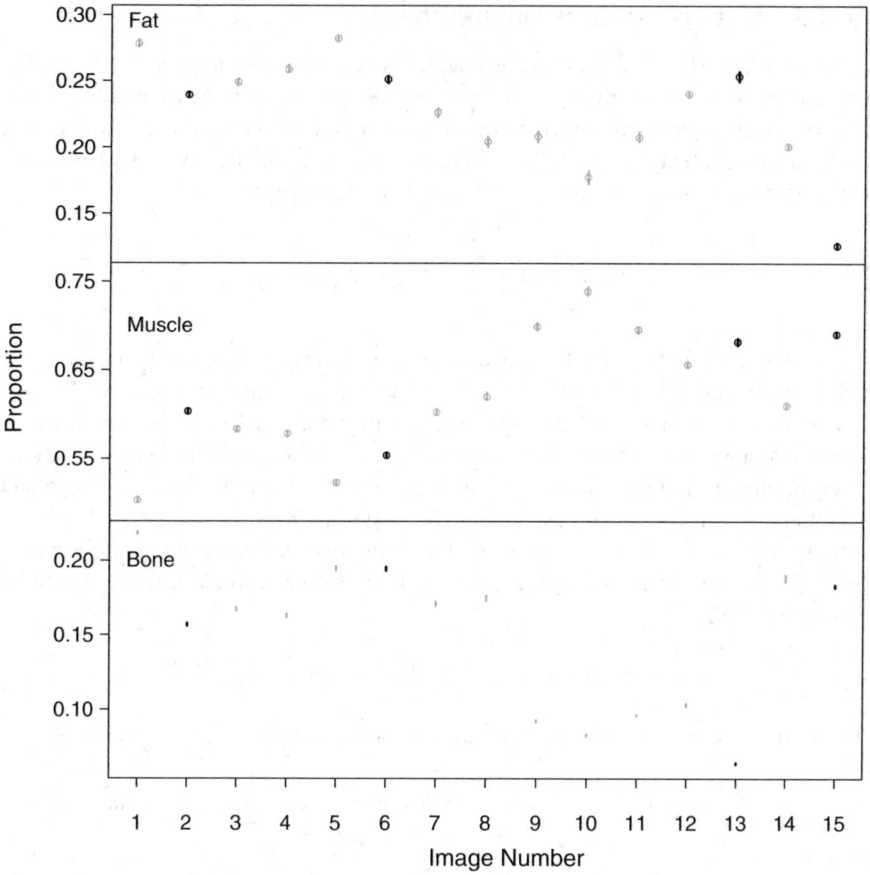

Figure 16.6 Estimated proportions and 95% credible intervals for each of the 15 CT scans taken on an individual sheep. The four images of Figure 16.2 are in bold, the interim scans are in grey. Top: Proportion estimate for fat tissue. Middle: Estimated proportion of muscle tissue. Bottom: Estimated proportion of bone.

The estimates of the probabilities (λ_j) and the number of pixels in each scan are then used to calculate the area of tissue in each of the CT scans. These 15 area estimates are then used to estimate the overall proportion of tissue types in the carcase (Table 16.3). Using the mixture model results, the Cavalieri method yields estimates of tissue proportions that are in close agreement with the findings of an extensive study published by Thompson and Atkins (1980). Additionally, a chemical analysis of this carcase indicated a fat percentage of 20.2% (4.47 kg), which is also in close agreement with these estimates. However, the results obtained using the boundary approach indicated that the proportion of fat and bone in this carcase is being overestimated quite substantially.

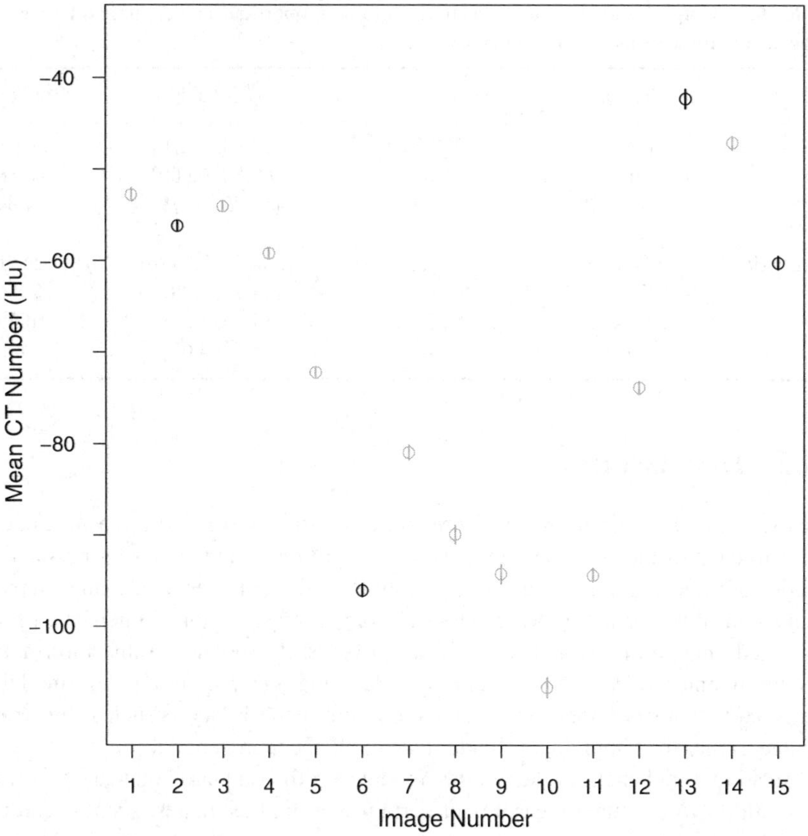

Figure 16.7 Mean CT numbers (95% CI) associated with pixels associated with fat tissue.

Figure 16.7 illustrates the mean CT number for the pixels allocated to fat tissue over the 15 scans. These smooth transitions between values reinforce the earlier observation from Figure 16.3 that the density of fat and muscle tissues seems to vary according to the region of the carcase. This provides further support for the observation that fixed boundaries may be unsuitable for analysing CT data, particularly when estimates of weight are a major outcome of interest. This idea is illustrated in Table 16.3, where it can be seen that the estimate of fat under the boundary allocation has values far exceeding the chemical analysis, the mixture model estimate and the findings of Thompson and Atkins (1980).

The weight of the carcase and individual tissues therein was also estimated at each of the last 1000 iterations of the Gibbs sampler (Table 16.3). A weight measurement of the carcase at slaughter was 25.4 kg, which again shows a close agreement with the estimates obtained using the mixture model approach, particularly given the error that is associated with manually editing images to extract the unwanted viscera from the live animal CT scans.

Table 16.3 Comparison of estimated carcase volume proportions and credible intervals using spatial mixture model and boundary techniques.

	Tissue	Spatial mixture	95% CI	Boundary
Percentage	Fat	23.03	[22.94, 23.10]	32.38
	Muscle	61.89	[61.80, 62.00]	42.80
	Bone	14.75	[14.72, 14.79]	24.84
Weight (kg)	Carcase	23.773	[23.771, 23.776]	24.30
	Fat	4.77	[4.75, 4.78]	7.87
	Muscle	14.35	[14.33, 14.38]	10.39
	Bone	4.64	[4.63, 4.65]	6.04

16.5 Discussion

The use of the Bayesian normal mixture model to estimate tissue volume and allocate pixels to each of the three tissue types is shown to be advantageous to the standard practice of imposing fixed boundaries to achieve these estimates. The change in mean Hounsfield units for the allocated tissues along carcase regions, together with the illustrated change in histogram densities, provides clear evidence that information may be misinterpreted if assessment is made using a standard boundary model. It seems likely that fixed boundaries need to be calibrated for factors such as breed, sex and age, rather then applying a standard set to all sheep in general.

An additional benefit of the mixture model over the standard boundary technique is the ability to calculate measures of certainty around estimates, such as credible intervals. The boundary method provides a point estimate only, hence giving no means of assessing uncertainty.

The computation of the mixture model is more burdensome than the boundary method. However, our current research is showing that, with deft programming and the exploitation of parallel computing and graphics processing unit (GPU) calculations, the proposed Gibbs sampler could be implemented for use in live animal experimental work, based on the current computing power available to most modern researchers.

References

Alston CL and Mengersen KL 2010 Allowing for the effect of data binning in a Bayesian Normal mixture model. *Computational Statistics & Data Analysis* **54**(4), 916–923.

Alston CL, Mengersen KL and Gardner GE 2009 A new method for calculating the volume of primary tissue types in live sheep using computed technology (CT) scanning. *Animal Production Science* **49**(11), 1035–1042.

Alston CL, Mengersen KL, Robert CP, Thompson JM, Littlefield PJ, Perry D and Ball AJ 2007 Bayesian mixture models in a longitudinal setting for analysing sheep CAT scan images. *Computational Statistics & Data Analysis* **51**(9), 4282–4296.

Alston CL, Mengersen KL, Thompson JM, Littlefield PJ, Perry D and Ball AJ 2004 Statistical analysis of sheep CAT scan images using a Bayesian mixture model. *Australian Journal of Agricultural Research* **55**(1), 57–68.

Alston CL, Mengersen KL, Thompson JM, Littlefield PJ, Perry D and Ball AJ 2005 Extending the Bayesian mixture model to incorporate spatial information in analysing sheep CAT scan images. *Australian Journal of Agricultural Research* **56**(4), 373–388.

Ball AJ, Thompson JM, Alston CL, Blakely AR and Hinch GN 1998 Changes in maintenance energy requirements of mature sheep fed at different levels of feed intake at maintenance, weight loss and realimentation. *Livestock Production Science* **53**, 191–204.

Brenøe UT and Kolstad K 2000 Body composition and development measured repeatedly by computer tomography during growth in two types of turkeys. *Poultry Science* **79**, 546–552.

Dempster AP, Laird NM and Rubin DB 1977 Maximum likelihood from incomplete data via the EM algorithm (with discussion). *Journal of the Royal Statistical Society, Series B* **39**(1), 1–38.

Diebolt J and Robert CP 1994 Estimation of finite mixture distributions through Bayesian sampling. *Journal of the Royal Statistical Society, Series B* **56**(2), 363–375.

Fullerton GD 1980 Fundamentals of CT tissue characterization. In *Medical Physics of CT and Ultrasound: Tissue Imaging and Characterization*. Medical Physics Monograph 6 (eds GD Fullerton and J Zagzebski J), pp. 125–162. American Institute of Physics, New York.

Gundersen HJG and Jensen EB 1987 The efficiency of systematic sampling in sterology and its prediction. *Journal of Microscopy* **147**(3), 229–263.

Gundersen HJG, Bendtsen TF, Korbo L, Marcussen N, Møller A, Nielsen K, Nyengaard JR, Pakkenberg B, Sørensen FB, Vesterby A and West MJ 1988 Some new, simple and efficient stereological methods and their use in pathological research and diagnosis. *APMIS: Acta Pathologica, Microbiologica et Immunologica Scandinavica* **96**(5), 379–394.

Jopson NB, Thompson JM and Fennessy PF 1997 Tissue mobilisation rates in male fallow deer (*Dama dama)* as determined by computed tomography: the effects of natural and enforced food restriction. *Animal Science* **65**, 311–320.

Kolstad K 2001 fat deposition and distribution measured by computer tomography in three genetic groups of pigs. *Livestock Production Science* **67**, 281–292.

Kvame T and Vangen O 2007 Selection for lean weight based on ultrasound and CT in a meat line of sheep. *Livestock Science* **106**, 232–242.

McLachlan G and Peel D 2000 *Finite Mixture Models*. John Wiley & Sons, Inc., New York.

Pearson K 1894 Contributions to the mathematical theory of evolution. *Philosophical Transactions of the Royal Society of London* **185**, 71–110.

Potts RB 1952 Some generalized order-disorder transitions. *Proceedings of the Cambridge Philosophical Society* **48**, 106–109.

Robert CP and Casella G 2004 *Monte Carlo Statistical Methods*, 2nd edn. Springer, New York.

Rydén T and Titterington DM 1998 Computational Bayesian analysis of hidden Markov models. *Journal of Computational & Graphical Statistics* **7**(2), 194–211.

Stanford DC 1999 Fast automatic unsupervised image segmentation and curve detection in spatial point patterns. PhD thesis. University of Washington.

Thompson J and Kinghorn B 1992 Catman – a program to measure CAT-Scans for prediction of body components in live animals. *Proceedings of the Australian Association of Animal Breeding and Genetics*, Vol. 10, pp. 560–564. AAAGB Distribution Service, University of New England, Armidale, NSW.

Thompson JM and Atkins KD 1980 Use of carcase measurements to predict percentage carcase composition in crossbred lambs. *Australian Journal of Agricultural Research and Animal Husbandry* **20**, 144–150.

West M 1992 Modelling with mixtures. In *Bayesian Statistics* (eds J Bernardo *et al.*), Vol 4, pp. 503–524. Oxford University Press, Oxford.

Winkler G 2003 *Image Analysis, Random Fields and Markov Chain Monte Carlo Methods: A Mathematical Introduction*, 2nd edn. Springer, Berlin.

17

Latent class models in medicine

Margaret Rolfe,[1] Nicole M. White[1,2] and Carla Chen[1]

[1] *Queensland University of Technology, Brisbane, Australia*
[2] *CRC for Spatial Information, Australia*

17.1 Introduction

Clustering based on patient-centred information is commonplace in studies of medicine. The ability to carry this out has wide-reaching implications, including the identification of meaningful clinical phenotypes or at-risk patients and a basis for reasoning observed differences in treatment outcome. Nevertheless, clustering is often complicated by the complexity and variability observed between subjects. In many cases, differences between subjects are thought to be attributed to an underlying unobservable process. To this end, the use of statistical models that include an unobservable or latent variable may assist in discovering and learning more about such underlying processes.

The use of latent variables for this purpose has found an abundance of applications in medicine. Latent class models to identify subgroups of patients with symptom profiles have been used in Alzheimer research (Walsh 2006); migraine symptom groupings (Chen *et al.* 2009); for trajectory response patterns to identify responders and non-responders in clinical trials for interstitial cystitis (Leiby *et al.* 2009); and identification of trajectories for positive affect and negative events following myocardial infarction (Elliott *et al.* 2005).

In this chapter, we showcase the flexibility of latent class models in subgroup identification given complex patient data. In doing so, two real-life case studies are presented, namely the potential presence of symptom subgroups in Parkinson's disease and in the response over time of cognitive abilities before and after

Case Studies in Bayesian Statistical Modelling and Analysis, First Edition. Edited by Clair L. Alston, Kerrie L. Mengersen and Anthony N. Pettitt.

chemotherapy treatment for early-stage breast cancer. This chapter is presented as follows. Section 17.2 introduces each case study, including motivation for the use of latent variable models. Section 17.3 presents the finite mixture model approach with specific models for each of the case studies and Sections 17.3.5 and 17.4 the computational implementation and the results of the case-study-specific finite mixture models.

17.2 Case studies

17.2.1 Case study 1: Parkinson's disease

Parkinson's disease (PD) is the second most diagnosed neurodegenerative disease worldwide, affecting an estimated 1-2% of persons over 60 globally. An incurable disease with no definitive diagnosis anti-mortem, PD is complicated by high heterogeneity in symptom presentation and progression between patients, not only in terms of principal motor symptoms but also in cognitive and psychiatric symptoms. In light of these observed differences, multiple phenotypes of PD are thought to exist. Consequently, much research has been invested in the identification of potential subtypes, in the hope that they provide insight into underlying pathophysiology and the causation of these variable phenotypes.

For this case study, we aim to identify PD phenotypes based on symptom information collected using the Unified Parkinson's Disease Rating Scale (UPDRS). Considered the gold standard for the assessment of PD, the UPDRS is a 42-item Likert-style questionnaire that assesses the presence and severity of symptoms related to PD and their impact on activities of daily living (ADL). Here, we consider a total of five symptoms assessed by the UPDRS: speech impairment (UDPRS item 18), postural instability (UPDRS item 30), duration of offs[1] (UPDRS item 39), tremor (UPDRS items 20 and 21); and rigidity (UPDRS item 22). Symptoms were coded as absent or present, with the exception of tremor and rigidity, which were coded as none/unilateral, bilateral asymmetric or bilateral symmetric, in an attempt to capture differences in lateral presentation of key motor symptoms. Data were collected on 370 subjects from three specialist movement disorder clinics in Brisbane, Australia, between 2002 and 2006, as part of the Queensland Parkinson's Project (Sutherland *et al.* 2009). Data collection resulted in 72 observed symptom profiles, indicative of the high levels of heterogeneity observed in PD. This case study was considered using latent class analysis by White *et al.* (2010), and is summarized in the chapter.

17.2.2 Case study 2: Cognition in breast cancer

Success in the treatment of breast cancer, especially when diagnosed in early stages of the disease, has resulted in an increased relative 5 year survivorship rate in the last quarter century from 74% in 1982 to nearly 90% in 2006 (Cancer Council Queensland 2008) in Queensland, Australia, while the annual incidence rate increased from 86.3 to 116.4 cases per 100 000 population for the same period. Similar increases in

[1] Offs is a term used to describe fluctuations in the effectiveness of medication.

incidence rates and 5 year survival rates have been experienced both nationally and internationally in Western countries. With this rapid increase in the number of breast cancer survivors, quality of life becomes an area of primary attention. Decline in cognitive function, also known as 'chemo-brain', is a frequently reported side effect for women undergoing adjuvant chemotherapy treatment for breast cancer, with estimates of women suffering from cognitive impairment after chemotherapy in the short term varying from 20% or 25% to 50% for women with moderate or severe impairment. The level of cognitive dysfunction has been shown to improve over time with a subset of women still below baseline levels at 12 months post-chemotherapy. However, in these studies the nature of this cognitive impairment has been described as subtle with deficit levels better than those required for a clinical impairment diagnosis.

Many cognitive domains of attention, concentration, verbal and visual memory and processing speed have been specifically indicated as areas of functional deficit. However, the domain of verbal memory was consistently identified by several studies as suffering compromise from chemotherapy treatment. Full references for breast cancer and cognitive decline described here can be found in Rolfe (2010). This case study concentrates on this area of verbal memory and on the identification of potential subgroups of women with differential responses over time.

For this study, we used data collected from the Cognition in Breast Cancer (CBC) study, which was a prospective longitudinal study undertaken to assess the impact of adjuvant chemotherapy on cognitive functioning in early-stage breast cancer patients drawn from hospitals throughout south-east Queensland, in women up to 2 years' post-chemotherapy. Neuropsychological testing was administered before commencement of chemotherapy (but after definitive surgery), and at 1, 6 and 18 months after completion of chemotherapy. Aspects of verbal memory considered here include verbal learning, immediate retention and delayed recall. These three aspects of verbal memory were measured by the administration of the Auditory Verbal Learning Test (AVLT) of Geffen *et al.* (1990) over four measurement occasions described above, with complete data for all four measurements available from 120 participants. For all three outcome measures, higher scores are indicative of better verbal memory. Figure 17.1 graphically depicts the mean verbal memory trajectory patterns.

We aim to identify subgroups of participants with distinct trajectory patterns, in particular the potential patterns of decline, recovery or no change in verbal memory function before and after chemotherapy treatment, as well as investigate ways to incorporate mediating variables on these response trajectories and/or in the probability of membership allocation.

17.3 Models and methods

In this section, we introduce two types of latent variable model that were used to analyse the two featured case studies. The first, the finite mixture model, is suited to the analysis of multivariate, cross-sectional data and is considered for Case study 1. The other, the trajectory mixture model, is suited to the analysis of data recorded longitudinally and was used to model the data presented in Case study 2.

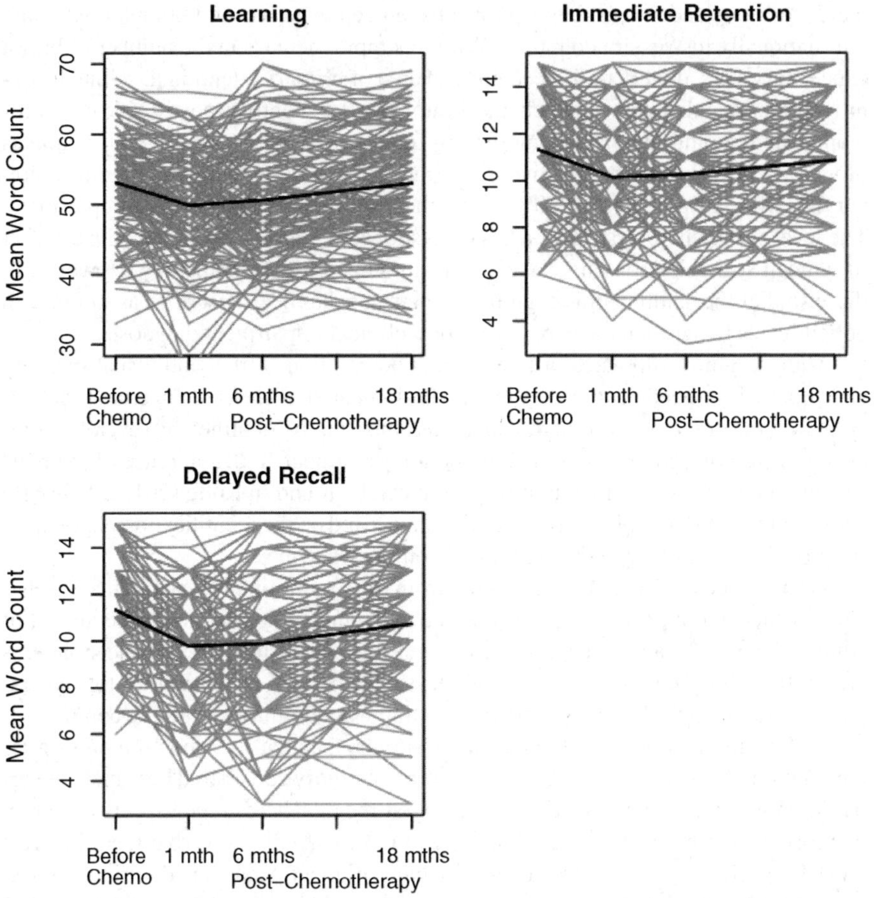

Figure 17.1 Sample mean scores for learning, immediate retention and delayed recall for four times (before chemotherapy, 1, 6 and 18 months post-treatment). Reproduced with permission from John Wiley from Rolfe, Margaret I., Mengersen, Kerrie L., Vearncombe, Katharine J., Andrew, Brooke, Beadle, Geoffrey F. (2011): Bayesian estimation of extent of recovery for aspects of verbal memory in women undergoing adjuvant chemotherapy treatment for breast cancer. *Journal of the Royal Statistical Society*, Series C (Applied Statistics) **60**(5), 655–674.

17.3.1 Finite mixture models

Central to the methodology of the aforementioned case studies and a common tool for data-driven clustering, the finite mixture model (FMM) describes multivariate data under the assumption that the population consists of K unobservable components or subgroups, where K is fixed a priori. This assumption implies that, for an FMM with K latent subgroups, the likelihood of a single observation $\mathbf{y}_i$ can be expressed as a sum of component-specific probability distributions, $f(y_i|\theta_k)$, each weighted by a component proportion p_k, subject to the constraint $\sum_{k=1}^{K} p_k = 1$. For a sample size of n subjects, the likelihood is given by

$$p(\mathbf{y}|p, \theta) = \prod_{i=1}^{n} \sum_{k=1}^{K} p_k f(\mathbf{y}_i|\theta_k). \tag{17.1}$$

As an unsupervised clustering tool, the FMM has the appealing interpretation that these components may correspond to complex, clinically meaningful patient subgroups, learned from the data. For example, in Case study 1, the latent subgroups may represent different phenotypes that may help explain the heterogeneity in observed symptom profiles.

The choice of probability density, $f(\mathbf{y}_i|\theta_k)$, depends on the data under study and, in this sense, the FMM is a flexible modelling tool given its ability to accommodate a variety of data types. When multiple outcomes are under consideration, $f(\mathbf{y}_i|\theta_k)$ can be replaced by a suitable multivariate distribution (e.g. multivariate normal) or, more commonly, a product of J distributions, such that Equation 17.1 now reads

$$p(\mathbf{y}|p, \theta) = \prod_{i=1}^{n} \sum_{k=1}^{K} p_k \prod_{j=1}^{J} f(y_{ij}|\theta_{jk}),$$

where J is the number of outcomes.

Typically, each $f(y_{ij}|\theta_k)$ is taken from the same parametric family. For Case study 1, responses to individual symptoms are treated as discrete data. Specifically, for symptoms that are either absent or present,

$$y_{ij} \sim \text{Bernoulli}(\theta_{jk})$$

and for all remaining symptoms a Multinomial distribution with $L = 3$ levels is assumed, leading to

$$y_{ij} \sim \text{Multinomial}(\theta_{jkl}); \qquad \sum_{l=1}^{L} \theta_{jkl} = 1.$$

Relating this back to the likelihood equation in Equation 17.1, we have

$$p(\mathbf{y}|p, \theta) = \prod_{i=1}^{n} \sum_{k=1}^{K} p_k \prod_{j=1}^{J^*} \theta_{jk}^{y_{ij}} (1 - \theta_{jk})^{1-y_{ij}} \prod_{j=J^*+1}^{J} \prod_{l=1}^{3} \theta_{jkl}^{\mathbf{I}_{y_{ij}=l}}$$

assuming the first J^* symptoms are coded as Bernoulli and the remaining $(J - J^*)$ as Multinomial.

The inclusion of a latent variable $\mathbf{C}$ in the FMM allows us to model each subject's membership to one of the K subgroups. Let C_i be a single-valued, discrete latent variable, taking values $k = 1, \ldots, K$, linked to the model's component proportions,

$$C_i \sim \text{Multinomial}(p_1, \ldots, p_K).$$

Consequently, p_k now has the interpretation of the probability that a randomly selected subject is assigned to subgroup k, $p_k = Pr(C_i = k)$.

The Bayesian specification of the above model involves the inclusion of prior distributions to express uncertainty about the component proportions $\mathbf{p}$ and component density parameters, $\Theta = \{\theta_{jk}, \theta_{jkl}\}$. In this chapter, we only consider conjugate priors for the model parameters. For the above model, priors take the following form:

$$\begin{aligned}\mathbf{p} &\sim \text{Dirichlet}(\gamma_1, \ldots, \gamma_K)\\ \theta_{jkl} &\sim \text{Dirichlet}(\eta_1, \ldots, \eta_L)\\ \theta_{jk} &\sim \text{Beta}(\alpha, \beta)\end{aligned}$$

where $(\boldsymbol{\gamma}, \boldsymbol{\eta}, \alpha, \beta)$ are the hyperparameters set by the user to express their level of uncertainty about the parameters of interest. For vague, non-informative priors, we choose $\boldsymbol{\gamma} = \mathbf{1}$, $\boldsymbol{\eta} = \mathbf{1}$, $\alpha = 1$ and $\beta = 1$.

From an inferential perspective, FMMs are appealing as they facilitate the expression of uncertainty in each subject's true subgroup membership. This is achieved upon calculating posterior probabilities of membership, which is a simple application of Bayes' rule. For each subject, these probabilities are given by

$$\begin{aligned}Pr(C_i = k|\mathbf{y}_i, \Theta) &= \frac{Pr(\mathbf{y}_i|C_i = k, \Theta)Pr(C_i = k)}{Pr(\mathbf{y}_i|\Theta)}\\ &= \frac{p_k p(\mathbf{y}_i|\Theta_k)}{\sum_{k=1}^{K} p_k p(\mathbf{y}_i|\Theta_k)}.\end{aligned}$$

In studies of patient classification, these probabilities may be used to explore (i) the strength of evidence for a single patient belonging to a single latent class and, in doing so, (ii) the possibility that the patient may convey qualities of multiple classes. These questions are explored for Case study 1 in Section 17.4.1, using information contained in the MCMC output (see Section 17.3.5 for details on computation).

17.3.2 Trajectory mixture models

Latent class models have been extended to incorporate longitudinally observed data and are known as growth mixture models (Muthen 2002; Muthen and Shedden 1999; Nagin 1999). These models assume that the trajectories of the observed data from n subjects are driven by an underlying subject-level latent (unobserved) growth process. The mean structure for this latent growth process itself depends on whether the subject belongs to one of K latent classes, where $K \ll n$. This approach has several useful features. When longitudinal measures are under consideration, the equation becomes

$$p(\mathbf{y}_{ti}|p, \theta) = \sum_{k=1}^{K} p_k \prod_{t=1}^{T} f(y_{ti}|\theta_{tik}),$$

where t denotes measurement or observation times and $t = 1, \ldots, T$. This is different from Equation 17.1 in that there is an additional layer of complexity with the inclusion of multiple observation times. Let y_{ti} be the response of individual i ($i = 1, 2, \ldots, n$) at time t ($t = 1, 2, 3, 4$). Then

$$\begin{aligned} y_{ti}|\eta_i &\sim \mathrm{N}(\Lambda_t \eta_i, \sigma_i^2) \\ \eta_i|C_i = k &\sim \mathrm{N}_m(\theta_k, \sigma^2) \\ C_i &\sim \mathrm{Multinomial}(p_1, \ldots, p_k) \\ p_k &\sim \mathrm{Dirichlet}(\gamma_1, \gamma_2, \ldots, \gamma_k) \\ \sigma_i = \sigma_k \mid C_i = k &\sim \mathrm{Uniform}(\alpha_1, \alpha_2). \end{aligned}$$

Here, C_i gives the latent class membership for subject i with $C_i = k$ if subject i belongs to class k ($k = 1, \ldots, K$). The subject-level variances σ_i^2 were set to be equal over time but able to vary across groups in this case study. Λ_t denotes the specification of the form or shape of the trajectory. The $\eta_i = \theta_k + b_i$ represent the trajectory parameters, where θ_k are the class-dependent model parameters and b_i the random components. For the linear trajectory there are two trajectory components and a variance component for $\Theta_k = \theta_{1k}, \theta_{2k}, \theta_{3k}$ for the intercept, slope and variance parameters for each of the $k = 1, \ldots, K$ groups. For a quadratic trajectory there are three trajectory components and a variance component for $\Theta_k = \theta_{1k}, \theta_{2k}, \theta_{3k}, \theta_{4k}$ for the intercept, slope, quadratic parameter and variance parameters for each of the $k = 1, \ldots, K$ groups. The parameters $(\boldsymbol{\gamma}, \boldsymbol{\alpha})$ are the hyperparameters set by the user to express their level of uncertainty about the parameters of interest. For vague, non-informative priors, we choose $\boldsymbol{\gamma} = \mathbf{1}$, $\alpha_1 = 0.01$ and $\alpha_2 = 10$.

Alternate trajectory parameterizations can include piecewise polynomial spines or change point models, where different polynomial functions are exhibited before and after the known or unknown change point. Bayesian applications of change point models include cognitive function in dementia sufferers (Hall *et al.* 2003), markers for ovarian cancer (Skates *et al.* 2001) and degree of cognitive recovery (Rolfe *et al.* 2011).

A two-piece linear process was used as the underlying growth trajectory profile for the longitudinal response. This allowed for the identification of two temporal responses, L_1 and L_2: for the treatment phase, baseline (T1) to 1 month post-chemotherapy (T2) and the recovery phase from 1 month (T2) to 18 months (T4) post-chemotherapy. The following is a two piecewise linear model for verbal memory trajectories where the first piece L_1 covers the period from baseline or initial time (T1) to 1 month after completion of chemotherapy (T2). Values for $L_1(t)$ were set as (0, 1, 1, 1) and $L_2(t)$ as (1, 1, 2, 4) for times $t = 1$ to 4 respectively.

If there are K groupings the trajectory mixture model can be written with superscript k indicating class or group k,

$$\lambda_{ti}^k = \beta_0^k + \beta_1^k L_1(t) + \beta_2^k L_2(t). \tag{17.2}$$

Estimates of probability of group membership are also obtained. If classes are well defined, then each subject will have a high probability of belonging to a single class.

The Bayesian latent class growth mixture model proposed is

$$\begin{aligned} y_{ti} &\sim \mathrm{N}(\lambda_{ti}^{k}, \sigma_i^2) \\ \lambda_{ti}^{k} \mid C_i = k &\sim \mathrm{N}_m(\beta_k^*, \sigma^2) \\ C_i &\sim \mathrm{Multinomial}(p_1, \ldots, p_k) \\ p_k &\sim \mathrm{Dirichlet}((1, 1, \ldots, 1)_k) \\ \sigma_i^2 = \sigma_k^2 \mid C_i = k &\sim \mathrm{Uniform}(\alpha_1, \alpha_2). \end{aligned}$$

Here, C_i gives the latent class membership for subject i with $C_i = k$ if subject i belongs to class k ($k = 1, \ldots, K$). The λ_{ti}^k are determined by class membership based on Equation 17.2.

The regression parameters β_1^k and β_2^k used non-informative prior distributions, namely N(0,1000) with β_0^k being ordered, $\beta_0^1 < \beta_0^2 < \beta_0^3$, so $\beta_0^2 = \beta_0^1 + \theta_1$, and $\beta_0^3 = \beta_0^1 + \theta_1 + \theta_2$ with θ_1, θ_2 restricted to positive values from N(0,1000) for $k = 2$ or $k = 3$.

Inclusion of covariates

Covariates can be included in the growth mixture models in two ways: as part of the trajectory model, that is as influencing the level of the trajectory; and/or as interactions with polynomial regression parameters (Colder *et al.* 2002; Nagin 2005). Alternatively, covariates can be included to influence the probability of group or class allocation, that is as predictors of class membership similar to parameters of a logistic regression (Elliott *et al.* 2005; Muthen 2004).

The addition of a time-invariant covariate, x_m, say, for class k (indicated by a superscript) is presented as follows:

$$\begin{aligned} \lambda_{ti}^k = & \beta_0^k + \beta_1^k L_1(t) + \beta_2^k L_2(t) \\ & + B_{x0m}^k x_{im} + B_{x1m}^k x_{im} L_1(t) + B_{x2m}^k x_{im} L_2(t), \end{aligned}$$

where $L_1(t)$ and $L_2(t)$ are functions of time.

Time-varying covariates, denoted by x_{ts}, say, can also be included as part of the trajectory process (Muthen 2004; Nagin 2005). Thus the model for the kth group which includes both a time-invariant covariate x_m and time-varying covariate x_{ts}, say, is

$$\begin{aligned} \lambda_{ti}^k = & \beta_0^k + \beta_1^k L_1(t) + \beta_2^k L_2(t) \\ & + B_{x0m}^k x_{im} + B_{x1m}^k x_{im} L_1(t) + B_{x2m}^k x_{im} L_2(t) + \kappa_{tk} x_{tis}. \end{aligned}$$

As before, the categorical latent variable C_i represents the unobserved subpopulation class membership for subject i, with $C_i = 1, 2 \ldots, K$.

A multinomial logistic regression model can specify the functional relationship between the probability of class membership p_k for the kth group, where $k = 1, \ldots, K$, and set of M covariates x_m, where $m = 1, \ldots, M$, and is estimated simultaneously with the trajectory parameters (Muthen 2004; Muthen and Shedden 1999; Nagin 2005).

The probability of class membership for class k can be defined as

$$Pr(p_k(x_m)) = \frac{\exp(\delta_{0k} + \delta_{1k}x_{i1} + \cdots + \delta_{mk}x_{im} + \cdots + \delta_{Mk}x_{iM})}{\sum_{k=1}^{K} \exp(\delta_{0k} + \delta_{1k}x_{i1} + \cdots + \delta_{mk}x_{im} + \cdots + \delta_{Mk}x_{iM})},$$

where the logistic regression parameters δ_{mK} are set to zero for last class K being set as the reference class.

For $K = 2$

$$\log\left[\frac{p_1(x_{mi})}{p_2(x_{mi})}\right] = \delta_{01} + \delta_{11}x_{1i} + \cdots + \delta_{m1}x_{mi} + \cdots + \delta_{M1}x_{Mi}.$$

In the Bayesian literature, a range of prior distributions have been used for the logistic regression parameters, and include non-informative normal prior distribution N(0,1000) (Congdon 2005), N(0,9/4) prior (Elliott *et al.* 2005; Leiby *et al.* 2009) and weakly informative Cauchy distribution priors with mean 0 and scale 2.5 (Gelman *et al.* 2008).

Care needs to be taken in the selection of covariates and where they are best used, with the trajectory or with the probability of class membership part of the model. Muthen indicates that the same time-invariant covariates can be used in both parts; however, this often results in the covariate being a significant contributor to the trajectory part of the model and non-significant contributor to group membership (Muthen and Shedden 1999; Nagin 2005). Nagin (2005) restricts covariates in the class prediction part of the model to those variables available at baseline for interpretive meaning.

Convergence

Gelman (1996) discusses possible ways to compare parallel chains, with the parameter–iteration plots of overlaid chains for each estimated model parameter used to visually assess their degree of separation. The Gelman–Rubin statistic, which is used to assess or monitor convergence (Gelman and Rubin 1992), takes a quantitative approach which separately monitors the convergence of all parameters of interest and is based on comparing the variance between multiple chains with the variances within each chain. Convergence is monitored by estimating the factor by which this scale parameter might shrink if sampling were to continue indefinitely. So for each estimated scalar parameter ψ the simulation draws are labelled as ψ_{ij} $(i = 1, \ldots, n; j = 1, \ldots, m)$ for m $(m > 1)$ simulated chains of length n (after discarding burn-in) and where the between- and within-sequence variances are B and W respectively:

$$B = \frac{n}{m-1}\sum_{j=1}^{m}(\bar{\psi}_{.j} - \bar{\psi}_{..})^2; \qquad \bar{\psi}_{.j} = \frac{1}{n}\sum_{i=1}^{n}\psi_{ij}, \bar{\psi}_{..} = \frac{1}{m}\sum_{i=1}^{m}\psi_{.j}$$

$$W = \frac{1}{m}\sum_{j=1}^{n} s_j^2; \qquad s_j^2 = \frac{1}{n-1}\sum_{i=1}^{n}(\psi_{ij} - \bar{\psi}_{.j})^2.$$

The marginal posterior variance of the estimated parameter ψ is denoted as $\widehat{var(\psi)}$ as a weighted average of B and W, so

$$\widehat{var(\psi)} = \left(\frac{n-1}{n}\right) W + \left(\frac{1}{n}\right) B,$$

which overestimates the marginal posterior variance assuming over-dispersion of the starting distribution but unbiased under stationarity. The potential scale reduction is estimated by $\hat{R} = \sqrt{\widehat{var}(\psi)/W}$ which tends to 1 as $n \rightarrow \infty$. If the potential scale reduction is high then further simulations are indicated, but if $\hat{R}$ is near 1, that is where $\hat{R} < 1.1$ is an indicator of convergence. In this chapter, visual assessment of multi-chain parameter simulation plots and the Gelman–Rubin diagnostic $\hat{R} < 1.1$ were the primary methods of assessing convergence.

17.3.3 Goodness of fit

When dealing with complex data, the true number of latent subgroups is unknown a priori. Therefore, from an inference perspective, the selection of the most appropriate value for K is an important task. To assist, there exist many goodness-of-fit criteria that measure a model's ability to adequately describe the data. In this section, we briefly outline a number of goodness-of-fit criteria commonly employed in finite mixture modelling.

The most commonly used assessments of model fit are in the form of information criteria. The general form of information criteria contain a likelihood-based term plus a complexity penalty. Table 17.1 summarizes a number of information criteria commonly applied in mixture modelling.

Table 17.1 Summary of the Akaike, Bayesian and deviance information criteria used in finite mixture modelling.

Criterion	Formula	Reference
AIC	$-2\log L + 2 * p$	Akaike (1987)
BIC	$-2\log L(\Theta) + p_K \log(n)$	Schwartz (1978)
DIC	$-2E(\log L) + p_D$	Celeux *et al.* (2006)
Adjusted BIC	$-2\log L(\Theta) + p_K \log((n+2)/24)$	Nylund *et al.* (2007)

Given models with different values for K, we seek to minimize one or more of these criteria, such that the value of K corresponding to this minimized value is taken as the most appropriate number of latent components to fit to the data. In recent studies assessing the ability of these and other criteria for detecting the true number of clusters (Nylund *et al.* 2007; Tofighi and Enders 2007), the BIC and adjusted BIC performed well.

Alternatively, the choice of K can be inferred via the use of model comparison criteria. Unlike the aforementioned goodness-of-fit criteria that are based purely on minimization, model comparison criteria are pairwise assessments of models, typically in the form of a ratio of evidence or support for one model over the other. In a Bayesian context, Bayes' factors (Kass and Raftery 1995) are a natural choice where, given models with K and K' components respectively,

$$B_{K',K} = \frac{m_{K'}(\mathbf{y})}{m_K(\mathbf{y})},$$

where $m_K(\mathbf{y})$ is the marginal likelihood, given by the integral $m_k(y) = \int_{\Theta_k} \pi(y|\Theta_k)\pi(\Theta_k)\,d\Theta_k$. Therefore, $B_{K',K}$ is interpreted as the ratio of the probability of $\mathbf{y}$ being generated by a mixture with K' components versus the probability of $\mathbf{y}$ being generated by K components. Given two models, values of $B_{K',K}$ in excess of 3 are generally taken as evidence for the K' component model, whereas values less than the reciprocal tend to support the K component model (Jeffreys 1961). Given the presence of the latent variable $\mathbf{C}$, $m_K(\mathbf{y})$ and therefore $B_{K',K}$ cannot be computed analytically. Nevertheless, a stable approximation to $B_{K',K}$ that makes use of existing MCMC output and is simple to implement was proposed by Chib (1995).

17.3.4 Label switching

Given the general representation of the FMM in Equation 17.1, a concern when it comes to modelling estimation is the invariance of the likelihood to permutations on K, known as the *label switching* problem. For MCMC in particular, the presence of label switching can have severe implications for posterior inference, as random permutations between MCMC iterations lead to the arbitrary labelling of subgroups and, consequently, the posterior distributions of component-specific parameters are affected. While we restrict our discussion here, the reader is directed to Jasra *et al.* (2005) for a more detailed treatment of label switching.

The simplest and perhaps most common approach to addressing the label switching problem is the placement of constraints on the prior. For instance, one may impose a strict ordering on the mixture component proportions, $\eta_1 < \eta_2 < \cdots < \eta_K$, to be implemented in either a within-MCMC or post-hoc fashion. With the main benefit this ordering being its simplicity of application, prior constraints may not be desirable in some situations. For example, in cases where two or more mixture components have similar estimates for select parameters, this forced ordering may distort the interpretation of the true underlying subgroups. Furthermore, for more complicated models, the choice of constraint is a cumbersome task. Frühwirth-Schnatter (2004) proposed a permutation MCMC sampler to encourage 'balanced' label switching, serving as a diagnostic approach to identify the most appropriate prior constraint or constraints.

In cases where the choice of prior constraint are of concern, an appealing alternative is the use of loss functions to 'untangle' posterior samples where label switching is likely to have occurred. For a model with latent labels C_i and component-specific parameters θ_{C_i}, this approach aims to select the permutation of C_i at each MCMC iteration, say ν, that minimizes a pre-specified loss,

$$L(a, \theta) = \min_{\nu} L(a, \nu(\theta)).$$

For clustering problems, this loss may refer to the Kullback–Leibler divergence (Stephens 2000) or a similar metric, including the canonical scalar product (Marin *et al.* 2005) and the sum of standardized errors over all model parameters. The loss function approach may be carried out either post hoc or, when computational storage is of concern, at the completion of each MCMC iteration (Celeux 1998).

17.3.5 Model computation

In this section, we outline the computation approach taken for each case study. In each case, Gibbs sampling was employed.

Case study 1

Model estimation for the FMM was implemented in MATLAB, with the Gibbs sampler summarized in Algorithm 18.

Algorithm 18: Gibbs sampler pseudo-code for Case study 1 (mixture of Bernoulli and multinomial items)

1. Draw initial values for $\mathbf{p}$, Θ from prior distributions.
2. Generate the latent classifications $C_i = k$ for $i = 1, \ldots, n$ and $k = 1, \ldots, K$ from its posterior conditional distribution:

$$Pr(C_i = k | p_k, \theta_{jk}, \theta_{jkl}, \mathbf{y}_i) \propto p_k f(\mathbf{y}_i | \theta_{jk}, \theta_{jkl}).$$

3. Simulate $\mathbf{p}$ from $D(N_1(\mathbf{C}) + \gamma_1, \ldots, N_K(\mathbf{C}) + \gamma_K)$, where $N_k(\mathbf{C})$ is the number of individuals assigned to component k.
4. Simulate θ_{jk} from $B(S_{jk}(\mathbf{C}) + \alpha, N_k(\mathbf{C}) - S_{jk}(\mathbf{C}) + \beta)$, where $S_{jk}(\mathbf{C})$ is the number of individuals with symptom j present, having been assigned to component k.
5. Simulate θ_{jkl} from $D(M_{jk1}(\mathbf{C}) + \eta_1, \ldots, M_{jkL}(\mathbf{C}) + \eta_L)$, where M_{jkl} equals the number of individuals with level l of symptom j, conditional on k.
6. Store current values for all parameters.
7. Return to step 2 and repeat for T iterations.

Based on current values for $\mathbf{p}$ and Θ, step 2 updates the posterior probabilities of membership for each individual. Sampling from these multinomial probabilities then provides an updated latent classification $\mathbf{C}$, then used to compute sufficient statistics required to update the component parameters. For $\mathbf{p}$ in step 3, we require the number of individuals assigned to each latent class, denoted by $N_k(\mathbf{C}), k = 1, \ldots, K$. For each θ_{jkl} in step 4, the term $S_{jk}(\mathbf{C})$ is defined as the number of subjects where symptom j is recorded as present, conditional on their assignment to latent class k. This may be conveniently represented by the following indicator function:

$$S_{jk}(\mathbf{C}) = \sum_{i=1}^{n} \mathbf{I}\left\{y_{ij} = 1 \cap C_i = k\right\}, \qquad \forall y_{ij} \in \{0, 1\}.$$

Finally, in step 5, we define $M_{jk1}(\mathbf{C})$ by the indicator function

$$M_{jkl}(\mathbf{C}) = \sum_{i=1}^{n} \left\{y_{ij} = l \cap C_i = k\right\}, \qquad \forall y_{ij} \in \{0, \ldots, L\},$$

which represents the number of subjects with level l of multinomial symptom j, again conditional on membership to the kth subgroup. This sufficient statistic is used to update each θ_{jkl}.

For Case study 1, Algorithm 18 was implemented for two chains, each consisting of T =100 000 iterations with the first 50 000 in each chain discarded as the burn-in period.

On completion of the above-defined Gibbs sampler, it may be of interest to assess a select subject's true subgroup membership. This can be achieved by computing the posterior probabilities of membership based on MCMC-based estimates of the component weights and parameters, for example the posterior means. In Section 17.4.1, these probabilities are explored for select symptom profiles, as a means of assessing the 'strength' of a subject's membership of a single subgroup or multiple subgroups.

Case study 2

WinBUGS and R2WinBUGS were used for the cognition trajectory mixture models.

Data are specified as $AVLT(i, j), i = 1, \ldots, N, j = 1, \ldots, T$; N is the number of subjects, T the number of measurement occasions, K the number of groups. Piecewise components are set as $pw1 = (0, 1, 1, 1)$, $pw2 = (1, 1, 2, 4)$; Dirichlet parameters are set as $\alpha = (1, 1)$ or $\alpha = (1, 1, 1)$ for $K = 2$ or 3.

Initial or starting values are specified for class membership $C(i)$, trajectory parameters $\beta_0^1, \beta_1^k, \beta_2^k, \theta^{k-1}, \sigma^k$, and the covariate trajectory parameters b_m^k or covariate logistic parameters δ_m^{k-1} for each chain. Two chains were used with 10 000 iterations of which 1000 were discarded as burn-in.

Table 17.2 Selected goodness-of-fit and classification criteria for FMM with $K = 1$ to $K = 4$ classes.

K	AIC	BIC_{adj}	$B_{K+1,K}$
1	3004.6	3009.8	9×10^{202}
2	2933.2	2944.3	0.03
3	2982.2	2999.2	10×10^{-78}
4	2964.0	2986.9	—

17.4 Data analysis and results

17.4.1 Case study 1: Phenotype identification in PD

Here, the main results for Case study 1 are summarized, with a description of the data provided in Section 17.2.1. Data for this case study were modelled with the FMM presented in Section 17.3.1, with three Bernoulli and two three-level Multinomial items. For a more detailed summary of results for this case study, see White *et al.* (2010).

Case study 1: Model selection

Beginning with goodness of fit, models with different values of K were fitted and evaluated using AIC, adjusted BIC and Bayes' factors, all featured in Section 17.3.3. The results of this analysis are presented in Table 17.2, for $K = 1$ to $K = 4$. For both AIC and adjusted BIC, values suggest an FMM with $K = 2$ or $K = 3$ classes providing the best fit to the data. However, the inclusion of Bayes' factors showed clear support for a two-class model, with $B_{3,2} = 0.03$.

Case study 2: Subgroup identification

Taking the $K = 2$ model, the subgroups identified had similar component weights, 0.48 for Class 1 and 0.52 for Class 2. For Bernoulli distributed items (speech, postural instability and offs), Table 17.3 contains the estimated class probabilities of presence and corresponding 95% credible intervals. For Class 1, we see that, out of all symptoms, speech impairment had the highest posterior probability of presence.

Table 17.3 Posterior means and 95% credible intervals for Bernoulli distributed symptoms.

	Speech	Postural instability	Offs
Class 1	0.53 (0.46, 0.67)	0.29 (0.14, 0.37)	0.24 (0.12, 0.32)
Class 2	0.88 (0.77, 0.94)	0.75 (0.58, 0.81)	0.63 (0.54, 0.79)

Table 17.4 Posterior means and 95% credible intervals for multinomial distributed symptoms.

		None unilateral	Bilateral asymmetric	Bilateral symmetric
Tremor	Class 1	0.63 (0.53, 0.74)	0.31 (0.22, 0.40)	0.06 (0.01, 0.11)
	Class 2	0.38 (0.30, 0.48)	0.42 (0.34, 0.50)	0.20 (0.13, 0.27)
Rigidity	Class 1	0.52 (0.40, 0.64)	0.40 (0.29, 0.50)	0.09 (0.01, 0.19)
	Class 2	0.16 (0.07, 0.25)	0.29 (0.18, 0.38)	0.56 (0.44, 0.70)

Conversely, Class 2 had high posterior probabilities of presence for all three symptoms.

For the Multinomial items, tremor and rigidity, estimates are summarized in Table 17.4. Here, Class 1 was characterized by none/unilateral forms of tremor and rigidity, with some evidence of none/unilateral and bilateral asymmetric tremor. This is in stark contrast to Class 2, described as containing patients with bilateral rigidity, in addition to one of none/unilateral or bilateral asymmetric tremor.

Taking the information presented in Tables 17.3 and 17.4, Class 1 was labelled the 'Less Affected' subgroup. This overall description was attributed to low posterior probabilities of presence for postural instability and offs, in addition to majority none or unilateral expressions of both tremor and rigidity. Conversely, Class 2 is labelled the 'More Affected' subgroup with reference to bilateral versions of tremor and rigidity, along with high probabilities of presence for speech impairment, postural instability and offs.

In Section 17.3.1, it was highlighted that a key strength of the FMM was its ability to quantify uncertainty in true subgroup membership. Exploring this notion further, Table 17.5 gives the posterior probabilities of membership for six different symptom profiles, based on parameter estimates for both the component weights and component-specific parameters.

Table 17.5 Posterior probabilities of membership for six observed symptom profiles, based on posterior mean estimates. Each symptom profile is of the form (tremor, rigidity, speech, postural instability, offs).

Symptom profile	$Pr(C_i = 1 \mid \mathbf{y}_i, \Theta)$	$Pr(C_i = 2 \mid \mathbf{y}_i, \Theta)$	Frequency
(1,2,1,0,0)	0.87	0.13	148
(2,3,1,1,1)	0.01	0.99	143
(2,2,0,1,0)	0.67	0.33	8
(2,2,1,1,0)	0.37	0.63	20
(1,2,1,1,0)	0.54	0.46	29
(1,3,1,0,0)	0.47	0.53	22

The comparison of different observed symptom profiles with regards to posterior probabilities of membership shows varying degrees of uncertainty. For the first two symptom profiles listed in Table 17.5, there is convincing evidence of membership of a single latent class: 0.87 to Class 1 for the first and 0.99 to Class 2 for the second symptom profile. Relative to others listed, we define these two profiles as corresponding to 'Distinct' phenotypes present in the data set, and, given their frequencies, as also the two most frequently observed symptom profiles. By comparison, the true subgroup membership of the final two profiles (1,2,1,1,0) and (1,3,1,0,0) is less clear, with close to equal probabilities of belonging to either Class 1 or Class 2. With this in mind, these symptom profiles are labelled as 'Overlapping', in light of their probable expression of features characteristic of both identified subgroups.

17.4.2 Case study 2: Trajectory groups for verbal memory

Only two- and three-class models were considered due to the possibility of small numbers for group membership given the relatively small total sample size n=120. The results of two- and three-class models fitted without the inclusion of covariates (unconditional) are presented in Table 17.6 and Table 17.7 respectively. All two- and three-class models attained convergence with the Gelman–Rubin diagnostic $\hat{R} < 1.1$ for all modelled parameters. Based on their smaller DIC values, the three-class models for both learning and immediate retention and the two-class model for delayed recall are preferred.

For all classes and all verbal memory outcomes, the first part of the piecewise function exhibited a negative slope (β_1), thus indicating a decline from baseline T1 (before chemotherapy treatment) to T2 (1 month post-chemotherapy), where β_1 measured the rate of decline. On the whole the second part of the piecewise function indicated a positive or flat slope (β_2), implying a varying rate of recovery or improvement.

For learning, the rate of decline, β_1, was steepest for the low class (Class 1), becoming less so as the baseline level of the response increased for the mid and high classes. Similarly the rate of recovery, β_2, increased from low to high classes. For immediate retention, the steepest decline, β_1, was for the mid class, which subsequently exhibited the strongest rate of improvement, β_2. However, the rate of improvement for the low class was relatively flat. For delayed recall, for the two classes, again the low class indicated the steepest decline, β_1, but the higher class exhibited the fastest rate of recovery.

Inclusion of covariates

As discussed earlier, covariates can be included in the growth mixture models in two ways: as part of the trajectory model, that is as influencing the level of the trajectory and/or as interactions with polynomial regression parameters; or alternatively as influencing the probability of group or class allocation, similar to the parameters of a logistic regression. Care is required if a specific covariate is included in both ways in a single model as lack of identifiability may occur. Using a common covariate

Table 17.6 Posterior means, SD and 95% credible interval (CI) for each two-class unconditional model for verbal memory outcomes.

	Class 1 Low posterior estimates				Class 2 High posterior estimates			
	Mean	SD	95% CI		Mean	SD	95% CI	
Learning								
β_0	47.970	0.922	46.19	49.79	56.520	0.860	54.83	58.22
β_1	−4.300	1.049	−6.390	−2.25	−2.43	0.973	−4.34	−0.54
β_2	0.850	0.367	0.130	1.57	1.34	0.353	0.65	2.03
σ	6.150	0.295	5.60	6.76	5.490	0.293	4.94	6.09
n	63.310	3.669	56	71	56.69	3.669	49	64
p	0.530	0.054	0.42	0.63	0.470	0.054	0.37	0.58
DIC	3101.4	p_D	48	10.91				
Immediate retention								
β_0	9.780	0.306	9.20	10.37	12.720	0.309	12.00	13.35
β_1	−1.450	0.340	−2.10	−0.80	−0.920	0.321	−1.5	−0.29
β_2	0.240	0.120	0.001	0.47	0.260	0.113	0.04	0.49
σ	2.050	0.099	1.90	2.25	1.660	0.111	1.40	1.88
n	67.370	5.270	57	77	52.630	5.270	43	63
p	0.560	0.062	0.44	0.68	0.440	0.062	0.32	0.56
DIC	2007.6	p_D	45.9					
Delayed recall								
β_0	9.780	0.301	9.200	10.360	12.730	0.303	12.000	13.330
β_1	−1.460	0.339	−2.100	−0.790	−0.920	0.318	−1.500	−0.290
β_2	0.240	0.119	0.005	0.470	0.260	0.112	0.038	0.480
σ	2.050	0.099	1.900	2.250	1.660	0.111	1.400	1.880
n	67.640	5.242	57	77	52.370	5.242	43	63
p	0.560	0.062	0.440	0.680	0.440	0.062	0.320	0.560
DIC	2008.2	p_D	46.5					

adjustment for all classes in the trajectory process may be a way to overcome this problem.

As an example, the covariate 'age at diagnosis' has been included (i) as moderating the level of the trajectory mixture process and (ii) as a predictor of class membership. The results for the three-class learning models with age centred at 49 years as the covariate are presented in Table 17.8.

The addition of the covariate, age, for learning was restricted to the three-class model as it was the preferred alternative when the unconditional model was fitted. The covariate, age, had a significant negative impact B_{age} on the trajectory intercept for low and mid classes, with a reduction of −0.19 (0.09) and −0.29 (0.09) for each year above 49 respectively. However, there was no impact of the age covariate on the trajectory intercept for the high class.

Table 17.7 Posterior means, SD and 95% credible interval (CI) for each three-class unconditional model for verbal memory outcomes.

	Class 1 Low posterior estimates				Class 2 Mid posterior estimates				Class 3 High posterior estimates			
	Mean	SD	95% CI		Mean	SD	95% CI		Mean	SD	95% CI	
Learning												
β_0	45.41	1.81	41.48	48.55	50.13	1.11	48.14	52.50	57.00	0.94	55.22	58.80
β_1	−5.74	1.93	−9.64	−2.01	−3.29	1.04	−5.32	−1.24	−2.39	1.04	−4.39	−0.32
β_2	0.71	0.72	−0.67	2.16	0.90	0.38	0.15	1.64	1.44	0.38	0.71	2.18
σ	6.44	0.58	5.39	7.66	4.81	0.42	4.02	5.63	5.31	0.33	4.68	5.95
n	24.08	5.90	13	37	47.03	6.56	34	60	48.89	5.83	36	58
p	0.20	0.06	0.10	0.33	0.39	0.07	0.26	0.53	0.41	0.06	0.27	0.52
DIC	3074.1	p_D	108.3									
Immediate retention												
β_0	8.61	0.60	7.33	9.70	10.76	0.40	10.03	12.00	13.08	0.33	12.44	14.00
β_1	−1.18	0.59	−2.30	0.01	−1.55	0.33	−2.21	−0.90	−0.74	0.36	−1.43	0.00
β_2	0.04	0.23	−0.43	0.46	0.36	0.12	0.11	0.60	0.22	0.12	−0.02	0.46
σ	1.97	0.17	1.65	2.30	1.72	0.12	1.48	1.90	1.50	0.13	1.22	1.80
n	25.37	6.72	14	40	55.60	6.58	42	68	39.03	6.21	26	50
p	0.22	0.07	0.10	0.36	0.46	0.07	0.32	0.59	0.33	0.07	0.20	0.45
DIC	1972.2	p_D	103.1									
Delayed recall												
β_0	8.80	0.63	7.46	9.91	10.43	0.39	9.69	11.23	12.91	0.31	12.28	13.51
β_1	−2.19	0.70	−3.62	−0.83	−1.59	0.40	−2.35	−0.77	−1.44	0.35	−2.11	−0.74
β_2	0.11	0.27	−0.49	0.59	0.33	0.14	0.06	0.59	0.43	0.12	0.20	0.66
σ	2.03	0.27	1.42	2.53	1.91	0.16	1.56	2.19	1.57	0.12	1.36	1.81
n	22.53	9.18	5	43	55.41	8.92	35	70	42.06	4.83	33	51
p	0.19	0.08	0.05	0.37	0.46	0.09	0.28	0.61	0.35	0.06	0.24	0.47
DIC	2065.8	p_D	129.6									

Table 17.8 Posterior means, SD and 95% credible interval (CI) for three-class age covariate models for learning.

	Class 1 Low posterior estimates				Class 2 Mid posterior estimates				Class 3 High posterior estimates			
	Mean	SD	95% CI		Mean	SD	95% CI		Mean	SD	95% CI	
Trajectory												
β_0	46.22	1.54	42.94	48.97	51.72	1.29	49.33	54.47	57.75	1.78	55.32	62.67
β_1	−5.20	1.64	−8.43	−1.93	−3.32	1.05	−5.38	−1.26	−2.05	1.19	−4.31	0.38
β_2	0.79	0.59	−0.40	1.93	0.99	0.38	0.23	1.73	1.37	0.44	0.44	2.20
σ	6.30	0.48	5.44	7.31	4.79	0.42	3.99	5.61	5.06	0.53	3.66	5.88
n	32.13	7.92	20	53	49.57	7.99	34	66	38.30	11.09	12	54
p	0.27	0.08	0.15	0.45	0.41	0.08	0.26	0.57	0.32	0.10	0.09	0.48
B_{age}	−0.19	0.09	−0.35	−0.04	−0.29	0.09	−0.44	−0.08	−0.13	0.09	−0.31	0.06
DIC	3110.5	p_D	147.3									
Probability												
β_0	45.24	1.71	42.00	48.28	50.23	1.03	48.00	52.37	57.06	0.92	55.00	58.86
β_1	−5.76	1.93	−9.70	−2.08	−3.34	1.03	−5.40	−1.35	−2.35	1.05	−4.40	−0.30
β_2	0.75	0.71	−0.64	2.17	0.90	0.38	0.16	1.64	6.35	0.58	5.30	7.58
σ	1.44	0.37	0.71	2.18	4.91	0.40	4.10	5.71	5.29	0.32	4.70	5.93
δ_1	−0.89	0.48	−1.90	−0.04	0.12	0.05	0.04	0.22				
δ_2	0.05	0.31	−0.56	0.67	0.06	0.04	0.00	0.13				
n	23.42	5.59	12.00	35.00	48.75	6.31	36.00	61.00	47.82	4.98	37.00	57.00
DIC	3080.1	p_D	113.3									

Table 17.9 Class membership for comparisons of the three classes of unconditional, trajectory covariate and probability covariate models for learning.

		Age in trajectory			Age in probability			
		Low	Mid	High	Low	Mid	High	*n*
Unconditional	Low	14	9	1	13	10	1	24
	Mid	15	32	9	9	35	12	56
	High	0	10	30	0	5	35	40
	n	29	51	40	22	50	48	120

		Age in probability			
		Low	Mid	High	*n*
Age in trajectory	Low	20	9	0	29
	Mid	2	38	11	51
	High	0	3	37	40
	n	22	50	48	120

When the covariate age was included in the probability of class membership, there was a significant positive effect in the odds of being in the mid class compared with the high class, but none with the odds for the low/high class comparison.

Class allocation was made using the posterior median of $C(i)$ for each individual i, for $i = 1$ to n. Even though class membership numbers were similar for the three classes of the unconditional (24,56,40), trajectory covariate (29,51,40) and the probability covariate (22,50,48) models, the membership composition differed markedly as can be seen in Table 17.9.

There was no departure from symmetry with the class allocation numbers for the unconditional model with each of the covariate models; however, divergence from symmetry occurred when the covariate in trajectory and covariate in probability models were compared. The inclusion of a covariate in the mixture trajectory model in either the trajectory part or in the probability of membership had a marked impact on class membership.

17.5 Discussion

In this chapter, we have highlighted the use of latent class models in modelling and gaining understanding of complex populations, often encountered in studies of medicine. For a flexible class of models, their utility was showcased by means of two real-life case studies: the first a study of symptom patterns in Parkinson's disease given cross-sectional data; and the second an investigation into modelling temporal changes in cognition measures for women undergoing chemotherapy for the treatment of breast cancer.

The use of these models in medicine is by no means restricted to the two featured case studies, with many other studies appearing in the literature. For example, Chen *et al.* (2009) also used FMM to analyse migraine symptom data and found that there are potentially three subtypes of migrainous headache. However, unlike Case study 1, the responses to the migraine symptoms are binary, therefore these authors assumed all components to have binomial distribution. Similarly, in Walsh (2006), the analysis of symptoms recorded at a single time point from patients with Alzheimer's disease revealed three latent subgroups with different dominant characteristics.

Likewise, trajectory mixture models similar to those outlined for Case study 2 have been used by Elliott *et al.* (2005), Leiby *et al.* (2009), Rolfe (2010) and Rolfe *et al.* (2011) for medical applications in a Bayesian framework; however, the tradition of trajectory mixture models or growth mixture models has been widely used in the social science area with the frequentist paradigm as seen in the work of Bengt Muthen and David Nagin and colleagues. The benefits of the Bayesian framework permit the ability to obtain distinct classes with relatively small subject numbers (less than 150), and in the ability to incorporate informative prior information in cases of extremely low subject numbers (Elliott *et al.* 2005).

While this chapter is intended as a general and practical introduction to latent class models, there are a number of extensions the analyst may wish to consider. Firstly, in both case studies presented, priors were conjugate and non-informative. For the former, this was done purely for ease of model estimation, as it permits the use of Gibbs sampling; however, one need not restrict oneself to the conjugate case. In this case, other methods of model estimation, for example the Metropolis–Hastings algorithm, could be implemented. Likewise, one may wish to incorporate informative belief or expert opinion into the model, in which case the prior distributions could be adjusted to reflect these beliefs. Secondly, with reference to Case study 1, the analysis of differences between subgroups with respect to external covariates could also have been explored. In a Bayesian setting, an approach to this type of analysis that fully captures uncertainty in a patient's subgroup membership is presented by White *et al.* (2010).

References

Akaike H 1987 Factor analysis and AIC. *Psychometrika* **52**, 317–332.

Cancer Council Queensland 2008 Queensland cancer statistics online. http://www.cancerqld.org.au/f/QCSOL/ (accessed 10 April 2009).

Celeux G 1998 Bayesian inference for mixtures: the label-switching problem. *COMPSTAT* **98**, 227–232.

Celeux G, Forbes F, Robert C and Titterington D 2006 Foundations of DIC. *Bayesian Analysis* **1**, 701–706.

Chen C, Keith J, Nyholt D, Martin N and Mengersen K 2009 Bayesian latent trait modeling of migraine symptom data. *Human Genetics Research* **126**, 277–288.

Chib S 1995 Marginal likelihood from the Gibbs output. *Journal of the American Statistical Association* **90**, 1313–1321.

Colder CR, Campbell RT, Ruel E, Richardson JL and Flay BR 2002 A finite mixture model of growth trajectories of adolescent alcohol use: predictors and consequences. *Journal of Consulting and Clinical Psychology* **70**(4), 976–985.

Congdon P 2005 *Bayesian Models for Categorical Data.* John Wiley & Sons, Ltd, Chichester.

Elliott M, Gallo J, Ten Have T, Bogner H and Katz Il 2005 Using a Bayesian latent growth curve model to identify trajectories of positive affect and negative events following myocardial infarction. *Biostatistics* **6**(1), 119–143.

Frühwirth-Schnatter S 2004 Estimating marginal likelihoods for mixture and Markov switching models using bridge sampling techniques. *Econometrics Journal* **7**, 143–167.

Geffen GM, Moar KJ, O'Hanlon AP, Clark CR and Geffen LB 1990 Performance measures of 16 to 86 year old males and females on the Auditory Verbal Learning Test. *The Clinical Neuropsychologist* **4**(1), 45–63.

Gelman A 1996 Inference and monitoring convergence. In *Markov chain Monte Carlo in Practice* (eds WR Gilks *et al.*), pp. 131–140. Chapman & Hall, London.

Gelman A and Rubin D 1992 Inference from iterative simulation using multiple sequences. *Statistical Science,* **7**, 457–511.

Gelman A, Jakulin A, Pittau M and Su Y 2008 A weakly informative default prior distribution for logistic and other regression models. *Annals of Applied Statistics* **2**(4), 1360–1383.

Hall CB, Ying J, Kuo L and Lipton RB 2003 Bayesian and profile likelihood change point methods for modeling cognitive function over time. *Computational Statistics & Data Analysis* **42**(1–2), 91–109.

Jasra A, Holmes C and Stephens D 2005 Markov chain Monte Carlo methods and the label switching problem in Bayesian mixture modeling. *Statistical Science* **20**, 50–67.

Jeffreys H 1961 *Theory of Probability*. Oxford University Press, New York.

Kass R and Raftery A 1995 Bayes factors. *Journal of the American Statistical Association* **90**(430), 773–795.

Leiby BE, Sammel MD, Ten Have TR and Lynch KG 2009 Identification of multivariate responders and non-responders by using Bayesian growth curve latent class models. *Journal of the Royal Statistical Society, Series C (Applied Statistics)* **58**(4), 505–524. Published online April 2009.

Marin J, Mengersen K and Robert C 2005 Bayesian modelling and inference on mixtures of distributions. In *Handbook of Statistics*, Vol. 25, pp. 459–507. Elsevier, Amsterdam.

Muthen B 2002 Beyond SEM: General latent variable modeling. *Behaviormetrika* **29**(1), 81–117.

Muthen B 2004 Latent variable analysis: growth mixture modeling and related techniques for longitudinal data. In *Handbook of Quantitative Methodology for the Social Sciences* (ed. Kaplan D), pp. 345–368. Sage, Newbury Park, CA.

Muthen B and Shedden K 1999 Finite mixture modeling with mixture outcomes using the EM algorithm. *Biometrics* **55**, 463–469.

Nagin DS 1999 Analyzing developmental trajectories: a semi-parametric, group-based approach. *Psychological Methods* **4**, 139–157.

Nagin DS 2005 *Group-Based Modeling of Development*. Harvard University Press, Cambridge, MA.

Nylund KL, Asparouhov T and Muthen BO 2007 Deciding on the number of classes in latent class analysis and growth mixture modeling: a Monte Carlo simulation study. *Structural Equation Modeling* **14**(4), 535–569.

Rolfe M 2010 *Bayesian models for longitudinal data.* PhD thesis. Queensland University of Technology.

Rolfe MI, Mengersen KL, Vearncombe KJ, Andrew B and Beadle GF 2011 Bayesian estimation of extent of recovery for aspects of verbal memory in women undergoing adjuvant chemotherapy treatment for breast cancer. *Journal of the Royal Statistical Society, Series C (Applied Statistics)* **60**, 655–674.

Schwartz G 1978 Estimating the dimension of a model. *Annals of Statistics* **6**, 461–464.

Skates SJ, Pauler DK and Jacobs IJ 2001 Screening based on the risk of cancer calculation from Bayesian hierarchical changepoint and mixture models of longitudinal markers. *Journal of the American Statistical Association* **96**(454), 429–439.

Stephens M 2000 Dealing with label switching in mixture models. *Journal of the Royal Statistical Society, Series B (Statistical Methodology)* **62**, 795–809.

Sutherland GT, Halliday G and Silburn PA 2009 Do polymorphisms in the familial Parkinsonism genes contribute to risk for sporadic Parkinson's disease?. *Movement Disorders* **24**, 833–838.

Tofighi D and Enders C 2007 Identifying the correct number of classes in a growth mixture model. In *Advances in Latent Variable Mixture Models* (eds GR Hancock and KM Samuelsen), pp. 317–342. Information Age, Greenwich, CT.

Walsh CD 2006 Latent class analysis identification of syndromes in Alzheimer's disease: a Bayesian approach. *Metodoloski Zvezki: Advances in Methodology and Statistics* **3**(1), 147–162.

White N, Johnson H, Silburn P, Mellick G, Dissanayaka N and Mengersen K 2010 Probabilistic subgroup identification using Bayesian finite mixture modeling: a case study in Parkinson's disease phenotype identification. *Statistical Methods in Medical Research. doi: 10.1177/0962280210391012.*

18

Hidden Markov models for complex stochastic processes: A case study in electrophysiology

Nicole M. White[1,4], Helen Johnson[1], Peter Silburn[2], Judith Rousseau[3] and Kerrie L. Mengersen[1]

[1]*Queensland University of Technology, Brisbane, Australia*
[2]*St. Andrew's War Memorial Hospital and Medical Institute, Brisbane, Australia*
[3]*Université Paris-Dauphine, Paris, France and Centre de Recherche en Économie et Statistique (CREST), Paris, France*
[4]*CRC for Spatial Information, Australia*

18.1 Introduction

Understanding the complexities of human physiology remains an exciting and challenging field in modern medicine. Of the many research streams in this field, a popular area is the study of action potentials (APs), or electrophysiology. Defined by a rapid rise and fall of electrical potential in an activated cellular membrane, an AP is visibly characterized by a unique waveform shape or trajectory, that is considered as an event separate from background noise. Biologically, APs play a central role in the activation of intracellular processes in the human body, including heart and muscle contraction, the release of insulin from the pancreas, facilitation of communication between neurons in the brain and motor sensory signals between the brain and muscles and tissues in the body. As such, the study of APs has the potential to gain understanding of

Case Studies in Bayesian Statistical Modelling and Analysis, First Edition. Edited by Clair L. Alston, Kerrie L. Mengersen and Anthony N. Pettitt.

these processes and how they are affected under different environmental, genetic and physical conditions.

The understanding of this and other complex phenomena involves the statistical analysis of stochastic processes. For what is essentially random behaviour over time and/or space, the collection of data on stochastic processes is carried out in a myriad of contemporary research areas, including but not limited to finance, economics, bioinformatics, signal processing and machine learning. Regardless of the origin of these data, their analyses centre on the uncovering of patterns or trends amid what is otherwise perceived as unpredictable behaviour or noise. For example, in speaker diarization (Gales and Young 2008; Tranter and Reynolds 2006), an important topic in speech processing, analysis aims to recognize not only voices in an audio recording originating from different speakers, but also transitions between speakers and the number of speakers, at the same time accounting for natural variation in an individual's tone, volume and pitch. Likewise, in bioinformatics, the alignment of multiple, highly variable DNA sequences is central to the identification of potential regions of functionality or structural importance (Eddy 1995; Holmes and Bruno 2001).

Of the statistical tools available for analysis, *hidden Markov models* (HMMs) have proven successful in light of their relative simplicity and flexibility in describing a wide variety of stochastic processes (Cappé *et al.* 2005; MacDonald and Zucchini 1997). A form of latent variable model, HMMs aim to describe the observed outputs of a stochastic process by a finite alphabet of unobservable, discrete-valued 'states', where different states are taken to represent different features of the process's behaviour and are inferred from the data. To relate these models to other chapters in this book, an HMM is a type of Bayesian network but may also be conceptualized as a finite mixture model for time-dependent observations. In this chapter, we aim to introduce these models to the reader and to demonstrate their applicability to a case study in electrophysiology. Specifically, we seek to apply an HMM to the identification and sorting of APs in extracellular recordings (White 2011).

18.2 Case study: Spike identification and sorting of extracellular recordings

In this chapter, we focus on the analysis of APs in the brain, made possible by the collection of extracellular recordings. Extracellular recordings consist of measurements of electrical potential discharged by either a single or multiple cells, in this case neurons, over time. An example of an extracellular recording is given in Figure 18.1.

From the definition of APs, their presence in Figure 18.1 is visualized as 'spikes', with an amplitude notably higher (or lower) than the background noise. That said, it is not clear which of these spikes genuinely correspond to APs and which merely form part of the background noise. For this reason, this area of research has witnessed substantial literature on statistical methods for reliable spike detection, from simple thresholding rules (Freeman 1971; Rizk and Wolf 2009; Thakur *et al.* 2007) to more adaptive alternatives, including nonlinear and wavelet-based detection methods (Kim and Kim 2000; Mtetwa and Smith 2006; Nenadic and Burdick 2005; Yang and Shamma 1988).

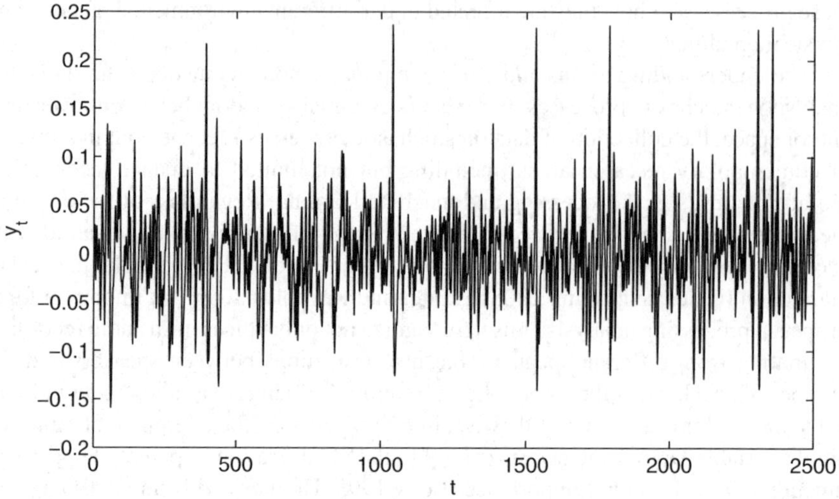

Figure 18.1 Example of a neural extracellular recording, to be analysed in Section 18.4.

In this chapter, HMMs are considered for the unsupervised detection of APs in extracellular recordings. By assuming the behaviour of an AP as a sequence of unobservable or latent states, HMMs model transitions between these states and therefore provide a means of unsupervised spike detection, without the need to set thresholds. In the next section, we first consider a single sequence HMM and how it may be specified for the spike detection problem. This model is then extended to multiple, independent HMMs, in an attempt to model spikes with different trajectories in the same recording. This concept was first proposed by Herbst *et al.* (2008), as a novel solution for the simultaneous identification and assignment of spikes to source cells, the latter commonly referred to as the 'spike sorting' problem. In this chapter, the model proposed by Herbst *et al.* (2008) is recast into the Bayesian framework, and is a summary of work presented in White (2011).

To explore these ideas, we consider a case study of extracellular recordings taken during deep brain stimulation, a popular treatment for advanced Parkinson's disease. Deep brain stimulation is a surgical procedure involving the placement of electrodes in an affected part of the brain, to provide a constant source of electrical stimulation. Best described as a 'brain pacemaker', this constant supply of electrical pulses has been consistently shown to alleviate symptoms associated with Parkinson's disease (Kleiner-Fisman *et al.* 2003; Krack *et al.* 2003; Kumar *et al.* 1998; Limousin *et al.* 1998). In this chapter, we model an extracellular recording taken at the subthalamic nucleus (STN), a popular surgical target for deep brain stimulation.

18.3 Models and methods

18.3.1 What is an HMM?

As described in Section 18.1, a HMM is defined by its use of latent or 'hidden' states to describe the behaviour of each value in an observed stochastic process, defined as

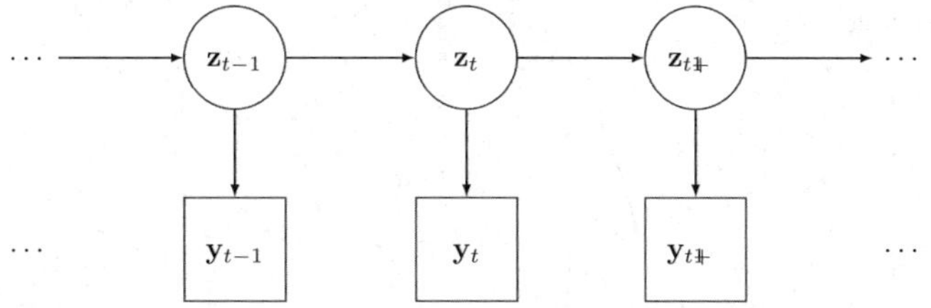

Figure 18.2 Directed acyclic graph of a simple HMM, where dependencies observed in the observed data $\mathbf{y}$ are attributed to a latent variable $\mathbf{z}$, modelled by a first-order Markov chain.

$\mathbf{y} = y_1, y_2, \ldots, y_T$ or $\mathbf{y}_{1:T}$. Given these data, the simplest HMM exists when $\mathbf{y}_{1:T}$ is modelled using a single latent sequence, say $\mathbf{z}_{1:T}$, taking a finite number of discrete values, $z_t \in (1, \ldots, G)$. This scenario is shown in Figure 18.2.

The elements of $\mathbf{z}_{1:T}$ in Figure 18.2 are connected due to their generation by a first-order Markov chain, whereby the distribution of z_t is dependent only on the state inferred at time $t-1$, as opposed to its entire history.

The transition from z_{t-1} to z_t is governed by a time homogeneous transition matrix, $\mathbf{Q}$, such that the distribution of each z_t is given by

$$z_t | z_{t-1}, \mathbf{Q} \sim \mathbf{Q}_{z_{t-1},:}$$

where $\mathbf{Q}_{z_{t-1},:}$ denotes the row of $\mathbf{Q}$ corresponding to state z_{t-1}. The elements of $\mathbf{Q}$ are defined as $q_{ij} = Pr(z_t = j | z_{t-1} = i)$. Note that as a result of introducing $\mathbf{z}_{1:T}$, the time dependence exhibited by $\mathbf{y}_{1:T}$ is now assumed to be completely attributed to the latent sequence.

Accompanying this transition matrix and completing the model is the specification of a distribution for y_t, conditional on z_t. Changes in the behaviour of the stochastic process are defined by changes in the parameters of the proposed distribution. For example, given Poisson distributed data, a suitable choice for y_t may be

$$y_t | z_t, \mathbf{Q}, \lambda \sim Poisson(\lambda_{z_t})$$

such that the behaviour of $\mathbf{y}_{1:T}$ is assumed to be driven by changes in the unknown rate, λ, which depends on $\mathbf{z}_{1:T}$.

From this basic definition, the remainder of this section aims to demonstrate how HMMs can be adapted to the problems of spike identification and spike sorting. The discussion begins with the application of the single HMM in Figure 18.2, where the specification of the transition matrix $\mathbf{Q}$ can be done in such a way as to model the dynamics of a single AP. This model is then extended to multiple independent latent sequences, commonly referred to as a *factorial* HMM (fHMM) (Ghahramani and Jordan 1997).

18.3.2 Modelling a single AP: Application of a simple HMM

Given the model defined in Section 18.3.1, we first consider modelling a single AP over time, given an extracellular recording consisting of observations $\mathbf{y}_{1:T}$, which are

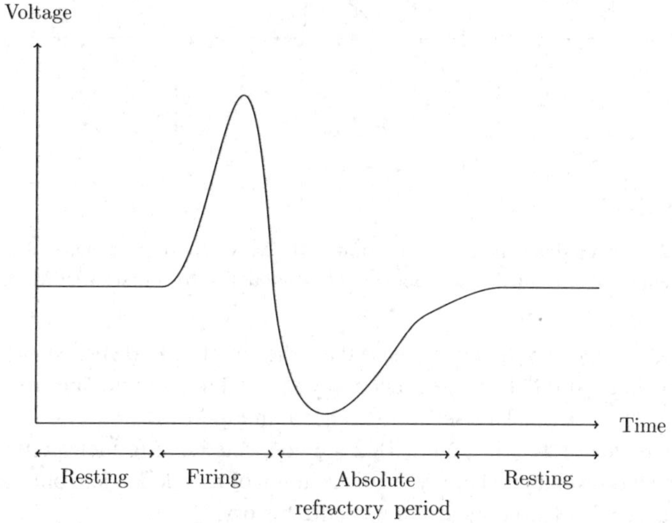

Figure 18.3 Schematic for the behaviour of an AP, where its behaviour can be divided into one of three phases: resting, firing and the refractory period.

subject to background noise. The latent variable z_t therefore represents the state of behaviour of the AP at time t.

Central to the modelling of this behaviour is the specification of the transition probability matrix $\mathbf{Q}$. To do this, we consider the behaviour of a typical AP over time. Explicitly, we acknowledge that, at any given time, an AP exists exclusively in one of three phases, depicted in Figure 18.3.

The first of these phases, *resting*, corresponds to times of equilibrium membrane potential and, therefore, the AP is indistinguishable from the background noise. When this equilibrium is interrupted and the potential rises above a certain threshold, the AP's trajectory is initiated, a state which we refer to as the *firing* state and which refers to the processes of depolarization and repolarization. Upon completion of the firing phase, the AP enters what is known as the *absolute refractory period*, before returning to the resting period.

During transitions between these phases it is assumed that, from the time of firing until the end of the absolute refractory period, it is biologically impossible for the AP to refire. Given this description, $\mathbf{Q}$ is given by

$$\mathbf{Q} = \begin{bmatrix} q_{11} & 1-q_{11} & 0 & 0 & \cdots & 0 \\ 0 & 0 & 1 & 0 & \ldots & 0 \\ 0 & 0 & 0 & 1 & \ldots & 0 \\ \vdots & \vdots & \vdots & \vdots & \vdots & \vdots \\ 0 & 0 & \ldots & \ldots & 0 & 1 \\ 1 & 0 & 0 & 0 & \ldots & 0 \end{bmatrix}_{G\times G}$$

where $z_t = 1$ denotes that the AP is in the resting phase at time t. The number of states G is chosen proportional to the sampling frequency of $\mathbf{y}_{1:T}$ and enforces the aforementioned assumption. For example, if an extracellular recording has a sampling rate of 20 kHz, $G = 20$ means that from the moment the AP fires, with probability $1 - q_{11}$, it is unable to refire for 1 ms.

Conditional on z_t, y_t is assumed normally distributed with unknown mean and common unknown variance, namely

$$y_t \mid z_t = g, \mathbf{Q}, \mu, \sigma^2 \sim N\left(\mu_g, \sigma^2\right), \qquad g = 1, \ldots, G. \tag{18.1}$$

This distributional form is chosen for two reasons. Firstly, the common variance term, σ^2, provides an estimate of the background noise. Secondly, by conditioning the unknown mean on z_t, we obtain an estimate of the average voltage across all defined states of the AP's behaviour. By ordering these means for 1 to G, the predicted average shape or trajectory of the AP is produced, commonly referred to as the spike template.

For all unknown parameters, prior distributions are conjugate and of the form

$$\sigma^2 \sim IG(\alpha, \beta) \tag{18.2}$$

$$\mu|\sigma^2 \sim N\left(b, \sigma^2\tau^2\right) \tag{18.3}$$

$$q_{11} \sim Beta(\gamma, \phi). \tag{18.4}$$

The inclusion of σ^2 in Equation 18.3 allows for the derivation of closed form full conditionals for model estimation, discussed further in Section 18.3.4.

In summary, by specifying an appropriate transition probability matrix, one is able to predict not only when the AP enters the firing state, but also its expected trajectory. This second property may become useful when multiple APs are present, addressed in the next section, as a possible solution to the spike sorting problem.

18.3.3 Multiple neurons: An application of a factorial HMM

In some cases, it is possible that an extracellular recording contains spikes from more than a single AP. With this comes the additional task of spike sorting, whereby identified spikes are classified by predicted trajectory, as this is indicative of their origin from different cells. In order to accommodate this general case, the current HMM must be extended to include N latent sequences, one for the behaviour of each distinct AP. This collection of sequences is denoted throughout this chapter by $\mathbf{z}_{1:T}^{1:N} = (\mathbf{z}_{1:T}^1, \ldots, \mathbf{z}_{1:T}^N)$.

The introduction of an additional $N - 1$, independent latent sequences results in the specification of a factorial HMM (fHMM) (Ghahramani and Jordan 1997), depicted in Figure 18.4.

The independence among latent sequences assumes that the behaviour of APs originating from different neurons is independent. As a consequence, the behaviour of each AP can be expressed by its own transition probability matrix, $\mathbf{Q}^n, n = 1, \ldots, N$.

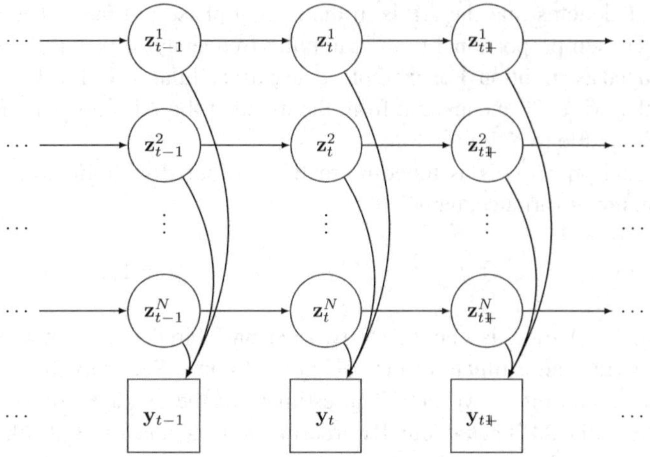

Figure 18.4 Factorial HMM for N independent APs.

Furthermore, it means that, for all t, the joint transition probability for $t-1 \rightarrow t$ is the product of individual AP transition probabilities,

$$Pr(z_t^{1:N} | z_{t-1}^{1:N}) = \prod_{n=1}^{N} q_{z_t^n, z_{t-1}^n}^n. \tag{18.5}$$

This assumption is chosen entirely for computational convenience, with model estimation described in Section 18.3.4.

We also assume for this model that the observed voltage at any given time is the sum of individual voltages from all APs. This is a biologically relevant assumption and allows for the possibility of overlapping APs, that is when more than one AP enters the firing state simultaneously or one AP fires during another AP's refractory or resting phase. Taking this second assumption into account, Equation 18.1 becomes

$$y_t | \mathbf{y}_{1:t-1}, z_t^{1:N}, \mathbf{Q} \sim N\left(\sum_{n=1}^{N} \mu_{z_t^n}^n, \sigma^2 \right)$$

where $\mu_{z_t^n}^n$ is the mean for the nth AP conditional on z_t^n. As a result of this modification to the distribution of y_t, each AP is defined by its own spike template, $\boldsymbol{\mu}^n = (\mu_1^n, \ldots, \mu_G^n)$, and common variance, σ^2, as per the single AP model. This model was first introduced by Herbst *et al.* (2008) in the maximum likelihood framework.

In this chapter, the extension of the model into the Bayesian framework requires the extension of priors in Equations 18.2–18.4 to accommodate the $N > 1$ case. While Equation 18.2 remains unchanged, this extension involves the hierarchical prior on μ defined by Equation 18.3 being replaced by an independence prior over n. Also, for each q_{11}^n, we specify a $Beta(\gamma, \phi)$ distribution.

18.3.4 Model estimation and inference

Estimation of any HMM, regardless of its structure, consists of two key tasks: the estimation of (i) the latent sequence/s and (ii) the parameters associated with each latent state. In a Bayesian framework, we can achieve these tasks using a combination of MCMC, in this case Gibbs sampling, and a unique algorithm for latent sequence estimation, henceforth referred to as the forward–backward algorithm. For the sake of brevity, we restrict our attention to the factorial HMM, as this model reduces to a single sequence HMM when $N = 1$.

The forward–backward algorithm

First developed by Rabiner (1989) in the classical statistical framework, the forward–backward algorithm aims to produce samples from the joint distribution of the latent sequence or sequences, represented by

$$\mathbf{z}_{1:T}^{1:N} \sim p(\mathbf{z}_{1:T}^{1:N} | \mathbf{y}_{1:T}, \mathbf{Q}^{1:N}, \boldsymbol{\Theta}). \tag{18.6}$$

Here, $\boldsymbol{\Theta}$ defines the parameter set $\{\mu, \sigma^2\}$. The generation of samples from Equation 18.6 is simplified by first re-expressing this joint distribution by the decomposition

$$\begin{aligned} p(\mathbf{z}_{1:T}^{1:N} | \mathbf{y}_{1:T}, \mathbf{Q}^{1:N}, \boldsymbol{\Theta}) =& p(\mathbf{z}_{T}^{1:N} | \mathbf{y}_{1:T}, \mathbf{Q}^{1:N}, \boldsymbol{\Theta}) \times \dots \\ & \times p(\mathbf{z}_{t}^{1:N} | \mathbf{z}_{t+1}^{1:N}, \dots, \mathbf{z}_{T}^{1:N}, \mathbf{y}_{1:T}, \mathbf{Q}^{1:N}, \boldsymbol{\Theta}) \times \dots \\ & \times p(\mathbf{z}_{1}^{1:N} | \mathbf{z}_{2}^{1:N}, \dots, \mathbf{z}_{T}^{1:N}, \mathbf{y}_{1:T}, \mathbf{Q}^{1:N}, \boldsymbol{\Theta}). \end{aligned} \tag{18.7}$$

This implies that simulation of $\mathbf{z}^{1:N}$ involves calculation of each probability mass function on the right hand side of Equation 18.7. This task is further simplified by the knowledge that the dependence exhibited by $\mathbf{z}^{1:N}$ is restricted to first order, as illustrated in both Figures 18.2 and 18.4. Chib (1996) showed that each of these terms, through the application of Bayes' rule, is given by

$$p(\mathbf{z}_{t}^{1:N} | \mathbf{z}_{t+1:T}^{1:N}, \mathbf{y}_{1:T}, \mathbf{Q}^{1:N}, \boldsymbol{\Theta}) = \frac{p(\mathbf{z}_{t}^{1:N} | \mathbf{y}_{1:t}, \mathbf{Q}^{1:N}, \boldsymbol{\Theta}) p(\mathbf{z}_{t+1}^{1:N} | \mathbf{z}_{t}^{1:N}, \mathbf{Q}^{1:N})}{\sum_{\mathbf{z}_{t+1}^{1:N}} p(\mathbf{z}_{t}^{1:N} | \mathbf{y}_{1:t}, \mathbf{Q}^{1:N}, \boldsymbol{\Theta}) p(\mathbf{z}_{t+1}^{1:N} | \mathbf{z}_{t}^{1:N}, \mathbf{Q}^{1:N})}. \tag{18.8}$$

with $p(\mathbf{z}_{t+1}^{1:N} | \mathbf{z}_{t}^{1:N}, \mathbf{Q}^{1:N})$ corresponding to the joint transition matrix probability defined in Equation 18.5. The term $p(\mathbf{z}_{t}^{1:N} | \mathbf{y}_{1:t}, \mathbf{Q}^{1:N}, \boldsymbol{\Theta})$ is computed recursively for each t, a process known as forward filtering. Beginning at $t = 1$, the recursive scheme consists of iterating between the two steps

$$p(\mathbf{z}_{t}^{1:N} | \mathbf{y}_{1:t}, \mathbf{Q}^{1:N}, \boldsymbol{\Theta}) = \frac{p(\mathbf{z}_{t}^{1:N} | \mathbf{y}_{1:t-1}, \mathbf{Q}^{1:N}, \boldsymbol{\Theta}) p(y_t | \Theta, z_t^{1:N})}{\sum_{z_t^{1:N}} p(\mathbf{z}_{t}^{1:N} | \mathbf{y}_{1:t-1}, \mathbf{Q}^{1:N}, \boldsymbol{\Theta}) p(y_t | \Theta, z_t^{1:N})} \tag{18.9}$$

$$p(\mathbf{z}_{t}^{1:N} | \mathbf{y}_{1:t-1}, \mathbf{Q}^{1:N}, \boldsymbol{\Theta}) = \sum_{z_{t-1}^{1:N}} p(\mathbf{z}_{t}^{1:N} | \mathbf{z}_{t-1}^{1:N}, \mathbf{Q}^{1:N}) p(\mathbf{z}_{t-1}^{1:N} | \mathbf{y}_{1:t-1}, \mathbf{Q}^{1:N}, \boldsymbol{\Theta}). \tag{18.10}$$

Upon computing Equation 18.9 for all t, sampling from the joint distribution of $\mathbf{z}^{1:N}$ is achieved via Equation 18.8. Explicitly, sampling is performed backwards in time and is thus referred to as backward smoothing. This process is outlined as follows:

$$\begin{aligned}
\mathbf{z}_T^{1:N} &\sim p(\mathbf{z}_T^{1:N}|\mathbf{y}_{1:T}, \mathbf{Q}^{1:N}, \boldsymbol{\Theta}) \\
\mathbf{z}_{T-1}^{1:N} &\sim p(\mathbf{z}_{T-1}^{1:N}|\mathbf{z}_T^{1:N}, \mathbf{y}_{1:T}, \mathbf{Q}^{1:N}, \boldsymbol{\Theta}) \\
&\vdots \\
\mathbf{z}_1^{1:N} &\sim p(\mathbf{z}_1^{1:N}|\mathbf{z}_2^{1:N}, \ldots, \mathbf{z}_T^{1:N}, \mathbf{y}_{1:T}, \mathbf{Q}^{1:N}, \boldsymbol{\Theta}). \qquad (18.11)
\end{aligned}$$

The forward–backward algorithm has found success in many areas of research, particularly in speech processing, and also appearing under the guise of the Baum–Welch (Baum *et al.* 1970) and Viterbi (Forney Jr 1973) algorithms. For a more comprehensive discussion of this algorithm in the Bayesian framework, see Frühwirth-Schnatter (2006).

Gibbs sampler

The choice of conjugate priors on all unknown parameters means that model estimation can be implemented using Gibbs sampling (Chib 1996; Geman and Geman 1984; Smith and Roberts 1993). For our model, the Gibbs sampler involves sampling from the following full conditionals:

$$\mathbf{z}^{1:N}|\mathbf{y}_{1:T}, \mathbf{Q}^{1:N}, \boldsymbol{\Theta} \qquad (18.12)$$

$$\mathbf{Q}^{1:N}|\mathbf{y}_{1:T}, \mathbf{z}^{1:N} \qquad (18.13)$$

$$\boldsymbol{\Theta}|\mathbf{y}_{1:T}, \mathbf{z}^{1:N}. \qquad (18.14)$$

Equation 18.12 is sampled from using the forward–backward algorithm described in the previous section, given current realizations of $\mathbf{Q}^{1:N}$ and $\boldsymbol{\Theta}$. Given $\mathbf{z}^{1:N}$, $\mathbf{Q}^{1:N}$ and $\boldsymbol{\Theta}$ are then updated via their corresponding full conditionals. For $\mathbf{Q}^{1:N}$, this involves updating each q_{11}^n, for $n = 1, \ldots, N$. These updates are in the form of beta distributions

$$q_{11}^n|\mathbf{z}^n \sim Beta(\gamma + m_{11}^n, \phi + 1)$$

where $m_{11}^n = \#(z_t^n = 1|z_{t-1}^n = 1)$, or the number of transitions where the latent sequence remains in the resting state. For Θ, we note that the posterior can be decomposed to give

$$p(\boldsymbol{\mu}, \boldsymbol{\sigma}^2|\mathbf{y}_{1:T}, \mathbf{z}^{1:N}) = p(\boldsymbol{\mu}|\boldsymbol{\sigma}^2, \mathbf{y}_{1:T}, \mathbf{z}^{1:N}) \times p(\boldsymbol{\sigma}^2|\mathbf{y}_{1:T}, \mathbf{z}^{1:N}).$$

The full conditional for $\boldsymbol{\mu}$ follows a multivariate normal distribution,

$$p(\boldsymbol{\mu}|\sigma^2, \mathbf{y}_{1:T}, \mathbf{z}^{1:N}) \sim MVN\left(\left(\frac{b}{\tau^2} + d(\mathbf{y})\right)^T \Sigma, \sigma^2\Sigma\right) \tag{18.15}$$

$$\mu = \left[\mu_1^1, \mu_2^1, \ldots, \mu_G^1, \mu_1^2, \ldots, \mu_G^N\right]^T$$

$$d(\mathbf{y}) = \left[d_{z_t^1=1}(\mathbf{y}), \ldots, d_{z_t^1=G}(\mathbf{y}), d_{z_t^2=1}(\mathbf{y}), \ldots, d_{z_t^N=G}(\mathbf{y})\right]^T$$

$$d_{z_t^n=g}(\mathbf{y}) = \sum_{t=1}^{T} \mathbf{I}\left\{z_t^n = g\right\} y_t.$$

Here, μ has been defined as a row vector, containing all possible combinations of $n = 1, \ldots, N$ and $g = 1, \ldots, G$. This allows for all mean parameters to be sampled in a single, vectorized step and is done so for computational convenience.

The remaining term, σ^2, is updated by an inverse-gamma distribution, having integrated out $\boldsymbol{\mu}$:

$$\sigma^2|\mathbf{y}_{1:T}, \mathbf{z}^{1:N} \sim IG\left(\alpha + \frac{T}{2}, \beta + \frac{1}{2}\left(\sum_{t=1}^{T} y_t^2 - d(\mathbf{y})^T \Sigma d(\mathbf{y})\right)\right).$$

The covariance matrix Σ contains entries in the form of counts and ordered identically to $d(\mathbf{y})$. The diagonal of the inverse of Σ is given by

$$diag(\Sigma^{-1}) = \frac{1}{\tau^2} + N(\mathbf{y}),$$

with $N(\mathbf{y})$ being a row vector of latent state counts for each action potential, written as

$$N(\mathbf{y}) = \left[N_{z^1=1}(\mathbf{y}), \ldots, N_{z^N=G}(\mathbf{y})\right]^T$$

$$N_{z^n=g}(\mathbf{y}) = \sum_{t=1}^{T} \mathbf{I}\left\{z_t^n = g\right\}.$$

Off diagonal elements of Σ^{-1} involve crosstabulated state counts for each pair of APs. The primary goal of model inference is to predict the latent sequences $\mathbf{z}^{1:N}$. In the Bayesian framework, posterior inference on each z_t^n given $\mathbf{y}_{1:T}$ is possible through the Monte Carlo estimate of the forward filter (Chib 1998). For D Gibbs iterations, this probability is given by

$$\begin{aligned} Pr(z_t^n|\mathbf{y}) &= \int p(z_t^n|y_{1:t-1}, \boldsymbol{\Theta}, Q^n) p(\Theta, Q^n|\mathbf{y})\, d\boldsymbol{\Theta}\, dQ^n \\ &= \frac{1}{D}\sum_{d=1}^{D} p(z_t^{n(d)}|\mathbf{y}_{1:t-1}, \boldsymbol{\Theta}^{(d)}, Q^{n(d)}) \end{aligned}$$

which is simply the average over Equation 18.10, marginalized over all action potentials $n' \neq n$.

18.4 Data analysis and results

In this section, we apply the model defined in Section 18.3.3 to both a simulated and real data set, the latter being extracellular recordings collected during deep brain stimulation, a popular treatment for patients diagnosed with advanced Parkinson's disease.

For each data set, the Gibbs sampler presented in the previous section was run for 10 000 iterations, discarding the first 5000 as the burn-in period. The results presented are based on the following choice of hyperparameters: $b = 0, \tau^2 = 10, \gamma = 10, \phi = 1$, $\alpha = 2.5$ and $\beta = 0.5s^2$, where s^2 is the sample variance of the observed data.

18.4.1 Simulation study

Data for this study were simulated assuming two distinct APs ($N = 2$) and a sampling frequency of 15 kHz, leading to the specification of 15 latent states for each AP ($G = 15$) or an AP duration of 1 ms. This data set is illustrated in Figure 18.5.

Spike onset locations were set at every 95th time step for the first AP, beginning at $t = 85$ and at every 122nd time step for the second AP, starting at $t = 453$. This resulted in one full overlap of the two APs (each AP fires simultaneously) and two partial overlaps (one AP fires during the refractory period of the other AP). From Figure 18.5, the presence of the second AP is clear, characterized by an amplitude between 0.2 and 0.3; however, the locations of the first AP are less clear and the locations of overlaps even less so.

For the Gibbs sampler in this study, two chains were run with different initial templates $(\mu_1, \ldots, \mu_G)$, for each AP. In the first instance, initial templates were set to the true, simulated template for each AP. For the second chain, each μ_g^n was randomly

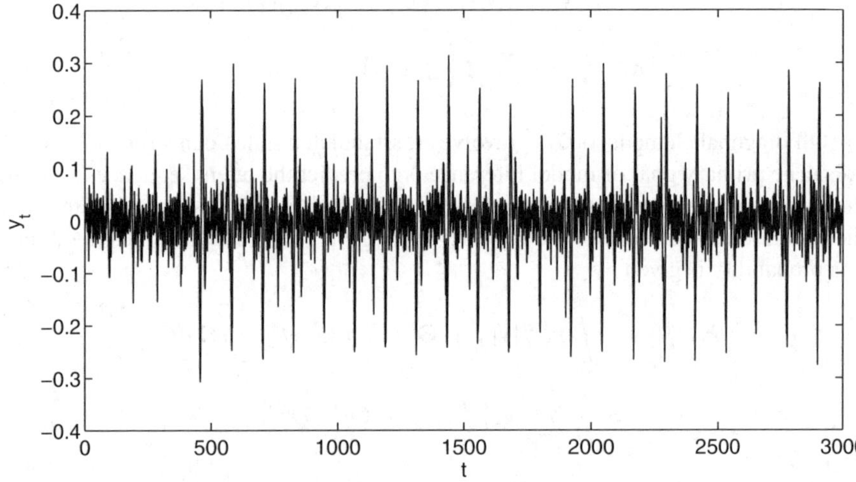

Figure 18.5 Simulated data set with $N = 2$ and $G = 15$.

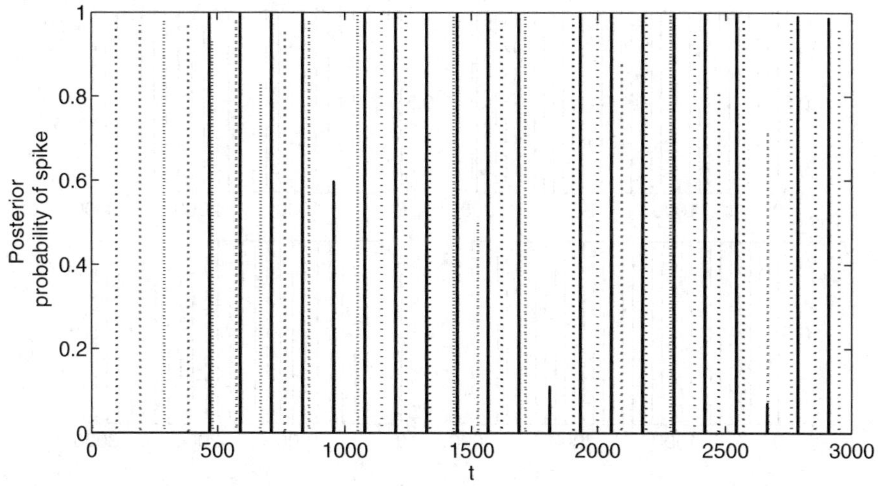

Figure 18.6 Plot of isolated firing probabilities for the simulated data set, under the assumption of normally distributed emissions.

drawn from a $N(0, \tau^2\sigma^2)$ distribution with each μ_1^n then set to zero to correspond to the assumed mean amplitude of the background noise. The comparison of inferences between these two chains did not result in any major discrepancies. Thus, the results that follow are based on MCMC output for the second chain.

The posterior probabilities of firing for isolated or non-overlapping spikes are summarized in Figure 18.6. Given each latent sequence was sampled backwards recursively, Figure 18.6 gives posterior probabilities for each AP of the form $Pr(z_t^n = G, z_t^{n'} = 1|\mathbf{y}_{1:T}, \boldsymbol{\mu}, \sigma^2, \mathbf{Q}^{1:N})$.

In Figure 18.6, there are the high posterior probabilities associated with the locations of each isolated AP. Assuming a probability threshold of 0.5, Figure 18.7 shows that posterior inference performed well with recovering the true locations of each AP. However, for the inferred locations of each AP, one sees some omissions when compared with the true sequences, which may be the result of overlaps.

To explore the possibility of overlapping APs being present, Figure 18.8 illustrates the posterior probabilities of partially overlapping APs over all locations, for each AP individually. In this case, these probabilities are of the form $Pr(z_t^n = G, z_t^{n'} \neq 1 \cap z_t^{n'} \neq G|\mathbf{y}_{1:T}, \boldsymbol{\mu}, \sigma^2, Q^{1:N})$, or the probability that the nth AP fires given the other AP is in the resting period.

Given Figure 18.8, two partially overlapping APs have been recovered, one for each AP. For the first AP, we see that it is initiated during the refractory period of the second AP at approximately $t = 1000$. Likewise, the second AP fires during the refractory period of the first AP later in the data set between $t = 2500$ and $t = 3000$. Comparing this result with the omissions in Figure 18.7, one sees these partial overlaps for two of the four omissions between the inferred and true latent sequences.

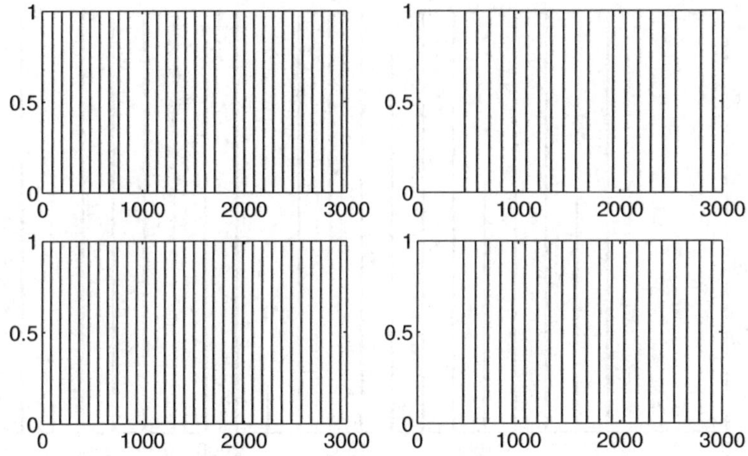

Figure 18.7 Comparison of true locations of spike onset (top row) versus inferred spikes given a posterior probability threshold of 0.5 (bottom row). Comparisons are given for each individual AP: first AP (left column) and second AP (right column).

For the identification of fully overlapping APs, summarized in Figure 18.8 by the posterior probability $Pr(z_t^{1:N} = G|\mathbf{y}_{1:T}, \boldsymbol{\mu}, \sigma^2, \mathbf{Q}^{1:N})$, the final omissions in Figure 18.7 are recovered, corresponding to the full overlap between $t = 1500$ and $t = 2000$.

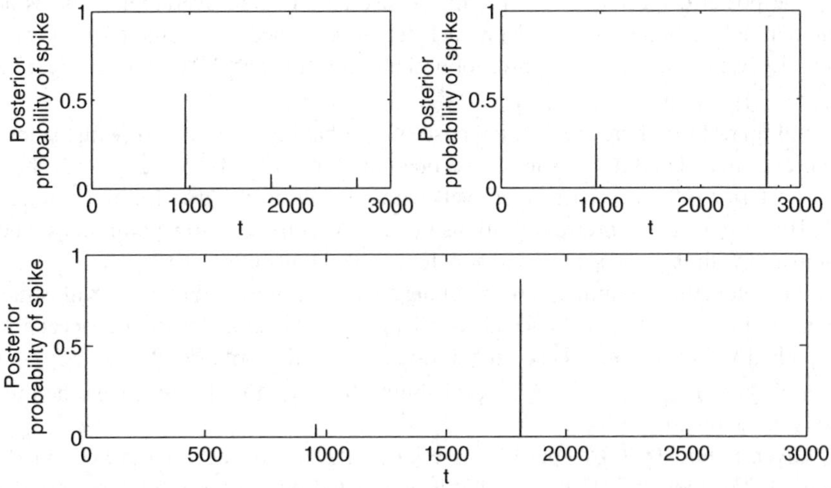

Figure 18.8 Plots of partially overlapping firing probabilities (top row) and simultaneous firing probabilities (bottom row) for the simulated data set, under the assumption of normally distributed emissions. Partially overlapping probabilities are given for neuron 1 (top left) and neuron 2 (top right).

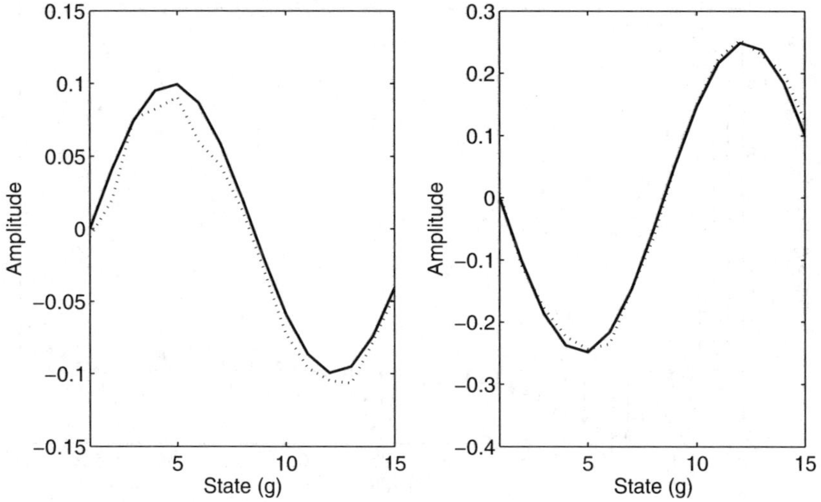

Figure 18.9 Plot of simulated templates (solid line) superimposed on predicted templates (dotted line), based on the approximate MAP.

Given the favourable performance of this modelling approach, Figure 18.9 compares the true versus predicted templates for each AP, given by $\mu_1^n, \ldots, \mu_G^n, n = 1, 2$. Given the presence of label switching throughout the course of the Gibbs sampler, predicted templates were constructed by choosing the estimates of $\boldsymbol{\mu}$ corresponding to the approximate *maximum a posteriori* (MAP) estimate. For each Gibbs iteration, the approximate MAP estimate was calculated by multiplying the observed data likelihood by the prior density $p(\mu^{1:N}, \sigma^2)$. Given these estimates for each AP, Figure 18.9 further supports the favourable performance of the model, with minimal discrepancies between true and predicted templates.

18.4.2 Case study: Extracellular recordings collected during deep brain stimulation

Results are now presented on an extracellular recording taken from the STN. Given the computational burden involved in estimating HMMs, analysis is restricted to $N = 2$ and to a small section of real-time recording of approximately 0.1 seconds. The data set is given in Figure 18.1. Potential extensions to allow for the feasible computation of larger data sets is left to the discussion.

To estimate the number of latent states G to model the trajectory of each AP, the sample autocorrelation function (ACF) over the entire data set was calculated and is summarized in Figure 18.10. G was then approximated by identifying the nearest lag corresponding to the ACF crossing zero for the second time, to take into account the biphasic nature of an AP, as illustrated by the schematic in Figure 18.3. This resulted in setting G equal to 30, representing a refractory period of approximately 1.3 ms.

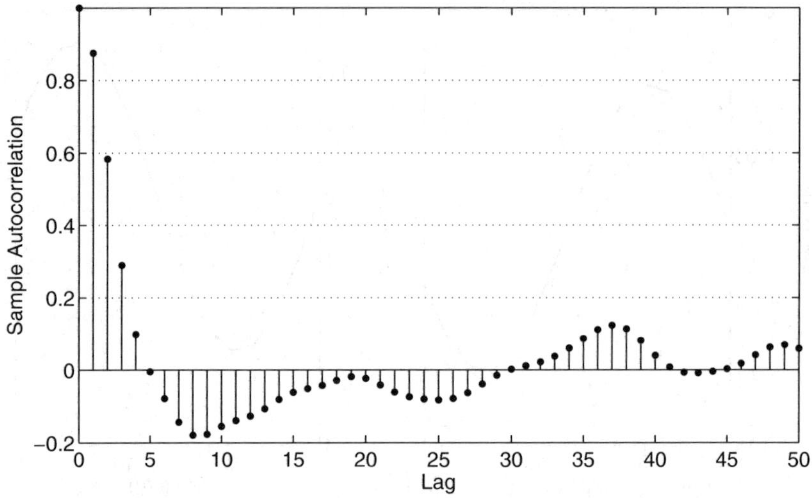

Figure 18.10 Sample autocorrelation function (ACF) up to and including lag 50 for the real data set given in Figure 18.1.

Based on the last 5000 iterations of the Gibbs sampler, Figure 18.11 summarizes the posterior distribution of the total number of spikes detected.

For both neurons, the posterior probability of remaining in the resting state was very high, with posterior expectations both equalling approximately 0.94. This result indicated that, for both modelled APs, spiking events were rare. Reviewing the number

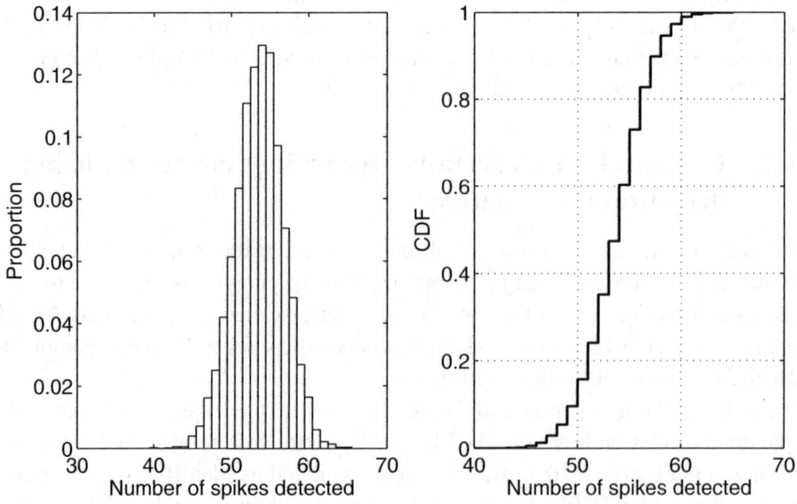

Figure 18.11 Summary of the total number of APs detected by a histogram (left) and empirical cumulative distribution function (CDF) (right).

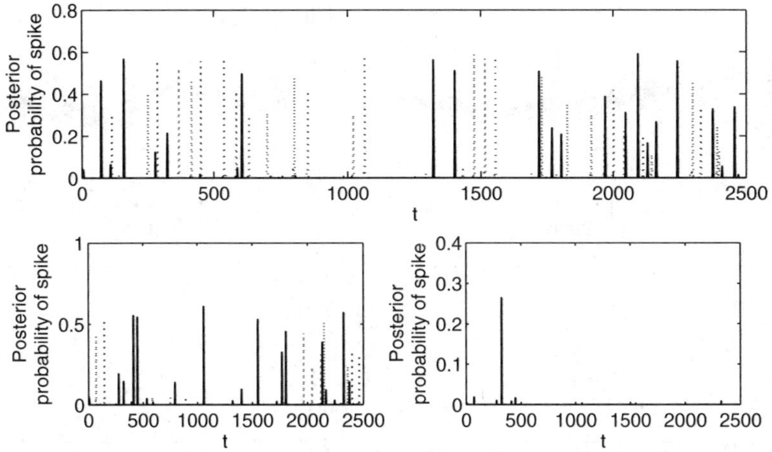

Figure 18.12 Posterior probability plots for isolated APs (top), partially overlapping APs (bottom left) and fully overlapping APs (bottom right).

of identified spikes in Figure 18.11, some variation was noted, with an expectation of 54 spikes and a 95% credible interval of between 47 and 60 spikes.

This analysis of real data, however, provided inferences that were not as conclusive as those obtained for the simulation study. In Figure 18.12, posterior inference for the chosen model was summarized for isolated, partially overlapping and fully overlapping APs. For isolated APs, the majority of non-zero posterior probabilities did not exceed 0.5 and therefore suggested substantial uncertainty about the presence of many APs. This result was somewhat expected, in part due to the higher level of background activity visible in Figure 18.1. Furthermore, investigation into this result revealed that many posterior probabilities regarding the locations of isolated APs were in fact shared between neighbouring locations. This suggested that the uncertainty in Figure 18.12 not so much suggested uncertainty in the presence of APs but more the uncertainty in the exact locations where APs were initiated. A similar statement was made for possible locations of partially overlapping APs, also summarized in Figure 18.12. That said, in inferring the most likely locations for each AP, it was thought feasible to set the posterior probability threshold lower than 0.5. Finally, in this case, there was no evidence of fully overlapping APs, indicated by the bottom right plot in Figure 18.12.

Concerns about the use of this model also arose following inference on the average templates for each of the two assumed APs. This inference is summarized in Figure 18.13.

In Figure 18.13, predicted APs given posterior probabilities of onset greater than 0.4 are given and compared with their predicted template. In the chosen model's favour, many APs assigned to the second neuron resemble the predicted average template well; however, there appeared to be a small subset of APs with notably higher amplitude. This raised concern that this subset in fact corresponded to a third

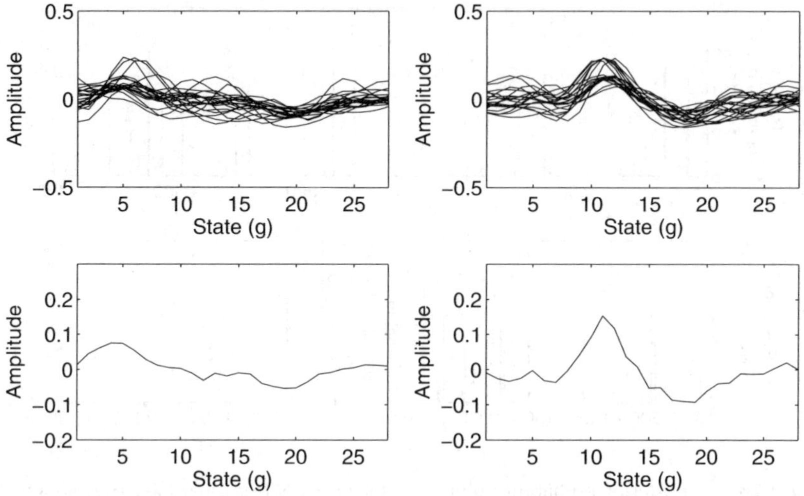

Figure 18.13 Identified APs given a probability threshold of 0.4 for the first (top left) and second AP (top right). The predicted templates based on the MAP estimate are also given for each AP (bottom left and right).

neuron and were thus incorrectly sorted. Representing a less favourable result, the predicted APs attributed to the first neuron did not appear to consistently follow the predicted average template, for either threshold. This was thought to be the result for two possible reasons. Firstly, given that the maximum and minimum amplitudes of the predicted template are closer to the background noise than the other template for the second AP, it was possible that the corresponding spikes were corrupted by noise to a greater degree. Secondly, and of greater concern, was that these spikes were false positives and did not correspond to true APs at all.

18.5 Discussion

In this chapter, we have highlighted the use of HMMs for modelling data-rich, complex stochastic processes. Furthermore, taking the simplest definition of an HMM, we have shown its flexibility in adapting to different modelling situations, in this case the extension of a single sequence HMM to a factorial setting, in an attempt to simultaneously identify and sort spikes arising from multiple APs.

In the simulation study, the model provided promising results and was able to correctly identify both APs and locations where they overlapped. However, in the analysis of the real data set, a number of key concerns arose, presenting opportunities for future research in this area and forming the focus of this discussion. Broadly speaking, these concerns could be categorized as either computational or methodological.

An analysis not considered in this chapter is that of model comparison. This comparison could take place for models with different values of N or, as explored in

White (2011), models with different distributional assumptions for y_t and/or different transition matrix specifications. Analyses of this form could be conducted with the use of goodness-of-fit criteria, such as the deviance information criterion (Celeux *et al.* 2006; Spiegelhalter *et al.* 2002), as is used in White (2011), or by Bayes' factors (Kass and Raftery 1995).

For many segments of real data analysed but not presented in this chapter, the specification of G also led to problems, in terms of the number of APs detected and the posterior probabilities of spikes occurring at a given location. Furthermore, for the real data analysis in Section 18.4.2, a number of different values for G were trialled. Although these additional analyses are not included in this chapter, it was seen that smaller values of G than the one chosen resulted in many false positives, compared with larger values of G that resulted in the detection of very few or no spikes in the observed data. The most likely reason for sensitivity regarding the choice of G is that, in recordings of multiple cells, the refractory period may vary considerably. It was this reason that motivated the alternative choice of the transition matrix; however, in both analyses presented, this extension did not appear to provide an improved fit to the data. For the real data set, evidence for the presence of two different refractory periods could be seen in the sample ACF, with its almost crossing zero for the second time at lag 20. To implement different refractory periods in the current model would involve the specification of identifiability constraints and would be a suitable extension for future work.

To conclude, while implementation of the proposed model in the Bayesian framework was appealing, in terms of the range of probability statements possible, the cost of model computation by MCMC was found to be a prohibitive factor, with computation taking approximately 2 hours per 0.1 seconds of real recording. That said, less expensive alternatives to MCMC would be desirable for future work in this area. Examples of viable alternatives to MCMC include variational Bayes (Ghahramani and Hinton 2000; McGrory and Titterington 2009; McGrory *et al.* 2011), approximate Bayesian computation (Jasra *et al.* 2010) and sequential Monte Carlo (Doucet *et al.* 2000) methods.

References

Baum L, Petrie T, Soules G and Weiss N 1970 A maximization technique occurring in the statistical analysis of probabilistic functions of Markov chains. *Annals of Mathematical Statistics* **41**, 164–171.

Cappé O, Moulines E and Rydén T 2005 *Inference in Hidden Markov Models*. Springer, Berlin.

Celeux G, Forbes F, Robert C and Titterington M 2006 Deviance information criteria for missing data models. *Bayesian Analysis* **1**(4), 651–674.

Chib S 1996 Calculating posterior distributions and modal estimates in Markov mixture models. *Journal of Econometrics* **75**(1), 79–97.

Chib S 1998 Estimation and comparison of multiple change-point models. *Journal of Econometrics* **86**(2), 221–241.

Doucet A, Godsill S and Andrieu C 2000 On sequential Monte Carlo sampling methods for Bayesian filtering. *Statistics and Computing* **10**, 197–208.

Eddy S 1995 Multiple alignment using hidden Markov models. *Proceedings of the Third AAAI International Conference on Intelligent Systems for Molecular Biology*, Vol. 3, pp. 114–120.

Forney Jr G 1973 The Viterbi algorithm. *Proceedings of the IEEE* **61**(3), 268–278.

Freeman J 1971 A simple multi-unit channel spike height discriminator. *Journal of Applied Physiology* **31**, 939–941.

Frühwirth-Schnatter S 2006 *Finite mixture and Markov switching models*. Springer, Berlin.

Gales M and Young S 2008 The application of hidden Markov models in speech recognition. *Foundations and Trends in Signal Processing* **1**(3), 195–304.

Geman S and Geman D 1984 Stochastic relaxation, Gibbs distributions, and the Bayesian restoration of images. *IEEE Transactions on Pattern Analysis and Machine Intelligence* **6**, 721–741.

Ghahramani Z and Hinton G 2000 Variational learning for switching state-space models. *Neural Computation* **12**(4), 831–864.

Ghahramani Z and Jordan M 1997 Factorial hidden Markov models. *Machine Learning* **29**(2), 245–273.

Herbst J, Gammeter S, Ferrero D and Hahnloser R 2008 Spike sorting with hidden Markov models. *Journal of Neuroscience Methods* **174**(1), 126–134.

Holmes I and Bruno W 2001 Evolutionary HMMs: a Bayesian approach to multiple alignment. *Bioinformatics* **17**(9), 803–820.

Jasra A, Singh S, Martin J and McCoy E 2010 Filtering via approximate Bayesian computation. *Statistics and Computing*. doi: 10.1007/s11222-010-9185-0.

Kass R and Raftery A 1995 Bayes factors. *Journal of the American Statistical Association* **90**(430), 773–795.

Kim K and Kim S 2000 Neural spike sorting under nearly 0-dB signal-to-noise ratio using nonlinear energy operator and artificial neural-network classifier. *IEEE Transactions on Biomedical Engineering* **47**, 1406–1411.

Kleiner-Fisman G, Fisman D, Sime E, Saint-Cyr J, Lozano A and Lang A 2003 Long-term follow up of bilateral deep brain stimulation of the subthalamic nucleus in patients with advanced Parkinson's disease. *Journal of Neurosurgery* **99**, 489–495.

Krack P, Batir A, Van Blercom N, Chabardes S *et al.* 2003 Five-year follow-up of bilateral stimulation of the subthalamic nucleus in advanced Parkinson's disease. *New England Journal of Medicine* **349**, 1925–1934.

Kumar R, Lozano A, Kim Y, Hutchison W, Sime E, Halket E and Lang A 1998 Double-blind evaluation of subthalamic nucleus deep brain stimulation in advanced Parkinson's disease. *Neurology* **51**, 850–855.

Limousin P, Krack P, Pollak P, Benazzouz A, Ardovin C, Hoffman D and Benabid A 1998 Electrical stimulation of the subthalamic nucleus in advanced Parkinson's disease. *New England Journal of Medicine* **339**, 1105–1111.

MacDonald IL and Zucchini W 1997 *Hidden Markov and Other Models for Discrete-Valued Time Series*. Chapman & Hall: New York.

McGrory C and Titterington D 2009 Variational Bayesian analysis for hidden Markov models. *Australian & New Zealand Journal of Statistics* **51**(2), 227–244.

McGrory C, White N, Mengersen K and Pettitt A 2011 A variational Bayes approach to fitting hidden Markov models in Parkinson's disease research. *Biometrics*, submitted.

Mtetwa N and Smith L 2006 Smoothing and thresholding in neuronal spike detection. *Neurocomputing* **69**, 1366–1370.

Nenadic Z and Burdick J 2005 Spike detection using the continuous wavelet transform. *IEEE Transactions on Biomedical Engineering* **52**, 74–87.

Rabiner L 1989 A tutorial on hidden Markov models and selected applications in speech recognition. *Proceedings of the IEEE* **77**(2), 257–286.

Rizk M and Wolf P 2009 Optimizing the automatic selection of spike detection thresholds using a multiple of the noise level. *Medical and Biological Engineering and Computing* **47**, 955–966.

Smith A and Roberts G 1993 Bayesian computation via the Gibbs sampler and related Markov chain Monte Carlo methods. *Journal of the Royal Statistical Society, Series B (Statistical Methodology)* **55**(1), 3–23.

Spiegelhalter D, Best N, Carlin B and van der Linde A 2002 Bayesian measures of model complexity and fit. *Journal of the Royal Statistical Society, Series B (Statistical Methodology)* **64**(4), 583–639.

Thakur P, Lu H, Hsiao S and Johnson K 2007 Automated optimal detection and classification of neural action potentials in extra-cellular recordings. *Journal of Neuroscience Methods* **162**, 364–376.

Tranter S and Reynolds D 2006 An overview of automatic speaker diarization systems. *IEEE Transactions on Audio, Speech, and Language Processing* **14**(5), 1557–1565.

White N 2011 *Bayesian mixtures for modelling complex medical data: a case study in Parkinson's disease*. PhD thesis. Queensland University of Technology.

Yang X and Shamma S 1988 A totally automated system for the detection and classification of neural spikes. *IEEE Transactions on Biomedical Engineering* **35**, 806–816.

19

Bayesian classification and regression trees

Rebecca A. O'Leary[1], Samantha Low Choy[2,4], Wenbiao Hu[3] and Kerrie L. Mengersen[2]

[1]*Department of Agriculture and Food, Western Australia, Australia*
[2]*Queensland University of Technology, Brisbane, Australia*
[3]*University of Queensland, Brisbane, Australia*
[4]*Cooperative Research Centre for National Plant Biosecurity, Australia*

19.1 Introduction

Classification and regression trees (CARTs) are binary decision trees, which are built by repeatedly splitting the predictor space according to splitting rules of the predictor variables in order to best classify or estimate a response variable (Breiman *et al.* 1984). These binary decision rules may be easily interpreted as a sequence of *if–then–else* statements. Each *if*-clause selects cases according to whether they fall below or above a threshold on a single predictor. The *then*-clause estimates the average response in these cases. This modelling approach facilitates the identification and description of complex nonlinear interactions between predictor variables, such as combinations of habitat variables describing an ecological niche, or gene–gene interactions that explain a disease. In contrast, with linear regression analysis (Chapter 4) usually only linear and low-order polynomial effects are examined, and it is often difficult to determine nonlinear interactions.

CARTs are popular because they are easy to interpret due to their binary nature and simplistic graphical output. Moreover, they have been shown to have good predictive power (Breiman 2001b; De'ath and Fabricius 2000). In a review paper,

Case Studies in Bayesian Statistical Modelling and Analysis, First Edition. Edited by Clair L. Alston, Kerrie L. Mengersen and Anthony N. Pettitt.

Breiman (2001b) gave CARTs a rating of A+ for interpretability and B for prediction. Following on from this letter-grade assessment, Fan and Gray (2005) gave the Bayesian approach to CARTs (BCARTs) an A+ for interpretability and B+ for prediction. A comparable method with respect to predictive capacity is a random forest (RF) (Breiman 2001a), but the interpretability of this method was rated much lower (as F) by Breiman (2001b). RF results in many trees, and provides the average prediction for the response for each case, but does not give an easy interpretable model.

The CART approach is becoming more common in the ecological and medical literature for prediction and exploratory modelling in a wide range of problems. In ecology, these include identifying the habitat preference of upland birds (Bell 1996), estimating the abundance of soft coral in the Great Barrier Reef (De'ath and Fabricius 2000), habitat suitability modelling for red deer (Debeljak *et al.* 2001), and describing the distribution of four vegetation alliances (Miller and Franklin 2002). In medicine, some examples include gene–gene interactions that explain human genetic diseases (Cordell 2009), epidemiological studies determining the risk factors for mortality and morbidity from specific diseases (BachurHarper 2001; Nelson *et al.* 1998, e.g.), and influenza treatment strategies (SmithRoberts 2002).

BCARTs were proposed by Chipman *et al.* (1998) and Denison *et al.* (1998). Buntine (1992) also suggested a Bayesian approach to classification trees. Their major focus of BCARTs inference was the posterior probability distribution, under non-informative priors. BCARTs have been further extended to include inference via a decision-theoretic approach (O'Leary 2008), informative priors (O'Leary *et al.* 2008), and overfitting of BCARTs which was assessed using cross-validation and posterior predictive distribution (Hu *et al.* 2011). O'Leary (2008) added a decision-theoretic layer to the BCART algorithm, which is an approach to identifying the best performing trees based on a selection criterion (cost function, e.g. false negative rate) tailored to the research goals. BCARTs have the ability to incorporate expert opinion or historical data into the prior model, and combine this information with the observed data to produce the trees (posterior distribution) (O'Leary *et al.* 2008). To date, there have been very few real applications of BCARTs published. Some examples include modelling multiple environmental stressors (eutrophication) to a regional lake (Lamon and Stow 2004), habitat suitability modelling of the threatened Australian brush-tailed rock wallaby *Petrogale penicillata* (O'Leary *et al.* 2008), identifying important genes for predicting acute lymphoblastic leukaemia (O'Leary *et al.* 2009), and identification and estimation of the spatial distribution of cryptosporidium (Hu *et al.* 2011). One reason for this slow uptake in the applications is the current lack of user-friendly software; this is slowly being addressed. In this chapter, we use the BCART software developed by O'Leary (2008), which is currently under development for wider distribution.

BCARTs explore the model space more fully than the traditional CART methods. In particular, the traditional CART algorithm starts at the top of the tree and at each splitting node selects the optimal split until the data cannot be partitioned any further. Thus, the choices of tree are constrained; specifically the choices of splitting rules at nodes down the tree are constrained by choices made at nodes above, and only get one optimal tree. Alternatively, BCARTs produce a large number of plausible trees with a wider variety of tree structures. In this chapter, BCART functionality is extended by closer examination of the decision-theoretic approach applied to assist in the selection of the 'best' tree and convergence of the algorithm. The BCART algorithm developed in this chapter is based on the model formulation and computational approach proposed by Denison *et al.* (1998) and is detailed by O'Leary (2008). The two case studies are described in Section 19.2 and the methodology of the BCART approach is detailed in Section 19.3. Results are provided in Section 19.5 and discussed in Section 19.6.

19.2 Case studies

In this chapter, we consider two case studies. The first is a well-known study that acts as a standard or reference data set. The second is a new problem based on real data.

19.2.1 Case study 1: Kyphosis

Kyphosis is a common condition in which there is curvature of the upper spine. This data set contains 81 cases of children who have had corrective spinal surgery (Chambers and Hastie 1992). The response variable is binary indicating whether kyphosis was present (1) or absent (0) after the operation. The covariates are age in months of the child (Age), number of vertebrae involved (Number) and the number of the first vertebra operated on (Start). After the operation, 17 of the 81 children had kyphosis. This data set is freely available in R, in library RPART (Therneau and Atkinson 2003, 1997). Based on exploratory data analysis, the variable providing the greatest difference between presence and absence of kyphosis is Start. All variables displayed weak pairwise correlation.

19.2.2 Case study 2: Cryptosporidium

Cryptosporidiosis is caused by *Cryptosporidium parvum*, a microscopic single cell parasite that can live in the intestines of humans. The data set considered here has been described fully by Hu *et al.* (2010); here we describe it briefly. This data set contains 1332 notified cryptosporidiosis cases in Queensland for the period of 1 January to 31 December 2001 from the Queensland Department of Health (Hu *et al.* 2009). The response variable is the incidence rate per 100 000 of cryptosporidium. The covariates are monthly mean maximum temperature (°C), month rainfall (mm) and social–economic index for areas (SEIFA), which were obtained for the same period from the Australian Bureau of Meteorology and the Australian Bureau of Statistics, respectively. In Hu *et al.* (2011), a two stage-model for predicting cryptosporidium rates was developed. Firstly, the presence and absence of cryptosporidium were modelled using Bayesian classification trees. Secondly, conditional on presence, the abundance was modelled using Bayesian regression trees. Here we just describe the second-stage model further; that is, we apply Bayesian regression trees to predict the spatial distribution of positive incidence rates (ignoring zeros) of the cryptosporidiosis infection using SEIFA and climate variables. For further details, including a comparison of BCARTs with a Bayesian spatial conditional autoregressive (CAR) model using cryptosporidium, see Hu *et al.* (2011).

Figure 19.1 shows scatterplots, histograms and correlations of the covariates and response variable for cryptosporidium. There are a large number of zero incidence

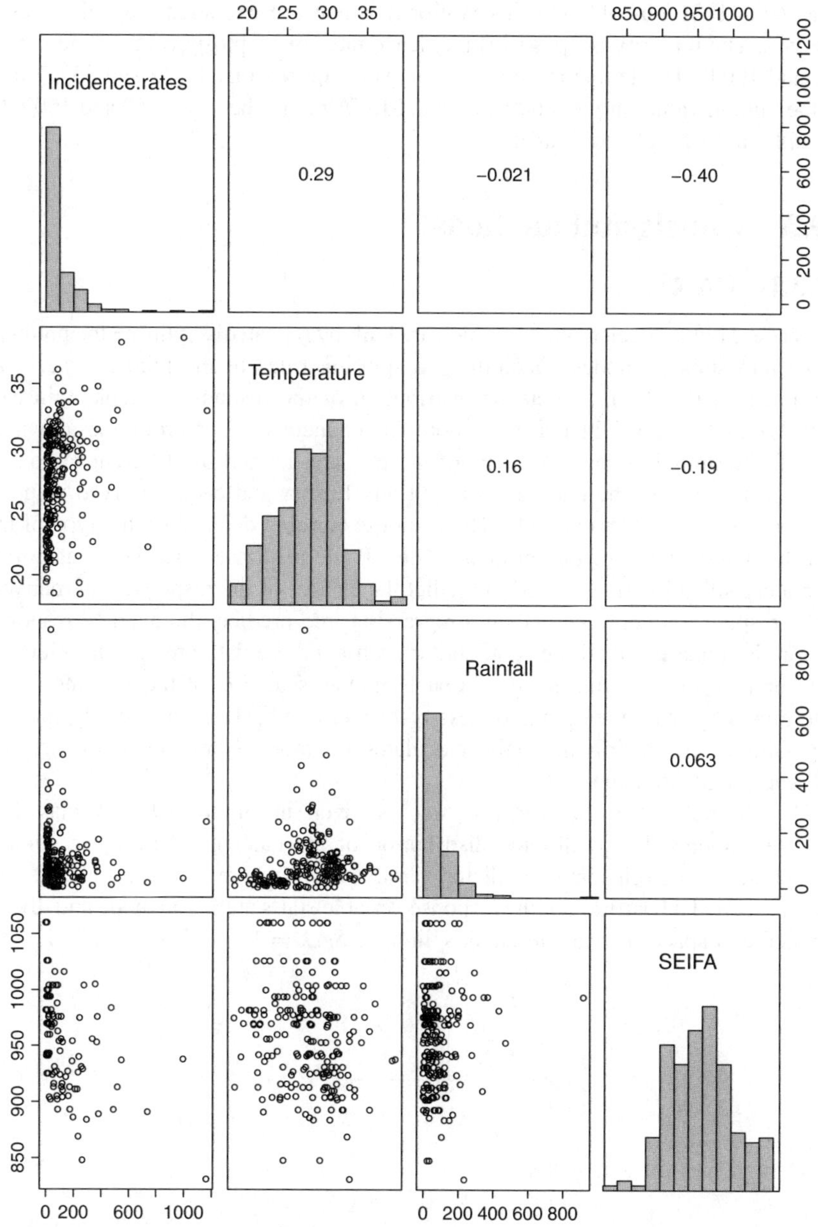

Figure 19.1 Scatterplots (lower panel), histograms (diagonal), correlations (upper panel) of the three covariates (temperature, rainfall and SEIFA) and response (incidence rate of cryptosporidium).

rates ($n = 1131$ out of 1332 observations) which are excluded from the present analysis. The majority of (positive) incidence rates of cryptosporidium are between 1.0 and 100.0. The temperature ranges between approximately 20 and 35°C; most of the rainfall values are less 600 mm; and SEIFA ranges between 850 and 1000. No variables are strongly correlated.

19.3 Models and methods

19.3.1 CARTs

CARTs are binary decision trees, which are built by repeatedly splitting the predictor space into subsets (nodes), according to splitting rules of the predictor variables (Breiman *et al.* 1984), so that the distribution of the response variable y in each subset (node) becomes increasingly more homogeneous as one progresses down the tree. The terminal nodes, or leaves, of the tree are defined by different regions of the partition, where the data are not split any further and each observed response is assigned to one terminal node. The response variable determines the type of tree and the homogeneity of the terminal nodes. If the response variable is categorical then a classification tree is used to predict the classes of the response; alternatively, if the response is continuous then a regression tree predicts the average response. The initial node is called the root, and at each node the data are split into left and right branches or splitting nodes. At each branch, selection of the splitting rule is determined by maximizing a goodness-of-fit measure such as an impurity function (e.g. Gini or information index) that calculates the impurity or diversity of each split (Therneau and Atkinson 1997).

An example of a classification tree is given in Figure 19.2. A classification tree models the conditional distribution of the categorical response variable $y_i \in \{1, \ldots, J\}$ given the L predictor variables $x_{i\ell}$ (where $\ell = 1, \ldots, L$) at sites $i = 1, \ldots, n$. The partition of the response variable starts at the root node and divides the predictor space at each internal or split node S_k, $k = 1, \ldots K - 1$, where K is the

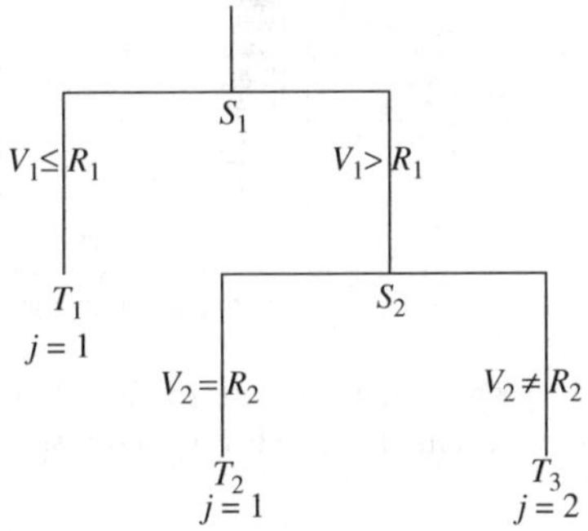

Figure 19.2 Example of a CART, including the parameters used to define the size and structure of the tree. This tree has three terminal nodes ($K = 3$).

size of the tree (number of terminal nodes). At each splitting node S_k, the partition is based on a splitting rule R_k of a variable V_k that divides the observations $\{y_i; y_i \in S_k\}$ into the left and right child nodes. Terminal nodes $T_1, \ldots, T_K$, also called leaves, are the bottom nodes in which the predictor space is not split any further. For classification trees each terminal node is usually classified into one of the J categories of the response variable. For regression trees, summary statistics of y given x are displayed (e.g. mean and standard deviation) at each terminal node. At each splitting node S_k, if the predictor V_k is continuous (e.g. S_1 in Figure 19.2) then the splitting rule R_k is based on a value a so $R_k = a$, where $\min(V_k) \leq a \leq \max(V_k)$. Thus the rule is $V_k \leq R_k$ which defines the observations in the left child node and $V_k > R_k$ defines those observations in the right child node. Alternatively, if it is categorical (e.g. S_2 in Figure 19.2) then R_k is based on a class subset c so $R_k = c$, where $c \subset \{\text{possible levels of} V_k\}$. Thus the rule $V_k \in c$ defines the observations in the left child node and $V_k \notin c$ defines those in the right child node. The parameter $\theta_k = \{R_k, S_k, V_k\}$ defines the parameter set of the kth node, so that $\theta_K = \{\theta_k, k = 1, \ldots K\}$.

19.3.2 Bayesian CARTs

Bayesian approaches to CARTs (BCARTs) were proposed by Chipman *et al.* (1998), Denison *et al.* (1998) and Buntine (1992), and further extended by O'Leary (2008) and O'Leary *et al.* (2008). These approaches have the potential to explore the model space more fully than the traditional CART methods. Although the model structure is (relatively) straightforward to formulate (see below), a fully Bayesian approach that evaluates the (posterior) probability over the whole model parameter space, including K, is a challenge.

Chipman *et al.* (1998) and Denison *et al.* (1998) both formulate the BCART model as follows. The joint distribution of (K, θ_K, y) is described by

$$p(K, \theta_K, y) = p(K)p(\theta_K|K)p(y|K, \theta_K),$$

where $p(K)$ is the prior probability for each model (defined by the number of terminal nodes), $p(\theta_K|K)$ is the prior probability of the parameter set θ_K given model K, and $p(y|K, \theta_K)$ is the likelihood of the data y given the model K and the corresponding parameter set θ_K. Inference about K and θ_K is derived from the joint posterior $p(K, \theta_K|y)$, and is factorized as

$$p(K, \theta_K|y) = p(K|y)p(\theta_K|K, y).$$

The likelihood $p(y|K, \theta_K)$ is the distribution of $p(y|\theta_K)$ over the terminal node T_k. As an example of a regression tree, if the data y are assumed to have a normal distribution, then $\psi_k = (\mu_k, \sigma_k^2)$ and the likelihood is

$$p(y|K, \theta_k) \propto \prod_{k=1}^{K} \left\{ \sigma_k^{-1} \exp\left[-\frac{1}{2\sigma_k^2} \left(\sum_{j \in \mathbf{T}_k} (y_j - \mu_k)^2 \right) \right] \right\}.$$

As an example of a classification tree, if the observations are assumed to have a multinomial distribution, so that if there are N categories, $\psi_k = (p_{k1}, \ldots, p_{kN})$, then the likelihood is

$$p(y|k, \theta_{(k)}) \propto \prod_{k=1}^{K} \prod_{j=1}^{N} (p_{kj})^{m_{kj}}, \tag{19.1}$$

where m_{kj} is the number of data points at terminal node k, which are classified into category j, and p_{kj} is the corresponding probability.

A conjugate Dirichlet prior can be adopted for p_{kj}; in the absence of other information, a uniform distribution may be assumed so that

$$\pi(p_{k1}, \ldots, p_{kJ}) = \mathrm{Dir}_{J-1}(p_{k1}, \ldots, p_{kJ}|1, \ldots, 1).$$

The prior for the model is $p(\theta_K|K)p(K)$, where $\theta_k = \{R_k, V_k, S_k\}$, so that

$$p(\theta_K|K)p(K) = p(R_k|V_k, S_k, K)p(V_k|S_k, K)p(S_k|K)p(K)p(\psi_k|V, S, K). \tag{19.2}$$

For a regression tree with a normal likelihood, a non-informative prior for $p(\psi_k|V, S, K)$ can be represented by a normal prior with a large variance for μ_k and a uniform prior with a large range for σ_k. For a classification tree with a multinomial likelihood, a non-informative prior for $p(\psi_k|V, S, K)$ can be represented by a Dirichlet prior for p_k with hyperparameters equal to 1.

A Dirichlet distribution may be specified for each of the possible splitting nodes S_k, variables V_k and splitting rules R_k at node S_k:

$$\begin{aligned} p(R_k|V_k, S_k, K) &= \mathrm{Dir}(R_k|\alpha_{R_1}, \ldots, \alpha_{R_K}), \\ p(V_k|S_k, K) &= \mathrm{Dir}(V_k|\alpha_{V_1}, \ldots, \alpha_{V_K}), \\ p(S_k|K) &= \mathrm{Dir}(S_k|\alpha_{S1}, \ldots, \alpha_{S_K}). \end{aligned} \tag{19.3}$$

Again, when no information is available on S_k, V_k and R_k, then a non-informative prior is used, with $\alpha_{V_1}, \ldots, \alpha_{V_K} = 1$, $\alpha_{R_1}, \ldots, \alpha_{R_K} = 1$ and $\alpha_{S_1}, \ldots, \alpha_{S_K} = 1$ in Equation (19.3).

The prior $p(K)$ may be assumed to be a truncated Poisson distribution with parameter λ (expected number of nodes in the tree),

$$p(k) \propto \frac{\lambda^k}{(\exp(\lambda) - 1)k!}, \qquad I_{0<k<K^*}. \tag{19.4}$$

Thus this prior places equal weight on trees of the same size. This prior imposes a left limit of $k = 0$ because the minimum model contains one terminal node. The value of λ is selected such that the size of the tree (number of terminal nodes K) is restricted to an interpretable size K^*. In the kyphosis and cryptosporidium case studies, there was no information available on the size of tree, thus $\lambda = 10$ (Denison *et al.* 1998).

Alternative distributions for the tree structure (size and shape) include a tree-generating process prior that can depend on both the tree size and shape (Chipman *et al.* 1998), and a 'pinball prior' (Wu *et al.* 2007).

In the present two case studies there was no information available about the model variables, so non-informative priors were adopted. In other situations, if such information is available, then informed priors may be used instead. For example, in an analysis of habitat suitability of a threatened species, O'Leary *et al.* (2008) discussed how to elicit from an expert the size of the tree, the relative importance of the variables, and the splitting rules for the most important variables. They also showed how to translate this information into priors and combine it with the data for Bayesian classification trees.

The sensitivity of the Bayesian CART model to the choice of priors has been investigated by O'Leary (2008) for classification trees. The sensitivity analysis involved the investigation of the hyperparameters of the priors for tree size (number of terminal nodes), splitting nodes, splitting variables and splitting rules. The results indicated that the posterior distribution is relatively robust to these priors except for extreme choices of the hyperparameters.

19.4 Computation

Buntine (1992) used a deterministic search algorithm, whereas Chipman *et al.* (1998) and Denison *et al.* (1998) developed stochastic search algorithms to identify good trees. Chipman *et al.* (1998) described a Metropolis–Hastings algorithm, while Denison *et al.* (1998) and O'Leary (2008) employed a reversible jump Markov chain Monte Carlo method (Green 1995). These algorithms use the following four steps or moves: addition of a splitting node (grow or birth move), deletion of a splitting node (prune or death move), change of variable or splitting rule at some node (change move). Chipman *et al.* (1998) also included a swap move in which the splitting rules among internal nodes (parent and child node pair) are swapped.

Below is an explanation of the BCART stochastic search algorithm, model diagnostics and identification of good trees; see O'Leary (2008) for further details.

19.4.1 Building the BCART model – stochastic search

In this chapter, the joint posterior distribution $p(\theta_K, K|y)$ is simulated using a reversible jump Markov chain Monte Carlo algorithm, developed from the general framework of Green (1995), to stochastically search over the space of possible trees. In this algorithm four moves are used to build the CART model:

1. Birth – addition of a splitting node. The prior $p(S_k|K)$ is used to select one of the existing terminal nodes T_k to become the new splitting node. Then a variable and splitting rule are chosen, using the priors $p(V_k|S_k, K)$ and $p(R_k|V_k, S_k, K)$, to split the observations y_i at this new node into two new terminal nodes.

2. Death – deletion of a splitting node. The prior $p(S_k|K)$ is used to select a current splitting node to be deleted and become one terminal node, but the selection is constrained to splitting nodes that have two terminal nodes.
3. Variable – change of variable at some splitting node. A splitting node is chosen by the prior $p(S_k|K)$ and its variable and splitting rule are changed using the prior $p(V_k|S_k, K)$ and $p(R_k|V_k, S_k, K)$.
4. Splitting rule – change of splitting rule at some splitting node. A splitting node is selected via the prior $p(S_k|K)$ and a new splitting rule is selected using the prior $p(R_k|V_k, S_k, K)$.

Given a model of size K, a birth move results in an increase in the number of terminal nodes from $m = K$ to $m = K + 1$ and a death move decreases this number to $m = K - 1$. Denoting the probability of selecting the birth move by b_K, the death move by d_K, the variable move by ν_K and the splitting rule move by ρ_K, then $\nu_K + \rho_K + b_K + d_K = 1$. Following Denison *et al.* (1998), these probabilities are set as follows:

$$\begin{aligned} \nu_k &= \rho_k, \\ b_k &= 2c \times \min\{1, p(K+1)/p(K)\}, \\ d_k &= c \times \min\{1, p(K)/p(K+1)\}, \end{aligned}$$

where the constant c is selected so $b_K + d_K \leq 0.75$.

Moreover, at the root of the tree ($K = 1$), $b_1 = 1$ and $d_1 = \nu_1 = \rho_1 = 0$, therefore ensuring that a birth move is always performed at this point. At each iteration, the selection of a move type is determined by the following constraints:

$$\begin{aligned} \text{birth move if} \quad & (u \leq b_K) \\ \text{death move if} \quad & (b_K < u \leq b_K + d_K) \\ \text{variable move if} \quad & (b_K + d_K < u \leq b_K + d_K + \nu_K) \\ \text{else splitting rule move} & \end{aligned}$$

where u is a random number generated from a uniform distribution $U(0, 1)$.

The probability α of accepting a move from model (K, θ) to a model (K^*, θ^*) is given generically (Green 1995) as

$$\alpha = \min\{1, (\text{likelihood ratio}) \times (\text{proposal ratio}) \times (\text{prior ratio})\}.$$

The probability of accepting a variable or splitting rule move (Green 1995) only includes the likelihood ratio term. For the birth move, after simple calculations, the proposal ratio multiplied by the prior ratio equals $(K_{\text{die}} + 1)/K$, where K_{die} is the number of possible splitting nodes (with two terminal nodes) that can be deleted. For the death move this fraction is inverted.

Acceptance of the current proposed model is restricted to $(|\mathbf{T}_k| > T_{\min})$; that is, $|\mathbf{T}_k|$, the number of data points classified into terminal node T_k, must be greater than the minimum number of data points $T_{\min}$. Here $T_{\min}$ was set to five (Becker *et al.* 1988).

This BCART algorithm was written in MATLAB, in which each move is a separate function. The pseudo-code is available on the book's website.

19.4.2 Model diagnostics and identifying good trees

Important issues of a BCART stochastic algorithm searching over the tree space are deciding on the stopping criterion, identifying good trees and model diagnostics. Denison *et al.* (1998) carried out a single long chain and used the stability of the posterior probability as the stopping rule. Chipman *et al.* (1998) performed multiple restarts and stopped the algorithm when it became trapped in a local posterior model. Their model diagnostic and stopping criteria of a BCART algorithm were just determined through stability of the posterior probability (Chipman *et al.* 1998; Denison *et al.* 1998). O'Leary (2008) implemented the reversible jump (RJ) algorithm of Denison *et al.* (1998), with two extensions: adding a decision-theoretic layer to identify the best performing trees; and including informative priors.

The selection criterion used by Denison *et al.* (1998) to identify good trees was the modal tree structure for each tree size. For classification trees, comparison between trees was achieved through the posterior probability, total number of misclassifications and the deviance (minus twice the log likelihood $p(y|K, \theta_K)$ given in Equation 19.1). Chipman *et al.* (1998) defined good trees as trees with the largest marginal likelihood and the lowest misclassification rate (for classification trees). Identification of trees with this criterion was determined by plotting the log marginal likelihood or misclassification rate of all trees against the number of terminal nodes. Trees were also compared using the diagnostic plots of the iteration number against log posterior, log likelihood and number of terminal nodes.

A confusion matrix (Table 19.1) can be calculated for each tree simulated from the posterior distribution. Then the 'best' tree can be defined as the one that minimizes one or more accuracy measures, depending on the aims of the study. For classification trees (in our case binary response variable, presence/absence), numerous accuracy measures can be calculated, for example Fielding and Bell (1997) and Pearce *et al.* (2001), which compares the observed versus predicted responses. Here the stopping criterion, the model diagnostics and identification of good classification trees were examined further through the investigation of several of these accuracy measures

Table 19.1 The confusion or loss matrix, where the response variable is binary (presence = 1, absence = 0).

		Observed (actual)		
		Presence (+ve)	Absences (−ve)	Total
Predicted	Presence (+ve)	A (true +ve)	B (false +ve)	(A+B)
	Absence (−ve)	C (false −ve)	D (true −ve)	(C+D)
	Total	(A+C)	(B+D)	N

(O'Leary 2008). In the kyphosis case study, interest centred on the correct prediction of presence of the Kyphosis disease. Therefore the most important aim was to minimize the misclassification of presences, the false negative rate, FNR $= C/(A + C)$. Since the data are limited, it was also important to minimize misclassification of absences (false positive rate, FPR $= C/(A + C)$) and the overall misclassification rate (MCR $= (B + C)/N$).

For a certain number of iterations (e.g. 300 000 iterations) after burn-in, a set SC_G of G good C classification trees is identified that have the lowest FNR, FPR and MCR. The variables and splitting rules at each splitting node of the trees in SC_G are examined. When the membership of SC_G and structure of the component trees have stabilized, this set of classification trees is declared 'good' and is examined further through diagnostic plots of the number of terminal nodes against log likelihood and log posterior probabilities. These probabilities can be compared with all proposed trees to confirm that this set of good trees SC_G has good values of goodness-of-fit measures, that is highest likelihood, maximum posterior probabilities and minimum deviance.

For each tree in the set of good classification trees SC_G the following summary statistics can be examined: tree structure (variables, splitting rules and number of terminal nodes), FNR, FPR, deviance and log likelihood, log posterior probability. From this set of good classification trees, depending on the aims of the analysis, two or three trees may be chosen as the 'best' trees, based on the modal tree structure (same-size tree with the same variables and splitting rules), lowest FNR, FPR and deviance, and the highest likelihood and posterior probability. Alternatively the 'best' tree could be selected from SC_G based on biological interpretability, expert judgement or other criteria.

For regression trees, the stopping criterion is based on posterior probabilities $p(K, \theta_K | y)$ and residual sums of squares (RSS)

$$\text{RSS} = \sum_{k=1}^{K} \sum_{j \in \mathbf{T}_k} (y_j - \overline{y}_{t_k})^2, \tag{19.5}$$

where $\overline{y}_{t_k}$ is the mean of response at terminal node t_k.

Therefore a set of good regression trees (SR_G), for a certain number of iterations (e.g. 300 000 iterations) after burn-in, is identified that have the lowest RSS and maximum posterior probabilities. Similar to classification trees, the variables and splitting rules at each splitting node of the trees in SR_G are investigated. Once the membership of SR_G and structure of the component trees have stabilized, this set of regression trees is declared 'good'.

The search space of the BCART algorithm can be examined to establish if it proposes tree solutions provided by the traditional recursive partitioning methods. Examples of packages in S-PLUS and R that perform frequentist recursive partitioning methods include `tree` and `RPART` (Recursive PARTitioning) respectively. The trees proposed by our BCART algorithm can also be compared with those identified using the Denison *et al.* (1998) algorithm.

19.5 Case studies – results

19.5.1 Case study 1: Kyphosis

Here we will compare the results of the kyphosis case study analysis (modelling the presence/absence of Kyphosis) between our BCART algorithm and that in Denison *et al.* (1998). Following on from Denison *et al.* (1998), the prior $p(K)$ was given a Poisson prior distribution with mean 10 in our BCART algorithm.

Our BCART algorithm identified a set of 22 good classification trees (SC_{22}) with the lowest FNR, FPR and MCR. (See the book's website for details.) The error rates of these trees were: FNR < 0.2, FPR < 0.3 and MCR < 0.3. The trees had four to seven terminal nodes and the majority had Start (number of the first vertebra operated on) as the first splitting variable V_1. For this set SC_{22}, the number of terminal nodes K was also plotted against the log likelihood and log posterior of tree structure given the data (see the book's website); as expected, trees with the maximum likelihood and posterior had the most terminal nodes.

Two of the best trees are depicted in Figure 19.3. The first tree (Figure 19.3a) has FNR of 0.1765 while the second tree (Figure 19.3b) has FNR of 0.0588. Since the aim of this case study is to achieve a tree with the lowest FNR, the second tree was selected as the 'best'. Figure 19.3a) was proposed nine times and accepted three times over the 300 000 iterations, thus showing the need for a very large number of iterations. The 'best' tree predicts three types of kyphosis presence (Figure 19.3b). These fall into two main groups, depending on whether the group has low or high Start. For people with a low Start value (less than or equal to 12), there are two presence groups and one absence group. Six people are correctly predicted to have no kyphosis after the operation and zero people with presence of Kyphosis, who have low Start, less than or equal to four Number of vertebrae and their age is less than or equal to 36 years old. People with the same low Start and low total Number of vertebrae but older than 36 are predicted to have kyphosis present after the operation, with 5 out of 12 people predicted correctly. Kyphosis is predicted as present for people with low Start but greater than four Number of vertebrae, in which 10 of 17 people are correctly predicted. There are three groups of people whose Start values here are greater than 12. People with Start less than or equal to 13 are predicted to have no kyphosis after the operation, with 11 of 12 correctly predicted. Absence of kyphosis is also predicted for people whose Start values are greater than 13 and less than 4 Number of vertebrae, with all 28 people correctly predicted. However, people with Start greater than 13 but Number of vertebrae less than 4 are predicted to have kyphosis present.

The model acceptance rate under the BCART algorithm was 41%, with birth move acceptance being 32%, death 32%, variable 13%, and splitting rule 23%. These acceptance rates for the four moves were fairly constant over the size of tree. Convergence plots of the BCART algorithm (see the book's website for details) indicated that the proposed iterations (thinned by 250) against FNR, FPR, MCR, log posterior tree structure, log likelihood and number of terminal nodes (K) show stability since the algorithm is varying locally within a defined region. Moreover, the diagnostic plots showed that the algorithm search space included trees with FNR and FPR both

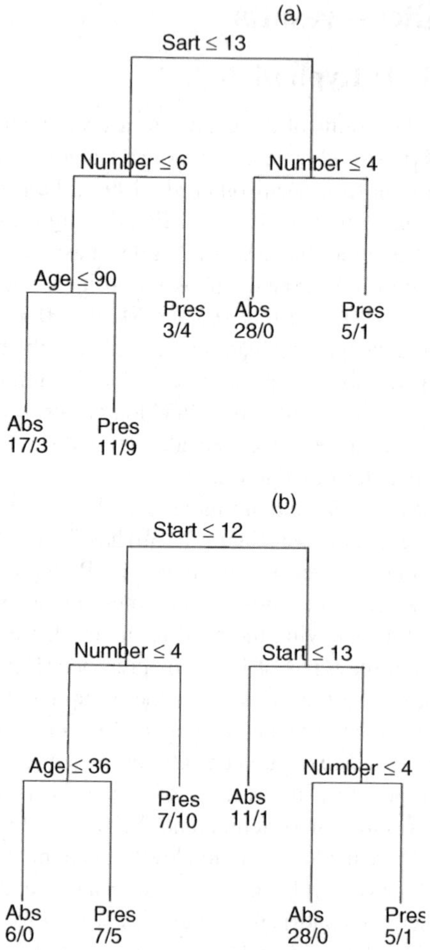

Figure 19.3 The two best trees for presence/absence of kyphosis identified from BCART.

ranging from 0.0 to 1.0 and MCR from 0.0 to 0.85. The maximum size tree this algorithm proposed was $K = 10$ terminal nodes. This indicates that the search space of this algorithm included trees of various sizes and trees with zero to total misclassification of presences and absences. Figure 19.4 shows a box-and-whisker plot of the log likelihood for all trees that were accepted at each size tree (number of terminal nodes). Trees that perform well have the highest likelihood, 4 to 10 terminal nodes and a log likelihood above −25.

Denison *et al.* (1998) stated that convergence was reached after only 20 000 iterations for the kyphosis data set. However, the results from our algorithm suggest, through the examination of the variables and splitting rules of the set of good classification trees SC_G, that convergence is reached after 300 000 iterations, with a burn-in of 20 000. As anticipated, the best tree identified using our algorithm and selection

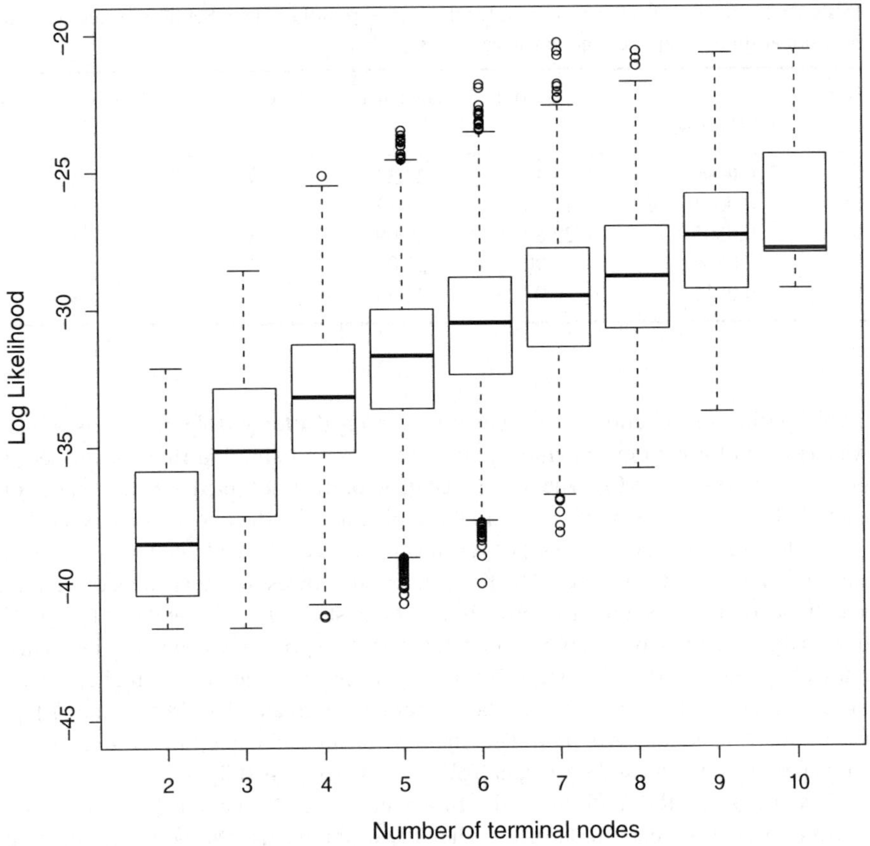

Figure 19.4 Boxplot of log likelihood for trees of each number of terminal nodes that were accepted for the kyphosis analysis.

method had a lower FNR than that selected by Denison *et al.* (FNR of 0.0588 and 0.294 respectively), but their selected tree had a lower FPR (0.2969 compared with 0.047) and MCR (0.247 compared with 0.099). Analysis of the kyphosis data set revealed that our BCART proposed trees were similar to the tree solutions from the traditional algorithms and the selected tree identified by Denison *et al.* (1998).

Trees were developed from the traditional CART methods, using the `RPART` package and `tree` function in R (see the book's website), and were compared with the proposed trees of the BCART algorithm. The BCART algorithm did not propose the same best tree solutions, but the tree from `RPART` without the age ≥ 111 splitting node was proposed four times and accepted twice.

19.5.2 Case study 2: Cryptosporidium

The Bayesian regression tree algorithm was applied to the cryptosporidium case study, in which the response variable is the positive incidence rates (ignoring zeros) of the

Table 19.2 Results of regression trees for the cryptosporidium data. Set of best trees (based on residual sums of squares and deviance) for Bayesian regression trees.

Tree	RSS (/1.0e-024)	Log posterior	Deviance	Size of tree	Number of trees
1	**0.0009**	**−18.15**	**28.37**	**4**	**1**
2	0.0051	−16.45	23.12	3	1
3	0.0505	−22.82	39.09	5	1
4	0.0535	−15.90	22.04	3	2
5	0.0816	−16.03	22.30	3	1

cryptosporidiosis infection. Convergence was reached after 300 000 iterations, which was assessed through examination of the variables and splitting rules of the set of good regression trees SR_G. The total percentage of trees accepted and acceptance of each of the four moves were similar to those obtained for the kyphosis analysis.

Table 19.2 summarizes the set of seven good trees (SR_7) with the lowest residual sums of squares and deviances. The first tree has the lowest residual sums of squares and the second highest posterior probability of tree structure, so is declared the 'best' tree. In this tree, displayed in Figure 19.5, there are four groups with positive incidence rates of cryptosporidium, ranging from low to high incidence. The highest mean incidence rate (144.83) occurs when the temperature is greater than 28.5°C and SEIFA is less than or equal to 1010.4. The smallest mean incidence rate is 4.73, when temperatures are at most 28.5°C and SEIFA is greater than 1039.8.

Traditional CARTs built using the `tree` and `RPART` functions (see the book's website) produced very similar trees. The only difference is that the `tree` function had one additional splitting node, SEIFA < 945.5, which is a child node of SEIFA ≥ 892 and Temperature ≥ 32.08. The 'best' tree obtained by BCART had smaller RSS and deviance than these two models.

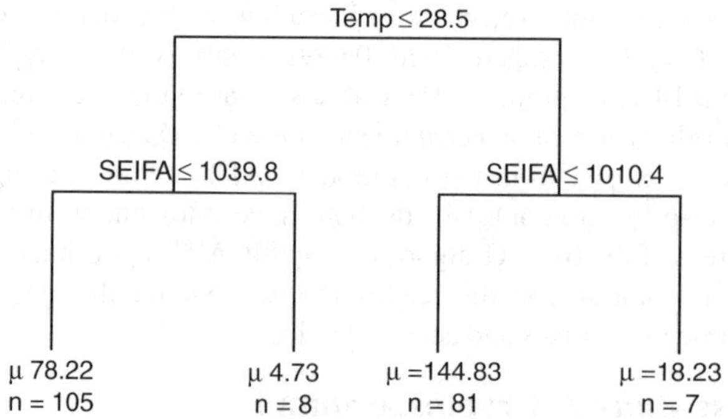

Figure 19.5 The best tree for positive incidence rates of cryptosporidium identified using BCART

19.6 Discussion

It can be argued that the BCART method explores the model space more fully than traditional CART. The latter method uses a recursive partitioning approach to build a binary decision tree, by repetitively splitting the data based on some splitting rule of the predictor variables (Breiman *et al.* 1984; De'ath and Fabricius 2000). This search algorithm starts at the top of the tree and at each splitting node selects the optimal split (using impurity functions such as information and Gini index) until the data cannot be partitioned any further. Thus, the choices of trees are constrained; specifically the choices of splitting rules at nodes down the tree are constrained by choices made at nodes above, and only get one optimal tree. Alternatively, BCART produces a large number of plausible trees with wider variety of tree structures with different variables, splitting rules and number of terminal nodes. At any splitting node, the variable and splitting rules are randomly selected from the prior and trees that perform well in terms of high likelihood and posterior probabilities are chosen. This stochastic search algorithm is based on the careful choice of model performance criteria to ensure a range of good models are selected. This choice of model performance criteria was the major algorithm design challenge addressed in this chapter.

For the kyphosis case study, trees developed using the traditional recursive partitioning method (implemented using `RPART` and `tree` functions in R) were also proposed by the BCART algorithm. The BCART algorithm and methodology we suggested for identifying good trees were able to find trees with good predictions of presence of kyphosis. For the traditional methods a higher weight was placed on the correct prediction of presences in the loss matrix. However, the BCART algorithm identified trees with better model performance in terms of lower FNR.

Hu *et al.* (2011) compared BCART with the Bayesian spatial conditional autoregressive (BCAR) model using the cryptosporidium case study. This analysis found that the nature and magnitude of the effect estimates were similar for the two methods, but the BCART model identified higher order interaction effects. Several second-order interactions and a third-order interaction suggested by the BCART analysis were also tested in BCAR, but only main effects remained significant. Overfitting of BCART for the cryptosporidium case study was also explored by Hu *et al.* (2011). A cross-validation approach was adopted by splitting the data into a training and test data set, using a stratified random sample to obtain an equivalent proportion of observed presences and absences. It was found that there was no evidence of overfitting in BCART, since there was little difference in the goodness-of-fit measures between the training and validation data sets.

As demonstrated by these two examples, BCART can assist in the variable selection problem, by identifying important main effects and nonlinear interactions of variables (Clark and Pregibon 1993; De'ath and Fabricius 2000; Guisan *et al.* 2002). In particular, BCART can identify variables that partition the response variable into more homogeneous groups. The selection of good trees can be based on different goodness of fit, or loss criterion, both through prior specification and post-hoc evaluation. Moreover, the Bayesian paradigm facilitates greater inferential capacity via the posterior densities and posterior parameter estimates as illustrated in the kyphosis and cryptosporidium examples.

References

Bachur RG and Harper MB 2001 Predictive model for serious bacterial infections among infants younger than 3 months of age. *Pediatrics* **108**, 311–316.

Becker R, Chambers JM and Wilks A 1988 *The New S Language*. Wadsworth, Belmont, CA.

Bell JF 1996 Application of classification trees to the habitat preference of upland birds. *Journal of Applied Statistics* **23**(2 and 3), 349–359.

Breiman L 2001a Random forests. *Machine Learning* **45**, 5–32.

Breiman L 2001b Statistical modelling: the two cultures. *Statistical Science* **16**(3), 199–231.

Breiman L, Friedman JH, Olshen R and Stone CJ 1984 *Classification and Regression Trees*. Wadsworth, Belmont, CA.

Buntine W 1992 Learning classification trees. *Statistics and Computing* **2**, 63–73.

Chambers JM and Hastie TJ 1992 *Statistical Models in S*. Wadsworth and Brooks/Cole, Pacific Grove, CA.

Chipman HA, George EI and McCulloch RE 1998 Bayesian CART model search. *Journal of the American Statistical Association* **93**(443), 935–960.

Clark L and Pregibon D 1993 Tree based models. In *Statistical Models in S* (eds JM Chambers and TJ Hastie TJ), pp. 377–420. Chapman & Hall/CRC, Boca Raton, FL.

Cordell H 2009 Detecting gene-gene interactions that underlie human diseases. *Nature Reviews Genetics* **10**, 392–404.

De'ath G and Fabricius KE 2000 Classification and regression trees: a powerful yet simple technique for ecological data analysis. *Ecology* **81**(11), 3178–3192.

Debeljak M, Dzeroski S, Jerina K, Kobler A and Adamic M 2001 Habitat suitability modelling for red deer (*Cervus elaphus L.*) in South-central Slovenia with classification trees. *Ecological Modelling* **138**(1-3), 321–330.

Denison D, Mallick B and Smith A 1998 A Bayesian CART algorithm. *Biometrika* **85**(2), 363–377.

Fan G and Gray JB 2005 Regression tree analysis using TARGET. *Journal of Computational and Graphical Statistics* **14**(1), 206–218.

Fielding AH and Bell JF 1997 A review of methods for the assessment of prediction errors in conservation presence/absence models. *Environmental Conservation* **24**(1), 38–49.

Green PJ 1995 Reversible jump Markov chain Monte Carlo computation and Bayesian model determination. *Biometrika* **82**, 711–732.

Guisan A, Edwards TC and Hastie T 2002 Generalized linear and generalized additive models in studies of species distributions: setting the scene. *Ecological Modelling* **157**, 89–100.

Hu W, Mengersen K and Tong S 2009 Spatial analysis of notified Cryptosporidiosis infections in Brisbane, Australia. *Annals of Epidemiology* **19**, 900–907.

Hu W, Mengersen K and Tong S 2010 Risk factor analysis and spatiotemporal CART model of cryptosporidiosis in Queensland, Australia. *BMC Infectious Diseases* **10**, 311.

Hu W, O'Leary R, Mengersen K and Low Choy S 2011 Bayesian classification and regression trees for predicting incidence of cryptosporidiosis. *PLoS ONE* **6**, e23903.

Lamon EC and Stow CA 2004 Bayesian methods for regional-scale eutrophication models. *Water Research* **38**, 2764–2774.

Miller J and Franklin J 2002 Modeling the distribution of four vegetation alliances using generalized linear models and classification trees with spatial dependence. *Ecological Modelling* **157**, 227–247.

Nelson L, Bloch D, Longstreth JW and Shi H 1998 Recursive partitioning for the identification of disease risk subgroups: a case-control study of subarachnoid hemorrhage. *Journal of Clinical Epidemiology* **51**, 199–209.

O'Leary R 2008 Informed statistical modelling of habitat suitability for rare and threatened species. PhD thesis. Queensland University of Technology.

O'Leary R, Murray J, Low Choy S and Mengersen K 2008 Expert elicitation for Bayesian classification trees. *Journal of Applied Probability and Statistics* **3**, 95–106.

O'Leary RA, Francis R, Carter K, Firth M, Kees U and de Klerk N 2009 A comparison of Bayesian classification trees and random forest to identify classifiers for childhood leukaemia. *18th World IMACS Congress and MODSIM09 International Congress on Modelling and Simulation. Modelling and Simulation Society of Australia and New Zealand and International Association for Mathematics and Computers in Simulation* (eds R Anderssen *et al.*), pp. 4276–4282.

Pearce J, Ferrier S and Scotts D 2001 An evaluation of the predictive performance of distributional models for flora and fauna in north-east New South Wales. *Journal of Environmental Management* **62**, 171–184.

Smith KJ and Roberts MS 2002 Cost-effectiveness of newer treatment strategies for influenza. *American Journal of Medicine* **113**, 300–307.

Therneau TM and Atkinson B 2003 RPART (Recursive Partitioning and Regression Trees).

Therneau TM and Atkinson EJ 1997 *An Introduction to Recursive Partitioning Using the RPART Routines*. Mayo Foundation.

Wu Y, Tjelmeland H and West M 2007 Bayesian CART: prior specification and posterior simulation. *Journal of Computational and Graphical Statistics* **16**(1), 44–66.

20

Tangled webs: Using Bayesian networks in the fight against infection

Mary Waterhouse[1,2] **and Sandra Johnson**[1]

[1] *Queensland University of Technology, Brisbane, Australia*
[2] *Wesley Research Institute, Brisbane, Australia*

20.1 Introduction to Bayesian network modelling

A Bayesian network (BN) is a probabilistic graphical model showing factors and their interactions, relating to a response of interest (Jensen and Nielsen 2007; Pearl 1988). The graphical model represents factors as nodes, and relationships between factors are shown as directed links which indicate the direction of the dependence between the nodes to form a directed acyclic graph (DAG) (Lauritzen and Sheehan 2003). An example of a DAG is shown in Figure 20.1. The link going from node C4 to node C2 indicates that node C4 is a parent of node C2. Similarly, node C2 is said to be a child of node C4. Apart from the parent–child relationship between two connected nodes, it is possible to distinguish three types of relationships between any three connected nodes in a BN (Korb and Nicholson 2010). These relationship types are serial, converging and diverging and are illustrated in Figure 20.1. The BN characteristics (*d-separation* and *Markov property*) inherent in these three relationships are important in their inferencing capabilities and simplification of probability calculations (Koller and Friedman 2009; Korb and Nicholson 2010). The relationships do not need to be

Case Studies in Bayesian Statistical Modelling and Analysis, First Edition. Edited by Clair L. Alston, Kerrie L. Mengersen and Anthony N. Pettitt.

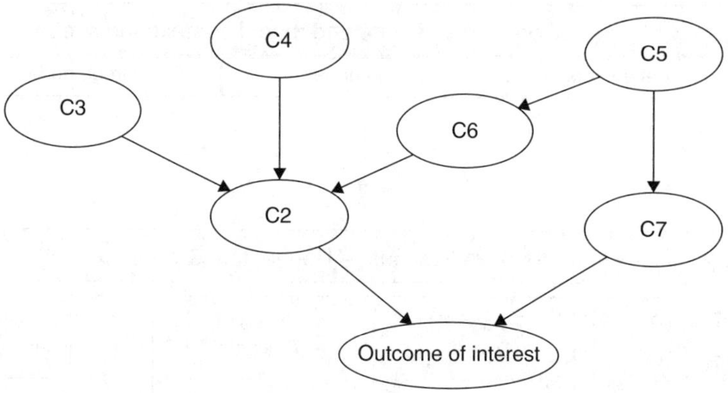

Figure 20.1 Example DAG showing the outcome of interest as the target node. One group of nodes (C2, C3, C4) represents a typical converging relationship, while another (C5, C6, C7) shows a diverging connection. Additionally nodes C5, C6 and C2 form a serial connection.

causal in nature; however, representing causality in the BN facilitates meaningful queries and conclusions to be drawn (Pearl 2000).

The graphical representation of the factors is combined with an underlying probabilistic framework that represents knowledge about the system, and captures the uncertainty associated with this knowledge through conditional probability distributions. There is a conditional probability table (CPT) for each node, including the outcome of interest. The CPT may be continuous or discrete; however, in practice continuous probability distributions are typically discretized into mutually exclusive states (Johnson *et al.* 2011) to utilize the full potential of BN software applications. Accordingly, the case study presented in this chapter has discrete nodes. The CPT of a node is determined by the states of the node and by the states of its parent nodes (Jensen and Nielsen 2007; Pearl 1988).

BNs are increasing in popularity and have been used extensively and successfully in many diverse areas (Johnson *et al.* 2011; Korb and Nicholson 2010), such as medical (Waterhouse *et al.* 2011), environmental (Donald *et al.* 2009), financial (Ammann and Verhofen 2007) and social science (Baumgartner *et al.* 2008). The majority of BNs are static, but object-oriented (OO) and dynamic BNs overcome many of the drawbacks of static BNs (Johnson and Mengersen 2011; Johnson *et al.* 2010a).

The variables (factors) in a BN may be at different temporal and spatial scales and the data represented in the network may originate from diverse sources such as empirical data, expert opinion and simulation outputs (Borsuk *et al.* 2006; Jensen and Nielsen 2007; McCann *et al.* 2006).

20.1.1 Building a BN

Constructing a BN is an iterative task, and a conceptual representation of an iterative approach to BN model development is illustrated in Figure 20.2. Essentially the iterative Bayesian network development cycle (IBNDC) consists of a

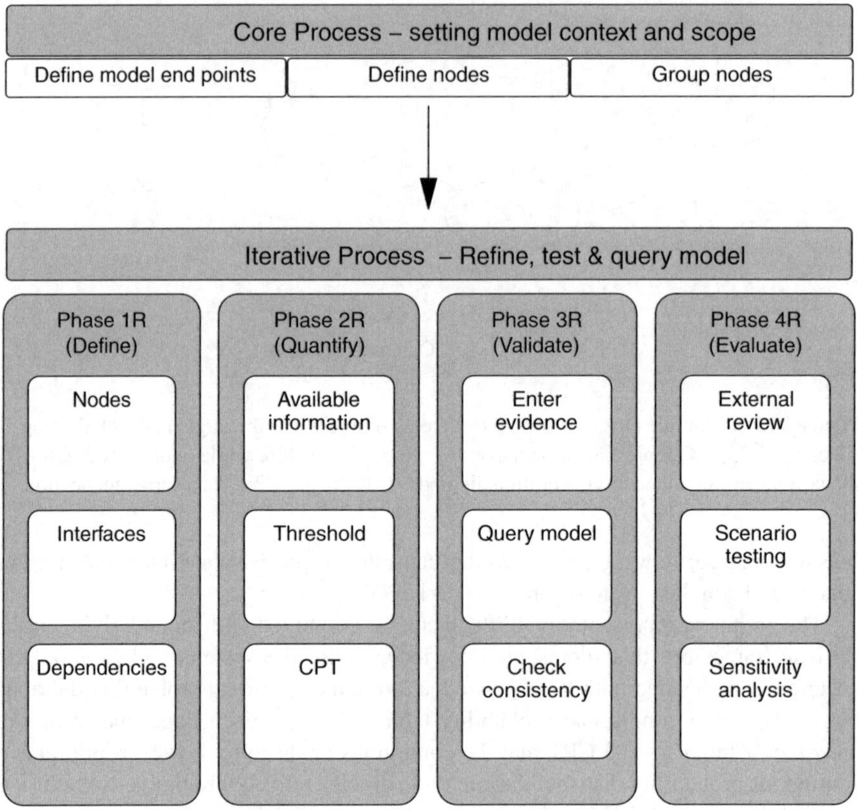

Figure 20.2 Conceptual representation of the iterative BN development cycle (IBNDC), showing key activities for each phase (Johnson *et al.* 2010b).

core process and an *iterative* process and is described in detail in Johnson *et al.* (2010b) and extended in Johnson and Mengersen (2011) for an OO approach to BN modelling.

Core process

During the core process, the response of interest is determined and then the factors known to affect this response are identified and grouped into roughly coherent subgroups.

Iterative process

The iterative process starts when the key factors identified in the core process are transferred into a BN modelling software package such as GeNIe, Netica or Hugin (Johnson *et al.* 2010b). During Phase 1R (Define Network) of the IBNDC each of the factors identified as being of interest in modelling the response is clearly defined. These definitions are entered into the software package, so that the model is 'self-documenting'. Careful documentation of factor definitions ensures consistency of interpretation of the factors during the numerous iterations of the IBNDC and prevents any dispute or uncertainty regarding each factor. During this phase the directed links are also reviewed to confirm that the direction of influence is correct and that all known dependencies have been represented. Some nodes and links may be added, deleted or modified as a result of these reviews (Johnson *et al.* 2010b).

In Phase 2R (Quantify) the states of discrete nodes and the distributions of continuous nodes are defined. The underlying CPTs are also populated during this phase and the existing documentation for the node is updated with information about the thresholds of the node's states and the information which was used to populate the CPTs. It is advisable to limit the number of states of discrete nodes and parent nodes to prevent unwieldy probability tables (Marcot *et al.* 2006).

Testing of the network is undertaken in Phase 3R (Validate) to check whether the network predictions and behaviour are consistent with current research, reflect known causal relationships and are robust to cross-validation. Any inconsistent behaviour requires the network to be carefully reassessed, which may result in changes to nodes, their states and/or thresholds, relationships between nodes and CPTs. Any changes resulting from this phase need to be addressed in Phases 1R and 2R in the next iteration (Johnson *et al.* 2010b).

In the final phase of each iteration, Phase 4R (Evaluate), evaluation of the network is performed in various ways. The most common forms are inference, scenario testing and sensitivity analysis. The latter can be used to determine the order in which factors influence a node of interest based on mutual information (MI) (Pearl 1988). The MI for two nodes, C and D say, is defined to be

$$I(\mathrm{C}, \mathrm{D}) = \sum_{d \in \mathrm{D}} \sum_{c \in \mathrm{C}} P(c, d) \log \frac{P(c, d)}{P(c)P(d)}.$$

It quantifies the extent to which a finding at one node reduces the uncertainty regarding the other node, by measuring the distance between the joint distribution, $P(\mathrm{C}, \mathrm{D})$, and the joint distribution assuming independence, $P(\mathrm{C})P(\mathrm{D})$. Higher values of $I(\mathrm{C}, \mathrm{D})$ indicate a stronger dependency between C and D, while $I(\mathrm{C}, \mathrm{D}) = 0$ if and only if C and D are independent.

For the last iteration of Phase 4R an external evaluation of the network by another expert panel should be conducted. This review may lead to additional changes to the BN, in which case another complete iteration of the four phases of the *iterative* process will be necessary.

20.2 Introduction to case study

Recent research estimates that health-care-acquired infections cost close to $1 billion per year in lost bed days in Australia (Graves *et al.* 2009). One cause of such infections is the organism methicillin-resistant *Staphylococcus aureus* (MRSA). MRSA is resistant to the standard treatments used to counter staphylococcal infections and it poses a serious risk to a patient's health and well-being. As such, limiting its transmission is a challenge facing hospitals worldwide.

MRSA is most commonly introduced to a hospital through the admission of a colonized or infected patient, and patient-to-patient transmission is primarily via the transiently colonized hands of health-care workers (Peacock *et al.* 1980). Many hospitals have screening programmes for new admissions and invest in campaigns to improve handwashing compliance (Grayson *et al.* 2008; Humphreys *et al.* 2009; Mears *et al.* 2009; Pittet *et al.* 2006). Ideally, patients whose carrier status is confirmed should be confined to isolation wards (Humphreys *et al.* 2009; Jernigan *et al.* 1996; Lucet *et al.* 2003; Mears *et al.* 2009). However, beds may not always be available for isolation, and increases in MRSA colonization have been observed when there is an isolation ward access block (Morton *et al.* 2009). Excessive overcrowding, in terms of high bed occupancy, and understaffing are believed to simultaneously encourage transmission and hinder control strategies due to: higher rates of interaction between nurses and patients; an increase in the movement of patients and staff between wards; lower levels of handwashing compliance; less thorough cleaning practices; and an increased burden on screening and isolation facilities (Andersen *et al.* 2002; Clements *et al.* 2008). In addition, the increased incidence of MRSA associated with overcrowding and understaffing serves to exacerbate these shortcomings, leading to a 'vicious cycle' (Clements *et al.* 2008). It therefore appears that the mechanisms behind MRSA transmission and containment have many confounding factors and control strategies may only be effective when used in combination.

Infectious diseases experts at a large tertiary-referral hospital in Brisbane, Australia, wanted to investigate the possible role of high bed occupancy on transmission of MRSA while simultaneously taking into account other risk factors (Waterhouse *et al.* 2011). Using a BN for this purpose is ideal, not only because it naturally facilitates the modelling of a complex system, but because it permits the combination of information from disparate sources, which is particularly useful when data are scarce (Uusitalo 2007). We use the case study to illustrate the use of the IBNDC modelling approach described in Section 20.1. We also demonstrate how the BN can be used to identify the most influential factors on MRSA transmission and to investigate scenarios.

In Section 20.3 we describe the model. The methods and results are presented in Sections 20.4 and 20.5 respectively. Section 20.6 is devoted to discussion.

20.3 Model

The BN model for MRSA was constructed using the IBNDC approach to guide model development and organization as described in Section 20.1. The final model is shown in Figure 20.3. In this case study the response of interest (target node), the presence/absence of MRSA isolate detected on swabs taken from patients, relates to the number of new in-hospital colonizations with multiresistant (mMRSA), non-multiresistant (nMRSA) or UK-MRSA. In this context MRSA is used as an umbrella term for all three strains.

The key factors (chance nodes) in the BN are loosely arranged in four clusters (subgroups). These are broadly classified as relating to handwashing behaviour, staffing adequacy, the hospital's physical capacity, and its screening for multiple

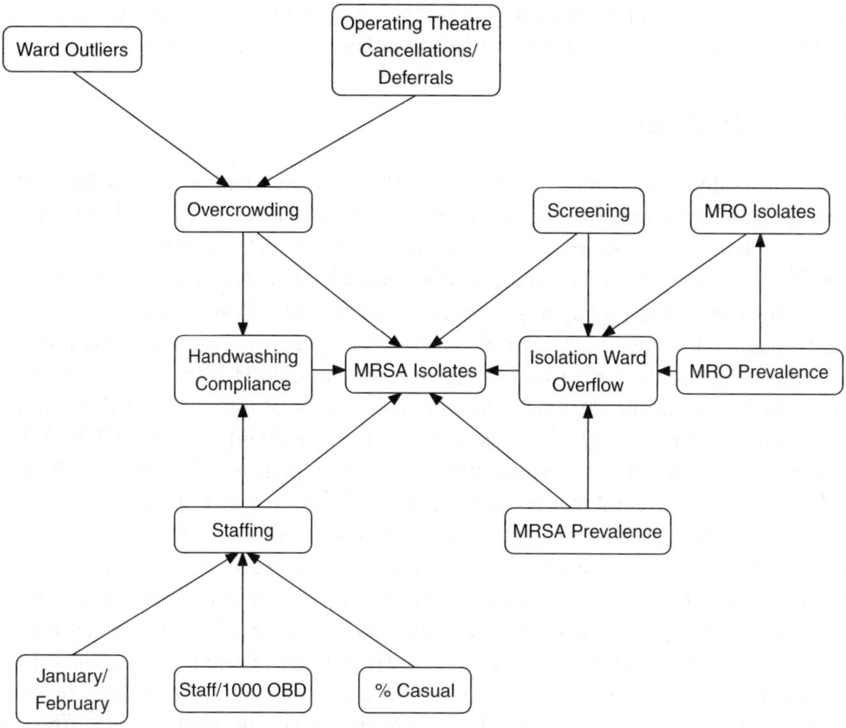

Figure 20.3 Structure of the BN used to model colonization with MRSA at a large public hospital in Brisbane. The outcome of interest, MRSA isolates, relates to the number of new in-hospital colonizations with MRSA. Copyright Elsevier.

antibiotic-resistant organisms (MROs) and isolation capacity. Since bed occupancy may be high throughout a study period, additional indicators have been used when defining the extent of overcrowding. These include the number of patients placed in outlying wards because their specialist home ward is full (ward outliers) and the number of cancellations and deferrals of operating theatre procedures due to unavailability of intensive care unit (ICU) beds. Nursing staff numbers per 1000 occupied bed-days (OBDs) and the proportion of non-permanent staff are included as indicators of staffing adequacy. Numbers of screenings for MROs, isolation ward overflow and the background MRO prevalence have been used to model the hospital's capacity to screen and isolate carriers.

Each node has two possible states. This choice was made for two pragmatic reasons. The first concerns the small data set. A node with n parents, all of which have two levels, will have a CPT with 2^n rows. Clearly the table quickly becomes large, leading to a situation in which the number of combinations to consider exceeds the available data. Secondly, the variability of several nodes was sufficiently small that additional levels would not add value to the modelling. Staffing is either satisfactory or unsatisfactory. January/February identifies early months of the year when there is an influx of new staff. Remaining nodes are either at a normal or high level.

20.4 Methods

Data were collected over the period 2008 to 2009 for the variables listed in Table 20.1. Average percentage bed occupancy relates to overcrowding in Figure 20.3. Remaining nodes in Figure 20.3 correspond to the variables in Table 20.1 in a natural way. Note that data were not collected for staffing. In some cases, data were available on a weekly basis. Otherwise, monthly aggregates were available. Means and standard deviations were calculated using all available data. Continuous data were categorized into two discrete levels based on quartiles derived from a subset of the 2008 records. The third quartile was the threshold used to classify observations as either 'normal' or 'high'.

When possible, the data were used to estimate the probabilities in the CPTs. When data were not available, the probabilities were estimated using simulation. Specifically, if a particular combination of the parents' levels was not observed during the monitoring period then parent values were randomly drawn from uniform distributions on the intervals of interest, and substituted into a regression model estimating the child node. A random component, drawn from a normal distribution with mean 0 and variance determined by the residual standard deviation of the regression model, was added to the regression output. We used 1000 sets of simulated parent values. The simulated responses for the child node were then categorized into discrete levels using the third-quartile thresholds, allowing estimation of the probability. Staffing on the other hand was populated on the basis of expert opinion.

Table 20.1 Summary of data collected over 2008–2009.

Factor	Collection frequency	*N*	Mean (SD)	Threshold
No. new in-hospital MRSA colonizations	Monthly	24	16.0 (5.5)	14.8/month
Handwashing compliance (%)	Monthly	24	46.5 (11.7)	50.0
No. nursing staff per 1000 OBDs	Monthly	24	95.9 (5.0)	94.2
Percentage non-permanent staff	Monthly	24	45.3 (1.1)	46.3
Average percentage bed occupancy	Monthly	24	92.3 (1.9)	94.0
No. ward outliers	Monthly	24	65.0 (9.8)	71.4/month
No. cancellations/deferrals of OT procedures	Monthly	24	46.6 (18.4)	68.3/month
IWO (daily average within weeks)	Weekly	101	6.0 (2.7)	8.3
Average MRSA prevalence	Weekly	101	51.0 (5.8)	52.5
No. screens for MRSA and other MROs	Weekly	101	854.8 (78.3)	868.8/week
No. MRO isolates	Weekly	101	1.5 (1.6)	1.8/week
Average MRO prevalence	Weekly	101	15.1 (3.3)	17.3

OBDs = Occupied Bed-Days.
OT = Operating Theatre.
IWO = Isolation Ward Overflow.
Reproduced with permission from Elsevier from M. Waterhouse, A. Morton, K. Mengersen, D. Cook, G. Playford. Role of overcrowding in meticillin-resistant Staphylococcus aureus transmission: Bayesian network analysis for a single public hospital. *The Journal of Hospital Infection* 2011; **78**(2), 92–96.

The BN was fitted using Netica (Norsys Software Corp. 2007) and WinBUGS (Lunn *et al.* 2000). The latter allowed for the construction of 95% credible intervals of the marginal probability for each node being at a specified level. When fitting the model in WinBUGS, we treated each stochastic node as a Bernoulli variable X with parameter p, where p followed a Beta(α, β) prior distribution. The parameters α and β were derived from the observed or simulated data. A Markov chain Monte Carlo method was used to establish the marginal probability distribution for each node. We performed 15 000 simulations, each of which provided an estimate of $Pr(X = 1)$. Within a single simulation, the estimate of $Pr(X = 1)$ is based on 50 000 draws of X. For a given draw, the values of α and β for a node with parents are chosen to reflect the randomly drawn states of its parent variables. For a more detailed discussion of this approach, refer to Donald *et al.* (2009).

20.5 Results

Sensitivity analysis results from the final iteration of Phase 4R are displayed in Table 20.2, which orders the factors from most to least influential, in terms of mutual information (MI), with respect to the number of MRSA isolates per month. Overcrowding is the most influential factor (MI = 0.034). The next most influential factor (MI = 0.0036), handwashing compliance, has about 1/10th the MI value of overcrowding. The results indicate that changes within the staffing adequacy cluster have very little effect on the response. Similarly, multiple antibiotic resistant organism (MRO) prevalence, MRO isolates and screening are effectively independent of MRSA isolates since the MI approaches zero for these nodes.

Table 20.2 also gives the probability that each factor is at the specified level, that is high or normal, under standard conditions. A 95% credible interval (CI) is given for each probability. Under standard conditions, the probability that there is a high level of MRSA isolates in a month is 0.536 (95% CI: 0.531, 0.540).

Deviations from standard conditions can be investigated by forcing one or more nodes to be at specified levels with probability 1. The effect of this 'evidence' is propagated throughout the network, with the probability of each node being at a specified level changing in response to the evidence. We are primarily interested in how evidence changes the probability of a high level of MRSA isolates. In particular, we are interested in the effect of changes to the levels of overcrowding and/or handwashing compliance, since these are identified as having the strongest association with the number of isolates. For example, we would like to know the probability of a high level of MRSA isolates if overcrowding is high with probability 1 and handwashing compliance is normal with probability 1. Results of this investigation are shown in Table 20.3.

The effect of overcrowding dominates that of handwashing compliance. In both cases, the effect is counter-intuitive. When evidence is entered for overcrowding alone, the probability of a high level of MRSA isolates is greater when overcrowding is at a normal level (0.594) than when it is at a high level (0.337). Similarly, when evidence is entered for handwashing compliance only, the probability of a high level of MRSA

Table 20.2 The probability, p, that the factor is at the specified level, with nodes listed from most to least influential with respect to MRSA isolates. The ranking of nodes is based on the mutual information.

Factor	Mutual information	Level	p (95% CI)
MRSA isolates	—	High	0.536 (0.531, 0.540)
Overcrowding	0.033 98	High	0.228 (0.225, 0.232)
Handwashing compliance	0.003 62	High	0.472 (0.468, 0.477)
Isolation ward overflow	0.002 81	High	0.241 (0.237, 0.244)
MRSA prevalence	0.001 27	High	0.459 (0.454, 0.463)
Operating theatre canc./def.	0.000 29	High	0.192 (0.189, 0.196)
Ward outliers	0.000 07	High	0.231 (0.227, 0.234)
Staffing	0.000 03	Unsat.	0.301 (0.297, 0.305)
MRO prevalence	0.000 01	High	0.202 (0.198, 0.205)
MRO isolates	0.000 01	High	0.406 (0.401, 0.409)
Staff/1000 OBDs	0.000 00	High	0.538 (0.534, 0.543)
Screening	0.000 00	High	0.364 (0.360, 0.369)
January/February	0.000 00	Yes	0.167 (NA)
% Casual nursing staff	0.000 00	High	0.154 (0.151, 0.157)

isolates is greater when compliance is at a high level (0.573) than when it is at a normal level (0.502). Under normal bed occupancy conditions, good handwashing compliance increases the probability of a high number of MRSA isolates from 0.574 to 0.672. Under conditions where the level of bed occupancy is known, increasing handwashing compliance from normal to high levels has very little effect. What is most striking is that a high level of bed occupancy is associated with a drop in the probability of a high level of new MRSA isolates regardless of whether handwashing compliance is at a normal or high level.

Table 20.3 The probability, p, that MRSA isolates is high for various handwashing compliance and overcrowding conditions. We use '—' to indicate that no evidence has been entered for a factor. For example, the first row considers the scenario where handwashing compliance is normal with probability 1, while overcrowding is unchanged from baseline levels.

Handwashing compliance	Overcrowding	p (95% CI)
Normal	—	0.502 (0.496, 0.508)
High	—	0.573 (0.566, 0.579)
—	Normal	0.594 (0.589, 0.599)
—	High	0.337 (0.329, 0.346)
Normal	Normal	0.574 (0.567, 0.581)
Normal	High	0.339 (0.329, 0.350)
High	Normal	0.612 (0.605, 0.619)
High	High	0.333 (0.318, 0.349)

The BN was subject to three broad types of validation in Phase 3R. The first involved iterative refinements in which nodes and/or links were added or removed. Decisions were made on the basis of relationships supported by the literature, the data and the opinions of an independent panel of infectious diseases physicians.

The second validation examined whether the BN's quantification produced robust predictions consistent with the observed data. We assessed robustness by quantifying the network using subsets of data and comparing the results obtained from these 'partial' networks with those from the complete model. In all cases, the importance of factors on MRSA isolates was found to largely replicate the results in Table 20.2. A notable exception was the screening node, based on the number of screens for MROs. The importance of screening varied quite markedly depending upon which reduced data set was used to estimate probabilities. On 2 (out of 24) occasions screening achieved its highest ranking of fourth most influential factor. In one case, it was the least influential factor. It is unclear whether this is practically significant.

Goodness of fit was assessed by comparing the BN's predictions to what was observed under particular conditions. This was achieved by entering evidence for selected nodes (other than the response) based on monthly observations. We defined the 'predicted level' of MRSA isolates to be the one that the model assigned highest probability. If this coincided with the observed level for that month, given the quartile-based threshold, the model and data were said to agree. For 2008, there was 50% agreement between the predicted and observed level of isolates. This increased to 91.7% for 2009.

The third type of validation involved comparing the BN's results with those obtained from standard statistical techniques based on the 'raw' data. Linear regression was used to relate the response to factors identified by the BN as being most influential, namely overcrowding and handwashing compliance. In line with the results from the BN, there was no significant interaction between overcrowding and handwashing compliance ($p = 0.211$). When the interaction term was omitted, overcrowding was significant ($p = 0.018$), but handwashing was not ($p = 0.829$). Regression analysis also confirmed the BN's findings of a negative association between overcrowding and MRSA isolates.

20.6 Discussion

The network, developed using data collected over two years, provides an enduring method for updating evidence. Its structure and underlying conditional probability distributions will continue to evolve as new data become available.

The results do not fit with our conventional views, suggesting that the meaning placed on previous associations between handwashing, overcrowding and rates of new MRSA isolates may not be universally applicable. Cunningham *et al.* (2006, 2005) attributed the increases they observed in MRSA transmission to high turnovers interfering with the ability of staff to prepare and clean rooms for incoming patients. In light of the results above, it therefore seems possible that high bed occupancy per se is not responsible for increased MRSA transmission, but rather that this is confounded by the consequences of overcrowding, for example interference with handwashing, cleaning and patient isolation. Furthermore, the relationship might be subject to tipping (boundary) points that depend upon the states of other factors. That is, the point at which overcrowding, say, becomes problematic is not fixed, but depends upon the overall state of the system. In such a multidimensional system, if most factors are at satisfactory levels then it would remain robust to deterioration in any one factor. For our data, we believe that the dependence of MRSA transmission rates on handwashing was relatively insensitive to changes in compliance because the system is operating far from the multidimensional boundary of 'good' or 'bad' conditions. Mathematical modelling suggests that problems will be encountered if handwashing compliance drops below 40% (McBryde *et al.* 2007) and that the benefits of increased compliance beyond 50% tend to be marginal (Beggs *et al.* 2009). The findings from the BN suggest that such static boundaries may be too simplistic.

The BN offers two obvious advantages over simpler modelling strategies. Firstly, many control strategies are thought to act synergistically. The BN allows these interactions to be explored more thoroughly through scenario analyses. Secondly, MI can be used to identify the order in which factors influence a given node, allowing the hospital to prioritize where attention should be focused.

Following the approach developed by Donald *et al.* (2009) CIs were constructed for the probability of MRSA isolates and the influential factors being high (Table 20.2). Similarly, the probability that MRSA isolates are high for a certain scenario is considered within the context of the CI around it (Table 20.3). We observed reasonably narrow CIs, which would suggest a high degree of confidence in the quantification of the measures. It is important to note that there may be additional unknown sources of uncertainty which are not explicitly modelled here. The subsequent inclusion of any additional uncertainties is likely to widen the CIs.

Using a BN allowed us to exploit various sources of information. This was valuable given our initial limited data, the effect of which was most acute for nodes with many parents. A node with n parents, all of which have two levels, will have a CPT with 2^n different conditioning values. Clearly the CPT quickly becomes complex, leading to a situation in which the number of combinations to consider exceeds the available data. In such cases, we used regression-based simulation to update prior probabilities. The CPT for the response, MRSA isolates, relied most heavily on simulation, with data observed for only 15 of the 64 possible combinations of parent levels. As more data become available, the simulation results will be replaced by data-driven estimates. Problems due to a node having many parents can be somewhat circumvented by the introduction of intermediate summary nodes.

Moreover, the structured and iterative nature of the IBNDC approach to BN model development was particularly well suited to the MRSA case study. Additional cycles of the *iterative* process will be invoked when the new data, currently being collected, are available to complement the initial data used for network quantification. These data can be used to improve the accuracy of node CPTs and to evaluate the model. The important concept is that the IBNDC approach to BN model development facilitates continual model improvement by learning from new data and research.

References

Ammann M and Verhofen M 2007 Prior performance and risk-taking of mutual fund managers: a dynamic Bayesian network approach. *Journal of Behavioral Finance* **8**(1), 20–34.

Andersen BM, Lindemann R, Bergh K, Nesheim BI, Syversen G, Solheim N and Laugerud F 2002 Spread of methicillin-resistant *Staphylococcus aureus* in a neonatal intensive unit associated with understaffing, overcrowding and mixing of patients. *Journal of Hospital Infection* **50**, 18–24.

Baumgartner K, Ferrari S and Palermo G 2008 Constructing Bayesian networks for criminal profiling from limited data. *Knowledge-Based Systems* **21**(7), 563–572.

Beggs CB, Shepherd SJ and Kerr KG 2009 How does healthcare worker hand hygiene behaviour impact upon the transmission of MRSA between patients? An analysis using a Monte Carlo model. *BMC Infectious Diseases* **9**, 64.

Borsuk ME, Reichert P, Peter A, Schager E and Burkhardt-Holm P 2006 Assessing the decline of brown trout (*Salmo trutta*) in Swiss rivers using a Bayesian probability network. *Ecological Modelling* **192**(1–2), 224–244.

Clements A, Halton K, Graves N, Pettitt A, Morton A, Looke D and Whitby MPH 2008 Overcrowding and understaffing in modern health-care systems: key determinants in methicillin-resistant *Staphylococcus aureus* transmission. *The Lancet Infectious Diseases* **8**, 427–434.

Cunningham J, Kernohan W and Rush T 2006 Bed occupancy, turnover intervals and MRSA rates in English hospitals. *British Journal of Nursing* **15**, 656–660.

Cunningham J, Kernohan W and Sowney R 2005 Bed occupancy and turnover interval as determinant factors in MRSA infections in acute settings in Northern Ireland. *Journal of Hospital Infection* **61**, 189–193.

Donald M, Cook A and Mengersen K 2009 Bayesian network for risk of diarrhoea associated with use of recycled water. *Risk Analysis* **49**(12), 1672–1685.

Graves N, Halton K, Paterson D and Whitby M 2009 Economic rationale for infection control in Australian hospitals. *Healthcare Infection* **14**, 81–88.

Grayson ML, Jarvie LJ, Martin R, Johnson PDR, Jodoin ME, McMullan C, Gregory RHC, Bellis K, Cunnington K, Wilson FL, Quin D and Kelly AM 2008 Significant reductions in methicillin-resistant *Staphylococcus aureus* bacteraemia and clinical isolates associated with a multisite, hand hygiene culture-change program and subsequent successful statewide roll-out. *Medical Journal of Australia* **188**, 633–640.

Humphreys H, Grundmann H, Skov R, Lucet JC and Cauda R 2009 Prevention and control of methicillin-resistant *Staphylococcus aureus*. *Clinical Microbiology and Infection* **15**, 120–124.

Jensen FV and Nielsen TD 2007 *Bayesian Networks and Decision Graphs*. Springer, New York.

Jernigan JA, Titus MG, Groeschel DH, Getchell-White SI and Farr BM 1996 Effectiveness of contact isolation during a hospital outbreak of methicillin-resistant *Staphylococcus aureus*. *American Journal of Epidemiology* **143**, 496–504.

Johnson S and Mengersen K 2011 Integrated Bayesian network framework for modelling complex ecological issues. *Integrated Environmental Assessment and Management*. doi: 10.1002/ieam.274.

Johnson S, Fielding F, Hamilton G and Mengersen K 2010a An integrated Bayesian network approach to *Lyngbya majuscule* bloom initiation. *Marine Environmental Research* **69**, 27–37.

Johnson S, Low-Choy S and Mengersen K 2011 Integrating Bayesian networks and geographic information systems: good practice examples. *Integrated Environmental Assessment and Management*. doi: 10.002/ieam.262.

Johnson S, Mengersen K, De Waal A, Marnewick K, Cilliers D, Houser AM and Boast L 2010b Modelling cheetah relocation success in southern Africa using an iterative Bayesian network development cycle. *Ecological Modelling* **221**(4), 641–651.

Koller D and Friedman N 2009 *Probabilistic Graphical Models: Principles and Techniques*. MIT Press, Cambridge, MA.

Korb KB and Nicholson AE 2010 *Bayesian Artificial Intelligence*, 2nd edn. Chapman & Hall/CRC, Boca Raton, FL.

Lauritzen SL and Sheehan NA 2003 Graphical models for genetic analyses. *Statistical Science* **18**(4), 489–514.

Lucet JC, Chevret S, Durand-Zaleski I, Chastang C and Regnier B 2003 Prevalence and risk factors for carriage of methicillin-resistant *Staphylococcus aureus* at admission to the intensive care unit – results of a multicenter study. *Archives of Internal Medicine* **163**, 181–188.

Lunn DJ, Thomas A, Best N and Spiegelhalter D 2000 WinBUGS – a Bayesian modelling framework: concepts, structure, and extensibility. *Statistics and Computing* **10**(4), 325–337.

Marcot BG, Steventon JD, Sutherland GD and McCann RK 2006 Guidelines for developing and updating Bayesian belief networks applied to ecological modeling and conservation. *Canadian Journal of Forest Research* **36**, 3063–3074.

McBryde ES, Pettitt AN and McElwain DLS 2007 A stochastic mathematical model of methicillin resistant *Staphylococcus aureus* transmission in an intensive care unit: predicting the impact of interventions. *Journal of Theoretical Biology* **245**, 470–481.

McCann RK, Marcot BG and Ellis R 2006 Bayesian belief networks: applications in ecology and natural resource management. *Canadian Journal of Forest Research* **36**(12), 3053–3062.

Mears A, White A, Cookson B, Devine M, Sedgwick J, Phillips E, Jenkinson H and Bardsley M 2009 Healthcare-associated infection in acute hospitals: which interventions are effective? *Journal of Hospital Infection* **71**, 307–313.

Morton A, Whitby M and Looke D 2009 Isolation ward access block. *Healthcare Infection* **14**, 47–50.

Norsys Software Corp. 2007 Netica 3.25.

Peacock JE, Marsik FJ and Wenzel RP 1980 Methicillin-resistant *Staphylococcus aureus* – introduction and spread within a hospital. *Annals of Internal Medicine* **93**, 526–532.

Pearl J 1988 *Probabilistic Reasoning in Intelligent Systems: Networks of Plausible Inference*. Morgan Kaufmann, San Mateo, CA.

Pearl J 2000 *Causality: Models, Reasoning, and Inference*. Cambridge University Press, Cambridge.

Pittet D, Allegranzi B, Sax H, Dharan S, Pessoa-Silva CL, Donaldson L and Boyce JM 2006 Evidence-based model for hand transmission during patient care and the role of improved practices. *The Lancet Infectious Diseases* **6**, 641–652.

Uusitalo L 2007 Advantages and challenges of Bayesian networks in environmental modelling. *Ecological Modelling* **203**(3–4), 312–318.

Waterhouse M, Morton A, Mengersen K, Cook D and Playford G 2011 Investigating the role of overcrowding in MRSA transmission using a Bayesian network: results for a single public hospital. *Journal of Hospital Infection* **78**, 92–96.

21

Implementing adaptive dose finding studies using sequential Monte Carlo

James M. McGree, Christopher C. Drovandi and Anthony N. Pettitt

Queensland University of Technology, Brisbane, Australia

21.1 Introduction

In sequential optimal design, data are observed temporally, which allows a design to be adjusted at certain stages of an experiment. Such an approach is particularly useful when nonlinear models are of consideration as an optimal design may be dependent upon the assumed prior information about, say, the true model and parameter values. As more data are observed, an informative experiment is achieved through well-informed interventions. The experimental conditions are determined by what the data reveal, removing a total reliance on elicited or prior information.

Bayesian statistics provide the methodology for incorporating uncertainty about the true model and parameter values when designing an experiment. All inferences about the unknown parameter are based on the posterior, a target distribution that can be difficult to sample from directly. Typically, Markov chain Monte Carlo (MCMC) techniques are employed to sample from such distributions. Unfortunately, MCMC techniques can be computationally inefficient in sequential design.

We propose to use a sequential Monte Carlo (SMC) approach to sample from target distributions in deriving experimental conditions. SMC algorithms have been

Case Studies in Bayesian Statistical Modelling and Analysis, First Edition. Edited by Clair L. Alston, Kerrie L. Mengersen and Anthony N. Pettitt.

shown to be useful for sampling from a sequence of target distributions that are in some sense connected (see Del Moral *et al.* (2006)). A set of weighted samples (particles) is generated from an easy-to-sample-from tractable distribution and the particles are propagated through the sequence of targets via reweighting, resampling and mutation steps. The eventual outcome is a weighted sample from the target of interest, usually the posterior distribution of the parameter given all the data.

We motivate our approach to sequential design through the consideration of phase I clinical trials. Such trials are generally the first human administration of a developmental compound. The primary goal is to determine safe (non-toxic) doses to take forward into phase II and phase III learning and confirming evaluations. The determination of safe doses has importance in drug development as relatively large doses could be potentially fatal for volunteers, and, conversely, taking forward relatively small doses may offer no therapeutic benefit. Although one could design to find efficacious doses, as in phase I/II clinical trials, see Dragalin and Fedorov (2006) and Denman *et al.* (2011), this is generally not of interest in phase I.

We focus on the general situation where there are only two possible outcomes after administration: whether a toxic event (or adverse reaction) was observed or not. The dose–response relationship is defined in a generalized linear model framework with the canonical link function where the predictor is a function of dosage. We assume that each volunteer receives a single dose, and the sequential nature of the trial means that data observed from all previous volunteers are available when selecting the next dosage to administer. Thus the sequence of targets is the posterior distributions of the parameter after each data point is observed. These targets are therefore built through data annealing, making this application suitable for an SMC approach.

As data are observed, current target distributions need to be updated to reflect the information gain. In SMC, this is achieved via a reweight step (avoiding a full MCMC approach). However, the continual reweighting of particles as more data become available can lead to highly variable and skewed particle weights reducing the effective sample size (ESS) of the particle set. This can be overcome by a resampling step which duplicates promising particles and eliminates those with negligible weight. The particle set can then be diversified via a mutation step. We base our mutation step on an MCMC kernel which is a Metropolis–Hastings procedure with proposals efficiently constructed from the previous target distribution. The resample and mutation steps are only performed when the ESS of the particle set becomes undesirably small. Hence, computationally expensive MCMC runs are only implemented at certain stages of the experiment (however, this step could be run in parallel for each particle on a graphics processing unit, for example).

For each volunteer in the trial, a decision about what dose to administer needs to be made. We will refer to this as the dose selection phase of the algorithm. This decision is based on the current target distribution and a chosen utility function. The utility function is generally chosen to reflect the experimenter's desires, such as minimizing the uncertainty about an estimate. We consider that a finite discrete set of doses is available (D_a) and the dose which optimizes the utility is chosen for administration. To evaluate the utility for a given dose, a new posterior sample needs to be formed for each possible outcome, then the expectation is approximated through consideration of the posterior predictive distribution of the data. Again, we avoid computationally expensive MCMC methods by implementing importance sampling (discussed in Section 21.3) in order to obtain weighted samples from each posterior, see McGree *et al.* (2012).

Bayesian sequential design procedures for phase I clinical trials have been considered previously. Here, we discuss some key publications. Whitehead and Brunier (1995) considered a dose finding problem for a drug intended for treatment of, for example, depression or arthritis. The drug is thought to possibly have mild adverse reactions such as a headache and/or a rash. The tolerated incidence of such reactions was set to 20% and interest was in determining what dosage of the drug yields this incidence. This will be referred to as the maximum tolerated dose (MTD). The Bayesian decision procedure developed was compared with the commonly used continual reassessment method (CRM) of O'Quigley *et al.* (1990). This comparison demonstrated the flexibility of a Bayesian decision procedure. Haines *et al.* (2003) considered c- and D-optimality within a Bayesian context to design phase I trials. Interest was in the estimation of the MTD, and the safety of volunteers was addressed by ensuring a low probability that an administered dose exceeds an acceptable level. More recently, Bayesian approaches have been adopted in clinical trials: see Zhou *et al.* (2008) who describe the implementation of methodology in a specific phase I trial for a new developmental compound.

This chapter is organized as follows. In Section 21.2 specifics about the model and priors on the parameter space are defined. The SMC procedure to simulate and evaluate utility functions is discussed in Section 21.3. This section details many of the important aspects of this research, including definitions of the Bayesian design criteria used for dose selection and the resampling techniques implemented. In Section 21.4, results of simulation studies are shown and subsequently discussed in Section 21.5.

21.2 Model and priors

We consider dose finding studies where volunteers may or may not suffer an adverse reaction from a drug. Under this assumption, the observation from the ith volunteer, y_i, is drawn from a logistic regression model, $y_i \sim Bernoulli(\pi_i)$, such that

$$\log\left(\frac{\pi_i}{1-\pi_i}\right) = \theta_0^T + \theta_1^T D_i, \tag{21.1}$$

where $\boldsymbol{\theta}^T = (\theta_0^T, \theta_1^T)'$ represents the true parameters of the dose–response relationship and D_i is the dose administered to the ith volunteer. We are required to sample from the posterior distribution of $\boldsymbol{\theta} = (\theta_0, \theta_1)'$ given all the data. We place an independent and uninformative normal prior on the parameters. The prior is uninformative with respect to the parameter ranges for flexible dose–response curves observed in practice. The prior is

$$p(\theta_0, \theta_1) = N(\theta_0; 0, 100)N(\theta_1; 0, 100),$$

which reflects our lack of knowledge of the parameters a priori. Of interest is the MTD, denoted by D^*, which is defined as the dose which has a probability p of experiencing an adverse effect. If p, θ_0^T and θ_1^T are known, the MTD can be calculated simply by

$$D^* = \frac{\log(p/(1-p)) - \theta_0^T}{\theta_1^T}.$$

The dose finding process benefits from the availability of several pseudo observations. These usually come in the form of expert elicited probabilities for a selection of doses. In our implementation, two doses, $D_{-1} = 0.9$ and $D_0 = 2.0$, were selected and the probabilities of toxicity are assumed to be elicited (denoted as δ_{-1} and δ_0, respectively). We trust that elicited probabilities are consistent with the true model, and our simulations which follow reflect this. The elicited data reduce the chance of selecting particularly poor doses early in the process when there is little information about the parameters. The effect of their presence is reduced as the sample size increases. As a safety precaution, the first volunteer is administered the minimal dose ($\min D = \min(D_a)$). This is the only safety provision considered in this chapter.

21.3 SMC for dose finding studies

21.3.1 Importance sampling

Before detailing SMC, it is useful to firstly describe importance sampling. For further details, the reader is referred to Robert and Casella (2004). Consider a target distribution $\pi(\boldsymbol{\theta})$ that is difficult to directly sample from but its density can be evaluated and is known up to a normalizing constant. Assume there is available another distribution, $\eta(\boldsymbol{\theta})$, called the importance distribution, from which it is straightforward to generate samples. We draw N samples from $\eta(.)$, $\boldsymbol{\theta}^{(1)}, \ldots, \boldsymbol{\theta}^{(N)}$, and weight these samples to reflect the distribution of interest with the formula

$$w(\boldsymbol{\theta}^{(i)}) = \frac{\pi(\boldsymbol{\theta}^{(i)})}{\eta(\boldsymbol{\theta}^{(i)})}.$$

The normalized weights, $W(\boldsymbol{\theta}^{(i)}) = w^{(i)}/\sum_{j=1}^{N} w(\boldsymbol{\theta}^{(j)})$ (referred to as the importance weights) together with the parameter samples from $\eta(.)$ form a weighted sample from $\pi(.)$. This weighted sample can be used to estimate various expectations under the distribution of interest such as $E_\pi(g(\boldsymbol{\theta}))$ since

$$\begin{aligned} E_\pi(g(\boldsymbol{\theta})) &= \frac{1}{\int_\gamma \pi(\gamma)d\gamma} \int_{\boldsymbol{\theta}} g(\boldsymbol{\theta})\pi(\boldsymbol{\theta})d\boldsymbol{\theta} \\ &= \frac{1}{\int_\gamma (\pi(\gamma)/\eta(\gamma))\eta(\gamma)d\gamma} \int_{\boldsymbol{\theta}} g(\boldsymbol{\theta})\frac{\pi(\boldsymbol{\theta})}{\eta(\boldsymbol{\theta})}\eta(\boldsymbol{\theta})d\boldsymbol{\theta} \end{aligned}$$

$$= \frac{1}{\int_{\gamma} w(\gamma)\eta(\gamma)d\gamma} \int_{\theta} g(\theta)w(\gamma)\eta(\theta)d\theta$$

$$\approx \sum_{i=1}^{N} \frac{w(\theta^{(i)})}{\sum_{j=1}^{N} w(\theta^{(j)})} g(\theta^{(i)})$$

$$= \sum_{i=1}^{N} W(\theta^{(i)})g(\theta^{(i)}).$$

The efficiency of a weighted sample can be measured through the ESS. The effective sample size provides us with how many independent and identically distributed samples are equivalent to the weighted sample. This can be approximated easily based on the normalized weights, ESS = $1/\sum_{i=1}^{N}(W^{(i)})^2$. This ESS ranges between 1 (when all but one particle has a non-negligible weight) and N (when the weights are all equal).

Unfortunately, the importance sampling method suffers from a number of drawbacks. Firstly, it is a requirement that the importance distribution covers the support of the target distribution so that the tails are not ignored. This condition is satisfied in this application as we bring in the data which make the targets increasingly informative. Additionally, importance sampling is usually inefficient in high dimensions because the importance distribution does not reflect the features of the unknown target (Robert and Casella 2004).

SMC methods can overcome the drawbacks of the basic importance sampling technique. In particular, the importance distribution is built sequentially so as the sequence of targets is traversed it more closely resembles the target of interest.

21.3.2 SMC

In sequential design, the sequence of targets (Chopin 2002; Del Moral *et al.* 2006) is given by

$$\pi_t(\theta|\mathbf{y}_{-1:t}) \propto p(\mathbf{y}_{-1:t}|\theta, \mathbf{D}_{-1:t})p(\theta), \text{ for } t = -1, \ldots, T,$$

where T is the number of volunteers, $p(\mathbf{y}_{-1:t}|\theta, \mathbf{D}_{-1:t})$ represents a probability mass function in this discrete data setting, $\mathbf{y}_{-1:t} = (\delta_{-1}, \delta_0, y_1, y_{2:t})$ and $\mathbf{D}_{-1:t} = (D_{-1}, D_0, \min D, \mathbf{D}_{2:t})$ (D_{-1}, D_0, δ_{-1} and δ_0 are as defined in Section 21.2). The initial importance distribution for π_{-1} is given by the prior, $p(\theta)$.

The general SMC procedure for dose finding is given in Algorithm 19. Initially, the particle set is drawn from the prior, and then the elicited data (and corresponding doses) are incorporated. As mentioned previously, the first volunteer receives $\min D$.

For any volunteer t, we have a particle set $\{\theta_t^{(i)}, W_t^{(i)}\}_{i=1}^{N}$. The reweighting step is performed to update the current target distribution with the information gained from the new data point. With the choice of an MCMC kernel, the incremental weights are

---

Algorithm 19: General dose finding procedure

---

1. %%%% Draw from prior
2. **for** $i = 1 : N$ **do**
3.   Draw $\theta_{-2}^{(i)} \sim p(\theta)$
4.   Let $W_{-2}^{(i)} = 1/N$
5. **end for**
6. %%%% Loop over all volunteers
7. **for** $t = 1 : T$ **do**
8.   **if** $t = 1$ **then**
9.   %%%% Doses for which data were elicited and the first volunteer receives the smallest dose
10.   Let $D_{-1} = 0.9$, $D_0 = 2.0$ and $D_1 = \min D$
11.   %%%% Elicited data
12.   Let $\delta_{-1} = p(Y = 1|\theta^T, D_{-1})$
13.   Let $\delta_0 = p(Y = 1|\theta^T, D_0)$
14.   **else**
15.   %%%% Dose selection phase (see Section 21.3.3)
16.   Evaluate ψ to find D_t
17.   **end if**
18.   Draw $y_t \sim \text{Bern}(p(Y = 1|\theta^T, D_t))$ %%%% Observation from true model
19.   Let $w_t^{(i)} = W_{t-1}^{(i)} f(y_t|\theta_{t-1}^{(i)})$, for $i = 1, \dots, N$
20.   Normalize weights $W_t^{(i)} = w_t^{(i)} / \sum_{j=1}^N w_t^{(j)}$
21.   Set $\theta_t^{(i)} = \theta_{t-1}^{(i)}$, for $i = 1, \dots, N$
22.   Let ESS $= 1 / \sum_{i=1}^N (W_t^{(i)})^2$
23.   **if** ESS $< NE$ **then**
24.   %%%% Resampling step
25.   Resample from $\{\theta_t^{(i)}, W_t^{(i)}\}_{i=1}^N \to \{\theta_t^{(i)}, 1/N\}_{i=1}^N$
26.   **for** $i = 1 : N$ **do**
27.   %%%% Mutation step
28.   Move $\theta_t^{(i)}$ with MCMC kernel R times with invariant distribution $\pi_t(\theta|y_{-1:t}, D_{-1:t})$
29.   **end for**
30.   **end if**
31. **end for**
32. $\hat{D}^* = \text{median}[D^*(\theta)|y_{-1:T}]$

---

simply the new target divided by the current. In an independent data setting (as is the case here), the incremental weights are then simply the likelihood of the introduced observation. Therefore the reweighting formula is given by

$$w_{t+1}^{(i)} = W_t^{(i)} p(y_{t+1}|\theta_t^{(i)}, D_{t+1}).$$

It is clear from the above weighting formula that the selection of an MCMC kernel results in an algorithm complexity of $\mathcal{O}(N)$.

As we move through the sequence of distributions, the importance weights tend to become more variable and skewed, sometimes known as an impoverishment of the particles. To prevent this from occurring, a resampling step is added to the algorithm and is performed when the ESS falls below a pre-defined threshold NE, for a suitably chosen E where $0 < E \leq 1$ ($E = 0.5$ is commonly used). This step has the effect of duplicating particles with high weight and discarding those with negligible weight, thereby focusing on the promising regions of the parameter space.

There are several resampling techniques available in the literature. Unfortunately, resampling causes an inflation in the variance of an estimator compared with the weighted sample. Each method maintains the expectation of an estimator but the different resampling methodologies vary in their ability to reduce this extra variation. Methods considered in this chapter include multinomial, residual and systematic resampling, see Kitagawa (1996) and Liu *et al.* (1998) for further details.

After the resampling step, the mutation step is performed. This generates new particles from the current set which reflect the current target distribution. This step is performed via an MCMC kernel. In our algorithm, a random walk Metropolis–Hastings procedure was used to accept or reject a proposed new particle. The variance of the random walk (or tuning parameter) was taken as the variance of each element in the particle set making it adaptive or self-tuning. This mutation step was performed R times to ensure a large proportion of unique particles is maintained throughout the experiment.

Once data from all T volunteers have been observed, the median of the posterior distribution of D^* is recorded as the estimate of D^* (denoted as $\hat{D}^*$). This estimate includes information, not only from the prior and observed data, but also from the elicited data.

21.3.3 Dose selection procedure

Integral in the sequential design algorithm is the dose selection phase. The aim of this phase is to choose the dose D which optimizes the expectation of some utility function. We start with a set of particles with corresponding weights for volunteer t denoted as $\{\boldsymbol{\theta}_t^{(i)}, W_t^{(i)}\}_{i=1}^N$. The expectation of the utility is evaluated by considering all possible outcomes for each dose. In our scenario, there are only two outcomes, $z \in \{0, 1\}$. For both outcomes of z for each dose $D \in D_a$, importance sampling is used to form a weighted sample from the posterior distribution of D^* based on the current particle set. The (weighted) criterion is evaluated for z and D, and the expectation of the utility over z is approximated by considering the predictive probability of z, given by $q_z(D) = p(z|\boldsymbol{y}_{-1:t}, D)$. This can be estimated straightforwardly from the importance weights. We approximate the expectation of the utility for all doses, and the dose corresponding to the optimized value is chosen for administration. See Algorithm 20 for more details.

Algorithm 20: Importance sampling algorithm for dose selection

1. For volunteer t, we have particles and weights $\{\boldsymbol{\theta}_t^{(i)}, W_t^{(i)}\}_{i=1}^N$
2. **for** $D \in D_a$ **do**
3. **for** $z \in \{0, 1\}$ **do**
4. %%%% Importance weights
5. $w_t^{(i)} = W_t^{(i)} p(y = z|\boldsymbol{\theta}_t^{(i)}, D)$, for $i = 1, \ldots, N$
6. %%%% Approximate predictive probability
7. $\hat{q}_z(D) = \sum_{i=1}^N w_t^i$
8. Normalize w_t^i
9. Evaluate weighted criterion $\psi(D, z)$
10. **end for**
11. %%%% Expectation of criterion
12. $\psi(D) = \sum_{z=0}^1 \hat{q}_z(D)\psi(D, z)$
13. **end for**
14. $D_t = \arg\min_{D \in D_a} \psi(D)$

Utility functions

There are many optimal design criteria (or utility functions) available in the literature, see Atkinson *et al.* (2007), and Bayesian optimal design was reviewed by Chaloner and Verdinelli (1995). In this chapter, two utility functions (ψ_1 and ψ_2) are explored. Both utilities are based on the posterior distribution (only), and we term these fully Bayesian design criteria. There is limited methodology in the literature relating to these criteria, particularly in the application to dose finding.

Posterior variance (VAR)

The utility function ψ_1 for individual $t + 1$ is defined as the dose D which minimizes the expectation of the posterior variance of D^*, that is the D which minimizes

$$\psi_1(D) = \sum_{z=0}^{1} q_z(D)\text{VAR}[D^*(\boldsymbol{\theta})|\mathbf{y}_{-1:t}, z],$$

for $t = 1, \ldots, T - 1$.

From Section 21.2, the functional form of D^* is essentially a ratio of two normally distributed random variables. In special cases, this ratio is a Cauchy distribution with undefined variance. The implication of this is that the distribution of D^* can be quite unstable and can often have outliers. Consequently, a variance may not be an appropriate summary of spread. McGree *et al.* (2012) instead considered a trimmed variance. However, we persist with variance and compare results with the interquartile range, which is a more robust measure of spread.

Posterior interquartile range

The utility function ψ_2 for individual $t+1$ is defined as the dose D which minimizes the expectation of the interquartile range (IQR) of D^*, that is the D which minimizes

$$\psi_2(D) = \sum_{z=0}^{1} q_z(D)\text{IQR}[D^*(\boldsymbol{\theta})|\boldsymbol{y}_{-1:t}, z],$$

for $t = 1, \ldots, T-1$.

This non-parametric measure of spread is preferred for skewed distributions and/or those with outliers. This can certainly be the case for the posterior distribution of D^*, particularly early on in the trial when relatively little information is available. The examples which follow will explore both utilities introduced in this section. The results will offer some insight into whether robust measures of spread are preferred when designing dose finding studies for the estimation of the MTD.

21.4 Example

In this example, we consider a dose finding study where the aim is to estimate as precisely as possible the dose D^* such that $P(Y = 1|D^*) = 0.20$. The adaptive nature of this implementation requires the generation of data from a 'true model' (as seen in Algorithm 19). We assume that the true probability of toxicity for a dose D is given by the logistic regression model with parameters $\boldsymbol{\theta}^T$, see Equation (21.1). Optimal designs within a nonlinear context generally require (and are dependent upon) the parameter set considered. Given this, we consider two different parameter sets for data generation, given by $\boldsymbol{\theta}^T = [-5, 5]$ and $[-3, 3]$ with $D_a = \{0.1, 0.2, \ldots, 2\}$. These lead to $D^* = 0.723$ and 0.538, respectively. In both cases, the only safety precaution used is the provision that the first individual receives the smallest dose. Further, we assume unbiased elicited data consistent with the true model for two doses D_{-1} and D_0.

The two fully Bayesian design criteria (see Section 21.3.3) are evaluated by comparing estimates of D^*. Given that estimates of D^* will (to a certain extent) depend upon the data generated, a large number (500) of dose finding trials were repeated for both utilities with the median of the posterior distribution of D^* recorded as $\hat{D}^*$, the estimate for D^*. This leads to a distribution of estimates of D^* which can be used to compare the utilities. Thus, under each of the two true models assumed, a study of $T = 100$ volunteers was simulated 500 times for both utilities each with the three resampling techniques discussed, and the estimates of D^* were then compared.

In regard to the specifics of the SMC algorithm, decisions needed to be made about how large the particle set (N) should be and at what level the ESS should be maintained (E). Both of these choices affect the run time of the algorithm, and the larger the values for both, the more stable the estimates of D^*. One of our aims of this research is not to show how to set up and run the most efficient SMC dose finding algorithm. Rather, our aims are to show how to apply SMC methodology to adaptive dose finding studies, compare the posterior variance and IQR as utility functions, and determine which resampling technique is preferred. Given this, we chose a sufficiently

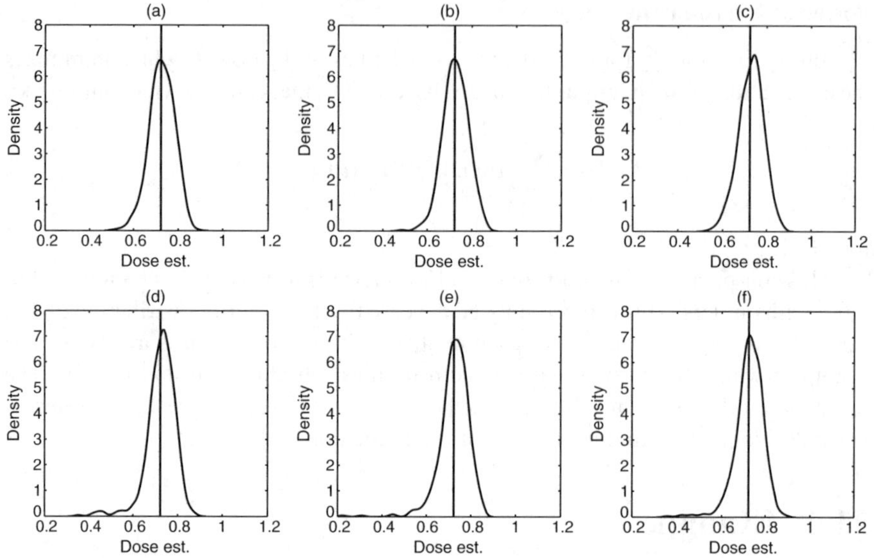

Figure 21.1 Distribution of $\hat{D}^*$ where $\boldsymbol{\theta}^T = [-5, 5]$, $P(Y = 1|D^*) = 0.20$ and MTD = 0.723 based on 100 patients. Rows relate to the variance and IQR utility (respectively) while the columns relate to the multinomial, residual and systematic resampling methods (respectively).

large particle set ($N = 100\,000$) and resampled and mutated when the ESS dropped below 75 000 ($E = 0.75$). The mutation step was run $R = 5$ times whenever the particle set was resampled.

We discuss the results for $\boldsymbol{\theta}^T = [-5, 5]$. Similar results were obtained when it was assumed that $\boldsymbol{\theta}^T = [-3, 3]$; these are reported in the Appendix to the chapter. The distributions of the estimates of D^* over the 500 runs given by both utilities and the three resampling techniques are shown in Figure 21.1 with summary statistics given in Table 21.1. The six distributions of estimates of D^* given in Figure 21.1 are visually quite similar. All seem relatively symmetric around the true MTD signified by the vertical line placed on $D^* = 0.723$. The rows relate to the posterior variance and IQR utilities, respectively. The main difference appears to be a longer left tail of the distributions for ψ_2 suggesting some small estimates were given under this criterion (and will inflate the standard deviation of the estimates).

Table 21.1 Summary statistics ($\boldsymbol{\theta}^T = [-5, 5]$, MTD = 0.723 and $T = 100$).

Criterion	Resample	Mean	Median	SD	IQR
ψ_1	Multinomial	0.727	0.728	0.057	0.078
ψ_1	Residual	0.727	0.728	0.057	0.076
ψ_1	Systematic	0.729	0.733	0.057	0.075
ψ_2	Multinomial	0.726	0.733	0.066	0.073
ψ_2	Residual	0.725	0.731	0.066	0.073
ψ_2	Systematic	0.726	0.731	0.064	0.074

The summary statistics for each of the distributions show that there is a slight tendency to overestimate the MTD for both utilities as the mean and median of the distributions are larger than D^*. This was also found in many of the examples considered in McGree *et al.* (2012). As suspected from the density plots, the standard deviations of the estimates of D^* are inflated when the IQR is used as a criterion. Similar summary IQRs are seen for all distributions, which highlights this aspect. However, overall the MTD was well estimated when using either criterion. Further, there appears to be no discernible differences in the three resampling techniques considered. None seemed to affect the location of the estimates nor do they seem to affect the variability, possibly due to the very large particle population size used. Hence, in this implementation, any resampling technique could be used.

21.5 Discussion

An SMC framework has been constructed to optimally design dose finding studies. This is a novel application of methodology, not only to dose finding, but to design generally. The iterative nature of dose finding trials naturally lends itself to the sequence of connected targets within SMC. The sequence of targets represents posterior distributions, and these form the prior information for dose selection for each volunteer in the trial. At each iteration, two utility functions were considered to select doses which yield data to potentially produce more informed dose selection for later volunteers. Trials were simulated so that the two utility functions and three resampling techniques could be evaluated and explored. The SMC algorithm proposed has computational advantages over a full MCMC implementation, and therefore the simulation properties of the utilities can be obtained in a more timely manner.

The two utilities compared were the posterior variance (ψ_1; suitable for symmetric distributions) and the posterior IQR (ψ_2; a more robust measure of spread). Despite the posterior distribution of D^* on occasions being skewed and having outliers, the posterior variance criterion performed well in comparison with the IQR. The posterior IQR consistently suffered from occasional small estimates of D^* which inflated the standard deviation of the estimate. Other than this, both criteria seemed to perform similarly in the two examples considered in this chapter. Further, three resampling techniques (multinomial, residual and systematic) were compared. None of these techniques produced largely different results, suggesting that this choice could be inconsequential for the large number of particles used ($N = 100\,000$).

An important part of design for dose finding studies is the safety of volunteers. In phase I trials, new experimental compounds that have never been trialled in humans before are used, so the potential for adverse reactions must be minimized. Safety approaches have been considered in the literature, see Zhou *et al.* (2008) and McGree *et al.* (2012). The only safety provision included in the examples was the requirement that the first individual receive the smallest dose, and obviously other provisions could be considered. Methods in the cited literature can easily be adopted within this SMC framework, and this is an obvious extension of our work.

Other extensions include the relaxation of the assumption that we know the model a priori, as in Toni and Stumpf (2010). It may not be the case that the logistic regression model is sufficient for the description of the dose–response relationship, and uncertainty at this level should be incorporated into the dose selection phase of our procedure. A set of models could be proposed and a model indicator could be included in the particle set. Then, one could 'learn' about the true model as more data become available. Difficulties to be overcome are how to propose new particles once the indicator has changed, and also how to adapt the mutation step. This is a research direction we hope to explore in future work.

References

Atkinson AC, Donev AN and Tobias RD 2007 *Optimum Experimental Designs, with SAS*, 2nd edn. Oxford University Press, New York.

Chaloner K and Verdinelli I 1995 Bayesian experimental design: a review. *Statistical Science* **10**, 273–304.

Chopin N 2002 A sequential particle filter method for static models. *Biometrika* **89**, 539–551.

Del Moral P, Doucet A and Jasra A 2006 Sequential Monte Carlo samplers. *Journal of the Royal Statistical Society, Series B (Statistical Methodology)* **68**(3), 411–436.

Denman NG, McGree JM, Eccleston JA and Duffull SB 2011 Design of experiments for bivariate binary responses modelled by Copula functions. *Computational Statistics & Data Analysis* **55**, 1509–1520.

Dragalin V and Fedorov V 2006 Adaptive designs for dose-finding based on efficacy-toxicity response. *Journal of Statistical Planning and Inference* **136**, 1800–1823.

Haines LM, Perevozskaya I and Rosenberger WF 2003 Bayesian optimal designs for phase I clinical trials. *Biometrics* **59**, 591–600.

Kitagawa G 1996 Monte Carlo filter and smoother for non-Gaussian nonlinear state space models. *Journal of Computational and Graphical Statistics* **5**, 1–25.

Liu J, Chen R and Wong W 1998 Rejection control and sequential importance sampling. *Journal of the American Statistical Association* **93**, 1022–1031.

McGree JM, Drovandi CC, Thompson MH, Eccleston JA, Duffull SB, Mengersen K, Pettitt AN and Goggin T 2012 Adaptive Bayesian compound designs for dose finding studies. *Journal of Statistical Planning and Inference* **142**, 1480–1492.

O'Quigley J, Pepe M and Fisher L 1990 Continual reassessment method: a practical design for phase I clinical trials in cancer. *Biometrics* **46**, 33–48.

Robert CP and Casella G 2004 *Monte Carlo Statistical Methods*, 2nd edn. Springer Science+Business Media, New York.

Toni T and Stumpf MPH 2010 Simulation-based model selection for dynamical systems in systems and population biology. *Bioinformatics* **26**, 104–110.

Whitehead J and Brunier H 1995 Bayesian decision procedures for dose determining experiments. *Statistics in Medicine* **14**, 885–893.

Zhou Y, Whitehead J, Korhonen P and Mustonen M 2008 Implementation of a Bayesian design in a dose-escalation study of an experimental agent in healthy volunteers. *Biometrics* **64**, 299–308.

21.A Appendix: Extra example

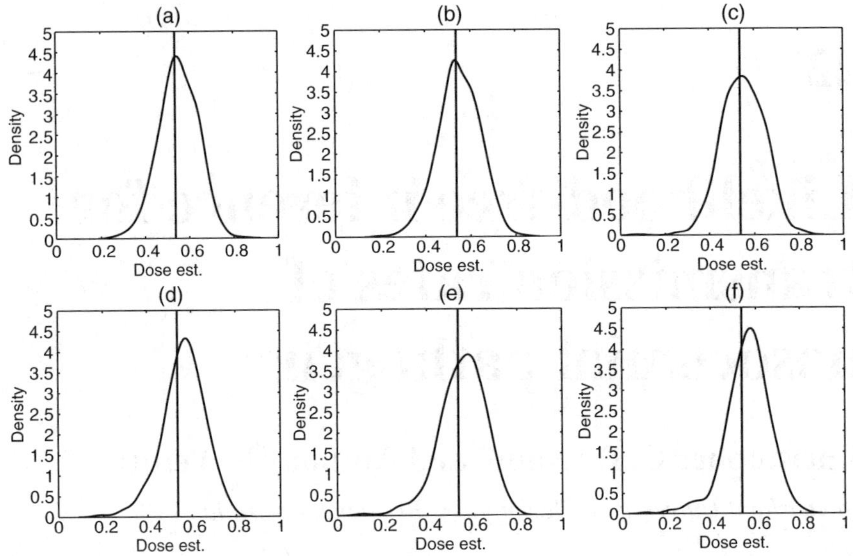

Figure 21.2 Distribution of $\hat{D}^*$ where $\boldsymbol{\theta}^T = [-3, 3]$, $P(Y = 1|D^*) = 0.20$ so that MTD = 0.538 based on 100 patients. Rows relate to the variance and IQR utility (respectively) while the columns relate to the multinomial, residual and systematic resampling methods (respectively).

Table 21.2 Summary statistics ($\boldsymbol{\theta}^T = [-3, 3]$, MTD = 0.538 and $T = 100$).

Criterion	Resample	Mean	Median	Var	IQR
ψ_1	Multinomial	0.552	0.552	0.087	0.123
ψ_1	Residual	0.548	0.546	0.092	0.122
ψ_1	Systematic	0.551	0.549	0.095	0.137
ψ_2	Multinomial	0.563	0.567	0.097	0.123
ψ_2	Residual	0.556	0.562	0.103	0.132
ψ_2	Systematic	0.564	0.568	0.098	0.117

22

Likelihood-free inference for transmission rates of nosocomial pathogens

Christopher C. Drovandi and Anthony N. Pettitt

Queensland University of Technology, Brisbane, Australia

22.1 Introduction

A Bayesian statistician is interested in obtaining the posterior distribution, $\pi(\theta|y)$, given by

$$\pi(\theta|y) \propto f(y|\theta)\pi(\theta),$$

where θ is the parameter of interest, y is observed data assumed to be drawn from $f(.|.)$ and $\pi(.)$ is the prior. Thus, Bayesian inferences are still heavily reliant on the availability of the likelihood function. However, it is becoming increasingly apparent that there are many statistical models that do not provide a computationally tractable likelihood function.

Fortunately a class of simulation methodologies popularly termed approximate Bayesian computation (ABC) can produce statistically valid inferences about the posterior distribution when the likelihood function is not computationally tractable.

Although ABC approaches do not necessitate the evaluation of the likelihood function, it is paramount that simulating data from the statistical model is relatively straightforward to perform. The initial need for a likelihood function is alleviated by simulating data from the model and searching for parameter values that produce

Case Studies in Bayesian Statistical Modelling and Analysis, First Edition. Edited by Clair L. Alston, Kerrie L. Mengersen and Anthony N. Pettitt.

simulated data close to the observed data. This assessment of closeness typically involves comparing a set of summary statistics.

Also often referred to as likelihood-free inference, ABC methods are becoming an increasingly important component of the statistician's toolbox since they allow for inferences on certain statistical models that were previously problematic. Furthermore, the approach allows for progressively more realistic models to be developed. For example, ABC is now widely applied in population genetics using a coalescent model, see for example Beaumont *et al.* (2002). An additional application is to Markov process models in epidemiology (Blum and Tran 2010) and biology (Drovandi and Pettitt 2011a). Such methods are suited to ABC as there is no closed form for the likelihood or numerical evaluation of the likelihood is too computationally expensive. Other models include Ising-type models (Grelaud *et al.* 2009) and quantile distributions (Allingham *et al.* 2009; Drovandi and Pettitt 2011b). Sisson and Fan (2010) provide a more comprehensive list of applications and models.

In this chapter, ABC is applied and used as an alternative to the likelihood-based approach of Drovandi and Pettitt (2008) to estimate the transmission rates of nosocomial pathogens within a hospital ward. Such estimates are important to help develop an understanding of the biological processes that may have generated the data and also to assess the impact of interventions (e.g. hand hygiene compliance) as per McBryde *et al.* (2007). Furthermore, in order to design hospital systems, it is critical to evaluate the effect of system changes on the incidence of MRSA infection and other similar adverse events. Hence, it is important for hospital administrators to know that the methods presented in this chapter suggest that transmission rates can be estimated accurately with quite coarse data and therefore allow detection of improvements from system changes. More details on this work can be found in Drovandi and Pettitt (2011c).

22.2 Case study: Estimating transmission rates of nosocomial pathogens

22.2.1 Background

Nosocomial (or hospital-acquired) infections are generally defined to be those that occur in patients who have been a resident inside a hospital for more than 48 hours and are due to various pathogens. Methicillin-resistant *Staphylococcus aureus* (MRSA) is an example of a hospital-acquired pathogen that continues to be of particular concern to patients and hospitals. Unfortunately, detection and eradication of the MRSA pathogen are now more difficult because it has become endemic in most hospitals (Evans and Brachman 1998; Harbarth 2006; Tiemersma *et al.* 2004). Moreover, the most effective infection control intervention is still unknown (Harbarth 2006). Complications particularly arise in the presence of hospital overcrowding and understaffing (Clements *et al.* 2008; Waterhouse *et al.* 2011).

A further complication to the containment of this pathogen is that it is resistant to certain antibiotics. MRSA is a strain of *Staphylococcus aureus* that has developed resistance to Methicillin, Cloxacillin and other related antibiotics (Ayliffe and English 2003, p. 228). Consequently, patients infected with MRSA can develop infections which are difficult and expensive to treat effectively. Patients most at risk are people with compromised immune systems, for instance long-term residents of a hospital or elderly people.

The pathogen can be spread among patients directly or indirectly, often indirectly via the hands of temporarily contaminated health-care workers. Transmission is often increased when health-care workers do not comply with hand hygiene protocol, which has shown to be the case for many health-care workers (Harbarth 2006). Other modes for transmission within the hospital involve contact with clothing, bed linen, and shared areas such as toilets etc. Nosocomial transmissions can also arise from outside the hospital, for example from colonized or infected patients who have been transferred from another hospital or new patients who unknowingly are carriers of the pathogen.

22.2.2 Data

The data available for analysis consist of 3329 patient records collected between 8 August 2001 and 3 March 2004 (inclusive) in an intensive care unit at the Princess Alexandra Hospital, Brisbane, Australia. These patients had a hospital length of stay of at least 48 hours. The patients were swabbed for MRSA on admission, discharge and twice weekly.

While individual patient data are available, data often take the form of daily/weekly/monthly prevalence or incidence counts as such data are reported routinely by hospitals. Prevalence refers to the current number of colonized patients while incidence refers to the number of new colonizations within a time frame. Here we use weekly incidence counts, which are shown in Figure 22.1. Here a new colonization is defined as a patient whose swab was negative on admission and positive during their stay in the ward. There are no covariates available and our modelling assumptions of these data are described later in the chapter.

22.2.3 Objective

The objective is to infer the transmission rates (i.e. health-care worker to patient and vice versa) from such data. All other modes of transmission are assumed negligible as the patients are not mobile.

22.3 Models and methods

22.3.1 Models

A two-compartment model was developed by McBryde *et al.* (2007) that models the number of colonized patients and health-care workers at time t, given by $Y_p(t)$ and $Y_h(t)$, respectively. The uncolonized states are not required in the model as it is

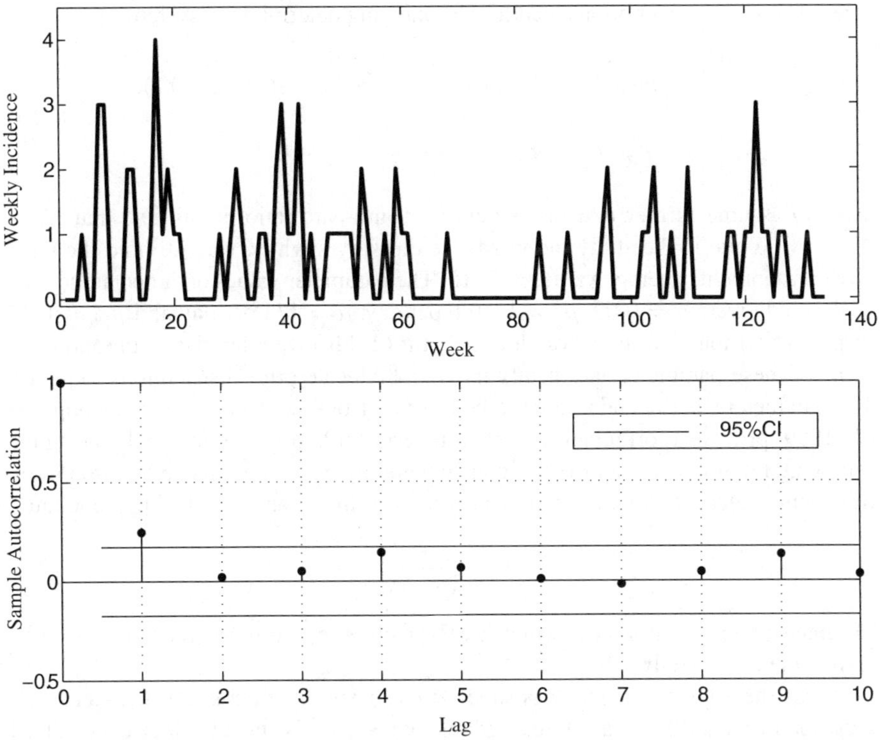

Figure 22.1 Top: Weekly incidence data collected from an ICU at the Princess Alexandra Hospital, Brisbane, between the dates of 8 August 2001 and 3 March 2004 (inclusive). Bottom: Sample autocorrelation function of the weekly incidence data. Reproduced with permission from de Guyter from Drovandi C.C. and Pettitt A.N. 2011c Using approximate Bayesian computation to estimate transmission rates of nosocomial pathogens. *Statistical Communications in Infectious Diseases* **3**(1), 2.

assumed that the number of patients and health-care workers is fixed at N_p and N_h respectively.

Given that the ward size is fixed, it is assumed that a discharged patient is immediately replaced with a new patient. Colonized and uncolonized patients are discharged at per-capita rates of μ' and μ respectively. The new patient to replace the discharged patient has a probability σ of being colonized.

Colonized health-care workers can become uncolonized due to a handwashing event that occurs at a rate of κ per number of colonized health-care workers. There is assumed to be no spontaneous decolonization of patients, so patients that are colonized before or during their stay remain so.

Finally, the colonized populations can increase by a transmission between a colonized patient and an uncolonized health-care worker and vice versa. The parameter c is a common contact rate. The probability of a transmission between colonized patient and uncolonized health-care worker and vice versa is given by p_{ph} and p_{hp} respectively. The rates of transmission are also dependent on the relevant population sizes. The above assumptions lead to the following deterministic system:

$$\begin{aligned}\frac{dY_p}{dt} &= cp_{hp}(N_p - Y_p)Y_h - \mu'(1-\sigma)Y_p + \mu\sigma(N_p - Y_p),\\ \frac{dY_h}{dt} &= cp_{ph}(N_h - Y_h)Y_p - \kappa Y_h.\end{aligned} \tag{22.1}$$

We will assume further that the patient to health-care ratio is unity so that $N_h = N_p = N$. While in the study the ward size did vary slightly, we will hold the ward size constant at its average value, $N = 15$. The parameter values obtained at the time of the study were $\sigma = 0.03$, $\mu' = 1/10.6$ per day, $\mu = 1/4$ per day and $p_{ph} = 0.13$ respectively (note that in Drovandi and Pettitt (2011c) we relax the assumption that some of these parameters are firmly fixed and allocate prior distributions to them). The parameters c, p_{hp} and κ need to be inferred from the data. However, Cooper *et al.* (1999) provide a formula to express κ in terms of known parameters by using the following relation between the pre-contact hand hygiene compliance, h (proportion of patient contacts that were preceded by handwashing), and the hand hygiene rate

$$h = \frac{\kappa}{\kappa + cN}. \tag{22.2}$$

The hand hygiene compliance, which is easy to observe, was measured to be $h = 0.59$ at the time of the study.

Since the population size N is small, it is appropriate to consider the stochastic variation in Y_p and Y_h as discrete, random counts. At this stage we introduce a third variable, $N(t)$, which we define to be the incidence, that is the number of new cases up to time t. We define a new colonized patient as one who becomes colonized as a result of transmission assumed to take place within the ICU. Furthermore, we introduce M, the maximum allowable count of incidence, in order to ensure a finite number of states. Empirically, there should be a very low probability of an incidence count higher than M. Using the model described in Equation (22.1), we can create the analogous discrete Markov process. Given that the current values of the states are $Y_p(t) = i$, $Y_h(t) = j$ and $N(t) = k$, and a small time interval Δ_t, the probabilities of various combinations of the states at time $t + \Delta_t$ are given by

$$\begin{aligned}P(Y_p = i+1, Y_h = j, N = k) &= \mu\sigma(N-i)\Delta_t + o(\Delta_t),\\ P(Y_p = i+1, Y_h = j, N = k+1) &= \phi_1(N-i)j\Delta_t + o(\Delta_t),\\ P(Y_p = i-1, Y_h = j, N = k) &= \mu'(1-\sigma)i\Delta_t + o(\Delta_t),\\ P(Y_p = i, Y_h = j+1, N = k) &= \phi_2(N-j)i\Delta_t + o(\Delta_t),\\ P(Y_p = i, Y_h = j-1, N = k) &= \frac{h\phi_2 N}{p_{ph}(1-h)} j\Delta_t + o(\Delta_t),\end{aligned} \tag{22.3}$$

where $\phi_1 = cp_{hp}$ and $\phi_2 = cp_{ph}$.

Such models can have computationally demanding likelihood functions since the transition matrix computation involves the matrix exponential of a large matrix (Sidje 1998) (here the matrix has $(N+1)^2(M+1)$ states). However, we consider an approximation to make the model more tractable by eliminating the colonized health-care workers using a so-called pseudo-equilibrium assumption demonstrated below. In this way we can compare the results between likelihood-based and likelihood-free inferences more easily.

McBryde *et al.* (2007) apply the pseudo-equilibrium approximation by assuming that the rate of change of the colonized health-care worker population is equal to zero (Drovandi and Pettitt (2008) demonstrate that this is a reasonable approximation in this application). Setting $dY_h/dt = 0$ in Equation (22.1) and making the substitution for κ given in Equation (22.2), we obtain the following steady state result denoted by $\bar{Y}_h$:

$$\bar{Y}_h = \frac{NY_p}{[hN/p_{ph}(1-h)] + Y_p}.$$

We can use this relation in the Markov process and the trivariate process in Equation (22.3) reduces to a bivariate process. Given that the current values of the states are $Y_p(t) = i$ and $N(t) = k$, and a small time interval Δ_t, the probabilities of various combinations of the states at time $t + \Delta_t$ are given by

$$\begin{aligned} P(Y_p = i+1, N = k) &= \mu\sigma(N-i)\Delta_t + o(\Delta_t), \\ P(Y_p = i+1, N = k+1) &= \phi_1(N-i)\bar{Y}_h\Delta_t + o(\Delta_t), \\ P(Y_p = i-1, N = k) &= \mu'(1-\sigma)i\Delta_t + o(\Delta_t). \end{aligned} \tag{22.4}$$

We see that there is now only a single parameter for inference, ϕ_1. Moreover, there are only $(N+1)(M+1)$ states of the chain, which provides substantial reduction in complexity.

22.3.2 Computing the likelihood

All of the models which we consider are continuous time models and we use continuous time Markov chain theory (Grimmett and Stirzaker 2001). We present here the likelihood computation for the trivariate process but similar steps are required for the likelihood of the pseudo-equilibrium model. If $\mathbf{P}_{\Delta_t}$ is a matrix of probabilities constructed from the relevant expressions given by Equation (22.3) the infinitesimal generator, $\mathbf{G}_\mathbf{C}$, is given by

$$\mathbf{G}_\mathbf{C} = \lim_{\Delta_t \to 0} \frac{1}{\Delta_t}\left(\mathbf{P}_{\Delta_t} - \mathbf{I}\right),$$

(Grimmett and Stirzaker 2001, p. 258).

The full trivariate process, Equation (22.3), contains an absorbing state, $N(t) = M$; however, the bivariate chain involving $Y_p(t)$ and $Y_h(t)$ remains irreducible. By denoting $(Y_p(t), Y_h(t), N(t) = n_t)$ to be equivalent to the events

$$(Y_p(t), Y_h(t), N(t) = n_t) \equiv \cup_{i,j=0,\ldots,N}(Y_p(t) = i, Y_h(t) = j, N(t) = n_t),$$

the initial distribution of the trivariate chain is given by

$$P((Y_p(0), Y_h(0), N(0) = 0); \ldots ; (Y_p(0), Y_h(0), N(0) = M)) = (\delta_{\mathbf{B}}, \mathbf{0}, \ldots, \mathbf{0}),$$

where $\delta_{\mathbf{B}}$ is the stationary distribution of the bivariate chain involving colonized patients and health-care workers. This can be obtained by solving $\delta_{\mathbf{B}}\mathbf{G}_{\mathbf{B}} = \mathbf{0}$ where $\mathbf{G}_{\mathbf{B}}$ is the generator matrix of the corresponding bivariate process. To obtain the probabilities at time t the initial distribution is multiplied by the matrix exponential of the trivariate generator, $\mathbf{G}_{\mathbf{C}}$, multiplied by the time interval

$$\begin{aligned}(\delta_{\mathbf{B}}, \mathbf{0}, \ldots, \mathbf{0})e^{t\mathbf{G}_{\mathbf{C}}} &= P((Y_p(t), Y_h(t), N(t) = 0); \ldots ; \\ &\quad (Y_p(t), Y_h(t), N(t) = M)|N(0) = 0).\end{aligned}$$

To compute the likelihood for the observation, the number of new colonization cases observed in $(0, t)$, denoted by n_t, we marginalize over all possible values of Y_p and Y_h:

$$P(N(t) = n_t|N(0) = 0) = \sum_{i=0}^{N}\sum_{j=0}^{N} P(Y_p(t) = i, Y_h(t) = j, N(t) = n_t|N(0) = 0).$$

The new initial distribution has to be found given that we have observed $N(t) = n_t$. The new initial distribution, $\delta_{\mathbf{B}}(n_t) = P(Y_p(t), Y_h(t)|N(t) = n_t, N(0) = 0)$, is easily calculated using Bayes' law. Its elements are given by

$$\delta_{B(ij)}(n_t) = \frac{P(Y_p(t) = i, Y_h(t) = j, N(t) = n_t|N(0) = 0)}{P(N(t) = n_t|N(0) = 0)}, \quad i, j = 0, \ldots, N.$$

To incorporate the fact that $N(t)$ is reset to zero at the beginning of each time interval, the new initial distribution for the next time interval is $(\delta_{\mathbf{B}}(n_t), \mathbf{0}, \ldots, \mathbf{0})$. This process is repeated for each observation. To calculate the likelihood for the pseudo-equilibrium bivariate approximation, Equation (22.4), to the trivariate process, the same principles as above (ignoring $Y_h(t)$) need to be applied. Assuming that $N = 15$ and $M = 4$, the number of states in the trivariate chain is 1260 while the pseudo-equilibrium chain has only 80. The matrix exponential for the trivariate model is still tractable but computationally demanding and unstable in some regions of the parameter space. If the time interval between observations is constant the matrix exponential only needs to be computed once, but requires computing for each unique time interval if this is not the case.

22.3.3 Model simulation

Given values for the parameters, it is relatively straightforward to simulate data from a Markov process model using Gillespie's algorithm (Gillespie 1977). This algorithm involves simulating the time until the next event with an exponential distribution and choosing an event type based on its relative hazards. Simulating for incidence has an additional step in that the incidence must be reset to zero at the time of each observation.

22.3.4 ABC

To avoid the computation of the likelihood, ABC introduces an auxiliary variable, x, which represents the simulated data. The joint approximate posterior distribution of the parameter and the auxiliary data is given by

$$\pi(x, \theta | y, \epsilon) \propto f(x|\theta)\pi(\theta)\phi(y|x, \theta, \epsilon) \tag{22.5}$$

(Luciani *et al.* 2009) where $f(x|\theta)$ is the likelihood function evaluated at the simulated data and $\phi(y|x, \theta, \epsilon)$ is a weighting function that assesses the similarity between x and y, giving high weight if they are close. Here ϵ is a tolerance specifying how close the simulated and observed data must be. Of interest then is the marginal approximate posterior distribution of the parameter which can be obtained by marginalizing over the simulated data

$$\pi(\theta | y, \epsilon) \propto \int_x f(x|\theta)\pi(\theta)\phi(y|x, \theta, \epsilon)dx.$$

ABC algorithms are designed to sample from either or both of the marginal and joint targets (Sisson *et al.* 2010).

Most weighting functions are of the form $\phi(y|x, \theta, \epsilon) = \phi(y|x, \epsilon)$ (Beaumont *et al.* 2002; Reeves and Pettitt 2005), producing a simple hierarchical structure $y \to x \to \theta$. There are several forms of the weighting function, but most involve a discrepancy function, $\rho(y, x)$, that provides an overall measure of the difference between observed and synthetic data sets. Typically, this function involves a comparison of a set of p summary statistics, $s(\cdot) = \{s_1(\cdot), \ldots, s_p(\cdot)\}$, of the observed and simulated data

$$\rho(y, x) = \|s(y) - s(x)\|,$$

for a suitably chosen norm. For example, the simplest and most commonly used weighting function is based on the indicator function

$$\phi(y|x, \epsilon) = 1(\rho(y, x) \leq \epsilon),$$

which is unity if the discrepancy function is less than or equal to a pre-defined tolerance ϵ. This is sometimes referred to as the uniform weighting function. Non-uniform weights can be obtained, see Del Moral *et al.* (2012) and Luciani *et al.* (2009), but the algorithm we use relies on the weighting function given above.

It is no surprise that there is some price to pay in terms of accuracy when the likelihood model is only used to simulate data. However, there are some scenarios in which there is no error due to ABC. For instance, it can be shown that if sufficient statistics are used and $\epsilon = 0$, the true posterior is produced. However, in most applications of ABC, sufficient statistics are not available. Such non-sufficiency in the summary statistics introduces one source of error. Secondly, regardless of the use of summary or sufficient statistics, it is generally impractical (and completely impossible for continuous data) to attempt to match these statistics exactly as the acceptance probabilities are too low (although see Beaumont *et al.* (2002) and Blum and François

(2010) for regression techniques that attempt to improve the ABC approximation after the tolerance is reached). The non-zero tolerance introduces a second source of error into the ABC approximation, but is necessary to ensure inferences can be obtained in reasonable time.

Tips and Tricks: Selecting optimal summary statistics reflects a trade-off between sufficiency and dimensionality. Typically we are after a low-dimensional set of summaries that convey most of the information in the observed data.

22.3.5 ABC algorithms

The first ABC algorithms to appear were rejection-based sampling algorithms. Such an algorithm was popularized by Beaumont *et al.* (2002). The algorithm involves generating a parameter from the prior, simulating a data set conditional on that parameter value, and accepting the parameter value if the discrepancy condition is satisfied. For most problems the method is too inefficient if the posterior is different from the prior. This is particularly relevant in ABC as simulation of a promising parameter value can still lead to rejection due only to the variability of the simulated data.

To help improve acceptance rates, Marjoram *et al.* (2003) proposed a Markov chain Monte Carlo approach to ABC (MCMC ABC), whereby a Markov chain is developed whose invariant distribution is the joint approximate posterior distribution in Equation (22.5). The proposal distribution is carefully selected so that evaluation of the likelihood is avoided in the Metropolis–Hastings ratio. More specifically, the proposal for (x^*, θ^*) based on current values (x, θ) is given by $q(x^*, \theta^*|x, \theta) = f(x^*|\theta^*)q(\theta^*|\theta)$.

However, recently developed algorithms based on sequential Monte Carlo (SMC; pioneered by Sisson *et al.* (2007)) have proved to be more efficient in the context of ABC. SMC methods are particularly suited to ABC, since a natural sequence of targets involves a non-increasing sequence of tolerances. More specifically, the sequence of joint targets is given by

$$\pi_t(\theta, x|y, \epsilon_t) \propto f(x|\theta)\pi(\theta)1(\rho(y, x) \leq \epsilon_t), \text{ for } t = 1, \ldots, T.$$

The SMC approach produces J weighted particles distributed according to each target in the sequence. Here we describe and use the SMC ABC replenishment algorithm of Drovandi and Pettitt (2011a). See Beaumont *et al.* (2009) and Del Moral *et al.* (2012) for other SMC ABC algorithms.

SMC involves an iterative process of reweighting, mutation and resampling steps. We base our mutation step on an MCMC kernel. If such a kernel is selected, the particle values do not get updated from one target to the next but are reweighted to reflect the new target. This incremental weight is simply the current target divided by the previous target, which is given by

$$\tilde{w}_t^i \propto \frac{1(\rho(x_t^i, y) \leq \epsilon_t)}{1(\rho(x_t^i, y) \leq \epsilon_{t-1})},$$

such that $W_t^i \propto \tilde{w}_t^i W_{t-1}^i$ (Chopin 2002). Clearly the current particle satisfies the tolerance ϵ_{t-1}, and the incremental weight will be proportional to 1 if it satisfies the tolerance ϵ_t. Otherwise, the incremental weight will be zero rendering the particle useless. Therefore after the reweighting step there will be at most J particles remaining with non-zero weight, referred to hereafter as 'alive' particles. To overcome degeneracy in this situation, the population size can be boosted back to J by resampling from the alive particles proportional to their weight. To ensure particle diversity the resampled particles can be moved according to an MCMC kernel that is invariant for the target involving ϵ_t (Chopin 2002).

The choice of an MCMC kernel results in an algorithm that is $O(J)$, at the expense of the creation of potentially duplicated particles that do not get perturbed by the MCMC kernel (Del Moral *et al.* 2006).

In an attempt to move the resampled particle with a probability close to $1 - c$ (with c set small), Drovandi and Pettitt (2011a) propose that the MCMC move step is iterated R_t times, where R_t is determined by monitoring the acceptance rate of the previous move phase, p_{t-1}^{acc}:

$$R_t = \frac{\log(c)}{\log(1 - p_{t-1}^{acc})}.$$

Moreover, the MCMC proposal distribution for the parameter at each iteration, $q_t(.|.)$, can be made adaptive in the spirit of Chopin (2002) since there are $J - \alpha J$ particles satisfying the current target at each iteration. For example, these particles can be used to estimate the covariance matrix required for a multivariate normal or t random walk proposal. Sisson and Fan (2010) show that this type of MCMC kernel in the ABC context satisfies the detailed balance condition.

The reweighting step lends itself naturally to the sequence of tolerances being determined dynamically. This can be achieved by sorting the particles by their discrepancies, ρ^i, and dropping a proportion of the particles, α, with the highest discrepancy. That is, the next tolerance, ϵ_t, is dynamically taken as the $(1 - \alpha)$th empirical quantile of the particles' discrepancies.

It is the fully adaptive nature of this algorithm that makes it so attractive. The only tuning parameters consist of ϵ_1, ϵ_T, α and c. A reasonable choice for c is 0.01 and a sensible choice for α is 0.5 (Drovandi and Pettitt 2011a). Furthermore, ϵ_1 can be chosen to achieve a particular acceptance rate in the initial rejection sampling phase. Finally, the stopping rule for the algorithm could be when the MCMC acceptance rate becomes intolerably low, which determines ϵ_T. The approach is shown in Algorithm 21.

It appears quite clear that the MCMC proposal in the SMC setup, q_t, will be far more efficient than the usual MCMC proposal, q, for a fixed tolerance as q_t will contain a closer to optimal random walk standard deviation and can incorporate the correlations between parameters. This algorithm could be considered as J MCMC kernels running in parallel, and this prevents the algorithm from becoming stuck in areas of low posterior probability and will have an improved chance of representing multimodal targets.

Algorithm 21: The SMC ABC replenishment algorithm of Drovandi and Pettitt (2011a)

1. Set J_a as the integer part of αJ.
2. Perform the rejection sampling algorithm with ϵ_1. This produces a set of particles $\{\theta^i, \rho^i\}_{i=1}^J$.
3. Sort the particle set by ρ and set $\epsilon_t = \rho^{J-J_a}$ and $\epsilon_{MAX} = \rho^J$. If $\epsilon_{MAX} \leq \epsilon_T$ then finish, otherwise go to 4.
4. Compute the tuning parameters of the MCMC kernel $q_t(.|.)$ using the particle set $\{\theta^i\}_{i=1}^{J-J_a}$.
5. **for** $j = J - J_a + 1$ **to** J **do**
6. Resample θ^j from $\{\theta^i\}_{i=1}^{J-J_a}$.
7. Move θ^j with an MCMC kernel using the proposal $q_t(.|.)$ for R_t iterations.
8. **end for**
9. Compute R_t based on the overall MCMC acceptance rate of the previous iteration and go to 3.

Tips and Tricks: Usually, simulating from the model is the most expensive operation in ABC algorithms as it can be called millions of times. Try writing this part in a lower level language (e.g. C or Fortran) and calling it into your main code that is written in a higher level language (e.g. MATLAB, R or Python).

22.4 Data analysis and results

We now compare the ABC approximation with that when the likelihood function is available. Firstly, we ran the SMC ABC replenishment algorithm with $J = 1000$, $\alpha = 0.5$ and $c = 0.01$, setting the target tolerance, $\epsilon_T = 0.04$ (we found that reducing the tolerance further had negligible impact on the approximate posterior). Secondly, we performed a normal random walk MCMC algorithm with a proposal standard deviation of 0.01 (tuned to ensure an acceptance rate of roughly 50%). We performed 11 000 iterations of this algorithm, discarding the first 1000 as burn-in and thinning out the resulting sample by a factor of 10, producing 1000 approximately independent draws. For this likelihood-based inference, we set $M = 4$ as the maximum allowable incidence value. In the parameter space that supports the data an incidence value above 4 is highly implausible.

The resulting inferences are comparable. The ABC analysis produced a median with a 95% credible interval of 0.041 (0.031, 0.054), while the equivalent results for the likelihood-based analysis were 0.040 (0.030, 0.052). A comparison of these posterior distributions is shown in Figure 22.2.

The closeness of the likelihood-free and likelihood-based inferences implies that the sum of the new cases over the time period is almost a sufficient statistic. This suggests that inferences will not be sensitive to the time grouping of the data (e.g. daily, weekly or monthly). This is consistent with the results of the time grouping sensitivity analysis in Drovandi and Pettitt (2008).

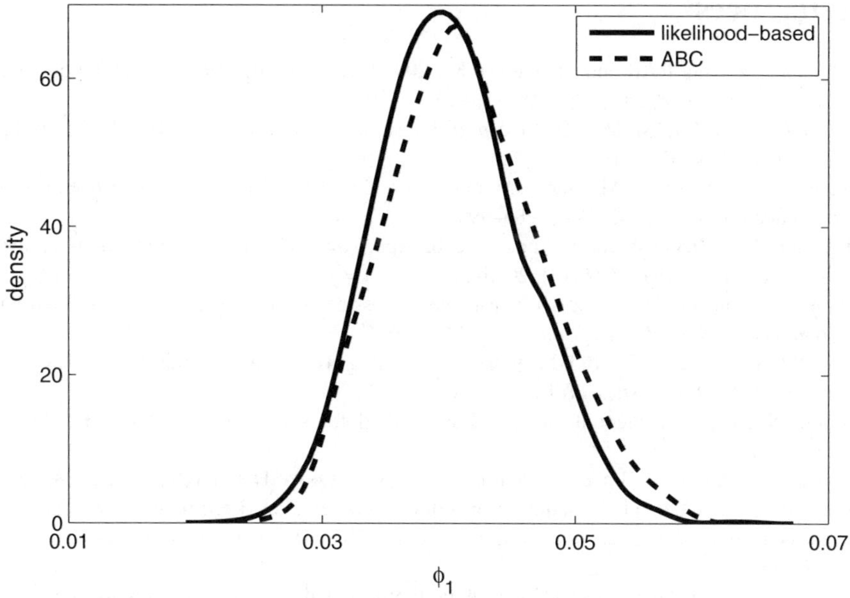

Figure 22.2 Posterior distributions for ϕ_1 of the bivariate model with likelihood-based inference (solid line) and ABC (dashed line). Reproduced with permission from de Guyter from Drovandi C.C. and Pettitt A.N. 2011c Using approximate Bayesian computation to estimate transmission rates of nosocomial pathogens. *Statistical Communications in Infectious Diseases* **3**(1), 2.

22.5 Discussion

Here we have presented a likelihood-free methodology to estimate the transmission rates of important nosocomial pathogens in a hospital ward. We considered a pseudo-equilibrium simplification so that inferences could be compared with likelihood-based results. There was an agreement with the two approaches with a slightly more precise inference when the likelihood was available, as expected.

The likelihood of the pseudo-equilibrium model is relatively straightforward to compute. However, likelihood evaluation of the full trivariate model is substantially more complicated as it involves calculating the exponential of a large matrix. Its computation is time consuming in a high-level language such as MATLAB and can be unstable in some regions of the parameter space. There would be extra motivation for an ABC approach if the time intervals between observations were not constant, since a matrix exponential would be required for each unique time difference. Furthermore, the likelihood would become even less tractable if the ward size increased and the incidence count was larger (see Drovandi *et al.* (2011) for an ABC example on macroparasite population evolution, where larger populations exist). Finally, in the likelihood-based analysis, an arbitrary upper limit for the incidence, M, required specification.

References

Allingham D, King RAR and Mengersen KL 2009 Bayesian estimation of quantile distributions. *Statistics and Computing* **19**, 189–201.

Ayliffe GAJ and English MP 2003 *Hospital Infection: From Miasmas to MRSA*. Cambridge University Press, Cambridge.

Beaumont MA, Cornuet JM, Marin JM and Robert CP 2009 Adaptive approximate Bayesian computation. *Biometrika* **96**(4), 983–990.

Beaumont MA, Zhang W and Balding DJ 2002 Approximate Bayesian computation in population genetics. *Genetics* **162**(4), 2025–2035.

Blum MGB and François O 2010 Non-linear regression models for approximate Bayesian computation. *Statistics and Computing* **20**(1), 63–73.

Blum MGB and Tran VC 2010 HIV with contact tracing: a case study in approximate Bayesian computation. *Biostatistics* **11**(4), 644–660.

Chopin N 2002 A sequential particle filter method for static models. *Biometrika* **89**(3), 539–551.

Clements A, Halton K, Graves N, Pettitt A, Morton A, Looke D and Whitby M 2008 Overcrowding and understaffing in modern health-care systems: key determinants in Methicillin-resistant *Staphylococcus aureus* transmission. *The Lancet Infectious Diseases* **8**(7), 427–434.

Cooper BS, Medley GF and Scott GM 1999 Preliminary analysis of the transmission dynamics of nosocomial infections: stochastic and management effects. *Journal of Hospital Infection* **43**, 131–147.

Del Moral P, Doucet A and Jasra A 2006 Sequential Monte Carlo samplers. *Journal of the Royal Statistical Society, Series B (Statistical Methodology)* **68**(3), 411–436.

Del Moral P, Doucet A and Jasra A 2012 An adaptive sequential Monte Carlo method for approximate Bayesian computation. *Statistics and Computing* **22**(5), 1009–1020.

Drovandi CC and Pettitt AN 2008 Multivariate Markov process models for the transmission of Methicillin-resistant *Staphylococcus aureus* in a hospital ward. *Biometrics* **64**(3), 851–859.

Drovandi CC and Pettitt AN 2011a Estimation of parameters for macroparasite population evolution using approximate Bayesian computation. *Biometrics* **67**(1), 225–233.

Drovandi CC and Pettitt AN 2011b Likelihood-free Bayesian estimation of multivariate quantile distributions. *Computational Statistics & Data Analysis* **55**(9), 2541–2556.

Drovandi CC and Pettitt AN 2011c Using approximate Bayesian computation to estimate transmission rates of nosocomial pathogens. *Statistical Communications in Infectious Diseases*. doi: 10.2202/1948-4690.1025.

Drovandi CC, Pettitt AN and Faddy MJ 2011 Approximate Bayesian computation using indirect inference. *Journal of the Royal Statistical Society, Series C (Applied Statistics)* **60**(3), 503–524.

Evans AS and Brachman PS 1998 *Bacterial Infections of Humans: Epidemiology and Control*. Plenum, New York.

Gillespie DT 1977 Exact stochastic simulation of coupled chemical reactions. *Journal of Physical Chemistry* **81**(25), 2340–2361.

Grelaud A, Robert C, Marin J, Rodolphe F and Taly JF 2009 ABC likelihood-free methods for model choice in Gibbs random fields. *Bayesian Analysis* **4**(2), 317–336.

Grimmett GR and Stirzaker DR 2001 *Probability and Random Processes*, 3rd edn. Oxford University Press, New York.

Harbarth S 2006 Control of endemic Methicillin-resistant *Staphylococcus aureus* – recent advances and future challenges. *Clinical Microbiology and Infectious Diseases* **12**, 1154–1162.

Luciani F, Sisson SA, Jiang H, Francis AR and Tanaka MM 2009 The epidemiological fitness cost of drug resistance in *Mycobacterium tuberculosis*. *Proceedings of the National Academy of Sciences* **106**(34), 14711–14715.

Marjoram P, Molitor J, Plagnol V and Tavaré S 2003 Markov chain Monte Carlo without likelihoods. *Proceedings of the National Academy of Sciences* **100**(26), 15324–15328.

McBryde ES, Pettitt AN and McElwain DLS 2007 A stochastic mathematical model of Methicillin resistant *Staphylococcus aureus* transmission in an intensive care unit: predicting the impact of interventions. *Journal of Theoretical Biology* **245**(3), 470–481.

Reeves RW and Pettitt AN 2005 A theoretical framework for approximate Bayesian computation. *Proceedings of the 20th International Workshop on Statistical Modelling, Sydney, Australia* (eds AR Francis *et al.*), pp. 393–396.

Sidje RB 1998 Expokit: a software package for computing matrix exponentials. *ACM Transactions on Mathematical Software (TOMS)* **24**(1), 130–156.

Sisson SA and Fan Y 2010 Likelihood-free Markov chain Monte Carlo. In *Handbook of Markov Chain Monte Carlo* (eds S Brooks *et al.*), pp 313–338. Chapman & Hall/CRC, Boca Raton, FL.

Sisson SA, Fan Y and Tanaka MM 2007 Sequential Monte Carlo without likelihoods. *Proceedings of the National Academy of Sciences* **104**(6), 1760–1765.

Sisson SA, Peters GW, Briers M and Fan Y 2010 A note on target distribution ambiguity of likelihood-free samplers. *arXiv:1005.5201v1*.

Tiemersma EW, Bronzwaer SLAM, Lyytikainen O, Degner JE, Schrijnemakers P, Bruinsma N, Monen J, Witte W and Grundmann H 2004 Methicillin-resistant *Staphylococcus aureus* in Europe, 1999–2002. *Emerging Infectious Diseases* **10**, 1627–1634.

Waterhouse M, Morton A, Mengersen K, Cook D and Playford G 2011 Role of overcrowding in Methicillin-resistant *Staphylococcus aureus* transmission: Bayesian network analysis for a single public hospital. *Journal of Hospital Infection* **78**, 92–96.

23

Variational Bayesian inference for mixture models

Clare A. McGrory[1,2]

[1]*Queensland University of Technology, Brisbane, Australia*
[2]*School of Mathematics, University of Queensland, St. Lucia, Australia*

23.1 Introduction

Variational Bayes (VB) is an approximate method for performing Bayesian inference. It is a non-simulation-based technique and as such it offers a practical and highly time-efficient alternative to using a Markov chain Monte Carlo (MCMC) based approach for Bayesian inference. In the VB approach, an approximation to the true posterior is derived; this results in a set of coupled update expressions for the variational posterior estimates of the model parameters which can be solved iteratively. Variational Bayesian inference has been used by researchers in machine learning since the late 1990s (Attias 1999; Jordan *et al.* 1999) and there are now many examples of variational approaches in the machine learning literature, for example Bishop (2006) and Winn and Bishop (2005). However, it has only been much more recently that VB has begun to gain popularity among statisticians as a tool for performing Bayesian analysis. Statistical articles have appeared describing the use of VB for mixture modelling (McGrory and Titterington 2007), hidden Markov chain modelling (McGrory and Titterington 2009), hidden Markov random field modelling for spatial data analysis (McGrory *et al.* 2009) and generalized linear mixed models (Ormerod and Wand 2011). Applications of VB include the analysis of functional magnetic resonance

Case Studies in Bayesian Statistical Modelling and Analysis, First Edition. Edited by Clair L. Alston, Kerrie L. Mengersen and Anthony N. Pettitt.

imaging (fMRI) data (Penny *et al.* 2003), genetics modelling (Huang *et al.* 2007) and human mobility pattern modelling (Wu *et al.* 2012).

In this chapter we will explain the key concepts of the VB approach and illustrate the ideas through application to the frequently encountered, but not straightforward task of fitting finite mixture models with unknown parameters. A finite mixture model is a linear combination of densities which are called the mixture components. Finite mixture models are often used in statistical applications because they are flexible and can represent very complex distributions. Inference for these models centres around estimating a suitable number of components for the model and estimating the parameters of these components. For mixture models, computing the posterior distribution required to infer these attributes of the model is a complex task. In the Bayesian literature, the most common approaches used for this have been based on the use of MCMC schemes. We refer the reader to Chapter 16 for a description of an MCMC-based approach to estimating mixture models. These schemes involve iteratively sampling from the posterior distribution until convergence is reached, and this could require tens of thousands of iterations. Therefore, such approaches can be highly computationally intensive and very time consuming, especially when there are large amounts of data to consider. For this reason, approximate methods such as VB are worthwhile exploring, particularly for tackling real-world problems.

It has been shown that the VB approach can be successfully applied to finite mixture modelling with unknown parameters and, very interestingly, it has been shown to lead to automatic model complexity selection (McGrory and Titterington 2007). More specifically, by using a VB algorithm we can effectively eliminate superfluous mixture components from the model during the iterative updating of the coupled expressions for the variational posterior estimates that are derived in the approximation. Since the model complexity and the parameters can be estimated simultaneously, VB might also be considered as an alternative to the transdimensional or reversible jump MCMC approach described in Richardson and Green (1997). We shall refer to this model complexity reduction feature as the elimination property of VB. When VB is used for hidden Markov modelling (McGrory and Titterington 2009), again this elimination feature has been observed. Currently there is no theoretical understanding of how this feature of the algorithm works.

In general, the theoretical properties of the variational approximation for mixture modelling have received less attention in the literature than its potential for use in practical applications. However, some results have been established. For example, Wang and Titterington (2006) demonstrated the asymptotic consistency of the VB approximation of the variational posterior modes of the parameters in mixtures of multivariate normal distributions with a fixed number of components. They also noted that the approximation is not biased for large enough samples, and proved that the variational Bayesian estimators for the model parameters will converge, at least locally, to the maximum likelihood estimators. A VB scheme also has the property that the variational approximation to the posterior will improve monotonically at each iteration. However, we must point out that the standard VB inference approach for Gaussian mixture modelling may be sensitive to the specific initialization of the hyperparameters in the prior and the initial allocation of observations to components used to start the algorithm (Wu *et al.* 2012). This means that for some data sets it will be possible to obtain slightly different fits by using different initialization settings for the variational algorithm, and there will be some cases where a poor choice for the initial settings might lead to convergence to a suboptimal final model (Wu *et al.* 2012).

One solution to this problem is to use model selection criteria to assess the automatic model choice made by the variational approximation in cases where different models are fitted for different initialization settings, see, for example, McGrory and Titterington (2007). However, more recently researchers are designing modified versions of the algorithm which incorporate component splitting in addition to the intrinsic elimination property described earlier in the chapter, see for example Wu *et al.* (2012) and references therein. These modified versions of the algorithm have an advantage over the approach of using the standard algorithm in conjunction with selection criteria, in that they provide a more fully automated approach; the user does not have to separately compute and compare selection criteria that can be arbitrarily chosen and which could be troublesome for certain applications.

In this chapter we will focus on the standard VB algorithm and its application to an animal science research case study.

23.2 Case study: Computed tomography (CT) scanning of a loin portion of a pork carcase

CT scanning is frequently used in animal science studies as it enables researchers to assess the composition of an animal carcase without having to resort to dissecting the animal. In this chapter we will examine data arising from a CT scan of a loin portion of a pork carcase; it is of interest to estimate the proportions of the carcase that are fat tissue, bone or muscle tissue.

Figure 23.1 shows the CT scan of the loin portion of a pork carcase that we will analyse. This representation is a 256-intensity-level greyscale image which is produced by binning the observed levels at each point on the image into 256 categories. Clearly, different types of body tissue and bone produce different intensity-level observations on the CT scan. Dark grey areas correspond to fat tissue, mid-grey areas correspond to muscle tissue, and white areas represent bone. By looking at the histogram of intensity levels from the CT scan image shown in Figure 23.2, we again see that there are three main levels of intensity observed. Based on anatomical knowledge of this part of the carcase, we would expect that the lower intensity levels would correspond to the fat tissue, the mid-levels of intensity to muscle, and the highest intensity levels to bone. While we can see an indication of these three different levels of intensity in the histogram, clearly there is some degree of overlap between them and varying spread. Mixture modelling is a very suitable way to represent these types of data as a combination of several components which will each correspond to one of the types of tissue or bone, and it provides a way of statistically allocating the observed CT intensity levels to one of these components. This then allows us to estimate the proportions of each type of tissue and bone present, which is our aim in this study.

Note that for the purpose of illustrating the variational approach in the context of finite mixture modelling, here we ignore any spatial association that may be present

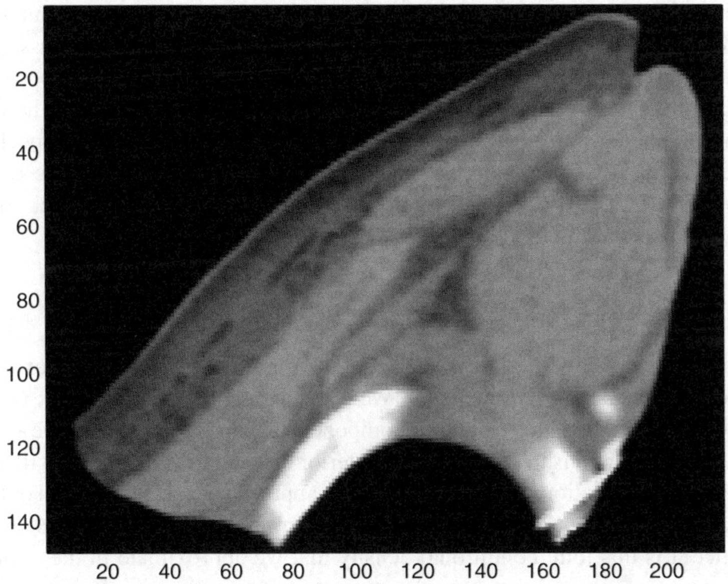

Figure 23.1 CT scan of the loin portion of a pork carcase. This representation is a 256-intensity-level greyscale image. Different types of body tissue and bone produce different intensity-level observations on the CT scan. Dark grey areas of the image correspond to fat tissue, mid-grey areas to muscle tissue, and white areas to bone.

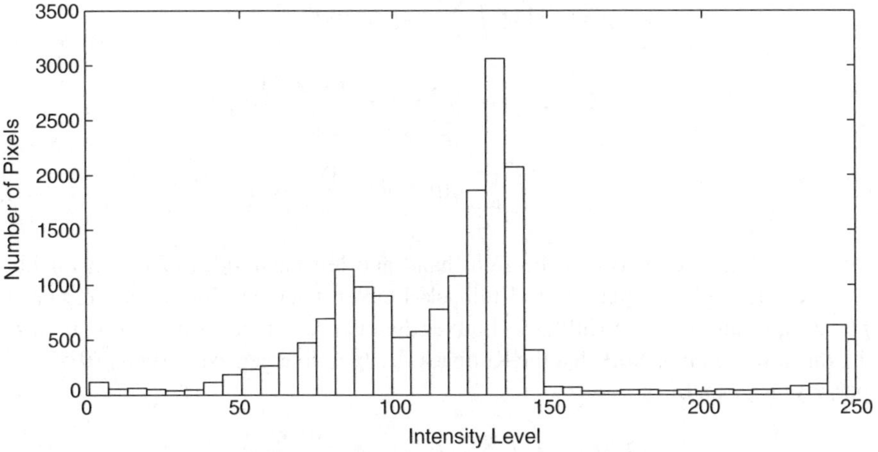

Figure 23.2 Histogram of intensity levels observed in the CT scan of the loin portion of a pork carcase shown in Figure 23.1.

in the data set and assume that a finite mixture model is adequate to represent the data. However, for an illustration of approaches that do take account of spatial association, we refer the reader to Chapter 16 which describes an MCMC-based approach for analysing this type of data via a spatial mixture model. Note also that the chapter describes another example of the use of CT technology in animal imaging; specifically, mixture modelling of CT scans of sheep is discussed. We also refer the reader to McGrory *et al.* (2009) which describes how to implement a VB approach to fit the same type of spatial model.

23.3 Models and methods

We denote our data by $\mathbf{y}$, and we shall assume a finite mixture model for our data whose components have parameters θ. We also have unobserved or latent variables, which we call $\mathbf{z}$, that indicate which component gave rise to each of the recorded observations. Our aim is to estimate the posterior distribution $p(\theta|\mathbf{y})$; we will achieve this by using the variational approach to find an approximation to it. We can obtain the posterior distribution $p(\theta|\mathbf{y})$ as the appropriate marginal distribution of $p(\theta, \mathbf{z}|\mathbf{y})$, and in practice it is this joint conditional density that we approximate in the variational Bayesian scheme. We approximate the complex joint conditional density, $p(\theta, \mathbf{z}|\mathbf{y})$, by another simpler distribution which we call the variational distribution, $q(\theta, \mathbf{z})$. Naturally this variational approximating distribution must be carefully chosen to give a good approximation and yet still be easily computable. In a variational Bayesian scheme $q(\theta, \mathbf{z})$ is chosen to give a tight lower bound on the marginal probability density $p(\mathbf{y})$ of $\mathbf{y}$. As we shall explain below, this is equivalent to minimizing the Kullback–Leibler divergence between $q(\theta, \mathbf{z})$ and $p(\theta, \mathbf{z}|\mathbf{y})$. By making use of Jensen's inequality we can obtain a tight lower bound for $p(\mathbf{y})$ as

$$\begin{aligned} \log p(\mathbf{y}) &= \log \int \sum_{\{\mathbf{z}\}} p(\mathbf{y}, \mathbf{z}, \theta) d\theta, \\ &= \log \int \sum_{\{\mathbf{z}\}} q(\theta, \mathbf{z}) \frac{p(\mathbf{y}, \mathbf{z}, \theta)}{q(\theta, \mathbf{z})} d\theta, \\ &\geq \int \sum_{\{\mathbf{z}\}} q(\theta, \mathbf{z}) \log \frac{p(\mathbf{y}, \mathbf{z}, \theta)}{q(\theta, \mathbf{z})} d\theta. \end{aligned} \tag{23.1}$$

Since the difference between the right hand and left hand sides of the inequality in Equation (23.1) is equal to the Kullback–Leibler divergence between $q(\theta, \mathbf{z})$ and $p(\theta, \mathbf{z}|\mathbf{y})$, minimizing the Kullback–Leibler divergence will correspond to maximizing this lower bound. Note that the Kullback–Leibler divergence between $q(\theta, \mathbf{z})$ and $p(\theta, \mathbf{z}|\mathbf{y})$ is given by

$$KL(q|p) = \int \sum_{\{\mathbf{z}\}} q(\theta, \mathbf{z}) \log \frac{q(\theta, \mathbf{z})}{p(\theta, \mathbf{z}|\mathbf{y})} d\theta. \tag{23.2}$$

We can minimize Equation (23.2), and hence maximize the lower bound, exactly by setting $q(\theta, \mathbf{z}) = p(\theta, \mathbf{z}|\mathbf{y})$, but we are seeking to simplify the necessary calculations so that they are feasible. This is achieved by assuming that the variational distribution $q(\theta, \mathbf{z})$ can be factorized over the unknown parameter θ and latent variables such that $q(\theta, \mathbf{z}) = q_\theta(\theta)q_\mathbf{z}(\mathbf{z})$. Maximizing the lower bound in this way produces the variational posteriors for $q_\theta(\theta)$ and $q_\mathbf{z}(\mathbf{z})$. Further factorization of $q_\theta(\theta)$ can take place as well. These assumptions result in a set of coupled expressions which can be solved through iteration. Typically the model to be estimated will be an exponential family model so that the usual conjugate priors can be used. This means that in the variational posterior the forms of the update expressions for model parameters will belong to the corresponding conjugate family.

Here we will describe how VB can be used in the case of a finite mixture model with an unknown number of components and unknown parameters. We follow McGrory and Titterington (2007) in our choice of model hierarchy and prior distributions. Consider a mixture of k univariate Gaussian distributions with unknown means, variances and mixing weights. Suppose we have a data set comprising observations $\mathbf{y} = (y_1, \ldots, y_n)$ which are assumed to have been drawn independently from the mixture distribution. Then the corresponding joint conditional density is of the form

$$\begin{aligned} p(y, z|\theta) &= \prod_{i=1}^{n}\prod_{j=1}^{k}\{\lambda_j N(y_i; \mu_j, \tau_j{}^{-1})\}^{z_{ij}}, \\ &= \prod_{i=1}^{n}\prod_{j=1}^{k}\left\{\lambda_j \sqrt{\frac{\tau_j}{2\pi}} e^{-\tau_j(y_i-\mu_j)^2/2}\right\}^{z_{ij}}. \end{aligned}$$

Here λ_j is the mixing weight associated with the jth component, $\lambda = (\lambda_1, \ldots, \lambda_k)$, and $\sum_{j=1}^{k}\lambda_j = 1$. The mixing weights represent the probabilities of component membership. The notation term $N(\cdot)$ corresponds to a univariate normal density for each component; μ_j is the mean of the jth component and τ_j is the precision of the jth component. Therefore, here the vector of unknown model parameters is $\theta = (\lambda, \mu, \tau)$. The $\{z_{ij}\}$ $(j = 1, \ldots, k)$ are the latent indicator variables which indicate which component each observation is from. Specifically, if data point y_i belongs to the lth component, then $z_{ij} = 1$ if $j = l$, and $z_{ij} = 0$ otherwise. We wish to estimate these latent indicators in our analysis.

As mentioned earlier, here we choose conjugate prior distributions as this simplifies the approximation. The prior for the mixing weights $\lambda = (\lambda_1, \ldots, \lambda_k)$ is Dirichlet, and the priors for the component means, $\mu = (\mu_1, \ldots, \mu_k)$, are independent normal priors, conditional on the precisions, $\tau = (\tau_1, \ldots, \tau_k)$. The precision matrices themselves are given independent gamma prior distributions. We have

$$\begin{aligned} p(\lambda) &= \mathrm{Dir}(\alpha_1{}^{(0)}, \ldots, \alpha_k{}^{(0)}), \\ p(\mu|\tau) &= \prod_{j=1}^{k} N(\mu_j; m_j{}^{(0)}, (\beta_j{}^{(0)}\tau_j)^{-1}), \\ p(\tau) &= \prod_{j=1}^{k} Ga\left(\tau_j|\frac{1}{2}\gamma_j{}^{(0)}, \frac{1}{2}\delta_j{}^{(0)}\right). \end{aligned}$$

The hyperparameters $\{\alpha_j^{(0)}\}$, $\{m_j^{(0)}\}$, $\{\beta_j^{(0)}\}$, $\{\gamma_j^{(0)}\}$ and $\{\delta_j^{(0)}\}$ are user-chosen values. This means that the resulting joint distribution of all of the variables is

$$p(y, z, \theta) \propto \prod_{j=1}^{k} \lambda_j^{\alpha_j^{(o)}-1+\sum_{i=1}^{n} z_{ij}} \prod_{j=1}^{k} \sqrt{\tau_j}^{(1+\sum_{i=1}^{n} z_{ij})} \tau_j^{\frac{1}{2}\gamma_j^{(0)}-1}$$
$$\times \exp\left\{-\frac{\tau_j}{2}\sum_{i=1}^{n} z_{ij}(y_i-\mu_j)^2\right\} \exp\left\{-\frac{\beta_j^{(0)}\tau_j}{2}(\mu_j - m_j^{(0)})^2 - \frac{1}{2}\delta_j^{(0)}\tau_j\right\}.$$

The VB posterior update expressions for the parameter values can then be derived algebraically by maximizing Equation (23.1) where the joint distribution is given by the above. It is assumed that $q(\theta) = q(\lambda)q(\mu)q(\tau)$; that is, we assume the same form as $p(\theta)$ and it turns out in the VB posteriors, the parameters they will have the following distributions:

$$q(\lambda) = Dir(\alpha_1, \ldots, \alpha_k),$$
$$q(\mu_j|\tau_j) = \prod_{j=1}^{k} N\left(m_j, \frac{1}{\beta_j\tau_j}\right),$$
$$q(\tau_i) = \prod_{j=1}^{k} Ga\left(\frac{1}{2}\gamma_j, \frac{1}{2}\delta_j\right).$$

The update expressions for the parameters in the VB posterior are given by

$$\alpha_j = \alpha_j^{(0)} + \sum_{i=1}^{n} q_{ij},$$
$$\beta_j = \beta_j^{(0)} + \sum_{i=1}^{n} q_{ij},$$
$$\gamma_j = \gamma_j^{(0)} + \sum_{i=1}^{n} q_{ij},$$
$$m_j = \frac{\beta_j^{(0)} m_j^{(0)} + \sum_{i=1}^{n} q_{ij} y_i}{\beta_j^{(0)} + \sum_{i=1}^{n} q_{ij}},$$
$$\delta_j = \delta_j^{(0)} + \sum_{i=1}^{n} q_{ij} y_i^2 + \beta_j^{(0)} m_j^{(0)2} - \beta_j m_j^2.$$

The term q_{ij} that is involved in each of these update expressions corresponds to the variational posterior expected probability that $z_{ij} = 1$; that is, it is the expected probability that observation i belongs to component j in the variational posterior. It is given by

$$q_{ij} = \exp\left\{\Psi(\alpha_j) - \Psi(\sum_{j\cdot} \alpha_{j\cdot}) + \frac{1}{2}\left[\Psi\left(\frac{1}{2}\gamma_j\right) - \log\frac{\delta_j}{2}\right]\right.$$
$$\left. - \frac{1}{2\beta_j} - \frac{\gamma_j}{2\delta_j}(y_i - m_j)^2\right\} / g_i,,$$

where g_i is a normalizing constant which ensures that $\sum_{j=1}^{k} q_{ij} = 1$, and $\Psi(\cdot)$ is the digamma function

$$\Psi(\alpha) = \frac{\partial\Gamma(\alpha)/\partial\alpha}{\Gamma(\alpha)} = \frac{\partial}{\partial\alpha}\ln\Gamma(\alpha).$$

The methodology outlined here extends straightforwardly to a mixture of multivariate normals. Again, conjugate priors are used in the multivariate case. The explicit forms of the priors and posterior update equations are given in the Appendix to this chapter; these are based on the approach for multivariate data as described by McGrory and Titterington (2007) and we refer the reader to that paper for further details. A detailed description of how the posterior update expressions are derived for the univariate and multivariate case, and for other mixtures of other exponential family models, is given in McGrory (2005).

Recall that the update expressions given above are coupled and must be solved iteratively. We coded a VB algorithm to iteratively solve these expressions in MATLAB. We outline the form of the algorithm in Algorithm 22.

Algorithm 22: Standard VB algorithm

Choose an appropriate initial number of components k
Set initial values for hyperparameters
Specify initial allocation of observations to components
while Not Converged **do**
  Update variational posterior expressions for model parameters
  Update variational posterior for the q_{ij}
  If any component has a weight $< \epsilon$ remove it from the model
  if the algorithm has converged **then**
    exit loop
  end if
end while

Since VB will lead to automatic selection of model complexity, that is the number of components in the model, we have to choose an initial value for the number of components k to begin the algorithm.

Tips and Tricks: Choosing an initial value for k that is too low may not allow enough scope for the algorithm to fit enough components. On the other hand, choosing an extremely large value for k will increase computation time, perhaps unnecessarily.

Notice that we must also choose some initial component allocation to start the algorithm. This is because we must initialize the q_{ij}, that is the probabilities of observation i being from each of the j components at the first iteration. Note that in subsequent iterations they will be updated based on current parameter estimates using the variational posterior update expression as described in Algorithm 22.

Tips and Tricks: One option is to use a clustering approach to choose initial allocations for the components, but simply using a random allocation in fact often works well. An advantage of a random allocation is that it avoids initializing the algorithm with strong prior information concerning which observations should belong to which components, as this, due to the elimination property of VB, could lead the early and permanent removal of components that might have represented the data well. However, caution should be exercised since in cases where the random initialization does not perform well, the resulting model fit may be very poor.

Given the allocations we must then set the initial q_{ij}.

Tips and Tricks: A straightforward approach which works well is to set the probability q_{ij} to be largest for the j corresponding to the initial allocation used and give smaller equal probabilities to the q_{ij} corresponding to the other j.

We also have to assess when the algorithm has converged and we can stop iteratively updating. This can be done in several ways.

Tips and Tricks: A commonly used approach is to compute the variational lower bound in Equation (23.1) at each iteration and stop once the bound is no longer being improved by successive updates. Another simpler method, which we use in this chapter, is to stop once estimates of the posterior means are no longer changed by iterative updates, up to some desired tolerance level.

Due to the component elimination property of VB, superfluous mixture components that are representing similar parts of the data set to other components will have their mixing weights move towards zero. This means that we can remove these components from the model and hence this leads to model complexity selection. We have to set a value for ϵ to specify a cutoff level at which a component is considered not to be useful. One reasonable choice would be to choose $\epsilon = 1$ as in McGrory and Titterington (2007), which would correspond to removing a component from the model if less than one observation is assigned to it under the VB approximation.

Tips and Tricks: In some cases it may make sense within the context of an application to choose the value of ϵ to be larger than 1. For instance, when the data set is very large, such as in the case study we consider, it might take many iterations for weightings of unnecessary components to become smaller than 1, and it may be clear that if their weighting reaches a certain cutoff higher than 1, then it will almost certainly drop further, leading to its eventual removal. In this case setting the cutoff level ϵ to be higher will save computation time.

23.4 Data analysis and results

We analysed the data shown in Figure 23.1 using the standard VB algorithm (see Algorithm 22). Recall that we are ignoring any spatial dependence here and assuming that a finite mixture model is adequate to represent the data. By fitting a mixture model to the data we aim to classify these observed values as belonging to one of the fitted components which each represent different types of tissue or bone. We initialized the standard VB algorithm with 15 components and used a random initial allocation of observations to these components. The data set was very large, comprising 17 514 observed intensity levels; however, we were able to carry out this VB analysis in a matter of minutes and a six-component mixture model was fitted to the data. A kernel density plot of the observed data with the variational posterior fitted model superimposed is shown in Figure 23.3, and we can see from this plot that the fitted model appears to represent the data well. Although both plots in Figure 23.3 are qualitatively similar, by fitting the mixture model we have quantifiably identified components, which is useful for analysing and classifying other observations obtained from the same source.

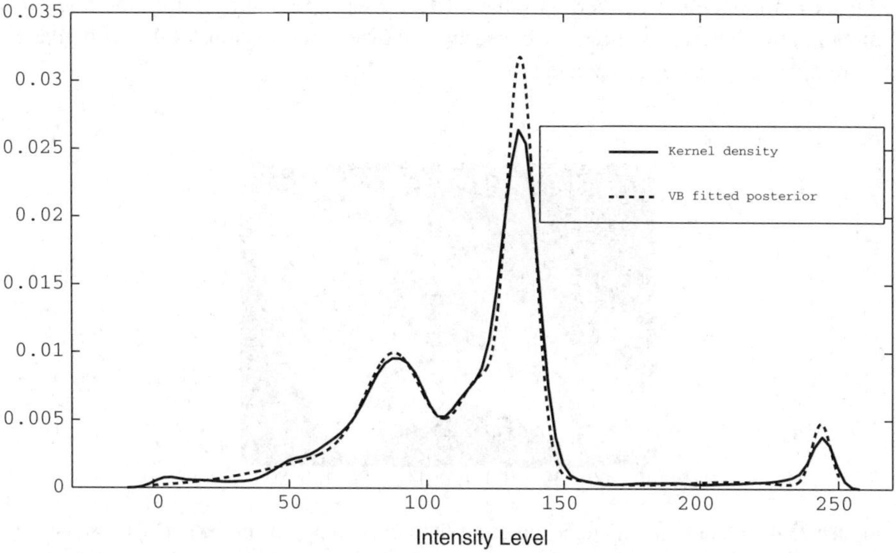

Figure 23.3 A kernel plot of the intensity levels recorded from the CT scan shown in Figure 23.1 (solid line) with the variational posterior fitted model superimposed (dashed line). We can see that by using the VB approach we have been able to fit the data well.

Table 23.1 VB posterior estimates of the component means and standard deviations and the mixing weights based on the six-component fit to the CT scan intensity levels. This fit was obtained using the standard VB algorithm initialized with 15 components.

Component type	Posterior means (μ)	Posterior standard deviations ($\sqrt{\tau^{-1}}$)	Posterior mixing weights (λ)
Fat	88.0364	10.0273	0.1762
Fat	88.3612	37.3851	0.2764
Muscle	119.2876	7.2415	0.1044
Muscle	134.2934	5.0746	0.3780
Bone	211.9350	19.9881	0.0255
Bone	243.8185	3.3884	0.0395

The numerical results obtained from the VB fit are given in Table 23.1. From these we can see that the first two components (corresponding to the lower mean intensity levels) represent the fat tissue, the third and fourth components (corresponding to the mid-range mean intensity levels) represent the muscle tissue, and the fifth and six components (corresponding to the higher mean intensity levels) represent bone. We can use the posterior parameter estimates for the mixing weights to provide an estimate of the proportions of each type of tissue or bone present in this section of the carcase. A graphical illustration of the variational posterior fit to the CT scan data is shown in Figure 23.4, where each pixel has been classified as belonging to one of the six components described in Table 23.1. We can see that by fitting the mixture model to the data we can now more clearly visualize which regions of the scan image represent each type of tissue and bone.

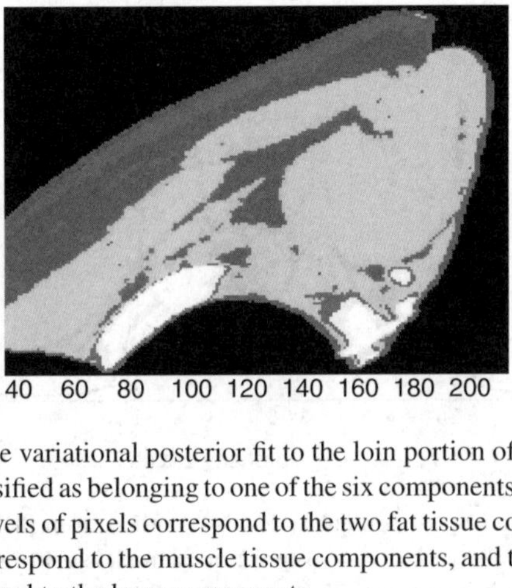

Figure 23.4 Plot of the variational posterior fit to the loin portion of the pork carcase where each pixel has been classified as belonging to one of the six components described in Table 23.1. The two darkest grey levels of pixels correspond to the two fat tissue components, the two mid-grey levels of pixels correspond to the muscle tissue components, and the two lightest coloured levels of pixels correspond to the bone components.

23.5 Discussion

We have seen that VB can be used to be provide a fast and efficient analysis of mixture data. The time efficiency is extremely useful in practical applications and is an advantage of using VB over a more computationally intensive MCMC-based approach. For example, in the case study we explored, we could easily use VB to analyse multiple similar images as part of a large study.

We also note again that, as with many approaches, if different initialization schemes are used before applying the standard VB algorithm, or, in some cases, if we choose different initial values for the number of components k, the final fit obtained may not be identical. As we have already mentioned, approaches for dealing with this include using model selection criteria to compare and select the final model, or perhaps using a splitting VB algorithm; see Wu *et al.* (2012) for an example of an application which makes use of both of these suggestions.

References

Attias H 1999 Inferring parameters and structure of latent variable models by variational Bayes. *Proceedings of the 15th Conference on Uncertainty in Artificial Intelligence*, pp. 21–30. Morgan Kaufmann, San Francisco.

Bishop CM 2006 *Pattern Recognition and Machine Learning*. Springer, New York.

Huang JC, Kannan A and Winn J 2007 Bayesian association of haplotypes and non-genetic factors to regulatory and phenotypic variation in human populations. *Bioinformatics* **23**(13), i212–i221.

Jordan MI, Ghahramani Z, Jaakkola TS and Saul LK 1999 An introduction to variational methods for graphical models. *Machine Learning* **37**, 183–233.

McGrory CA 2005 Variational approximations in Bayesian model selection. PhD thesis. University of Glasgow.

McGrory CA and Titterington DM 2007 Variational approximations in Bayesian model selection for finite mixture distributions. *Computational Statistics & Data Analysis* **51**, 5352–5367.

McGrory CA and Titterington DM 2009 Bayesian analysis of hidden Markov models using variational approximations. *Australian and New Zealand Journal of Statistics* **51**(2), 227–244.

McGrory CA, Titterington DM, Reeves R and Pettitt AN 2009 Variational Bayes for estimating the parameters of a hidden Potts Markov random field. *Statistics and Computing* **19**, 329–340.

Ormerod JT and Wand MP 2011 Gaussian variational approximate inference for generalized linear mixed models. *Journal of Computational and Graphical Statistics*. doi: 10.1198/jcgs.2011.09118.

Penny WD, Kiebel SJ and Friston KJ 2003 Variational Bayesian inference for fMRI time series. *NeuroImage* **19**(3), 727–741.

Richardson S and Green PJ 1997 On Bayesian analysis of mixtures with an unknown number of components (with discussion). *Journal of the Royal Statistical Society, Series C* **59**, 731–792.

Wang B and Titterington DM 2006 Convergence properties of a general algorithm for calculating variational Bayesian estimates for a normal mixture model. *Bayesian Analysis* **1**, 625–650.

Winn J and Bishop CM 2005 Variational message passing. *Journal of Machine Learning Research* **6**, 661–694.

Wu B, McGrory CA and Pettitt AN 2012 A new variational Bayesian algorithm with application to human mobility pattern modelling. *Statistics and Computing* **22**, 185–203.

23.A Appendix: Form of the variational posterior for a mixture of multivariate normal densities

Here we again follow McGrory and Titterington (2007) in specifying our model structure and we refer to that article for further details and simulation studies. The mixture of multivariate normal distributions has the form

$$p(\mathbf{y}, \mathbf{z}|\theta) = \prod_{i=1}^{n}\prod_{j=1}^{k}\{\lambda_j N_d(y_i; \mu_j, \mathbf{T}_j^{-1})\}^{z_{ij}}.$$

Here λ_j is the mixing weight associated with the jth component. The mixing weights represent the probabilities of component membership and they sum to 1. $N_d(\cdot)$ corresponds to a multivariate normal density component with dimensionality d; μ_j denotes the mean of the jth component and $\mathbf{T}_j$ denotes the precision matrix of the jth component. The $\{z_{ij}\}$ $(j = 1, \ldots, k)$ are the latent indicator variables which indicate which component each observation is from. Specifically, if data point y_i belongs to the lth component, then $z_{ij} = 1$ if $j = l$, and $z_{ij} = 0$ otherwise.

We use conjugate prior distributions: the prior for the mixing weights is Dirichlet; the priors for the component means, $\mu = (\mu_1, \ldots, \mu_k)$, are independent multivariate normal priors, conditional on the precision matrices $\mathbf{T} = (T_1, \ldots, T_k)$; the precisions themselves are given independent Wishart prior distributions.

$$\begin{aligned}
p(\lambda) &= Dir(\lambda; {\alpha_1}^{(0)}, \ldots, {\alpha_k}^{(0)})\\
p(\mu|\mathbf{T}) &= \prod_{j=1}^{k} N_d(\mu_j; {m_j}^{(0)}, ({\beta_j}^{(0)}T_j)^{-1})\\
p(\mathbf{T}) &= \prod_{j=1}^{k} W(T_j; {v_j}^{(0)}, {\Sigma_j}^{(0)}).
\end{aligned}$$

The hyperparameters $\{\alpha_j^{(0)}\}$, $\{m_j^{(0)}\}$, $\{\beta_j^{(0)}\}$, $\{v_j^{(0)}\}$ and $\{\Sigma_j^{(0)}\}$ are user-chosen values.

In the VB posterior, the parameters have the following distributions:

$$\begin{aligned}
q_\lambda(\lambda) &= Dir(\lambda; \alpha_1, \ldots, \alpha_k),\\
q_{\mu|\mathbf{T}}(\mu|\mathbf{T}) &= \prod_{j=1}^{k} N_d(\mu_j; m_j, (\beta_j T_j)^{-1}),\\
q_{\mathbf{T}}(\mathbf{T}) &= \prod_{j=1}^{k} W(T_j; v_j, \Sigma_j),
\end{aligned}$$

where we have the following VB posterior update expressions for the parameter values:

$$\alpha_j = \alpha_j{}^{(0)} + \sum_{i=1}^{n} q_{ij}$$

$$\beta_j = \beta_j{}^{(0)} + \sum_{i=1}^{n} q_{ij}$$

$$m_j = \frac{\beta_j{}^{(0)} m_j{}^{(0)} + \sum_{i=1}^{n} q_{ij} y_i}{\beta_j}$$

$$\Sigma_j = \Sigma_j{}^{(0)} + \sum_{i=1}^{n} q_{ij} y_i y_i{}^T + \beta_j{}^{(0)} m_j{}^{(0)} m_j{}^{(0)T} - \beta_j m_j m_j{}^T$$

$$v_j = v_j{}^{(0)} + \sum_{i=1}^{n} q_{ij}.$$

The term q_{ij} that is involved in each of these update expressions corresponds to the variational posterior expected probability that $z_{ij} = 1$; that is, it is the expected probability that observation i belongs to component j in the variational posterior. It is given by

$$q_{ij} = \frac{\exp\{\mathbf{E}(\log \lambda_j) + \frac{1}{2}\mathbf{E}(\log |T_j|) - \frac{1}{2}\mathrm{tr}(\mathbf{E}(T_j)(y_i - m_j)(y_i - m_j)^T + \beta_j{}^{-1})\mathbf{I}_d\}}{g_i}$$

where g_i is a normalizing constant. The expectations required to compute the q_{ij} are given by

$$\mathbf{E}(\mu_j) = m_j$$

$$\mathbf{E}(T_j) = v_j \Sigma_j{}^{-1}$$

$$\mathbf{E}(\log |T_j|) = \sum_{s=1}^{d} \Psi\left(\frac{v_j + 1 - s}{2}\right) + d \log(2) - \log |\Sigma_j|$$

$$\mathbf{E}(\log(\lambda_j)) = \Psi(\hat{\alpha}_j) - \Psi(\hat{\alpha}.),$$

where Ψ is the digamma function and $\hat{\alpha}. = \sum_j \hat{\alpha}_j$.

24

Issues in designing hybrid algorithms

Jeong E. Lee[1], Kerrie L. Mengersen[2] and Christian P. Robert[3]

[1]*Auckland University of Technology, New Zealand*
[2]*Queensland University of Technology, Brisbane, Australia*
[3]*Université Paris-Dauphine, Paris, France and Centre de Recherche en Économie et Statistique (CREST), Paris, France*

24.1 Introduction

In the Bayesian community, an ongoing imperative is to develop efficient algorithms for the more diverse and often complicated problems encountered in practice. There has been a substantial amount of progress on the fundamental ideas of designing efficient computational algorithms and the theoretical properties of these methods of simulation. The Markov chain Monte Carlo (MCMC) methods, originally proposed by Metropolis *et al.* (1953) and Hastings (1970), are designed to generate Markov chains with a given stationary distribution and various types of algorithms based on MCMC techniques have been proposed in the literature. Each algorithm has different strengths and weaknesses, and can be evaluated with respect to its ability to meet different criteria based on specified statistical properties. It is therefore natural that the question arises whether a better scheme can be developed by combining the best aspects of existing algorithms.

The concept of the hybrid algorithm was introduced by Tierney (1994). A hybrid algorithm can be designed from different perspectives, based on a variety of algorithms

Case Studies in Bayesian Statistical Modelling and Analysis, First Edition. Edited by Clair L. Alston, Kerrie L. Mengersen and Anthony N. Pettitt.

to combine and the way in which they are combined. The primary motivation is to propose an efficient algorithm that overcomes identified weaknesses in the individual algorithms. Tierney (1994) outlined the theoretical and practical parts of existing algorithms, and the Gibbs sampler and the Metropolis algorithm were compared.

In this chapter, we extend those parts proposed by Tierney (1994) from both practical and theoretical perspectives, and focus on three criteria: statistical efficiency, applicability and implementation. These are now described in more detail.

Statistical efficiency: This criterion is subject to the statistical properties of the generated Markov chain, such as the rate of convergence and the mixing speed of chains. Although classical studies ensure that the MCMC method guarantees convergence to the target, this simply may not be enough to guarantee convergence in real time. The work in recent years has thus focused on theoretical developments of MCMC samplers to maximize efficiency. Established results for MCMC algorithms include geometric ergodicity for Metropolis algorithms (Jarner and Hansen 2000; Mengersen and Tweedie 1996), and exponential ergodicity for the Langevin algorithm (Roberts and Tweedie 1996a). When the chain is updated via an accept–reject setup, the efficiency has been characterized by deriving optimal acceptance rates (Gelman *et al.* 1997; Roberts and Rosenthal 2001; Roberts and Tweedie 1996b).

There also has been considerable attention on the mixing of a simulated Markov chain. A chain with a slower rate of mixing is more strongly dependent on the initial value and may easily become stuck in certain parts of the state space for multimodal problems. This becomes a severe problem in generating reliable samples from the target as the dimension increases. Mechanisms such as the repulsive proposal (Mengersen and Robert 2003), geometric determinant proposal (Mengersen and Robert 2003), and delayed rejection (Tierney and Mira 1999) are proposed to improve the rate of mixing and will be examined in more detail later.

For the importance-sampling-based algorithms the convergence of estimates to their expected values is guaranteed as the number of particles increases. In practice limited numbers of particles are used and the degeneracy of importance weights is an unavoidable drawback. Attempts to limit degeneracy and reduce the variance can be found in Doucet *et al.* (2000) among others.

Applicability: In general a common feature of popular algorithms is that they are applicable in a diverse range of realistic settings. For example, the Gibbs sampler (Geman and Geman 1984) and the Metropolis–Hastings algorithm (MHA) (Hastings 1970; Metropolis *et al.* 1953) enjoy universal popularity because of their simple setting and, in the latter case, flexible choice of the proposal distribution (Besag *et al.* 1995; Besag and Green 1993; Geyer 1992). Well-known hybrid approaches based on the MHA are the delayed rejection algorithm (DRA) (Tierney and Mira 1999), the reversible jump MCMC for variable dimension models (RJMCMC) (Green 2001; Green and Mira 2001), the Metropolis adjusted Langevin algorithm (MALA) (Roberts and Tweedie 1996a), and the pinball sampler (PS) (Mengersen and Robert 2003).

Implementation: This criterion addresses computational issues such as the level of difficulty in coding, memory storage required and the demand on computation in real time.

In the MCMC literature, theoretical studies of statistical efficiency (first criterion above) are typically based on the number of iterations, not actual CPU time. From the perspective of practical implementation, the efficient sampling performance of an algorithm in real time (seconds) involves both the statistical efficiency and CPU time consumed in implementing the algorithm. Thus optimal performance may involve a compromise between the increase in computing workload and the gain in statistical efficiency.

In this chapter, we focus on selected hybrid approaches: the delayed rejection algorithm (DRA) with different types of moves for the proposal, the Metropolis adjusted Langevin algorithm (MALA) and the Metropolis–Hastings algorithm (MHA) with the repulsive proposal (RP), and make comparisons between the hybrid algorithms and the component algorithms on which they are based with respect to the three criteria defined above. The performance of hybrid algorithms is compared using two case studies: a simulated data set from a mixture of normal distributions and an aerosol particle size data set.

The simulated data set is created by generating random samples from a target distribution that comprises two well-separated modes, described by a mixture of two-dimensional normal distributions

$$\pi = \frac{1}{2}\mathbf{N}([0, 0]^T, I_2) + \frac{1}{2}\mathbf{N}([5, 5]^T, I_2).$$

The expectations of the two coordinates are $\mathbf{E}_\pi(\theta_1) = \mathbf{E}_\pi(\theta_2) = 2.5$. A normal proposal distribution $q_1(\theta, \cdot) = \mathbf{N}(\theta, sI_2)$, $s > 0$, is used for the MHA, the RP, and as the first-stage proposal for the DRA.

Aerosols are small particles in suspension in the atmosphere and have both direct and indirect effects on the Earth's climate. Aerosol size distributions describe the number of particles observed to have a certain radius, for various size ranges, and are studied to understand the aerosol dynamics in environmental and health modelling. The data represented in Figure 24.1 comprise a full day of measurements, taken

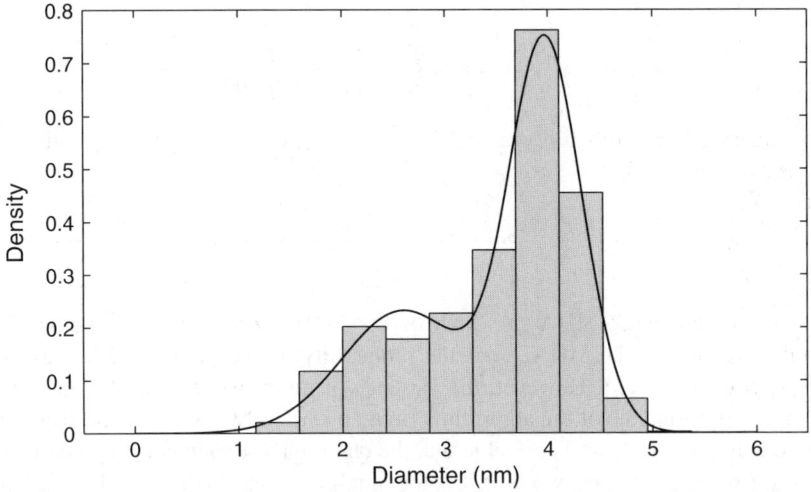

Figure 24.1 Histogram of the aerosol diameter data set, along with empirical density via the MHA (solid line).

at 10 minute intervals. The data set was collected at a boreal forest measurement site at Hyytiälä, Finland (Nilsson and Kulmala 2006); a random subsample of 2000 observation from the total of 19 998 measurements was taken for this case study.

24.2 Algorithms and hybrid approaches

A considerable amount of attention has been devoted to the MHA which was developed by Metropolis *et al.* (1953) and subsequently generalized by Hastings (1970); see Chapter 2 for further details. The advantages of the MHA are that it is straightforward to implement and is flexible with respect to the proposal density. However, it may take an unrealistically long time to explore high-dimensional state spaces and tuning parameters involved with proposal density could be difficult (Robert and Casella 2004).

The MHA associated with a stationary density π and a Markov kernel q_1 produces a Markov chain $\{\theta^{(t)}\}_{t=0}^{T}$ via the procedure described in Algorithm 23. The Markov chain $\{\theta^{(t)}\}_{t=0}^{T}$ converges to the target π at a given rate (Mengersen and Tweedie 1996; Roberts and Tweedie 1996b). A useful feature of the MHA is that it can be applied even when the target is only known up to a normalizing constant through the ratio π/q_1.

Algorithm 23: Metropolis–Hastings algorithm (MHA)

Generate the initial states from an initial distribution p, $\theta^{(0)} \sim p(\cdot)$.
for $t = 1$ to T **do**
Generate a candidate state φ from $\theta^{(t-1)}$ with some distribution q_1, $\varphi \sim q_1(\theta^{(t-1)}, \cdot)$.
Calculate an acceptance probability α_1

$$\alpha_1(\theta^{(t-1)}, \varphi) = \min\left\{1, \frac{\pi(\varphi)}{\pi(\theta^{(t-1)})}\frac{q_1(\varphi, \theta^{(t-1)})}{q_1(\theta^{(t-1)}, \varphi)}.\right\}$$

Accept φ with probability $\alpha_1(\theta^{(t-1)}, \varphi)$. If φ is accepted, set $\theta^{(t)} = \varphi$. Otherwise set $\theta^{(t)} = \theta^{(t-1)}$.
end for

The random walk MHA in which $q_1(\varphi, \theta^{(t-1)}) = q_1(\theta^{(t-1)}, \varphi)$ is perhaps the simplest version of the MHA, since the probability of acceptance reduces to $\alpha_1 = \min\{1, \pi(\varphi)/\pi(\theta^{(t-1)})\}$. However, the variance of the proposal distribution is critical to the performance of the algorithm (Gelman *et al.* 1997; Roberts and Rosenthal 2001). If it is too large, it is possible that the chain may remain stuck at a particular value for many iterations, while if it is too small the chain will tend to make small jumps and move inefficiently through the support of the target distribution.

24.2.1 Particle system in the MCMC context

The MHA, like most MCMC algorithms to date, produces a single Markov chain simulation and monitoring focuses on convergence of the chain to the target. An alternative particle system based on importance sampling (IS) can be used to approximate the posterior expectation of estimates of interest. The key idea of the particle system is to represent the required distribution by a set of random samples $(\varphi_1^{(t)}, \varphi_2^{(t)}, \ldots, \varphi_N^{(t)})$ with associated importance weights and to compute estimates based on these samples and weights.

In the MCMC context, instead of using the IS the whole random vector $(\theta_1^{(t)}, \theta_2^{(t)}, \ldots, \theta_N^{(t)})$ is resampled at each iteration according to a Markovian updating process. Mengersen and Robert (2003) showed that an MCMC algorithm generates the production of samples of size N from π and a product distribution $\pi^{\otimes N}$ is not fundamentally different from the production of a single output from π.

The simplest version of a particle system in an MCMC context is a parallel run of independent MCMC algorithms. In this type of implementation the result and chains are considered individually rather than as a whole vector. The parallel system can be found in the literature on coupling methods for Markov chains (Breyer and Roberts 2002). When the posterior distribution evolves with time, hybrid algorithms can be constructed by combining the parallel MCMC and the sequential Monte Carlo (SMC) algorithm (Andrieu *et al.* 2010; Del Moral *et al.* 2006). Mengersen and Robert (2003) proposed updating schemes, the repulsive proposal and the geometric determinant proposal to allow interaction between particles. These schemes are in essence hybrid algorithms, since they combine existing algorithms, and will be examined in detail later.

24.2.2 MALA

The random walk MHA has an advantage in that it moves independently from the shape of π and is easy to implement. However, an algorithm that is more specifically tailored to π may converge more rapidly. This leads to the MALA, which applies a Langevin diffusion approximation using the information about the target density (in the form of the derivative of $\log \pi$) to the MHA structure (Besag and Green 1993). Assuming that the target distribution π is everywhere non-zero and differentiable so that the derivative of $\log \pi$ is well defined, φ is generated from

$$q_1(\theta^{(t-1)}, \cdot) = \mathbf{N}\left(\theta^{(t-1)} + \frac{1}{2} h \nabla \log \pi(\theta^{(t-1)}), h\right) \tag{24.1}$$

where h is a scaling parameter. The proposal is accepted with probability $\alpha_1(\theta^{(t-1)}, \varphi)$, as per Algorithm 23. As with the traditional Euler method, scaling the step size h is important. If h is too large or too small, the chain may never converge to the target in real time (Roberts and Stramer 2002). Atchadé (2006) showed that h can be tuned adaptively.

The idea of shaping the candidate density based on the target was introduced long ago by Doll *et al.* (1978) and in the probabilistic literature has been studied by, among others, Roberts and Tweedie (1996a), Roberts and Tweedie (1996b), Stramer and Tweedie (1999a) and Stramer and Tweedie (1999b). In discrete time space, Roberts and Tweedie (1996a) proved that the MALA is exponentially ergodic when the tails of the target density are heavier than Gaussian. This is an advantage of the MALA over the MHA that is geometrically ergodic. With a fast convergence rate the algorithm may be efficient for sampling within a single mode, but it can still fail to explore more than a single mode of the distribution. Moreover, in practice $\nabla \log \pi$ can be expensive to compute and sometimes the first partial derivative form is not straightforward to evaluate.

As an attempt to cross low-probability regions between modes, Skare *et al.* (2000) used a smoothed Langevin proposal based on a function approximation and Celeux *et al.* (2000) demonstrated on mixtures of distributions a tempering scheme using Langevin algorithms.

24.2.3 DRA

In the regular MHA (Algorithm 23), whenever a proposed candidate is rejected, the chain retains the same position so that $\theta^{(t)} = \theta^{(t-1)}$. This increases the autocorrelation along the realized path and thus increases the variance of the estimates. Tierney and Mira (1999) proposed the DRA to ameliorate this effect. The central idea behind this algorithm is that if the candidate is rejected, a new candidate is proposed with an acceptance probability tailored to preserve the stationary distribution. Thus at the tth iteration, if a candidate φ is rejected with probability $\alpha_1(\theta^{(t-1)}, \varphi)$, a proposal ϑ is constructed from a possibly different proposal density $q_2(\theta^{(t-1)}, \varphi, \cdot)$ and ϑ is accepted with probability

$$\alpha_2(\theta^{(t-1)}, \varphi, \vartheta) = \min\left\{1, \frac{\pi(\vartheta)q_1(\vartheta, \varphi)q_2(\vartheta, \varphi, \theta^{(t-1)})[1 - \alpha_1(\vartheta, \varphi)]}{\pi(\theta^{(t-1)})q_1(\theta^{(t-1)}, \varphi)q_2(\theta^{(t-1)}, \varphi, \vartheta)[1 - \alpha_1(\theta^{(t-1)}, \varphi)]}\right\}.$$

The algorithm can be extended to allow proposals to be made after multiple rejections (Mira 2001).

This scheme does not guarantee an automatic increase in the acceptance probability. However, once a better candidate is accepted, it induces an increase in the rate of acceptance and a reduction of the autocorrelation in the constructed Markov chain.

The drawback of the DRA is that the amount of computation and program workload increases geometrically with the number of delayed rejection attempts. The number of attempts may depend on the type of problem, model and core questions to be answered. Green and Mira (2001) concluded that the workload required for three or more attempts was not worthwhile based on their limited experiments.

Since the proposal distributions (q_1 and q_2) are not constrained to be the same, various types of proposal distributions can be constructed using information from rejected proposals. Green and Mira (2001) developed a hybrid algorithm by applying the DRA to the reversible jump algorithm for transdimensional problems, and Mengersen and Robert (2003) suggested the proposal scheme in a two-dimensional space. We consider here the geometric deterministic proposal suggested by Mengersen and Robert (2003) and the Langevin proposal originally suggested by Mira *et al.* (2004).

Geometric deterministic proposal (Mengersen and Robert 2003): The idea is to push particles away using symmetry with respect to a line defined by the closest particle and rejected particle. Suppose that the proposal, $\varphi_i \sim q_1(\theta_i, \cdot)$, is rejected. The second proposal ϑ_i is generated by taking the reflection of θ_i with respect to the line connecting θ^* and φ_i where θ^* is the closest particle to φ_i among $\theta_1, \ldots \theta_{i-1}, \theta_{i+1}, \ldots, \theta_N$. This is the so-called pinball effect.

Langevin proposal: If the current proposal φ is rejected, the second proposal ϑ is generated according to

$$q_2(\theta^{(t-1)}, \varphi, \cdot) = \mathbf{N}\left(\varphi + \frac{1}{2}h\nabla \log \pi(\varphi), h\right).$$

24.2.4 PS

Mengersen and Robert (2003) introduced a hybrid algorithm, the PS, which is a combination of the MHA, delayed rejection mechanism, particle system and repulsive proposal. It is an updating system for a particle system that is based on a standard random walk with correction to avoid the immediate vicinity of other particles.

Unlike traditional particle systems, neither importance sampling schemes nor weights are used. At each iteration, the whole vector $(\theta_1^{(t)}, \theta_2^{(t)}, \ldots, \theta_N^{(t)})$ is updated. Thus the algorithm produces iid samples from the target, which is an advantage over MCMC methods. Moreover, at each vector update the repulsive mechanism discourages particles from being too close and hence avoids possible degeneracy problems.

We now study the properties of the repulsive proposal density in order to understand the mechanics of the PS.

Repulsive proposal (RP): The RP suggested by Mengersen and Robert (2003) uses the following pseudo distribution:

$$\pi^R(\theta_i) \propto \pi(\theta_i) \prod_{j \neq i} e^{-\xi/\pi(\theta_j)\|\theta_i - \theta_j\|^2} \tag{24.2}$$

where ξ is a tempering factor.

The pseudo distribution, π^R, is derived from the distribution of interest π by multiplying exponential terms that create holes around the other particles θ_j $(i \neq j)$. These thus induce a repulsive effect around the other particles. The factor $\pi(\theta_j)$ in the exponential moderates the repulsive effect in zones of high probability and enhances it in zones of low probability. If the tempering factor, ξ, is large, π^R is dominated by the repulsive term and becomes very different from the target distribution π. If ξ is small, there is a negligible repulsive effect on π^R, so $\pi^R \approx \pi$.

The MHA is easily adapted to include the RP through two acceptance steps, firstly with π^R and secondly with the true posterior π to ensure convergence to the target distribution. The procedure is described in Algorithm 24.

Algorithm 24: Metropolis–Hastings algorithm with the repulsive proposal

Generate initial particles from an initial distribution p, $\theta^{(0)}_{1,\dots,N} \sim p(\cdot)$.
for $t = 1, \dots, T$ **do**
  for $i = 1, \dots, N$ **do**
    Generate $\varphi_i \sim q_1(\theta_i^{(t-1)}, \cdot)$.
    Propose step
    Determine the probability using π^R,

$$\alpha_1^*(\theta_i^{(t-1)}, \varphi_i) = \min\left\{1, \frac{\pi^R(\varphi_i)}{\pi^R(\theta_i^{(t-1)})} \frac{q_1(\varphi_i, \theta_i^{(t-1)})}{q_1(\theta_i^{(t-1)}, \varphi_i)}\right\}.$$

    Generate $r \sim U[0, 1]$. If $r < \alpha_1^*$, go to *Correction step*. Otherwise, set $\theta_i^{(t)} = \theta_i^{(t-1)}$.
    Correction step
    Implement a final Metropolis acceptance probability calibrated for the target distribution π,

$$\alpha_1(\theta_i^{(t-1)}, \varphi_i) = \min\left\{1, \frac{\pi(\varphi_i)}{\pi(\theta_i^{(t-1)})} \frac{\pi^R(\theta_i^{(t-1)})}{\pi^R(\varphi_i)}\right\}.$$

    Accept φ_i, with probability α_1. If φ_i is accepted, set $\theta_i^{(t)} = \varphi_i$; otherwise set $\theta_i^{(t)} = \theta_i^{(t-1)}$.
  end for
end for

The tempering factor ξ can be calibrated during the simulation against the number of particles N and the acceptance rate. The value should be tuned such that proposals are easily rejected if they are generated from low-probability regions or are too close to existing particles.

Borrowing the idea from the DRA, two or more attempts at updating may be pursued using the information from rejected proposals, and the delayed rejection step can easily be included in this algorithm.

The two-stage PS algorithm for generating from a target distribution π is given in Algorithm 25. Note that other proposals can be considered for q_2 for different problems. In particular, some stochasticity could augment the deterministic proposal described in this algorithm.

24.2.5 Population Monte Carlo (PMC) algorithm

The PMC algorithm by Cappé *et al.* (2004) is an iterated IS scheme (Algorithm 26). The major difference between the PMC algorithm and earlier work on particle systems is that it allows adaptive proposals to draw samples from a target that does not evolve with time.

Algorithm 25: Pinball sampler (PS)

Generate initial particles from an initial distribution p, $\theta^{(0)}_{1,\ldots,N} \sim p(\cdot)$.
for $t = 1, \ldots, T$ **do**
  for $i = 1, \cdots, N$ **do**
    First stage
    Generate $\varphi_i \sim q_1(\theta_i^{(t-1)}, \cdot)$.
    Generate $r_1 \sim U[0, 1]$. If $r_1 < \alpha_1^*(\theta_i^{(t-1)}, \varphi_i)$, accept φ_i as in the *Correction step*. Either $r_1 > \alpha_1^*(\theta_i^{(t-1)}, \varphi_i)$ or φ_i is rejected with π, go to the *Second stage*.
    Second stage
    Draw a candidate $\vartheta_i \sim q_2(\theta_i^{(t-1)}, \varphi_i, \cdot)$. Here the q_2 is the geometric deterministic proposal.
    Generate $r_2 \sim U[0, 1]$. If $r_2 < \alpha_2^*(\theta_i^{(t-1)}, \varphi_i, \vartheta_i)$, accept ϑ_i as in the *Correction step*.
  end for
end for

Algorithm 26: Population Monte Carlo (PMC) algorithm

Generate initial particles from an initial distribution p, $\theta^{(0)}_{1,\ldots,N} \sim p(\cdot)$.
for $t = 1$ to T **do**
  Construct the importance function $g^{(t)}$ and sample $\varphi_1^{(t)}, \ldots, \varphi_N^{(t)} \sim g^{(t)}(\cdot)$.
  Compute the importance weight

$$\omega_i^{(t)} \propto \frac{\pi\left(\varphi_i^{(t)}\right)}{g^{(t)}\left(\varphi_i^{(t)}\right)}.$$

  Normalize the importance weights to sum to 1.
  Resample with replacement N particles $\{\theta_i^{(t)}; i = 1, \ldots, N\}$ from the set $\{\varphi_i^{(t)}; i = 1, \ldots, N\}$ according to the normalized importance weights.
end for

The advantage of the PMC over the MCMC algorithms is that the adaptive perspective can be achieved easily. Doucet *et al.* (2007) proposed a D-kernel PMC that automatically tunes the scaling parameters and induces a reduction in the asymptotic variance via a mixture of importance functions.

Resampled particles according to the importance weights are relatively more informative than those that are not resampled and this available information can be used to construct the g at the next iteration. A common choice for g is a set of individual normal distributions for $\theta_1, \ldots, \theta_N$ (Cappé *et al.* 2004; Doucet *et al.* 2007).

As with other IS-based algorithms, the downside of the PMC algorithm is that degeneracy may occur due to instability of the importance weight for the multimodal target as the dimension increases. The degeneracy is observed as the importance weight is concentrated on very few particles or a single particle. Theoretically this phenomenon can be avoided when the number of particles increases exponentially with the dimension (Li *et al.* 2005) but this usually causes storage problems.

24.3 Illustration of hybrid algorithms

In this section we examine the hybrid methods described in Section 24.2 and study how individual components of the algorithms influence the properties of the overall algorithm. Algorithms are programmed using MATLAB version 7.0.4 software. The investigation is first carried out via simulation of the hybrid MCMC algorithms using a toy example and the sensitivity of a tempering factor of the RP is given in Section 24.3.1.

24.3.1 Simulated data set

We consider two simulation studies. The first study focuses on the performance of the algorithms in terms of statistical efficiency and relative cost of computation. The second study focuses on the mobility of the chains in a special setup where initial values are generated from a certain part of state space and demonstrates the ability of the algorithm to detect the two modes of the target distribution.

Performance study

The comparison of the performance of algorithms is based on 100 replicated simulations of each algorithm. Each simulation result is obtained after running the algorithm for the same amount of time (1200 seconds) using 10 particles. This allows a comparison of the algorithms in a more realistic way by fixing the running time and observing the consistency of estimates throughout the replications.

For each simulation, the performance is defined as the accuracy of H which is the mean of θ_1, and the variance (σ_H^2), the efficiency of moves indicated by the rate of acceptance (A), the computational demand indicated by the total number of iterations (T), the mixing speed of the chain estimated by the integrated autocorrelation time ($\tau_\pi(H)$), and the loss in efficiency due to the correlation along the chain by the effective sample size (ESS_H) where $ESS_H = (T - T_0)/\tau_\pi(H)$ and T_0 is the length of the burn-in. The term τ_π indicates the number of correlated samples with the same variance-reducing power as one independent sample, and is a measure of the autocorrelation in the chain. The ESS_H can be interpreted as the number of equivalent iid samples from the target distribution for 1200 seconds of running and provides a practical measure for the comparison of algorithms in terms of both statistical and computational efficiency. Except for T and A, all measurements are estimated after ignoring the first 500 iterations ($T_0 = 500$).

Based on the 100 replicates, these measures are summarized by taking the respective averages ($\bar{H}$, $\bar{\sigma}_H^2$, $\bar{A}$, $\bar{T}$, $\bar{\tau}_\pi(H)$ and $\overline{ESS}_H$) and mean square error (MSE). The MSE measures the consistency of performances over 100 replicates and the

Table 24.1 Comparison of performance in the two-dimensional example with the proposal variance $s = 4$. The values in parentheses are the MSE of measures.

	$\overline{T}$	$\bar{A}$	$\bar{H}$	$\bar{\sigma}_H^2$	$\bar{\tau}_\pi(H)$	$\overline{ESS}_H$
True value			2.5			
MHA^s	1.0376×10^6	0.30	2.4979	7.2523	294.7127	3507.2
	(456.0547)	($< 10^{-4}$)	(0.0032)	(0.0013)	(0.9197)	(10.8990)
MHA	2.4259×10^6	0.30	2.5002	0.7250	296.5254	8181.7
	(1.2×10^4)	($< 10^{-4}$)	(0.0007)	(0.0005)	(0.5736)	(43.1207)
MHA^{RP}	1.2172×10^5	0.29	2.4904	0.7213	266.5477	371.2440
($\xi = 10^{-5}$)	(581.89)	($< 10^{-4}$)	(0.0036)	(0.0028)	(2.4224)	(3.8696)
DRA	1.0012×10^6	0.49	2.4978	0.7187	196.8483	5086.9
	(24.8503)	($< 10^{-4}$)	(0.0009)	(0.0006)	(0.5143)	(13.1896)
DRA^{LP}	1.9914×10^5	0.34	2.5010	0.7223	253.4174	786.2
(h=4)	(1031.5)	($< 10^{-4}$)	(0.0020)	(0.0018)	(1.3601)	(6.0554)
DRA^{Pinball}	6.4729×10^5	0.61	2.5025	0.7235	239.1775	2708.4
	(3091.8)	($< 10^{-4}$)	(0.0010)	(0.0009)	(0.8232)	(17.3921)
MALA	2.0415×10^5	0.29	2.4978	0.7265	828.4097	505.0
($h = 4$)	(241.9826)	($< 10^{-4}$)	(0.0138)	(0.0068)	(56.5102)	(60.2510)
PS	9.5226×10^4	0.56	2.4904	0.7213	266.5477	371.2440
($\xi = 10^{-5}$)	(891.984)	(0.0067)	(0.0268)	(0.0104)	(6.8292)	(8.2292)

stability of the algorithm. For instance, a small MSE can be interpreted to indicate that the algorithm will produce a reliable output with a given expected statistical efficiency. The performance of the single and parallel chain implementation of the MHA, the MHA with the RP, the DRA using three moves for the second-step proposal, the MALA using a traditional proposal and smoothed Langevin proposal, and the PS is summarized in Table 24.1 and Table 24.2 for two different sizes of scaling parameters, $s = 4, 2$ and $h = 4, 2$, respectively. All chains converged to the target reasonably well based on informal diagnostics. The performance of each algorithm is discussed below.

MHA: It is noticeable that due to the parallel computing the MHA takes the shortest CPU time per iteration and produces the largest sample size, 20 times more proposals than the single chain, MHA^s. The improvement in bias of samples from the parallel chains is not observable compared with a single chain. To optimize the performance, the proposal variance s is usually tuned such that $A \approx 0.3$ (Roberts and Rosenthal 2001) and for this example $s = 4$ is the optimal value.

MALA: It is known that the MALA often performs poorly for multimodal problems, as the chain is likely pulled back towards the nearest mode and may become stuck as can be seen in Skare *et al.* (2000). This is observed by a larger MSE of measures compared with other MCMC algorithms in general. When h is too large, samples are easily trapped in a certain part of the state space and τ_f is very large. For a smaller h, a very small τ_f is induced and the MSE of $\hat{H}$ remains relatively large.

Table 24.2 Comparison of performance in the two-dimensional example with the proposal variance $s = 2$. The values in parentheses are the MSE of measures.

	$\overline{T}$	$\bar{A}$	$\bar{H}$	$\bar{\sigma}_H^2$	$\bar{\tau}_\pi(H)$	$\overline{ESS}_H$
True value			2.5			
MHA^s	1.037×10^6	0.43	2.4999	7.2473	770.6577	1339.7
	(346.7434)	$(< 10^{-4})$	(0.0053)	(0.0014)	(1.7493)	(3.0647)
MHA	2.481×10^6	0.43	2.4992	0.7253	872.3864	2845.7
	(2.52×10^4)	$(< 10^{-4})$	(0.0012)	(0.0009)	(2.5346)	(29.9846)
MHA^{RP}	1.2294×10^5	0.42	2.4952	0.7230	786.3059	157.7317
$(\xi = 10^{-5})$	(1.37×10^3)	(0.0016)	(0.0062)	(0.0044)	(9.7029)	(2.3711)
DRA	1.0012×10^5	0.63	2.4983	0.7148	700.9815	1429.6
	(41.2523)	$(< 10^{-4})$	(0.0015)	(0.0013)	(2.6121)	(5.2876)
DRA^{LP}	2.2856×10^5	0.61	2.5021	0.7235	553.5437	413.6313
$(h = 2)$	(1.77×10^3)	$(< 10^{-4})$	(0.0031)	(0.0024)	(3.5946)	(4.0817)
DRA^{Pinball}	6.8763×10^5	0.68	2.4990	0.7218	674.7575	1020.0
$(\xi = 10^{-5})$	(5.21×10^3)	$(< 10^{-4})$	(0.0021)	(0.0015)	(3.0249)	(8.6733)
MALA	1.9467×10^5	0.67	2.4985	0.6983	37.9879	5738.4
$(h = 2)$	(1.51×10^3)	$(< 10^{-4})$	(0.0166)	(0.0131)	(1.340)	(196.05)
PS	9.3550×10^4	0.64	2.4971	0.7148	679.0638	141.1047
$(\xi = 10^{-5})$	(972.6976)	(0.0061)	(0.0240)	(0.0074)	(11.1321)	(2.6741)

MHA with RP (MHA^{RP}): It can be seen that the repulsive effect induces a fast mixing chain. The RP reduces the autocorrelation along the chain by pushing particles apart around a neighbourhood. Its effect is more obvious when the proposal variance is relatively small. The RP imposes an additional tuning parameter ξ on the MHA^{RP} compared with the MHA. The algorithm is illustrated with $\xi = 10^{-4}$, and approximately 99.5% of proposals that are accepted with π^R are accepted with π. The choice of ξ is critical in implementing the RP and we will discusses this matter later. The drawback of the MHA^{RP} is that it is expensive to compute. For D-dimensional problems with N particles and an arbitrary $N \times D$ matrix $\theta^{(t)}$ the repulsive part of a pseudo distribution adds $O(ND)$ operations. This algorithm also can be unstable and, as for the MALA, the MSE of measures tends to be large in some circumstances. As an illustration, even with ξ carefully tuned, we observed six exceptional measures out of 100 replicates using $s = 2, 4$. A very low rate of acceptance, a high τ_π and an extremely small ESS indicate that the chain is stuck in a certain region and does not adequately explore the state space. However, these instances occurred relatively rarely.

DRA: By making the second attempt to move instead of remaining in the current state, algorithms with a delayed rejection mechanism produce less correlated samples and have a higher rate of acceptance. The demand in computation is also increased and differs with different types of moves. We examined three types of moves for the second-step proposal: the normal random walk as for the first proposal (DRA), the geometric deterministic proposal (DRA^{GP}) and the Langevin proposal (DRA^{LP}).

With the geometric deterministic proposals, over 60% of proposed moves were accepted regardless of the value of the proposal variance. The geometric deterministic proposal pushes away from the closest particle; however, with a moderate repulsive probability proposed particles may remain in the neighbourhood of existing particles, in which case it is highly likely to be accepted. Generally the Langevin proposal seems a suitable choice for the second-stage proposal in terms of the improvement in the mixing speed of chains when the proposal variance is not relatively large. However, it requires a greater amount of computation than other types of proposal, and despite a diminished τ_{π} the ESS is relatively small. In this case the loss in the computational efficiency overwhelms the gain in the statistical efficiency. Given the nature of the Langevin diffusion, tuning of h is essential. The normal random walk for a second-stage proposal is faster to compute and results in improved mixing with a sufficiently flexible choice of the proposal variance. As seen in Table 24.1, when particles can move around the state space efficiently, neither the deterministic proposal nor the Langevin diffusion approximation will induce a substantive reduction in the correlation of the samples.

PS: The PS produces less correlated samples than the MHA and has a high rate of acceptance due to the DRA and repulsive effect. When the size of random walk is sufficiently large, the standard normal random walk proposal is better than the repulsive effect, and this concurs with the simulation result of the MHA^{RP}. It seems that the statistical efficiency of the PS is mostly affected by the DRA. The major drawback of the PS is a large computational workload due to the repulsive effect term, and the smallest number of samples among the MCMC simulations is produced for the same computational cost.

Ability to detect modes

We now consider a burn-in period in which chains are still dependent on initial states. We examine the selected hybrid approaches to study how quickly chains move from initial values and find other local modes. We run algorithms in the same manner with initial particles from one particular mode $\mathbf{N}([0, 0], I_2)$, $s = 2$ for the random walk, and $h = 2$ for the Langevin proposal. We considered the total number of trials in which the second mode $\mathbf{N}([5, 5], I_2)$ was detected within the first 50 iterations during 400 repeated simulations. The allocation of particles was determined by the distance from the centre of the mode. Approximately 86.5% of the MALA simulations failed to detect both modes; 72.5% of the DRA^{LP} trials found the second mode within 50 iterations, compared with only 50.8% for the MHA.

24.3.2 Application: Aerosol particle size

The data set y containing 2000 observations is described with a mixture of two normal distributions

$$\lambda \mathbf{N}(\mu_1, \sigma_1^2) + (1 - \lambda)\mathbf{N}(\mu_2, \sigma_2^2)$$

where $0 \leq \lambda \leq 1$. The unknown parameters ($\mu_1, \mu_2, \sigma_1, \sigma_2$ and λ) are estimated using the missing data representation (Marin and Robert 2007) with relatively vague priors

$$\mu_1, \mu_2 \sim \mathbf{N}(Mean(y), Var(y)), \qquad \sigma_1, \sigma_2 \sim \mathbf{G}(2, 2), \qquad \lambda \sim U[0, 1].$$

In light of the high dimensionality of the problem we use block updating in which an acceptance probability is based on the ratio of the full conditional distributions. In addition, this allows the use of the geometrical deterministic proposal.

The MHA and four hybrid algorithms (the MALA, DRA, DRA with the Langevin proposal and PS) were run using 10 particles for 4200 seconds. The sets of the parameters ($[\mu_1, \mu_2]$, $[\sigma_1, \sigma_2]$ and λ) were updated in turn. Details of implementing the algorithms are as follows.

MHA and DRA: Proposals are generated using the normal random walk with a covariance matrix of $25 \times 10^{-4} I_2$ for $[\mu_1, \mu_2]$, $2.25 \times 10^{-4} I_2$ for $[\sigma_1, \sigma_2]$ and a variance of 10^{-4} for λ. These updates are used for DRA^{LP} and PS as the first proposal scheme.

MALA and DRA with Langevin proposal: At the second stage $[\mu_1, \mu_2]$, $[\sigma_1, \sigma_2]$ and λ are updated using the Langevin proposal with h_μ, h_σ and h_λ respectively,

$$h_\mu = 25 \times 10^{-4} I_2\,, \qquad h_\sigma = 2.25 \times 10^{-4} I_2\,, \qquad h_\lambda = 10^{-4}\,.$$

PS: The two-dimensional updates $[\mu_1, \mu_2]$ and $[\sigma_1, \sigma_2]$ are implemented using the PS algorithm and λ is sampled using the MHA. Based on the ratio of approximate normalizing constant test, the tempering factor $\xi = 10^{-48}$ is chosen for both μ and σ.

The results of the MCMC simulations are summarized in Table 24.3. Since the two modes are fairly clearly separated, label switching was not observed in any of the simulations (Geweke 2007). Overall the estimates of the five parameters were very similar and the 95% credible intervals of parameters obtained from all algorithms overlapped each other. The $\bar{\tau}_\pi$ denotes the average of the τ_π of all parameters and $ESS = (T - T_0)/\bar{\tau}_\pi$ ($T_0 = 1000$).

In general, the features of the hybrid algorithms shown from this simulation are similar to those observed using the toy example in Section 24.3.1. The MHA demands the least CPU time per iteration. The Langevin proposal was very effective in reducing τ_f and the DRA did not improve the mixing as it did in the toy example. By the property

Table 24.3 Performance of selected hybrid algorithms (MALA, DRA^{LP} and DRA) and the MHA.

	T	$\bar{\tau}_\pi$	ESS
MALA	46 071	65.6773	686.2
DRA^{LP}	87 613	114.1824	758.5
DRA	125 020	130.7077	948.8
MHA	203 118	185.4259	1090.0

Table 24.4 The relative performance rating of the hybrid algorithms to the MHA and PMC.

Algorithm	Statistical eff.			Computation		Applicability		
	EPM	CR	RC	CE	SP	FH	CP	Mode
MALA	1	1	1	−2	−1	0	−1	Single
MHA^{RP}	0	1	0	−2	−1	−1	−2	Both
DRA	1	1	0	−1	−1	0	0	Both
DRA^{LP}	2	1	0	−2	−1	0	0	Both
DRA^{GP}	2	1	0	−2	−1	0	0	Both

of the Langevin diffusion, particles are pushed back to the mode and consequently can explore a single mode more quickly.

24.4 Discussion

In this chapter we have considered the problem of designing hybrid algorithms by studying selected hybrid methods: the DRA with three types of move for the second-stage proposal, the MALA, and the MHA with RP. Using a two-dimensional example, each algorithm was evaluated and the relative contributions of individual components to the overall performance of the hybrid algorithm were estimated. The performance was defined by the accuracy of estimation, the efficiency of the proposal move, the demand in computation, the rate of mixing of the chains, and the ability of the algorithm to detect modes in a special setup. The algorithms were also applied to a real problem of describing an aerosol particle size distribution. We observed that the results of this analysis largely coincided with those obtained using a toy example.

Based on these simulation studies, we subjectively rated the performance of the hybrid MCMC algorithms relative to the MHA. The results are presented in Table 24.4. Following the three criteria for an efficient algorithm defined in Section 24.1, we considered the efficiency of the proposal move (EPM), the correlation reduction of a chain (CR), the rate of convergence (RC), the cost effectiveness (CE), the simplicity of programming (SP), the flexibility of hyperparameters (FH), the consistency of performance (CP) and the preference between a single mode and a multimodal problem (Mode). As the algorithm improves with respect to the criterion, the rating increases positively, with zero indicating that there is no substantive difference to the MHA. It can be seen that the DRA and DRA^{LP} are sensible algorithms to use for multimodal problems and the MALA for unimodal problems.

We identified the following overall issues to be considered in designing hybrid algorithms:

1. Combining features of individual algorithms may lead to complicated characteristics in a hybrid algorithm. For instance, it was seen that the performance of some algorithms was sensitive with respect to the value of the tuning parameters such as s, h and ξ, and this also applied to the hybrid algorithms. For the MHA with the

repulsive effect the improvements from the statistical perspective were sensitive to both the size of the normal random walk (a well-known property of the MHA) and the tempering factor. Moreover, some inconsistency of estimates throughout replicates occurred, albeit rarely.

2. Each individual algorithm may have a strong individual advantage with respect to a particular performance criterion, but this does not guarantee that the hybrid method will enjoy a joint benefit of these strategies. The contribution of some components may be insignificant, and certain components may dominate the character of the hybrid method. This can be seen with hybrid versions of the DRA as the delayed rejection scheme most strongly contributes to statistical efficiency.
3. From the perspectives of applicability and implementation, the combination of algorithms may add complexity in setup, programming and computational expense. These phenomena are easily observed as all hybrid approaches considered in this chapter were computationally demanding, although the magnitude of the demand differed with the types of move and techniques. In practice it is important to be aware of these issues to optimize the performance in real time so that the improvement in one criterion does not become negligible due to the drawback in computation. This was observed, for example, with the DRA^{LP}.

These general considerations in building an efficient algorithm are easily applied to existing hybrid methods. The adaptive MCMC algorithm is another good example. Its motivation is to automate and improve the tuning of the algorithm by learning from the history of the chain itself. It can be easily coded and is statistically efficient without a strenuous increase in computational expense. Roberts and Rosenthal (2007) and Rosenthal (2008) demonstrated that an algorithm can learn and approach an optimal algorithm via automatic tuning of scaling parameters to optimal values. After a sufficiently long adaptation period it will converge much faster than a non-adapted algorithm. However, this fast convergence feature increases the risk that chains may converge to the wrong values (Robert and Casella 2004; Rosenthal 2008). Moreover, since proposals are accepted using the history of the chain, the sampler is no longer Markovian and standard convergence techniques cannot be used. The adaptive MCMC algorithms converge only if the adaptations are done at regeneration times (Brockwell and Kadane 2005; Gilks *et al.* 1998) or under certain technical types of the adaptation procedure (Andrieu and Moulines 2006; Atchadé and Rosenthal 2005; Haario *et al.* 2001; Roberts and Rosenthal 2007). Overall the adaptive MCMC algorithm can be efficient from the statistical perspective and cost effective only when it is handled with care. In other words, these advantages come with a reduction in robust reliability which may affect its applicability.

References

Andrieu C and Moulines E 2006 On the ergodicity properties of some adaptive MCMC algorithms. *Annals of Applied Probability* **16**(3), 1462–1505.

Andrieu C, Doucet A and Holenstein R 2010 Particle Markov chain Monte Carlo methods. *Journal of the Royal Statistical Society, Series B* **72**(3), 269–342.

Atchadé Y 2006 An adaptive version for the Metropolis adjusted Langevin algorithm with a truncated drift. *Methodology and Computing in Applied Probability* **8**, 235–254.

Atchadé Y and Rosenthal JS 2005 On adaptive Markov chain Monte Carlo algorithms. *Methodology and Computing in Applied Probability* **11**(5), 815–828.

Besag J, Green P, Hidgon D and Mengersen K 1995 Bayesian computation and stochastic systems. *Statistical Science* **10**(1), 3–41.

Besag JE and Green PJ 1993 Spatial statistics and Bayesian computation. *Journal of the Royal Statistical Society, Series B* **55**, 25–38.

Breyer LA and Roberts GO 2002 New method for coupling random fields. *LMS Journal of Computation and Mathematics* **5**, 77–94.

Brockwell AE and Kadane JB 2005 Identification of regeneration times in MCMC simulation, with application to adaptive schemes. *Journal of Computational and Graphical Statistics* **14**, 436–458.

Cappé O, Guillin A, Marin JM and Robert CP 2004 Population Monte Carlo. *Journal of Computational and Graphical Statistics* **13**(4), 907–929.

Celeux G, Hurn M and Robert CP 2000 Computational and inference difficulties with mixture posterior distributions. *Journal of the American Statistical Association* **95**, 957–970.

Del Moral PD, Doucet A and Jasra A 2006 Sequential Monte Carlo samplers. *Journal of the Royal Statistical Society, Series B* **68**, 411–436.

Doll JD, Rossky PJ and Friedman HL 1978 Brownian dynamics as smart Monte Carlo simulation. *Journal of Chemical Physics* **69**, 4628–4633.

Doucet R, Godsill S and Andrieu C 2000 On sequential Monte Carlo sampling methods for Bayesian filtering. *Statistics and Computing* **10**, 197–208.

Doucet R, Guillin A, Marin JM and Robert C 2007 Minimum variance importance sampling via population Monte Carlo. *ESAIM: Probability and Statistics* **11**, 427–447.

Gelman, A, Gilks WR and Roberts GO 1997 Weak convergence and optimal scaling of random walk Metropolis algorithms. *Annals of Applied Probability* **7**(1), 110–120.

Geman S and Geman D 1984 Stochastic relaxation, Gibbs distributions, and the Bayesian restoration of images. *IEEE Transactions on Pattern Analysis and Machine Intelligence* **PAMI-6**, 721–741.

Geweke J 2007 Interpretation and inference in mixture models: simple MCMC works. *Computational Statistics & Data Analysis* **51**, 3529–3350.

Geyer CJ 1992 Practical Markov chain Monte Carlo. *Statistical Science* **7**(4), 473–483.

Gilks WR, Roberts GO and Sahu SK 1998 Adaptive Markov chain Monte Carlo. *Journal of the American Statistical Association* **93**, 1045–1054.

Green PJ 2001 Reversible jump Markov chain Monte Carlo computation and Bayesian model determination. *Biometrika* **82**(4), 711–732.

Green PJ and Mira A 2001 Delayed rejection in reversible jump Metropolis-Hastings. *Biometrika* **88**, 1035–1053.

Haario H, Saksman E and Tamminen J 2001 An adaptive Metropolis algorithm. *Bernoulli* **7**, 223–242.

Hastings WK 1970 Monte Carlo sampling methods using Markov chains and their applications. *Biometrika* **57**, 97–109.

Jarner SF and Hansen E 2000 Geometric ergodicity of Metropolis algorithms. *Stochastic Processes and their Applications* **21**, 341–361.

Li B, Bengtsson T and Bickel P 2005 Curse-of-dimensionality revisited: collapse of importance sampling in very large scale systems. Technical Report 696, Department of Statistics, UC-Berkeley.

Marin JM and Robert CP 2007 *Bayesian Core: A Practical Approach to Computational Bayesian Statistics*. Springer, Berlin.

Mengersen KL and Robert CP 2003 IID sampling with self-avoiding particle filters: the pinball sampler. *Bayesian Statistics 7* (eds JM Bernardo *et al.*). Oxford University Press, Oxford.

Mengersen KL and Tweedie RL 1996 Rates of convergence of the Hastings and Metropolis algorithms. *Annals of Statistics* **24**(1), 101–121.

Metropolis N, Rosenbluth AW, Rosenbluth MN, Teller AH and Teller E 1953 Equations of state calculations by fast computing machines. *Journal of Chemical Physics* **21**, 1087–1091.

Mira A 2001 On Metropolis-Hastings algorithms with delayed rejection. *Metron – International Journal of Statistics* **59**, 231–241.

Mira A, Bressanini D, Morosi G and Tarasco S 2004 Delayed rejection variational Monte Carlo. Technical Report 2, Faculty of Economics of the University of Insubria.

Nilsson ED and Kulmala M 2006 Aerosol formation over the boreal forest in Hyytiälä, Finland: monthly frequency and annual cycles – the roles of air mass characteristics and synoptic scale meteorology. *Atmospheric Chemistry and Physics Discussions* **6**, 10425–10462.

Robert C and Casella G 2004 *Monte Carlo Statistical Methods*, 2nd edn. Springer, Berlin.

Roberts GO and Rosenthal JS 2001 Optimal scaling for various Metropolis-Hastings algorithms. *Statistical Science* **16**(4), 351–367.

Roberts GO and Rosenthal JS 2007 Coupling and ergodicity of adaptive Markov chain Monte Carlo algorithms. *Journal of Applied Probability* **44**(2), 458–475.

Roberts GO and Stramer O 2002 Langevin diffusion and Metropolis-Hastings algorithms. *Methodology and Computing Applied Probability* **4**, 337–357.

Roberts GO and Tweedie RL 1996a Exponential convergence of Langevin diffusions and their discrete approximations. *Bernoulli* **2**, 341–364.

Roberts GO and Tweedie RL 1996b Geometric convergence and central limit theorems for multi-dimensional Hastings and Metropolis algorithms. *Biometrika* **83**, 95–110.

Rosenthal JS 2008 Adaptive Markov chain Monte Carlo algorithms. *International Society for Bayesian Analysis, 9th World Meeting*, Keynote Talk.

Skare O, Benth FE and Frigessi A 2000 Smoothed Langevin proposals in Metropolis-Hastings algorithms. *Statistics & Probability Letters* **49**, 345–354.

Stramer O and Tweedie RL 1999a Langevin-type models I: Diffusions with given stationary distributions and their discretization. *Methodology and Computing in Applied Probability* **1**, 283–306.

Stramer O and Tweedie RL 1999b Langevin-type models II: Self-targeting candidates for MCMC algorithms. *Methodology and Computing in Applied Probability* **1**(3), 307–328.

Tierney L 1994 Markov chains for exploring posterior distributions. *Annals of Statistics* **22**, 1701–1762.

Tierney L and Mira A 1999 Some adaptive Monte Carlo methods for Bayesian inference. *Statistics in Medicine* **18**, 2507–2515.

25

A Python package for Bayesian estimation using Markov chain Monte Carlo

Christopher M. Strickland[1], Robert J. Denham[2], Clair L. Alston[1] and Kerrie L. Mengersen[1]

[1]Queensland University of Technology, Brisbane, Australia
[2]Department of Environment and Resource Management, Brisbane, Australia

25.1 Introduction

The most common approach currently used in the estimation of Bayesian models is Markov chain Monte Carlo (MCMC). **PyMCMC** is a Python module that is designed to simplify the construction of Markov chain Monte Carlo (MCMC) samplers, without sacrificing flexibility or performance. Python has extensive scientific libraries, is easily extensible, and has a clean syntax and powerful programming constructs, making it an ideal programming language to build an MCMC library; see van Rossum (1995) for further details on the programming language Python. **PyMCMC** contains objects for the Gibbs sampler, Metropolis–Hastings (MH), independent MH, random walk MH, orientational bias Monte Carlo (OBMC) as well as the slice sampler; see for example Robert and Casella (1999) for details on standard MCMC algorithms. The user can simply piece together the algorithms required and can easily include their own modules, where necessary. Along with the standard algorithms,

Case Studies in Bayesian Statistical Modelling and Analysis, First Edition. Edited by Clair L. Alston, Kerrie L. Mengersen and Anthony N. Pettitt.

PyMCMC includes a module for Bayesian regression analysis. This module can be used for the direct analysis of linear models, or as a part of an MCMC scheme, where the conditional posterior has the form of a linear model. It also contains a class that can be used along with the Gibbs sampler for Bayesian variable selection.

The flexibility of **PyMCMC** is important in practice, as MCMC algorithms usually need to be tailored to the problem of interest in order to ensure good results. Issues such as block size and parameterization can have a dramatic effect on the convergence of MCMC sampling schemes. For instance, Lui *et al.* (1994) show theoretically that jointly sampling parameters in a Gibbs scheme typically leads to a reduction in correlation in the associated Markov chain in comparison with individually sampling parameters. This is demonstrated in practical applications in Carter and Kohn (1994) and Kim *et al.* (1998). Reducing the correlation in the Markov chain enables it to move more freely through the parameter space and as such enables it to escape from local modes in the posterior distribution. Parameterization can also have a dramatic effect on the convergence of MCMC samplers; see for example Gelfand *et al.* (1995), Sahu and Roberts (1997), Pitt and Shephard (1999), Robert and Mengersen (1999), Schnatter (2004) and Strickland *et al.* (2008), who show that the performance of the sampling schemes can be improved dramatically with the use of efficient parameterization.

PyMCMC aims to remove unnecessary repetitive coding and hence reduce the chance of coding error, and, importantly, greatly speed up the construction of efficient MCMC samplers. This is achieved by taking advantage of the flexibility of `Python`, which allows for the implementation of very general code. Another feature of `Python`, which is particularly important, is that it is also extremely easy to include modules from compiled languages such as `C` and `Fortran`. This is important to many practitioners who are forced, by the size and complexity of their problems, to write their MCMC programs entirely in compiled languages, such as `C/C++` and `Fortran`, in order to obtain the necessary speed for feasible practical analysis. With`Python`, the user can simply compile `Fortran` code using a module called **F2py** (Peterson 2009), or inline `C` using **Weave**, which is a part of **Scipy** (Oliphant 2007), and use the subroutines directly from `Python`. **F2py** can also be used to directly call `C` routines with the aid of a `Fortran` signature file. This enables the use of **PyMCMC** and `Python` as a rapid application development environment, without compromising on performance, by requiring only very small segments of code written in a compiled language. It should be mentioned that for most reasonably sized problems **PyMCMC** is sufficiently fast for practical MCMC analysis without the need for specialized modules.

Figure 25.1 is a flow chart that depicts the structure of **PyMCMC**. Essentially, the implementation of an MCMC sampler can be seen to centre around the class `MCMC`, which acts as a container for various algorithms that are used in sampling from the conditional posterior distributions that make up the MCMC sampling scheme.

The structure of the chapter is as follows. In Section 25.2 the algorithms contained in **PyMCMC** and the user interface are described. This includes the Gibbs sampler, the Metropolis-based algorithms and the slice sampler. Section 25.2 also contains a description of the Bayesian regression module. Section 25.3 contains three empirical examples that demonstrate how to use **PyMCMC**. The first example demonstrates how to use the regression module for the Bayesian analysis of the linear model. In

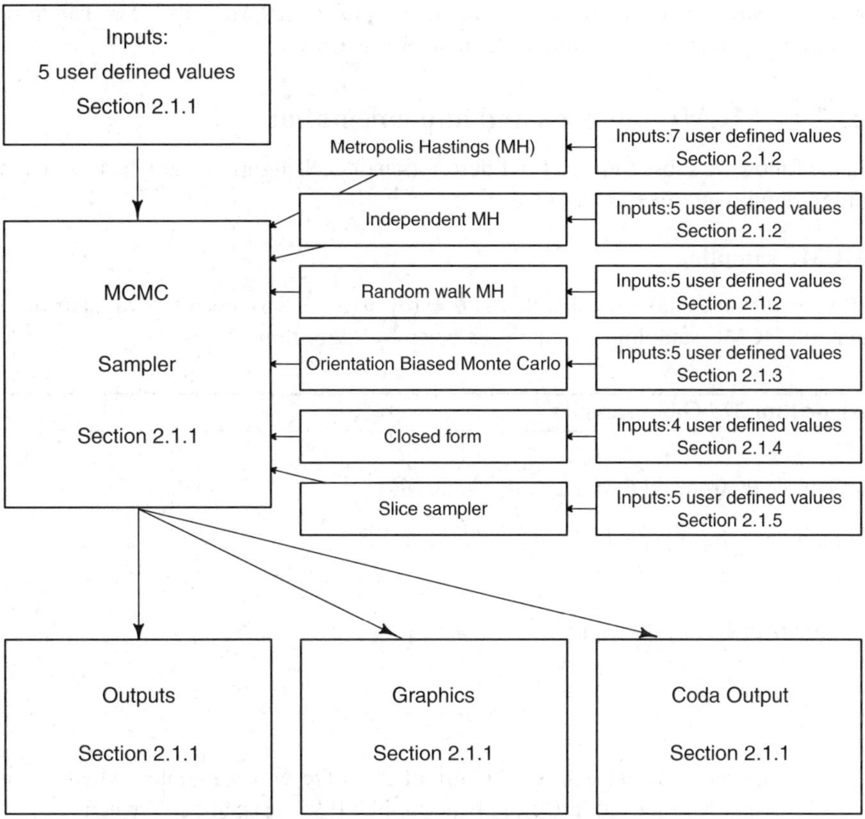

Figure 25.1 Flow chart illustrating the implementation of **PyMCMC**.

particular, the stochastic search variable selection algorithm, see George and McCulloch (1993) and Marin and Robert (2007), is used to select a set of 'most likely models'. The second example demonstrates how to use **PyMCMC** to analyse the log-linear model and the third example demonstrates how to use **PyMCMC** to analyse a linear model with first-order autoregressive errors. Section 25.4 contains a discussion on the efficient implementation of the MCMC algorithms using **PyMCMC**. Section 25.5 describes how to use **PyMCMC** interactively with R and Section 25.6 concludes.

25.2 Bayesian analysis

Bayesian analysis quantifies information about the unknown parameter vector of interest, $\boldsymbol{\theta}$, for a given data set, $\boldsymbol{y}$, through the joint posterior probability density function (pdf), $p(\boldsymbol{\theta}|\boldsymbol{y})$, which is defined such that

$$p(\boldsymbol{\theta}|\boldsymbol{y}) \propto p(\boldsymbol{y}|\boldsymbol{\theta}) \times p(\boldsymbol{\theta}), \tag{25.1}$$

where $p(\boldsymbol{y}|\boldsymbol{\theta})$ denotes the pdf of $\boldsymbol{y}$ given $\boldsymbol{\theta}$ and $p(\boldsymbol{\theta})$ is the prior pdf for $\boldsymbol{\theta}$. The most common approach used for inference about $\boldsymbol{\theta}$ is MCMC.

25.2.1 MCMC methods and implementation

In the following subsections, a brief description of each algorithm and the associated programming interface is included.

MCMC sampling

If we partition $\boldsymbol{\theta}$ into s blocks, that is $\boldsymbol{\theta} = (\boldsymbol{\theta}_1, \boldsymbol{\theta}_2, \ldots, \boldsymbol{\theta}_s)^T$, then the jth step for a generic MCMC sampling scheme is as given by Algorithm 27.

Algorithm 27: Gibbs sampler

1. Sample $\boldsymbol{\theta}_1^j$ from $p\left(\boldsymbol{\theta}_1|\boldsymbol{y},\boldsymbol{\theta}_2^{j-1}, \boldsymbol{\theta}_3^{j-1}, \ldots, \boldsymbol{\theta}_s^{j-1}\right)$.
2. Sample $\boldsymbol{\theta}_2^j$ from $p\left(\boldsymbol{\theta}_2|\boldsymbol{y},\boldsymbol{\theta}_1^{j}, \boldsymbol{\theta}_3^{j-1}, \boldsymbol{\theta}_4^{j-1}, \ldots, \boldsymbol{\theta}_s^{j-1}\right)$.

   ⋮

s. Sample $\boldsymbol{\theta}_s^j$ from $p\left(\boldsymbol{\theta}_s|\boldsymbol{y},\boldsymbol{\theta}_1^{j}, \boldsymbol{\theta}_2^{j}, \ldots, \boldsymbol{\theta}_{s-1}^{j}\right)$.

An important special case of Algorithm 27 is the Gibbs sampler, which is an algorithm that is proposed in Gelfand and Smith (1990). Specifically, when each of $\boldsymbol{\theta}_1, \boldsymbol{\theta}_2, \ldots, \boldsymbol{\theta}_s$ is sampled from a closed form then this algorithm corresponds to that of the Gibbs sampler. **PyMCMC** contains a class that facilitates the implementation of Algorithm 27, in which the user must define functions to sample from each block, that is a function for each of $\boldsymbol{\theta}_i$, for $i = 1, \ldots, s$. These functions may be defined using the Metropolis-based or slice sampling algorithms that are part of **PyMCMC**. The class is named `MCMC` and the following arguments are required in the initialization of the class:

`nit`: The number of iterations.
`burn`: The burn-in length of the MCMC sampler.
`data`: A dictionary (`Python` data structure) containing any data, functions or objects that the user would like to have access to when defining the functions that are called from the Gibbs sampler.
`blocks`: A list (`Python` data structure) containing functions that are used to sample from the full conditional posterior distributions of interest.
`**kwargs` Optional arguments:
`loglike` A tuple (a `Python` data structure) containing a function that evaluates the log-likelihood, number of parameters and the name of the data set. For example, `loglike = (loglike, nparam, 'yvec')`. If this is defined

then the log-likelihood and the Bayesian information criterion (BIC) will be reported in the standard output.

`transform` A dictionary, where the keys are the names of the parameters and the associated values are functions that transform the iterates stored in the MCMC scheme. This can be useful when the MCMC algorithm is defined under a particular parameterization, but where it is desirable to report the results under a different parameterization.

Several functions are included as a part of the class:

`sampler()`: Used to run the MCMC sampler.

`get_mean_cov(listname)`: Returns the posterior covariance matrix for the parameters named in `listname`, where `listname` is a list that contains the parameter names of interest.

`get_parameter(name)`: Returns the iterates for the named parameter including the burn-in.

`get_parameter_exburn(name)`: Returns the iterates for the named parameter excluding the burn-in.

`get_mean_var(name)`: Returns the estimate from the MCMC estimation for the posterior mean and variance for the parameter defined by `name`.

`set_number_decimals(num)`: Sets the number of decimal places for the output.

`output(**kwargs)`: Used to produce output from the MCMC algorithm.

`**kwargs`: Optional arguments that control the output.

`parameters`: A dictionary, list or string specifying the parameters that are going to be presented.

- If a string is passed (e.g. `parameters = 'beta'`), all elements of that parameter are given.
- If a list (e.g. `parameters = ['alpha', 'beta']`), all elements of each parameter in the list are given.
- If a dictionary (e.g. `parameters = {'alpha': {'range': range(5)}}`), then there is the possibility to add an additional argument `'range'` that tells the output to only print a subset of the parameters. The above example will print information for `alpha[0], alpha[1], ..., alpha[4]` only.

`custom`: A user-defined function that produces custom output.

`filename`: A filename to which the output is printed. By default output will be printed to `stdout`.

`plot(blockname, **kwargs)`: Create summary plots of the MCMC sampler. By default, a plot of the marginal posterior density, an autocorrelation function (ACF) plot and a trace plot are produced for each parameter in the block. The plotting page is divided into a number of subfigures. By default, the number of columns is approximately equal to the square root of the total number of subfigures divided by the number of different plot types. Arguments to `plot` are:

`blockname`: The name of the parameter, for which summary plots are to be generated.

`**kwargs`: An optional dictionary (`Python` data structure) containing information to control the summary plots. The available keys are summarized below:

`elements`: A list of integers specifying the elements that will be plotted. For example, if the blockname is `'beta'` and $\boldsymbol{\beta} = (\beta_0, \beta_1, \ldots, \beta_n)$ then you may specify elements as `elements = [0,2,5]`.

`plottypes`: A list giving the type of plot for each parameter. By default the plots are `'density'`, `'acf'` and `'trace'`. A single string is also acceptable.

`filename`: A string providing the name of an output file for the plot. As a plot of a block may be made up of a number of subfigures, the output name will be modified to give a separate filename for each subfigure. For example, if the filename is passed as `'plot.png'`, and there are multiple pages of output, it will produce the files `plot001.png`, `plot002.png`, etc. The type of file is determined by the extension of the filename, but the output format will also depend on the plotting backend being used. If the filename does not have a suffix, a default format will be chosen based on the graphics backend. Most backends support png, pdf, ps, eps and svg (see the documentation for **Matplotlib** for further details: http://matplotlib.sourceforge.net).

`individual`: A Boolean option. If true, then each subplot will be done on an individual page.

`rows`: Integer specifying the number of rows of subfigures on a plotting page.

`cols`: Integer specifying the number of columns of subfigures on a plotting page.

`CODAoutput(**kwargs)`: Outputs the results in a format suitable for reading in with the statistical package Convergence Diagnostic and Output Analysis (**CODA**) (Plummer *et al.* 2006). By default, there will be two files created, `coda.txt` and `coda.ind`.

`**kwargs`: An optional dictionary controlling the **CODA** output.

`filename`: A string to provide an alternative filename for the output. If the file has an extension this will form the basis for the data file and the index file will be named by replacing the extension with `ind`. If no extension is in the filename then two files will be created and named by adding the extensions `.txt` and `.ind` to the given filename.

`parameters`: A string, a list or a dictionary that specifies the items written to file. It can be a string such as `'alpha'` or it can be a list (e.g., `['alpha','beta']`) or it can be a dictionary (e.g. `{'alpha':{'range':[0,1,5]}}`. If you supply a dictionary the key is the parameter name. It is also permissible to have a range key with a range of elements. If the range is not supplied it is assumed that the user wants all of the elements.

`thin`: Integer specifying how to thin the output. For example, if `thin = 10`, then every 10th element will be written to the **CODA** output.

MH

A particularly useful algorithm that is often used as a part of MCMC samplers is the MH algorithm (Algorithm 28); see for example Robert and Casella (1999). This algorithm is usually required when we cannot easily sample directly from $p(\boldsymbol{\theta}|\mathbf{y})$; however, we have a candidate density $q(\boldsymbol{\theta}|\mathbf{y}, \boldsymbol{\theta}^{j-1})$, which in practice is close to $p(\boldsymbol{\theta}|\mathbf{y})$ and is more readily able to be sampled. The MH algorithm at the jth iteration for $j = 1, 2, \ldots, M$ is given by the steps in Algorithm 28.

Algorithm 28: Metropolis–Hastings

---

1. Draw a candidate $\boldsymbol{\theta}^*$ from the density $q\left(\boldsymbol{\theta}|\mathbf{y}, \boldsymbol{\theta}^{j-1}\right)$.
2. Accept $\boldsymbol{\theta}^j = \boldsymbol{\theta}^*$ with probability equal to

$$\min\left\{1, \frac{p\left(\boldsymbol{\theta}^*|\mathbf{y}\right)}{p\left(\boldsymbol{\theta}^{j-1}|\mathbf{y}\right)} \Big/ \frac{q\left(\boldsymbol{\theta}^*|\mathbf{y}, \boldsymbol{\theta}^{j-1}\right)}{q\left(\boldsymbol{\theta}^{j-1}|\mathbf{y}, \boldsymbol{\theta}^*\right)}\right\}.$$

3. Otherwise $\boldsymbol{\theta}^j = \boldsymbol{\theta}^{j-1}$.

PyMCMC includes a class for the MH algorithm, which is called `MH`. To initialize the class the user needs to define the following:

`func`: User-defined function that returns a sample for the parameter of interest.
`actualprob`: User-defined function that returns the log probability of the parameters of interest evaluated using the target density.
`probcandprev`: User-defined function that returns the log of $q(\boldsymbol{\theta}^*|\mathbf{y}, \boldsymbol{\theta}^{j-1})$.
`probprevcand`: User-defined function that returns the log of $q(\boldsymbol{\theta}^{j-1}|\mathbf{y}, \boldsymbol{\theta}^*)$.
`init_theta`: Initial value for the parameters of interest.
`name`: The name of the parameter of interest.
`**kwargs`: Optional arguments:
- `store`
  - `'all'` (default) – stores every iterate for the parameter of interest. This is required for certain calculations.
  - `'none'` – does not store any of the iterates from the parameter of interest.
- `fixed_parameter` – is used if the user wants to fix the parameter value that is returned. This is used for testing MCMC sampling schemes. This command will override any other functionality.

Independent MH

The independent MH is a special case of the MH described in Algorithm 28. Specifically, the independent MH algorithm is applicable when we have a candidate density $q(\boldsymbol{\theta}|\mathbf{y}) = q(\boldsymbol{\theta}|\mathbf{y}, \boldsymbol{\theta}^{j-1})$. The independent MH algorithm at the jth iteration for $j = 1, 2, \ldots, M$ is given by Algorithm 29.

Algorithm 29: Independent MH algorithm

1. Draw a candidate $\boldsymbol{\theta}^*$ from the density $q(\boldsymbol{\theta}|\mathbf{y})$.
2. Accept $\boldsymbol{\theta}^j = \boldsymbol{\theta}^*$ with probability equal to

$$\min\left\{1, \frac{p(\boldsymbol{\theta}^*|\mathbf{y})}{p(\boldsymbol{\theta}^{j-1}|\mathbf{y})} \Big/ \frac{q(\boldsymbol{\theta}^*|\mathbf{y})}{q(\boldsymbol{\theta}^{j-1}|\mathbf{y})}\right\}.$$

3. Otherwise accept $\boldsymbol{\theta}^j = \boldsymbol{\theta}^{j-1}$.

PyMCMC contains a class for the independent MH algorithm, named `IndMH`. To initialize the class the user needs to define the following:

`func`: A user-defined function that returns a sample for the parameter of interest.
`actualprob`: A user-defined function that returns the log probability of the parameters of interest evaluated using the target density.
`candpqrob`: A user-defined function that returns the log probability of the parameters of interest evaluated using the candidate density.
`init_theta`: Initial value for the parameters of interest.
`name`: Name of the parameter of interest.
`**kwargs`: Optional arguments:
  `store`
    `'all'` (default) – stores every iterate for the parameter of interest. This is required for certain calculations.
    `'none'` – does not store any of the iterates from the parameter of interest.
  `fixed_parameter` – is used if the user wants to fix the parameter value that is returned. This is used for testing MCMC sampling schemes. This command will override any other functionality.

Random walk MH

A useful and simple way to construct an MH candidate distribution is via

$$\boldsymbol{\theta}^* = \boldsymbol{\theta}^{j-1} + \boldsymbol{\varepsilon}, \tag{25.2}$$

where $\boldsymbol{\varepsilon}$ is a random disturbance vector. If $\boldsymbol{\varepsilon}$ has a distribution that is symmetric about zero then the MH algorithm has a specific form that is referred to as the random walk MH algorithm. In this case, note that the candidate density is both independent of $\mathbf{y}$ and, due to symmetry, $q(\boldsymbol{\theta}^*|\boldsymbol{\theta}^{j-1}) = q(\boldsymbol{\theta}^{j-1}|\boldsymbol{\theta}^*)$. The random walk MH algorithm at the jth iteration for $j = 1, 2, \ldots, M$ is given by Algorithm 30.

Algorithm 30: Random walk MH

1. Draw a candidate $\boldsymbol{\theta}^*$ from Equation (25.2) where the random disturbance $\boldsymbol{\varepsilon}$ has a distribution symmetric about zero.
2. Accept $\boldsymbol{\theta}^j = \boldsymbol{\theta}^*$ with probability equal to

$$\min\left\{1, \frac{p\left(\boldsymbol{\theta}^*|\mathbf{y}\right)}{p\left(\boldsymbol{\theta}^{j-1}|\mathbf{y}\right)}\right\}.$$

3. Otherwise accept $\boldsymbol{\theta}^j = \boldsymbol{\theta}^{j-1}$.

A typical choice for the distribution of $\boldsymbol{\varepsilon}$ is a normal distribution, that is $\boldsymbol{\varepsilon} \sim \text{iid}\mathcal{N}(0, \boldsymbol{\Omega})$, where the covariance matrix $\boldsymbol{\Omega}$ is viewed as a tuning parameter. **PyMCMC** includes a class for the random walk MH algorithm, named `RWMH`. The class `RWMH` is defined assuming $\boldsymbol{\varepsilon}$ follows a normal distribution. Note that more general random walk MH algorithms could be constructed using the MH class. To initialize the class the user must specify the following:

`post`: A user-defined function for the log of full conditional posterior distribution for the parameters of interest.
`csig`: Scale parameter for the random walk MH algorithm.
`init_theta`: Initial value for the parameter of interest.
`name`: Name of the parameter of interest.
`kwargs`: Optional arguments:
- `store`
  - `'all'` (default) – stores every iterate for the parameter of interest. This is required for certain calculations.
  - `'none'` – does not store any of the iterates from the parameter of interest.
- `fixed_parameter` – is used if the user wants to fix the parameter value that is returned. This is used for testing MCMC sampling schemes. This command will override any other functionality.
- `adaptive` - `'GFS'`: Then the adaptive random walk MH algorithm of Garthwaite *et al.* (2010) will be used to optimize $\boldsymbol{\Omega}$.

OBMC

The multiple try Metropolis (Liang *et al.* 2000) generalizes the MH algorithm to allow for multiple proposals. The OBMC algorithm is a special case of the multiple try Metropolis that is applicable when the candidate density is symmetric. The OBMC algorithm at iteration j is given in Algorithm 31.

Algorithm 31: Orientational bias Monte Carlo

1. Draw L candidates $\boldsymbol{\theta}_l^*$, $l = 1, 2, \ldots, L$, independently from the density $q(\boldsymbol{\theta}|\mathbf{y}, \boldsymbol{\theta}^{j-1})$, where $q(\boldsymbol{\theta}|\mathbf{y}, \boldsymbol{\theta}^{j-1})$ is a symmetric function.
2. Construct a *probability mass function (pmf)* by assigning to each $\boldsymbol{\theta}_l^*$ a probability proportional to $p(\boldsymbol{\theta}_l^*|\mathbf{y})$.
3. Select $\boldsymbol{\theta}^{**}$ randomly from this discrete distribution.
4. Draw $L-1$ reference points $\mathbf{r}_l$, $l = 1, 2, \ldots, L-1$, independently from $q(\boldsymbol{\theta}|\mathbf{y}, \boldsymbol{\theta}^{**})$ and set $\mathbf{r}_L = \boldsymbol{\theta}^{j-1}$.
5. Accept $\boldsymbol{\theta}^j = \boldsymbol{\theta}^{**}$ with probability equal to
$$\min\left\{1, \frac{\sum_{l=1}^{L} p\left(\boldsymbol{\theta}_l^*|\mathbf{y}\right)}{\sum_{l=1}^{L} p\left(\mathbf{r}_l|\mathbf{y}\right)}\right\}.$$
6. Otherwise accept $\boldsymbol{\theta}^j = \boldsymbol{\theta}^{j-1}$.

PyMCMC implements a special case of the OBMC algorithm, for which the candidate density is multivariate normal, making it a generalization of the random walk MH algorithm. The class for the OBMC algorithm is named `OBMC`. To initialize the class the user must specify the following:

`post`: A user-defined function for the log of the full conditional posterior distribution for the parameters of interest.
`ntry`: Number of candidates, L.
`csig`: A scale parameter for the OBMC algorithm.
`init_theta`: Initial value for the parameter of interest.
`**kwargs`: Optional arguments:
- `store`
  - `'all'` (default) – stores every iterate for the parameter of interest. This is required for certain calculations.
  - `'none'` – does not store any of the iterates from the parameter of interest.
- `fixed_parameter` – is used if the user wants to fix the parameter value that is returned. This is used for testing MCMC sampling schemes. This command will override any other functionality.

Closed form sampler

A class is included so that the user can specify a function to sample the parameters of interest when there is a closed form solution. The name of the class is `CFsampler`. To initialize the class the user must specify the following:

`func`: User-defined function that samples from the posterior distribution of interest.
`init_theta`: Initial value for the unknown parameter of interest.

`name`: The name of the parameter of interest.
`**kwargs`: Optional parameters:
  `store`
    `'all'` (default) – stores every iterate for the parameter of interest. This is required for certain calculations.
    `'none'` – does not store any of the iterates from the parameter of interest.
  `fixed_parameter` – is used if the user wants to fix the parameter value that is returned. This is used for testing MCMC sampling schemes. This command will override any other functionality.

Slice sampler

The slice sampler is useful for drawing values from complex densities; see Neal (2003) for further details. The required distribution must be proportional to one or a multiple of several other functions of the variable of interest:

$$p(\theta) \propto f_1(\theta) f_2(\theta) \cdots f_n(\theta).$$

A set of values from the distribution is obtained by iteratively sampling a new value, ω, from the vertical *slice* between 0 and $f_i(\theta)$, and then sampling a value for the parameter θ from the horizontal *slice* that consists of the set of possible values of θ, for which the previously sampled $\omega \leq p(\theta)$. This leads to the slice sampler algorithm, which can be defined at iteration j using Algorithm 32.

Algorithm 32: Slice sampler

1. For $i = 1, 2, \ldots, n$, draw $\omega_i \sim \text{Unif}[0, f_i(\theta^{j-1})]$.
2. Sample $\theta^j \sim \text{Unif}[A]$ where

$$A = \{\theta : f_1(\theta) \geq \omega_1 \in f_2(\theta) \geq \omega_2 \in \cdots \in f_n(\theta) \geq \omega_n\}.$$

In cases where the density of interest is not unimodal, determining the exact set A is not necessarily straightforward. The *stepping out* algorithm of Neal (2003) is used to obtain the set A. This algorithm is applied to each of the n slices to obtain the joint maximum and minimum of the slice. This results in a sampling interval that is designed to draw a new θ^j in the neighbourhood of θ^{j-1} and may include values outside the permissible range of A. The user is required to define an estimated typical slice size (ss), which is the width of set A, along with an integer value (N), which limits the width of any slice to $N \times ss$. The stepping out algorithm is given in Algorithm 33.

Algorithm 33: Stepping out

1. Initiate lower bound (LB) and upper bound (UB) for slice defined by set A.
   - $U \sim$ Unif(0, 1):
     - $LB = \theta^{j-1} - ss \times U$;
     - $UB = LB + ss$.
2. Sample $V \sim$ Unif(0, 1).
3. Set $J =$ Floor($N \times V$).
4. Set $Z = (N - 1) - J$.
5. Repeat while $J > 0$ and $\omega_i < f_i(LB) \forall i$:
   - $LB = LB - ss$;
   - $J = J - 1$.
6. Repeat while $Z > 0$ and $\omega_i < f_i(UB) \forall i$:
   - $UB = UB + ss$;
   - $Z = Z - 1$.
7. Sample $\theta^j \sim$ Unif(LB, UB).

The value of θ^j is accepted if it is drawn from a range $(LB, UB) \in A$. If it is outside the allowable range due to the interval (LB, UB) being larger in range than the set A, we then invoke a shrinkage technique to resample θ^j and improve the sampling efficiency of future draws, until an acceptable θ^j is drawn. The shrinkage algorithm is implemented as in Algorithm 34, repeating this algorithm until exit conditions are met.

Algorithm 34: Shrinkage

1. $U \sim$ Unif(0, 1).
2. $\theta^j = LB + U \times (UB - LB)$:
   - If $\omega_i < f_i(\omega_i) \forall i$, accept θ^j and exit.
   - Else if $\theta^j < \theta^{j-1}$, set $LB = \theta^j$ and return to step 1.
   - Else set $UB = \theta^j$ and return to step 1.

PyMCMC includes a class for the slice sampler named `SliceSampler`. To initialize the class the user must define the following:

`func`: A k-dimensional list containing the set of log functions.
`init_theta`: An initial value for θ.
`ssize`: A user-defined value for the typical slice size.
`sN` - An integer limiting slice size to $N \times ss$.
`**kwargs`: Optional arguments:

`store`

`'all'` (default) – stores every iterate for the parameter of interest. This is required for certain calculations.

`'none'` – does not store any of the iterates from the parameter of interest.

`fixed_parameter` – is used if the user wants to fix the parameter value that is returned. This is used for testing MCMC sampling schemes. This command will override any other functionality.

25.2.2 Normal linear Bayesian regression model

Many interesting models are partly linear for a subset of the unknown parameters. As such, drawing from the full conditional posterior distribution for the associated parameters may be equivalent to sampling the unknown parameters in a standard linear regression model. **PyMCMC** includes several classes that aid in the analysis of linear or partly linear models. In particular the classes `LinearModel`, `CondRegressionSampler`, `CondScaleSampler` and `StochasticSearch` are useful for this purpose. These classes are described in this section. For the standard linear regression model, see Zellner (1971), assume the $(n \times 1)$ observational vector, $\boldsymbol{y}$, is generated according to

$$\boldsymbol{y} = \boldsymbol{X\beta} + \boldsymbol{\varepsilon}; \quad \boldsymbol{\varepsilon} \sim N(\boldsymbol{0}, \sigma^2 \boldsymbol{I}), \tag{25.3}$$

where $\boldsymbol{X}$ is an $(n \times k)$ matrix of regressors, $\boldsymbol{\beta}$ is a $(k \times 1)$ vector of regression coefficients and $\boldsymbol{\varepsilon}$ is a normally distributed random variable with a mean vector $\boldsymbol{0}$ and an $(n \times n)$ covariance matrix, $\sigma^2\boldsymbol{I}$. Assuming that both $\boldsymbol{\beta}$ and σ are unknown, then the posterior distribution for Equation (25.3) is given by

$$p(\boldsymbol{\beta}, \sigma | \boldsymbol{y}, \boldsymbol{X}) \propto p(\boldsymbol{y}|\boldsymbol{X}, \boldsymbol{\beta}, \sigma) \times p(\boldsymbol{\beta}, \sigma), \tag{25.4}$$

where

$$p(\boldsymbol{y}|\boldsymbol{X}, \boldsymbol{\beta}, \sigma) \propto \sigma^{-n} \exp\left\{-\frac{1}{2\sigma^2}(\boldsymbol{y} - \boldsymbol{X\beta})^T(\boldsymbol{y} - \boldsymbol{X\beta})\right\} \tag{25.5}$$

is the joint pdf for $\boldsymbol{y}$ given $\boldsymbol{X}$, $\boldsymbol{\beta}$ and σ, and $p(\boldsymbol{\beta}, \sigma)$ denotes the joint prior pdf for $\boldsymbol{\beta}$ and σ.

A class named `LinearModel` is defined to sample from the posterior distribution in Equation (25.4). One of four alternative priors may be used in the specification of the model. The default choice is Jeffreys' prior. Denoting the full set of unknown parameters as $\boldsymbol{\theta} = (\boldsymbol{\beta}^T, \sigma)^T$, then Jeffreys' prior is defined such that

$$p(\boldsymbol{\theta}) \propto |I(\boldsymbol{\theta})|^{-1/2}, \tag{25.6}$$

where $I(\boldsymbol{\theta})$ is the Fisher information matrix for $\boldsymbol{\theta}$. For the normal linear regression model in Equation (25.3), given the assumption that β and σ are a priori independent, Jeffreys' prior is flat over the real-number line for β, and σ is distributed such that

$$p(\sigma) \propto \frac{1}{\sigma}. \tag{25.7}$$

See Zellner (1971) for further details on Jeffreys' prior. Three alternative informative prior specifications are allowed, namely the normal–gamma, the normal–inverted-gamma and Zellner's g-prior; see Zellner (1971) and Marin and Robert (2007) for further details. The normal–gamma prior is specified such that

$$\boldsymbol{\beta}|\kappa \sim N(\underline{\boldsymbol{\beta}}, \underline{\boldsymbol{V}}^{-1}), \quad \kappa \sim G\left(\frac{\underline{\nu}}{2}, \frac{\underline{S}}{2}\right), \tag{25.8}$$

where $\kappa = \sigma^{-2}$ and $\underline{\beta}$, $\underline{\boldsymbol{V}}$, $\underline{\nu}$ and $\underline{S}$ are prior hyperparameters that take user-defined values. For the normal–gamma prior, `LinearModel` produces estimates for $(\kappa, \boldsymbol{\beta}^T)^T$ rather than $(\sigma, \boldsymbol{\beta}^T)$. The normal–inverted-gamma prior is specified such that

$$\boldsymbol{\beta}|\sigma^{-2} \sim N(\underline{\boldsymbol{\beta}}, \underline{\boldsymbol{V}}^{-1}), \quad \sigma^{-2} \sim IG\left(\frac{\underline{\nu}}{2}, \frac{\underline{S}}{2}\right), \tag{25.9}$$

where $\underline{\beta}$, $\underline{\boldsymbol{V}}$, $\underline{\nu}$ and $\underline{S}$ are prior hyperparameters, which take values that are set by the user. Zellner's g-prior is specified such that

$$\boldsymbol{\beta}|\sigma \sim N\left(\underline{\boldsymbol{\beta}}, g\sigma^2\left(X^T X\right)^{-1}\right), \quad p(\sigma) \propto \sigma^{-1}, \tag{25.10}$$

where $\underline{\boldsymbol{\beta}}$ and g are hyperparameters with values that are specified by the user. To initialize the class `LinearModel` the user must specify the following:

`yvec`: One-dimensional **Numpy** array containing the data.
`xmat`: Two-dimensional **Numpy** array containing the regressors.
`**kwargs`: Optional arguments:
`prior`: A list containing the name of the prior and the corresponding hyperparameters. For example,
`prior = ['normal_gamma', betaubar, Vubar, nuubar, Subar]`,
`prior = ['normal_inverted_gamma',betaubar, Vubar, nuubar, Subar]` and
`prior = ['g_prior', betaubar, g]`.
If none of these options are chosen or they are misspecified then the default prior will be Jeffreys' prior.

`LinearModel` contains several functions that may be of interest to the user. In particular:

`sample()`: Returns a sample of σ and $\boldsymbol{\beta}$ from the joint posterior distribution for the normal–inverted-gamma prior, Jeffreys' prior and Zellner's g-prior. If the normal–gamma prior is specified then `sample()` returns κ and $\boldsymbol{\beta}$.
`update_yvec(yvec)`: Updates yvec in `LinearModel`. This is often useful when the class is being used as a part of the MCMC sampling scheme.
`update_xmat(xmat)`: Updates xmat in `LinearModel`. This is often useful when the class is being used as a part of the MCMC sampling scheme.

`loglike(scale, beta)`: Returns the log-likelihood.
`posterior_mean()`: Returns the posterior mean for the scale parameter (either σ or κ depending on the specified prior) and $\boldsymbol{\beta}$.
`get_posterior_covmat()`: Returns the posterior covariance matrix for $\boldsymbol{\beta}$.
`bic()`: Returns the BIC; see Kass and Raftery (1995) for details.
`plot(**kwargs)`: Produces standard plots. Specifically the marginal posterior density intervals for each element of $\boldsymbol{\beta}$ and for the scale parameter (σ or κ).
`residuals`: Returns the residual vector from the regression analysis. The residuals are calculated with $\boldsymbol{\beta}$ evaluated at the marginal posterior mean.
`output`: Produces standard output for the regression analysis. This includes the means, standard deviations and highest posterior density (HPD) intervals for the marginal posterior densities for each element of $\boldsymbol{\beta}$ and for the scale parameter (σ or κ). The output also reports the log-likelihood and the BIC.

In MCMC sampling schemes it is common that for a subset of the unknown parameters of interest the full conditional posterior distribution will correspond to that of a linear regression model, where the scale parameter is known. For the linear regression model specified in Equation (25.3) the posterior distribution for the case that σ is known is as follows:

$$p(\boldsymbol{\beta}|\mathbf{y}, \mathbf{X}, \boldsymbol{\beta}, \sigma) \propto p(\mathbf{y}|\mathbf{X}, \boldsymbol{\beta}, \sigma) \times p(\boldsymbol{\beta}), \tag{25.11}$$

where $p(\mathbf{y}|\mathbf{X}, \boldsymbol{\beta}, \sigma)$ is described in Equation (25.5) and $p(\boldsymbol{\beta})$ is the prior pdf for $\boldsymbol{\beta}$. To sample from Equation (25.11) a class named `CondRegressionSampler` can be used. The user may specify one of three alternative priors. The default prior is Jeffreys' prior, which for $\boldsymbol{\beta}$ is simply a flat prior over the real-number line. A normally distributed prior for $\boldsymbol{\beta}$ is another option, and can be specified such that

$$\boldsymbol{\beta} \sim N\left(\underline{\boldsymbol{\beta}}, \mathbf{V}^{-1}\right).$$

The user may also specify their a priori beliefs using Zellner's g-prior, where

$$\boldsymbol{\beta}|\sigma \sim N\left(\boldsymbol{\beta}, g\sigma^2 \mathbf{X}^T \mathbf{X}\right).$$

To initialize the class the user must specify the following:

`yvec`: A one-dimensional **Numpy** array containing the data.
`xmat`: A two-dimensional **Numpy** array containing the regressors.
`**kwargs`: Optional arguments:
  `prior` – a list containing the name of the prior and the corresponding hyperparameters. For example,
  `prior=['normal', betaubar, Vubar]` or `['g_prior', betaubar, g]`.
  If none of these options are chosen or they are misspecified then the default prior will be Jeffreys' prior.

`CondRegressionSampler` contains several functions that may be of interest to the user. In particular:

`sample(sigma)`: Returns a sample of $\boldsymbol{\beta}$ from the posterior distribution specified in Equation (25.11).

`get_marginal_posterior_mean()`: Returns the marginal posterior mean for Equation (25.11).

`get_marginal_posterior_precision()`: Returns the marginal posterior precision for the linear conditional posterior distribution specified in Equation (25.11).

`update_yvec(yvec)`: Updates yvec in `CondRegressionSampler`. This is often useful when the class is being used as a part of an MCMC sampling scheme.

`update_xmat(xmat)`: Updates xmat in `CondRegressionSampler`. This is often useful when the class is being used as a part of an MCMC sampling scheme.

Many Bayesian models contain linear components with unknown scale parameters, hence a class has been specified named `CondScaleSampler`, which can be used to individually sample scale parameters from their posterior distributions. In particular, we wish to sample from

$$p(\sigma|\mathbf{y}, \boldsymbol{\theta}), \tag{25.12}$$

where $\boldsymbol{\theta}$ is the set of unknown parameters of interest excluding σ. The user may choose to use one of three priors. Jeffreys' prior, which for σ given the posterior in Equation (25.12) is as follows:

$$p(\sigma) \propto \frac{1}{\sigma}.$$

The second option is to specify an inverted-gamma prior, such that

$$\sigma \sim IG\left(\frac{\nu}{2}, \frac{S}{2}\right).$$

Alternatively, the user may specify a gamma prior for $\kappa = 1/\sigma^2$, where

$$\kappa \sim G\left(\frac{\nu}{2}, \frac{S}{2}\right).$$

To initialize the class `CondScaleSamper` the user may first specify the following:

`**kwargs`: Options arguments:

`prior` – list containing the name of the prior and the corresponding hyperparameters. For example,
`prior=['gamma',nuubar,subar]` or
`prior=['inverted-gamma', nuubar, subar]`. If no prior is specified Jeffreys' prior is used.

PyMCMC also includes another class that can be used for the direct analysis of the linear regression model. The class is called `StochasticSearch` and can be used in conjunction with the class `MCMC`, for the purpose of variable selection.

The stochastic search algorithm can be used for variable selection in the linear regression model. Given a set of k possible regressors there are 2^k models to choose from. The stochastic search algorithm, as proposed by George and McCulloch (1993), uses the Gibbs sampler to select a set of 'most likely' models. The stochastic search algorithm is implemented in the class `StochasticSearch`. The specific implementation follows Marin and Robert (2007). The algorithm introduces the vector $\boldsymbol{\gamma}$, which is used to select the explanatory variables that are to be included in the model. In particular, $\boldsymbol{\gamma}$ is defined to be a binary vector of order k, whereby the inclusion of the ith regressor implies that the ith element of $\boldsymbol{\gamma}$ is a one, while the exclusion of the ith regressor implies that the ith element is zero. It is assumed that the first element of the design matrix is always included and should typically be a column of ones which is used to represent the constant or intercept in the regression. The algorithm specified to sample $\boldsymbol{\gamma}$ is a single move Gibbs sampling scheme; for further details see Marin and Robert (2007).

To use the class `StochasticSearch` the user must specify their a priori beliefs that the unknown parameters of interest, $(\sigma, \boldsymbol{\beta}^T)^T$, are distributed following Zellner's g-prior, which is described in Equation (25.10). `StochasticSearch` is designed to be used in conjunction with the MCMC sampling class. To initialize `StochasticSearch` the user must specify the following:

`yvec`: One-dimensional **Numpy** array containing the dependent variable.
`xmat`: Two-dimensional **Numpy** array containing the regressors.
`prior`: A list with the structure `[betaubar, g]`.

The class `StochasticSearch` also contains the following function:

`sample_gamma(store)`: Returns a sample of $\boldsymbol{\gamma}$. The only argument to pass into the function `sample_gamma` is the storage dictionary that is passed by default to each of the classes called from the class `MCMC` in **PyMCMC**.

25.3 Empirical illustrations

PyMCMC is illustrated though three examples. Specifically, a linear regression example with variable selection, a loglinear example and a linear regression model with first-order autoregressive errors. For each example, the model of interest is specified, then the code used for estimation is shown, following which a brief description of the code is given. Each example uses the module for **PyMCMC**, along with the `Python` libraries **Numpy**, **Scipy** and **Matplotlib**; see Oliphant (2007) and Hunter (2007) for further details of these `Python` libraries. Example 3 further uses the library **Pysparse** (Geus 2011).

25.3.1 Example 1: Linear regression model – variable selection and estimation

The data used in this example are a response of crop yield modelled using various chemical measurements from the soil. As the results of the chemical analysis of soil cores are obtained in a laboratory, many input variables are available and the data analyst would like to determine the variables that are most appropriate to use in the model.

The normal linear regression model in Equation (25.3) is used for the analysis. To select the set of 'most probable' regressors we use the stochastic search variable selection approach described in Section 25.2.2 using a subset of 19 explanatory variables.

In addition to **PyMCMC**, this example uses functions from **Numpy**, so the first step is to import the relevant packages:

```
import os
from numpy import loadtxt, hstack, ones, random, zeros,
    asfortranarray, log
from pymcmc.mcmc import MCMC, CFsampler
from pymcmc.regtools import StochasticSearch,
    LinearModel
```

The remaining code is typically organized so that the user-defined functions are at the top and the main program is at the bottom. We begin by defining functions that are called from the class `MCMC`, all of which take the argument `store`. In this example there is one such function:

```
def samplegamma(store):
    return store['SS'].sample_gamma(store)
```

In this case, `samplegamma` simply uses the pre-defined function available in the `StochasticSearch` class to sample from γ.

The data and an instance of the class `StochasticSearch` are now initialized:

```
data = loadtxt('yld2.txt')
yvec = data[:, 0]
xmat = data[:, 1:20]
xmat = hstack([ones((xmat.shape[0], 1)), xmat])
data ={'yvec':yvec, 'xmat':xmat}

prior = ['g_prior', zeros(xmat.shape[1]), 100.]
SSVS = StochasticSearch(yvec, xmat, prior);
data['SS'] = SSVS
```

Note that data is augmented to include the class instance for StochasticSearch. In this example we use Zellner's g-prior with $\underline{\boldsymbol{\beta}} = 0$ and $g = 100$. The normal–inverted-gamma prior could also be used.

The next step is to initialize γ and set up the appropriate sampler for the model. In this case, as the full conditionals are of closed form, we can use the CFsampler class:

```
initgamma = zeros(xmat.shape[1], dtype ='i')
initgamma[0] = 1
simgam = CFsampler(samplegamma, initgamma, 'gamma')
```

The required arguments to CFSampler are a function that samples from the posterior (samplegamma), the initial value for γ (initgamma) and the name of the parameter of interest ('gamma'). The single argument to samplegamma, store, is a dictionary (Python data structure) that is passed to all functions that are called from the MCMC sampler. The purpose of store is to contain all the data required to define functions that sample from, or are used in the evaluation of, the posterior distribution. For example, we see that samplegamma accesses store['SS'], which contains the class instance for StochasticSearch. In addition to all the information contained in data, store contains the value of the previous iterate for each block of the MCMC scheme. It is possible, for example, to access the current iteration for any named parameter (in this case 'gamma') from any of the functions called from MCMC by using store['gamma']. This feature is not used in this example, but can be seen in the following examples.

The actual sampling is done by setting a random seed and running the MCMC sampler:

```
random.seed(12346)
ms = MCMC(20000, 5000, data, [simgam])
ms.sampler()
```

The MCMC sampler is initialized by setting the number of iterations (20 000), the burn-in (5000), providing the data dictionary data and a list containing the information used to sample from the full conditional posterior distributions. In this case, this list consists a single element, simgam, the CFsampler instance.

The default output from the MCMC object is a summary of each parameter providing the posterior mean, posterior standard deviation, 95% credible intervals and inefficiency factors. This can output directly to the screen, or be captured in an output file. A sample of this output, giving only the first four elements of γ, is

```
ms.output()

--------------------------------------------------------

The time (seconds) for the MCMC sampler = 7.4
```

```
Number of blocks in MCMC sampler = 1

               mean        sd   2.5%   97.5%   IFactor
gamma[0]          1         0      1       1        NA
gamma[1]   0.0929      0.29      0       1      3.75
gamma[2]   0.0941     0.292      0       1      3.54
gamma[3]   0.0939     0.292      0       1      3.54
```

In this case, the standard output is not all that useful as a summary of the variable selection. A more useful output, giving the 10 most likely models ordered by decreasing posterior probabilities, is available using the output function in the `StochasticSearch` class. This can be called by using the `custom` argument to `output`:

```
ms.output(custom = SSVS.output)

Most likely models ordered by decreasing posterior
    probability

-------------------------------------------------------
probability | model
-------------------------------------------------------
0.09353333  | 0, 12
0.0504      | 0, 11, 12
0.026       | 0, 10, 12
0.01373333  | 0, 9, 12
0.01353333  | 0, 8, 12
0.013       | 0, 4, 12
0.01293333  | 0, 12, 19
0.01206667  | 0, 7, 12
0.01086667  | 0, 11, 12, 17
0.01086667  | 0, 12, 17
-------------------------------------------------------
```

As indicated earlier, the MCMC analysis is conducted using 20 000 iterations, of which the first 5000 are discarded. The estimation takes 7.4 seconds in total. The results indicate that a model containing variable 12 along with a constant (indicated in the table by 0) is the most likely (prob = 0.09). Furthermore, variable 12 is contained in each of the 10 most likely models, indicating its strong association with crop yield.

Following the variable selection procedure, we may wish to fit a regression to the most likely model. This can be done using the `LinearModel` class. Firstly, we extract the explanatory variables from the most likely model

```
txmat = SSVS.extract_regressors(0)
```

and fit the model using the desired prior

```
g_prior = ['g_prior', 0.0, 100.]
```

```
breg = LinearModel(yvec,txmat,prior = g_prior)
```

The output summarizes the fit:

```
breg.output()

             ----------------------------------------------------
                     Bayesian Linear Regression Summary
                                  g_prior
             ----------------------------------------------------
                    mean             sd            2.5%          97.5%
beta[0]          -0.1254         0.1058         -0.3361        0.08533
beta[1]           0.7587         0.0225          0.7139         0.8035
  sigma           0.3853        0.03152              NA             NA

loglikelihood = -4.143
log marginal likelihood = nan
BIC  = 21.32
```

The parameter associated with variable 12, $\hat{\beta}_1$, is estimated as a positive value, 0.7587, with a 95% credible interval [0.7139, 0.8035]. We note that zero is not contained in the credible interval, hence the crop yield increases with higher values of variable 12. The marginal posterior densities (Figure 25.2) can be created using

```
breg.plot()
```

This clearly shows this effect is far from zero.

25.3.2 Example 2: Loglinear model

The data analysed in this example are the number of nutgrass shoots counted, in randomly scattered quadrats, at weekly intervals, during the growing season. A loglinear model is used:

$$\log(\text{count}) = \beta_0 + \beta_1 \text{week} \tag{25.13}$$

where the intercept, β_0, is expected to be positive in value, as nutgrass is always present in this study site, and β_1 is also expected to be positive as the population of nutgrass increases during the growing season.

For the loglinear model, see Gelman *et al.* (2004), the ith observation y_i, for $i = 1, 2, \ldots, n$, is generated as follows:

$$p(y_i|\mu_i) = \frac{\mu_i^{y_i} \exp(-\mu_i)}{\mu_i!}, \tag{25.14}$$

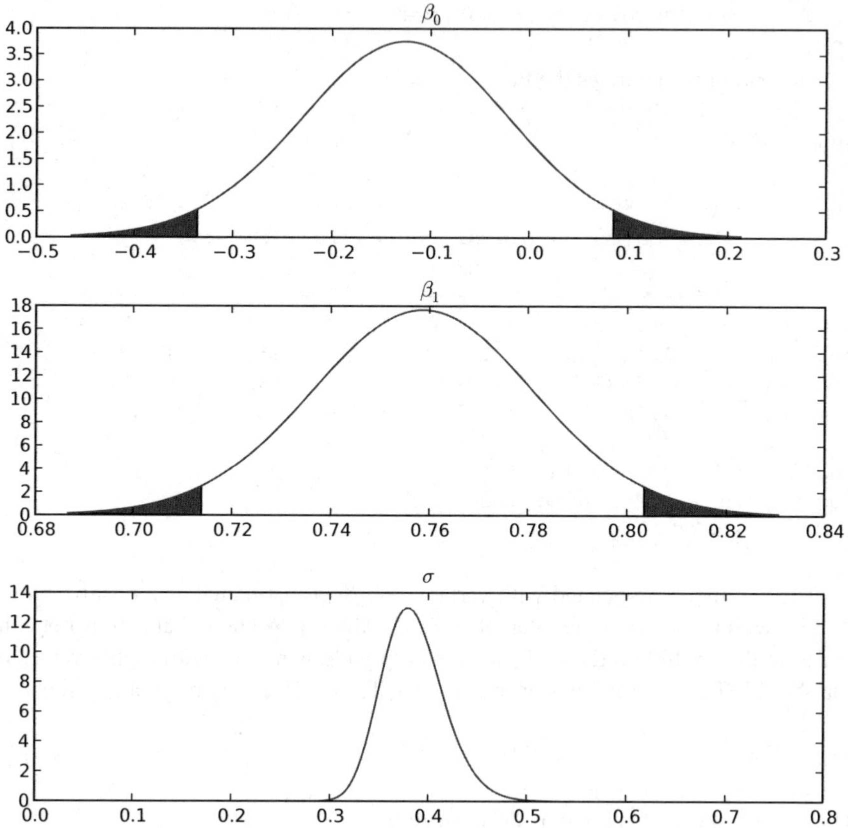

Figure 25.2 Marginal posterior density plots for the regression coefficients in Example 1.

with

$$\log(\mu_i) = \boldsymbol{x}_i^T \boldsymbol{\beta},$$

where $\boldsymbol{x}_i^T$ is the ith row of the $(n \times k)$ matrix $\boldsymbol{X}$.

The joint posterior distribution for the unknown parameter $\boldsymbol{\beta}$ is given by

$$p(\boldsymbol{\beta}|\boldsymbol{y}, \boldsymbol{X}) \propto p(\boldsymbol{y}|\boldsymbol{\beta}, \boldsymbol{X}) \times p(\boldsymbol{\beta}), \tag{25.15}$$

where $p(\boldsymbol{y}|\boldsymbol{\beta}, \boldsymbol{X})$ is the joint pdf for $\boldsymbol{y}$ conditional on the $\boldsymbol{\beta}$ and $\boldsymbol{X}$, and $p(\boldsymbol{\beta})$ denotes the prior pdf for $\boldsymbol{\beta}$. From Equation (25.14) it is apparent that

$$p(\boldsymbol{y}|\boldsymbol{\beta}, \boldsymbol{X}) = \prod_{i=1}^{n} \frac{\mu_i^{y_i} \exp(-\mu_i)}{\mu_i!}.$$

A priori we assume that

$$\boldsymbol{\beta} \sim N(\underline{\boldsymbol{\beta}}, \boldsymbol{V}^{-1}).$$

To sample from Equation (25.15), a random walk MH algorithm is implemented, where the candidate $\boldsymbol{\beta}^*$, at each iteration, is sampled following

$$\boldsymbol{\beta}^* \sim N\left(\boldsymbol{\beta}^{j-1}, \boldsymbol{\Omega}\right), \tag{25.16}$$

where

$$\boldsymbol{\beta}^0 = \boldsymbol{\beta}_{nls} = \arg\min\left(\boldsymbol{y} - \exp\left(\boldsymbol{X}\boldsymbol{\beta}\right)\right)^2$$

and

$$\boldsymbol{\Omega}^{-1} = -\sum_{i=1}^{n} \exp\left(\boldsymbol{x}_i^T \boldsymbol{\beta}_{nls}\right) \boldsymbol{x}_i \boldsymbol{x}_i^T.$$

The example code for **PyMCMC** uses two `Python` libraries, **Numpy** and **Scipy**, which the user must have installed to run the code.

As before, the code begins by importing the required packages:

```
import os
from numpy import random, loadtxt, hstack, ones, dot,
    exp, zeros, outer, diag
from numpy import linalg
from pymcmc.mcmc import MCMC, RWMH, OBMC
from pymcmc.regtools import LinearModel
from scipy.optimize.minpack import leastsq
```

The import statements are followed by the definition of a number or required functions, as below.

A function `minfunc` used in the nonlinear least squares routine:

```
def minfunc(beta, yvec, xmat ):
    return yvec - exp(dot(xmat, beta))
```

A function `prior` to evaluate the log of the prior pdf $\boldsymbol{\beta}$ and a function `logl` defining the log-likelihood function:

```
def prior(store):
    mu = zeros(store['beta'].shape[0])
    Prec = diag(0.005 * ones(store['beta'].shape[0]))
    return -0.5 * dot(store['beta'].transpose(),
        dot(Prec, store['beta']))

def logl(store):
    xbeta = dot(store['xmat'], store['beta'])
    lamb = exp(xbeta)
    return sum(store['yvec'] * xbeta - Lamb)
```

As in the variable selection example, `store` is a `Python` dictionary used to store all the information of interest that needs to be accessed by functions that are called from the MCMC sampler. For example, the function `logl` uses `store['beta']` which provides access to the vector $\boldsymbol{\beta}$.

A function `posterior` which evaluates the log of the posterior pdf for $\boldsymbol{\beta}$:

```
def posterior(store):
    return logl(store) + prior(store)
```

A function `llhessian` which returns the Hessian for the loglinear model:

```
def llhessian(store, beta):
    nobs = store['yvec'].shape[0]
    kreg = store['xmat'].shape[1]
    lamb = exp(dot(store['xmat'], beta))
    sum = zeros((kreg, kreg))
    for i in xrange(nobs):
        sum = sum + lamb[i] * outer(store['xmat'][i],
            store['xmat'][i])
    return -sum
```

Following the function definitions, the main program begins with the command to set the random seed

```
random.seed(12345)
```

and set up the data:

```
data = loadtxt('count.txt', skiprows = 1)
yvec = data[:, 0]
xmat = data[:, 1:data.shape[1]]
xmat = hstack([ones((data.shape[0], 1)), xmat])
data ={'yvec':yvec, 'xmat':xmat}
```

Bayesian regression is used to initialize the nonlinear least squares algorithm:

```
bayesreg = LinearModel(yvec, xmat)
sig, beta0 = bayesreg.posterior_mean()
```

The function `leastsq` from **Scipy** is used to perform the nonlinear least squares operation:

```
init_beta, info = leastsq(minfunc, beta0, args
    = (yvec, xmat))
data['betaprec'] =-llhessian(data, init_beta)
scale = linalg.inv(data['betaprec'])
```

Initialize the random walk MH algorithm:

```
samplebeta = RWMH(posterior, scale, init_beta, 'beta')
```

Finally, set up and run the sampling scheme. Note that the sampling algorithm is run for 20 000 iterations and the first 4000 are discarded. The MCMC scheme has only one block and is an MH sampling scheme:

```
ms = MCMC(20000, 4000, data, [samplebeta],
      loglike = (logl, xmat.shape[1], 'yvec'))
ms.sampler()
```

A summary of the model fit can be produced using the `output` function:

```
ms.output()
```

```
--------------------------------------------------------

The time (seconds) for the Gibbs sampler =  7.47
Number of blocks in Gibbs sampler =  1

              mean          sd       2.5%      97.5%     IFactor
beta[0]       1.14      0.0456       1.05       1.23        13.5
beta[1]      0.157     0.00428      0.148      0.165        12.2
Acceptance rate  beta  =  0.5625
BIC =  -7718.074
Log likelihood =  3864.453
```

It can be seen from the output that estimation is very fast (7.5 seconds), and that both β_0 and β_1 are positive values, with $\hat{\beta}_0 = 1.14$ [1.05, 1.23] and $\hat{\beta}_1 = 0.157$ [0.148, 0.165].

Summary plots can also be produced (Figure 25.3):

```
ms.plot('beta')
```

The marginal posterior densities of these estimates (Figure 25.3) confirm that both estimates are far from zero. The ACF plots (Figure 25.3) and low inefficiency factors (see Chib and Greenberg (1996) for details on inefficiency factors) of 13.5 ($\hat{\beta}_0$) and 12.2 ($\hat{\beta}_1$) show that autocorrelation in the sample is relatively low for an MCMC sampler.

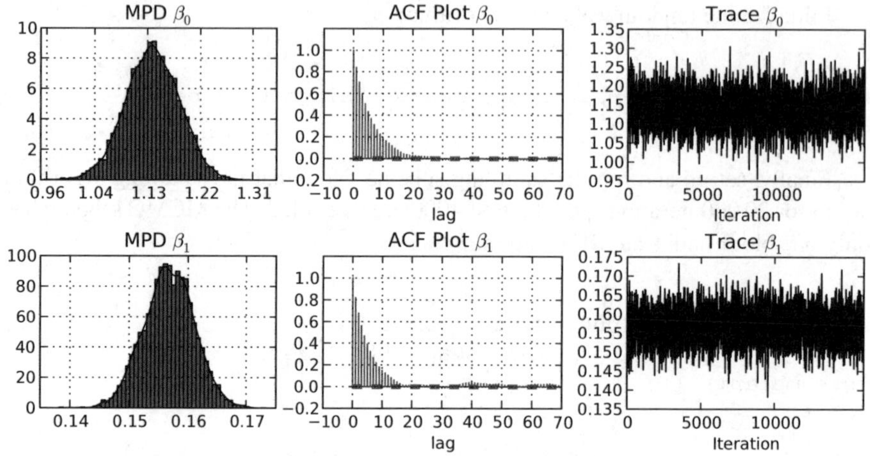

Figure 25.3 Plots of the marginal posterior density, autocorrelation and trace plots for the MCMC estimation of the loglinear model in Example 2. Note that summaries are calculated after removing the burn- in.

25.3.3 Example 3: First-order autoregressive regression

The final example demonstrates first-order autoregressive regression using a simulated data set, with 1000 observations and three regressors.

The linear regression model, with first-order autocorrelated serial correlation in the residuals, see Zellner (1971), is defined such that the tth observation, y_t, for $t = 1, 2, \ldots, n$, is

$$y_t = \boldsymbol{x}_t^T \boldsymbol{\beta} + \varepsilon_t, \qquad (25.17)$$

with

$$\varepsilon_t = \rho \varepsilon_{t-1} + \nu_t; \quad \nu_t \sim \text{iid } N(0, \sigma^2), \qquad (25.18)$$

where $\boldsymbol{x}_t$ is a $(k \times 1)$ vector of regressors, $\boldsymbol{\beta}$ is a $(k \times 1)$ vector of regression coefficients, ρ is a damping parameter and ν_t is an independent and identically normally distributed random variable with a mean of 0 and a variance of σ^2. Under the assumption that the process driving the errors is stationary, that is $|\rho| < 1$, and assuming that the process has been running since time immemorial, then Equations (25.17) and (25.18) can be expressed as

$$\boldsymbol{y} = \boldsymbol{X\beta} + \boldsymbol{\varepsilon}; \quad \boldsymbol{\varepsilon} \sim N\left(\boldsymbol{0}, \kappa^{-1}\boldsymbol{\Omega}^{-1}\right), \qquad (25.19)$$

where

$$\boldsymbol{\Omega} = \begin{bmatrix} 1 & -\rho & 0 & 0 & \cdots & 0 \\ -\rho & 1+\rho^2 & -\rho & 0 & \ddots & 0 \\ 0 & -\rho & 1+\rho^2 & \ddots & \ddots & \vdots \\ 0 & 0 & \ddots & \ddots & -\rho & 0 \\ \vdots & \vdots & \ddots & -\rho & 1+\rho^2 & -\rho \\ 0 & 0 & \cdots & 0 & -\rho & 1 \end{bmatrix}.$$

Further, if we factorize $\boldsymbol{\Omega} = \boldsymbol{L}\boldsymbol{L}^T$, using the Cholesky decomposition, it is straightforward to derive $\boldsymbol{L}$, where

$$\boldsymbol{L} = \begin{bmatrix} 1 & -\rho & 0 & 0 & \cdots & 0 \\ 0 & 1 & -\rho & \ddots & \ddots & 0 \\ 0 & 0 & \ddots & \ddots & \ddots & \vdots \\ \vdots & \ddots & \ddots & \ddots & \ddots & 0 \\ \vdots & \ddots & \ddots & \ddots & 1 & -\rho \\ 0 & 0 & \cdots & \cdots & 0 & \sqrt{1-\rho^2} \end{bmatrix}.$$

Pre-multiplying Equation (25.19) by $\boldsymbol{L}$ gives

$$\tilde{\boldsymbol{y}} = \tilde{\boldsymbol{X}}\boldsymbol{\beta} + \tilde{\boldsymbol{\varepsilon}}, \tag{25.20}$$

where $\tilde{\boldsymbol{y}} = \boldsymbol{L}^T\boldsymbol{y}$, $\tilde{\boldsymbol{X}} = \boldsymbol{L}^T\boldsymbol{X}$ and $\tilde{\boldsymbol{\varepsilon}} = \boldsymbol{L}^T\boldsymbol{\varepsilon}$. Note that $\boldsymbol{\varepsilon} \sim N(0, \kappa^{-1}\boldsymbol{I})$.

The joint posterior distribution for the full set of unknown parameters is

$$p(\boldsymbol{\beta}, \kappa, \rho|\boldsymbol{y}) \propto p(\boldsymbol{y}|\boldsymbol{\beta}, \kappa, \rho) \times p(\boldsymbol{\beta}, \kappa) \times p(\rho), \tag{25.21}$$

where $p(\boldsymbol{y}|\boldsymbol{\beta}, \kappa, \rho)$ is the joint pdf of $\boldsymbol{y}$ conditional on $\boldsymbol{\beta}$, κ and ρ, $p(\boldsymbol{\beta}, \kappa)$ is the joint prior pdf for $\boldsymbol{\beta}$ and κ, and $p(\rho)$ denotes the prior density function for ρ. The likelihood function, which is defined following Equations (25.17) and (25.18), is defined as follows:

$$\begin{aligned} p(\boldsymbol{y}|\boldsymbol{\beta}, \kappa, \rho) &\propto \kappa^{n/2}|\boldsymbol{\Omega}|^{1/2} \exp\left\{-\frac{\kappa}{2}(\boldsymbol{y} - \boldsymbol{X}\boldsymbol{\beta})^T \boldsymbol{\Omega}(\boldsymbol{y} - \boldsymbol{X}\boldsymbol{\beta})\right\} \\ &= \kappa^{n/2}|\boldsymbol{\Omega}|^{1/2} \exp\left\{-\frac{\kappa}{2}\left(\tilde{\boldsymbol{y}} - \tilde{\boldsymbol{X}}\boldsymbol{\beta}\right)^T \left(\tilde{\boldsymbol{y}} - \tilde{\boldsymbol{X}}\boldsymbol{\beta}\right)\right\} \\ &= \kappa^{n/2}\left(1-\rho^2\right)^{1/2} \exp\left\{-\frac{\kappa}{2}\left(\tilde{\boldsymbol{y}} - \tilde{\boldsymbol{X}}\boldsymbol{\beta}\right)^T \left(\tilde{\boldsymbol{y}} - \tilde{\boldsymbol{X}}\boldsymbol{\beta}\right)\right\}. \end{aligned} \tag{25.22}$$

For the analysis a normal–gamma prior is assumed for $\boldsymbol{\beta}$ and κ, such that

$$\boldsymbol{\beta}|\kappa \sim N\left(\underline{\boldsymbol{\beta}}, \kappa^{-1}\right), \quad \kappa \sim G\left(\frac{\underline{\nu}}{2}, \frac{\underline{S}}{2}\right). \tag{25.23}$$

It follows from Equations (25.22) and (25.23) that sampling $\boldsymbol{\beta}$ and κ conditional on ρ is simply equivalent to sampling from a linear regression model with a normal–gamma prior. A beta prior is assumed for ρ, thereby restricting the autocorrelation of the time series to be both positive and stationary. Specifically

$$\rho \sim \mathcal{B}e\left(\alpha, \beta\right).$$

An MCMC sampling scheme, for the posterior distribution in Equation (25.21), defined at iteration j is as follows:

1. Sample $\boldsymbol{\beta}^{(j)}, \kappa^{(j)}$ from $p(\boldsymbol{\beta}, \kappa|\mathbf{y}, \rho^{(j-1)})$.
2. Sample $\rho^{(j)}$ from $p(\rho|\mathbf{y}, \boldsymbol{\beta}, \kappa)$.

The code for this model follows the structure of the previous examples, and begins by importing the required packages:

```
from numpy import random, ones, zeros, dot, hstack,
    eye, log
from scipy import sparse
from pysparse import spmatrix
from pymcmc.mcmc import MCMC, SliceSampler, RWMH, OBMC,
    MH, CFsampler
from pymcmc.regtools import LinearModel
```

As this example uses simulated data, the first function is used to derive these data:

```
def simdata(nobs, kreg):
    xmat = hstack((ones((nobs, 1)), random.randn(nobs,
        kreg - 1)))
    beta = random.randn(kreg)
    sig = 0.2
    rho = 0.90
    yvec = zeros(nobs)
    eps = zeros(nobs)
    eps[0] = sig ** 2 / (1. - rho ** 2)
    for i in xrange(nobs - 1):
        eps[i + 1] = rho * eps[i] + sig * random.randn(1)
    yvec = dot(xmat, beta) + eps
    return yvec, xmat
```

Next, a function is defined to calculate $\tilde{\mathbf{y}}$ and $\tilde{\mathbf{X}}$:

```
def calcweighted(store):
    nobs = store['yvec'].shape[0]
    store['Upper'].put(-store['rho'], range(0, nobs - 1),
        range(1, nobs))
    store['Upper'].matvec(store['yvec'],
        store['yvectil'])
    for i in xrange(store['xmat'].shape[1]):
        store['Upper'].matvec(store['xmat'][:, i],
            store['xmattil'][:, i])
```

Note that $\mathbf{L}^T$ is updated based on the latest iteration in the MCMC scheme. Further, $\mathbf{L}^T$ is stored in the Python dictionary `store` and is accessed using the key `'Upper'`. It is stored in sparse matrix format using the library **Pysparse**.

Next a function is defined and used to sample $\boldsymbol{\beta}$ from its conditional posterior distribution:

```
def WLS(store):
    calcweighted(store)
    store['regsampler'].update_yvec(store['yvectil'])
    store['regsampler'].update_xmat(store['xmattil'])
    return store['regsampler'].sample()
```

Now functions are defined to evaluate the log-likelihood (`loglike`), the log of the prior pdf for ρ (`prior_rho`) and the log of the posterior pdf for ρ (`post_rho`):

```
def loglike(store):
    nobs = store['yvec'].shape[0]
    calcweighted(store)
    store['regsampler'].update_yvec(store['yvectil'])
    store['regsampler'].update_xmat(store['xmattil'])
    return store['regsampler'].loglike(store['sigma'],
        store['beta'])

def prior_rho(store):
    if store['rho'] > 0. and store['rho'] < 1.0:
        alpha = 1.0
        beta = 1.0
        return (alpha - 1.) * log(store['rho'])
                + (beta - 1.) * log(1.-store['rho'])
    else:
        return -1E256

def post_rho(store):
    return loglike(store) + prior_rho(store)
```

The main program begins by setting the seed and constructing the Python dictionary data, which is used to store information that will be passed to functions that are called from the MCMC sampler:

```
random.seed(12345)
nobs = 1000
kreg = 3
yvec, xmat = simdata(nobs, kreg)
priorreg = ('g_prior', zeros(kreg), 1000.0)
regs = LinearModel(yvec, xmat, prior = priorreg)
data ={'yvec':yvec, 'xmat':xmat, 'regsampler':regs}
U = spmatrix.ll_mat(nobs, nobs, 2 * nobs - 1)
U.put(1.0, range(0, nobs), range(0, nobs))
data['yvectil'] = zeros(nobs)
data['xmattil'] = zeros((nobs, kreg))
data['Upper'] = U
```

Initial values for σ and β are set using Bayesian regression:

```
bayesreg = LinearModel(yvec, xmat)
sig, beta = bayesreg.posterior_mean()
```

The parameters σ and β are jointly sampled using the closed form class:

```
simsigbeta = CFsampler(WLS, [sig, beta], ['sigma',
    'beta'])
```

The parameter ρ is sampled using the slice sampler, with initial value 0.9:

```
rho = 0.9
simrho = SliceSampler([post_rho], 0.1, 5, rho, 'rho')
```

In this example, there are two blocks in the MCMC sampler. The code

```
blocks = [simrho, simsigbeta]
```

constructs a Python list that contains the class instances that define the MCMC sampler. In particular, it implies that the MCMC sampler consists of two blocks. Further, ρ will be the first element sampled in the MCMC scheme.

Finally, the sampler can be initialized and sampling undertaken:

```
loglikeinfo = (loglike, kreg + 2, 'yvec')
ms = MCMC(10000, 2000, data, blocks, loglike
      = loglikeinfo)
ms.sampler()
```

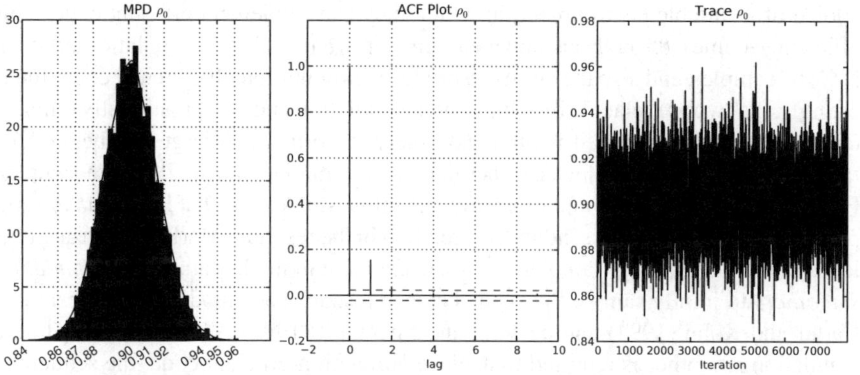

Figure 25.4 Marginal posterior density, autocorrelation function and trace plot based on the MCMC analysis for Example 3. Note that summaries are calculated after removing the burn-in.

The MCMC sampler is run for 10 000 iterations and the first 2000 are discarded. A summary can be generated by the `output` command:

```
ms.output()

The time (seconds) for the Gibbs sampler =  27.72
Number of blocks in Gibbs sampler =  2

              mean        sd       2.5%     97.5%    IFactor
beta[0]     -0.523    0.0716     -0.653    -0.373        3.5
beta[1]       1.85   0.00508       1.84      1.86       3.56
beta[2]      0.455   0.00505      0.445     0.465       3.75
  sigma      0.217   0.00489      0.207     0.226        3.5
    rho      0.901    0.0155      0.872     0.932       3.67

Acceptance rate  beta  =  1.0
Acceptance rate  sigma  =  1.0
Acceptance rate  rho  =  1.0
BIC =  -331.398
Log likelihood =   182.969
```

The total time of estimation is approximately 28 seconds. From the inefficiency factors it is clear that the algorithm is very efficient.

Summary plots generated using `ms.plot('rho')` (Figure 25.4) provide the marginal posterior density, autocorrelation plot and trace plot for the iterates.

25.4 Using PyMCMC efficiently

The fact that MCMC algorithms rely on a large number of iterations to achieve reasonable results and are often implemented on very large problems limits thepractitioner's choice of a suitable environment, in which they can implement efficient code. This efficiency comes through an understanding of what makes a simulation-efficient MCMC sampler, and also the ability to produce computationally efficient code. Interestingly, the two are related. To achieve both simulation and computationally efficient code in MCMC samplers it is often extremely important that large numbers of parameters are sampled in blocks, rather than the alternative of a single move sampler. From the perspective of simulation efficiency it is well known that individually sampling correlated parameters induces correlation in the resultant Markov chain and thus leads to a poorly mixing sampler. A classic example in the literature uses *simulation smoothers* to jointly sample the state vector in a state space model; see for example Carter and Kohn (1994) and de Jong and Shephard (1995). While implementing a simulation smoother is required to achieve simulation-efficient code, the sequential nature of their implementation often renders higher level languages impractical for large problems and thus forces the analyst to write their entire code in a lower level language. This is an inefficient use of time as usually only a small percentage of code needs to be optimized. This drawback is easily circumvented in **PyMCMC** as `Python` makes it easy to use a lower level language to write the specialized module and use the functions directly from `Python`. This ensures **PyMCMC**'s modules can be used for rapid development from `Python` and lower level languages are only resorted to when necessary.

This section aims to provide guidelines for producing efficient code with **PyMCMC**. We discuss alternative external libraries that are available to the user for producing efficient code using **PyMCMC**. Despite the ease of writing specialized modules, this should not be the first resort of the user. Instead, one should ensure that the `Python` code is as efficient as possible using the resources available with `Python`.

Arguably, the first thing the user of **PyMCMC** should concentrate on when optimizing their **PyMCMC** code is to ensure they use as many inbuilt functions and libraries as possible. As most high-performance libraries are written in `C` or `Fortran` this ensures that computationally expensive procedures are computed using code from compiled languages. `Python` users, and hence **PyMCMC** users, have an enormous resource of scientific libraries available to them as a result of the popularity of `Python` in the scientific community. Two of the most important libraries for most users will quite possibly be **Numpy** and **Scipy**. Making use of such libraries is one of the best ways of avoiding large loops in procedures that are called from the MCMC sampler. If a large loop is used inside a function that is called from inside the MCMC sampler then this could mean that a large proportion of the total computation is being done by `Python`, rather than a library that was generated from optimized compiled code. This can have a dramatic effect on the total computation time. As a simple and somewhat trivial example we modify Example 2 from Section 25.3.2 so that a loop is explicitly used to calculate the log likelihood:

```
def logl(store):
    suml=0.0
    for i in xrange(store['yvec'].shape[0]):
```

```
        xbeta=dot(store['xmat'][i,:],store['beta'])
        suml=suml+store['yvec'][i] * xbeta - Exp(xbeta)
    return suml
```

While the two functions to calculate the log-likelihood are mathematically equivalent, the one with the explicit loop is substantially slower. Specifically, the time taken for the MCMC sampler went from 7.3 seconds to 130.19 seconds. As such, this minor modification leads to an approximate 18-fold decrease in the speed of the program.

If the use of an inbuilt function is not possible and the time taken from the program is unacceptable then there are several alternative solutions available to the user. One such solution is to use the package **Weave**, which is a part of the **Scipy** library, to write inline C code which will accelerate the problem area in the code. An example is given below:

```
def logl(store):
    code = """
double sum = 0.0, xbeta;
for(int i=0; i<nobs; i++){
xbeta = 0.0;
for(int j=0; j<kreg; j++){
                          xbeta += xmat(i,j) * beta(j);
                        }
sum += yvec(i) * xbeta - exp(xbeta);
}
return_val = sum;
    """
    yvec = store['yvec']
    xmat = store['xmat']
nobs, kreg = xmat.shape
beta = store['beta']
    return weave.inline(code,['yvec','xmat', 'beta',
        'nobs','kreg'], compiler='gcc',type_converters=
            converters.blitz)
```

The total time taken for the **Weave** version is 4.33 seconds. The reason for the speed increase over the original version that uses **Numpy** functions is that the **Weave** version avoids the construction of temporary matrices that are typically a by-product of overloaded operators.

Another alternative, which is our preferred approach, is to use the Python module **F2py**; see Peterson (2009) for further details. **F2py** allows for the seamless integration of Fortran and Python code. Following on and using the same trivial example we use the following Fortran77 code. This example requires that the user have basic linear algebra subprograms (**BLAS**) and preferably also the automatically tuned linear algebra software (**ATLAS**).

```
c     fortran 77 code used to calculate the likelihood
c     of a loglinear model. Subroutine uses BLAS.

      subroutine logl(xb,xm,bv,yv,llike,n,k)
      implicit none
      integer n, k, i, j
      real*8 xb(n),xm(n,k), bv(k), yv(n), llike
      real*8 alpha, beta

cf2py intent(in,out) logl
cf2py intent(in) yv
cf2py intent(in) bv
cf2py intent(in) xmat
cf2py intent(in) xb

      alpha=1.0
      beta=0.0
      call dgemv('n',n,k,alpha,xm,n,bv,1,beta,xb,1)

      llike=0.0
      do i=1,n
          llike=llike+yv(i)*xb(i)-exp(xb(i))
      enddo
      end
```

In UNIX-type environments, such as Linux and OSX, the code is compiled with the following command (Windows users, see Section 25.4.1):

```
f2py -c loglinear.f -m loglinear -lblas -latlas
```

The loglinear library can then be imported as a `Python` module, and the function `logl` accessed as a standard `Python` function:

```
import loglinear
print loglinear.logl.__doc__

logl - Function signature:
  llike = logl(xb,xm,bv,yv,llike,[n,k])
Required arguments:
  xb : input rank-1 array('d') with bounds (n)
  xm : input rank-2 array('d') with bounds (n,k)
  bv : input rank-1 array('d') with bounds (k)
  yv : input rank-1 array('d') with bounds (n)
  llike : input float
Optional arguments:
  n := len(xb) input int
  k := shape(xm,1) input int
Return objects:
  llike : float
```

Note that this function requires as input xb a rank 1 array of length n. We add this to the data dictionary:

```
data['xb']=zeros(yvec.shape[0])
```

The array data['xb'] is a work array used for the calculation of $\boldsymbol{X\beta}$. As it is stored in the Python dictionary data, it is only created once rather that each time the function logl is called.

It is useful to pass arrays stored in column major order to **F2py** functions since this is what is used in Fortran, rather than the Python default, which is row major order, the default for the C programming language. This can be achieved using the **Numpy** function asfortranarray:

```
data['xmatf']=asfortranarray(xmat)
```

If store['xmat'] were passed to the function loglinear.logl then **F2py** would automatically produce a copy and convert it to column major order each time the function logl is called.

The function logl can now be rewritten as

```
import loglinear
def logl(store):
loglike=array(0.0)
return loglinear.logl(store['xb'],store['xmatf'],
          store['beta'],store['yvec'],loglike)
```

The total time for the MCMC sampler when using **F2py** is 4.03 seconds. This is slightly faster than the version that uses **Weave**, where most likely the small gain can be attributed to the use of **ATLAS**.

The user has many other choices available to them for writing specialized extension modules. For example, if it is the preference of the user it is not much more difficult to use **F2py** to compile procedures written in C, which then can be used directly from Python. Another popular library that can be used to marry C, as well as C++, code with Python is **SWIG**. In our opinion **SWIG** is more difficult than **f2py** for complicated examples. The user may also opt to manually call C and C++ routines using Python and **Numpy**'s C application interface. Another option for C++ users is to use **Boost** Python. These alternative approaches are beyond the scope of this chapter.

25.4.1 Compiling code in Windows

The previous section described how one might go about using **PyMCMC** efficiently. To do this, access to a compiler is necessary. Under most flavours of UNIX, this should pose no problem, but under Microsoft Windows this can be more difficult.

This section provides some brief guidelines to an approach we found workable under Windows.

In order to run **PyMCMC**, `Python`, **Numpy** and **Scipy** are all required, but to have a reasonable developer experience under Windows, we suggest a few additional packages, all of which are freely available:

- **mingw** (http://www.mingw.org/), which provides, among other things, the GNU compiler suite. The user should choose at least **gcc** and **g++**.
- **msys** (http://www.mingw.org/wiki/MSYS), which provides a set of GNU utilities commonly found on Linux. This will make building and compiling code more manageable under Windows.
- **ipython** (http://ipython.scipy.org/moin/), an interactive interface to `Python`, which can be used as an alternative to the idle interface that is distributed with `Python`.
- **pyreadline** (http://ipython.scipy.org/moin/PyReadline/Intro), which provides Windows readline capabilities for **IPython**.
- **gfortran** (http://gcc.gnu.org/wiki/GFortranBinaries), which provides a native Windows `Fortran` compiler.

Once these additional utilities are installed, it should be possible to compile code in different languages. To test that **Weave** works as expected, make sure that the **mingw** bin directory is in your path, and try the following code:

```
import scipy.weave
a=100
scipy.weave.inline('printf("a=%d\\n",a);',['a'],
    verbose=1)
```

The output should be similar to

```
In [4]: scipy.weave.inline('printf("a=%d\\n",a);',['a'],
    verbose=1)
<weave: compiling>
No module named msvccompiler in numpy.distutils; trying
    from distutils
Compiling code...
Found executable c:\mingw\bin\g++.exe
finished compiling (sec):  2.73600006104
a=100
```

F2py requires a `Fortran` compiler. To set this up under Windows, follow the instructions at http://www.scipy.org/F2PY_Windows, and make sure the simple example provided works on your system. The examples presented above additionally require **BLAS** or **ATLAS** to be available. This can be built under Windows (see instructions at http://www.scipy.org/Installing_SciPy/Windows, for example). To check that **F2py** and **ATLAS** are installed correctly, save the following code as, for example, blas_eg.f90:

```
subroutine dgemveg()
  REAL*8 X(2, 3) /1.D0, 2.D0, 3.D0, 4.D0, 5.D0, 6.D0/
  REAL*8 Y(3) /2.D0, 2.D0, 2.D0/
  REAL*8 Z(2)
  CALL DGEMV('N', 2, 3, 1.D0, X, 2, Y, 1, 0.D0, Z, 1)
  PRINT *, Z
end subroutine dgemveg
```

Set the location of your **ATLAS** libraries appropriately, ensure also that **gfortran** is in your path and compile. The following provides a template:

```
ATLAS_LIB_DIR="/d/tmp/pymcmc_win_install/BUILDS/lib"
export PATH=${PATH}:\
/c/Program\ Files/gfortran/libexec/gcc/i586-pc
    -mingw32/4.6.0:\
/c/Program\ Files/gfortran/bin:/c/python26
python /c/Python26/Scripts/f2py.py -c -m foo \
   --fcompiler=gfortran \
   blas_eg.f90 -L${ATLAS_LIB_DIR}-lf77blas-latlas-lg2c
```

This should produce a Python dll (foo.pyd), which can be imported into Python:

```
import foo
dir(foo)
print foo.__doc__
foo.dgemveg()
```

25.5 PyMCMC interacting with R

There are many functions from the R statistical language (R Development Core Team 2010) that can be useful in Bayesian analysis. The **RPy2** (Gautier 2011) Python library can be used to integrate R functions into **PyMCMC** programs. These can be accessed in **PyMCMC** through the **RPy2** Python library. As an example, consider the loglinear model described in Section 25.3.2. The random walk MH requires the specification of a candidate density function (Equation 25.2) and an initial value. The R functions glm and summary.glm can be used to set this to the maximum likelihood estimate $\hat{\beta}$ and the unscaled estimated covariance matrix of the estimated coefficients. The relevant code is summarized below:

```
import rpy2.robjects as robjects

def initial_values(yvec,xmat):
```

```
    ry = robjects.FloatVector(yvec)
    rv = robjects.FloatVector(xmat[:,1:].flatten())
    rx = robjects.r['matrix'](rv, nrow=xmat.shape[0],
                              byrow=True)
    robjects.globalenv['y'] = ry
    robjects.globalenv['x'] = rx
    mod = robjects.r.glm("y~x", family="poisson")
    init_beta =  array(robjects.r.coefficients(mod))
    modsummary = robjects.r.summary(mod)
    scale = array(modsummary.rx2('cov.unscaled'))
    return init_beta,scale


random.seed(12345)
data=loadtxt('count.txt',skiprows=1)
yvec=data[:,0]
xmat=data[:,1:data.shape[1]]
xmat=hstack([ones((data.shape[0],1)),xmat])

data={'yvec':yvec,'xmat':xmat}

init_beta,scale=initial_values(yvec,xmat)

samplebeta=RWMH(posterior,scale,init_beta,'beta')
ms=MCMC(20000,4000,data, [samplebeta],loglike=
    (logl,xmat.shape[1],'yvec'))
ms.sampler()
ms.CODAoutput(filename="loglinear_eg", parameter="beta")
```

It may also be useful to take advantage of the many MCMC analysis functions in R and associated packages. To facilitate this, **PyMCMC** includes a **CODA** (Plummer *et al.* 2006) output format which can easily be read into R for further analysis. A sample R session after **PyMCMC** might look like

```
library(coda)
aa <- read.coda("loglinear_eg.txt","loglinear_eg.ind")
plot(aa)
summary(aa)
raftery.diag(aa)
xyplot(aa)
densityplot(aa)
acfplot(aa,lag.max=500)
```

25.6 Conclusions

In this chapter, we describe the Python software package **PyMCMC**. **PyMCMC** takes advantage of the flexibility and extensibility of Python to provide the user with a code-efficient way of constructing MCMC samplers. The **PyMCMC** package includes classes for the MCMC sampler MH, independent MH, random walk MH,

OBMC and slice sampling algorithms. It also contains an inbuilt module for Bayesian regression analysis. We demonstrate **PyMCMC** using an example of Bayesian regression analysis with stochastic search variable selection, a loglinear model and also a time series regression analysis with first-order autoregressive errors. We demonstrate how to optimize **PyMCMC** using **Numpy** functions, inline `C` code using **Weave** and `Fortran77` using **F2py**, where necessary. We further demonstrate how to call `R` functions using **RPy2**.

25.7 Obtaining PyMCMC

The source code for **PyMCMC** is held in a **git** repostitory, and can be cloned by

```
git clone
https://bitbucket.org/christophermarkstrickland/pymcmc.git
```

As an alternative, for UNIX-based operating systems, a pre-packaged source distribution is available at

```
https://bitbucket.org/
christophermarkstrickland/pymcmc/downloads/pymcmc-1.0.tar.gz.
```

For most users, installation should require only

```
python setup.py install
```

More detailed installation instructions, including information on building for Microsoft Windows or Macintosh systems, are included in the `INSTALL` file in the source distribution. Additionally, binaries are available for Mac and Windows systems from

```
https://bitbucket.org/christophermarkstrickland/pymcmc/
 wiki/installing.
```

References

Carter C and Kohn R 1994 On Gibbs sampling for state space models. *Biometrika* **81**, 541–553.

Chib S and Greenberg E 1996 Markov chain Monte Carlo simulation methods in econometrics. *Econometric Theory* **12**, 409–431.

de Jong P and Shephard N 1995 The simulation smoother for time series models. *Biometrika* **82**, 339–350.

Garthwaite PH, Fan Y and Scisson SA 2010 Adaptive optimal scaling of Metropolis-Hastings algorithms using the Robbins-Monroe process. Technical Report, University of New South Wales.

Gautier L 2011 Rpy2: a simple and efficient access to R from Python. (accessed 6 March 2011).

Gelfand AE and Smith AFM 1990 Sampling-based approaches to calculating marginal densities. *Journal of the American Statistical Association* **85**, 398–409.

Gelfand AE, Sahu SK and Carlin BP 1995 Efficient parametrisations for normal linear mixed models. *Biometrika* **65**(3), 479–488.

Gelman A, Carlin JB, Stern HS and Rubin DB 2004 *Bayesian Data Analysis*. Chapman & Hall/CRC, Boca Raton, FL.

George EI and McCulloch RE 1993 Variable selection via Gibbs sampling. *Journal of the American Statistical Association* **88**, 881–889.

Geus R 2011 Pysparse. (accessed 6 March 2011).

Hunter JD 2007 Matplotlib: a 2D graphics environment. *Computing in Science and Engineering* **9**, 90–95.

Kass RE and Raftery AE 1995 Bayes factors. *Journal of the American Statistical Association* **90**, 773–795.

Kim S, Shephard N and Chib S 1998 Stochastic volatility: likelihood inference and comparison with arch models. *Review of Economic Studies* **65**(3), 361–393.

Liang F, Lui JS and Wong WH 2000 The use of multiple-try method and local optimization in Metropolis sampling. *Journal of the American Statistical Association* **95**, 121–134.

Lui JS, Wong WH and Kong A 1994 Covariance structure of the Gibbs sampler with applications to the comparisons of estimators and augmentations schemes. *Journal of the Royal Statistical Society, Series B* **57**(1), 157–169.

Marin JM and Robert CP 2007 *Bayesian core*. Springer, New York.

Neal RM 2003 Slice sampling. *Annals of Statistics* **31**(3), 705–741.

Oliphant TE 2007 Python for scientific computing. *Computing in Science and Engineering* **9**, 10–20.

Peterson P 2009 F2py: a tool for connecting Fortran and Python programs. *International Journal of Computational Science and Engineering* **4**, 296–605.

Pitt M and Shephard N 1999 Analytic convergence rates and parameterisation issues for the Gibbs sampler applied to state space models. *Journal of Time Series Analysis* **20**, 63–85.

Plummer M, Best N, Cowles K and Vines K 2006 CODA: convergence diagnosis and output analysis for MCMC. *R News* **6**(1), 7–11.

R Development Core Team 2010 *R: A Language and Environment for Statistical Computing*. R Foundation for Statistical Computing, Vienna.

Robert CP and Casella G 1999 *Monte Carlo Statistical Methods*. Springer, New York.

Robert CP and Mengersen KL 1999 Reparameterisation issues in mixture modelling and their bearing on MCMC algorithms. *Computational Statistics & Data Analysis* **29**(3), 325–343.

Sahu SK and Roberts GO 1997 Updating schemes, correlation structure, blocking and parameterisation for the Gibbs sampler. *Journal of the Royal Statistical Society, Series B* **59**, 291–317.

Schnatter SF 2004 *Efficient Bayesian Parameter Estimation for State Space Models Based on Reparameterisations, State Space and Unobserved Component Models: Theory and Applications*. Cambridge University Press, Cambridge.

Strickland CM, Martin GM and Forbes CS 2008 Parameterisation and efficient MCMC estimation of non-Gaussian state space models. *Computational Statistics & Data Analysis* **52**, 2911–2930.

van Rossum G 1995 Python tutorial. Technical Report cs-r9526. Centrum voor Wiskunde en Informatica (CWI), Amsterdam.

Zellner A 1971 *An Introduction to Bayesian Inference in Econometrics*. John Wiley & Sons, Inc., New York.

Index

Case Studies in Bayesian Statistical Modelling and Analysis, First Edition. Edited by Clair L. Alston, Kerrie L. Mengersen and Anthony N. Pettitt.

WILEY SERIES IN PROBABILITY AND STATISTICS

ESTABLISHED BY WALTER A. SHEWHART AND SAMUEL S. WILKS

Editors: David. J. Balding, Noel A.C. Cressie, Garrett M. Fitzmaurice, Harvey Goldstein, Iain M. Johnstone, Geert Molenberghs, David W. Scott, Adrian F. M. Smith, Ruey S. Tsay, Sanford Weisberg

Editors Emeriti: Vic Barnett, Ralph A. Bradley, J. Stuart Hunter, J.B. Kadane, David G. Kendall, Jozef L. Teugels

The ***Wiley Series in Probability and Statistics*** is well established and authoritative. It covers many topics of current research interest in both pure and applied statistics and probability theory. Written by leading statisticians and institutions, the titles span both state-of-the-art developments in the field and classical methods.

Reflecting the wide range of current research in statistics, the series encompasses applied, methodological and theoretical statistics, ranging from applications and new techniques made possible by advances in computerized practice to rigorous treatment of theoretical approaches.

This series provides essential and invaluable reading for all statisticians, whether in academia, industry, government, or research.

† ABRAHAM and LEDOLTER · Statistical Methods for Forecasting
AGRESTI · Analysis of Ordinal Categorical Data, *Second Edition*
AGRESTI · An Introduction to Categorical Data Analysis, *Second Edition*
AGRESTI · Categorical Data Analysis, *Second Edition*
ALSTON, MENGERSEN and PETTITT (editors) · Case Studies in Bayesian Statistical Modelling and Analysis
ALTMAN, GILL, and McDONALD · Numerical Issues in Statistical Computing for the Social Scientist
AMARATUNGA and CABRERA · Exploration and Analysis of DNA Microarray and Protein Array Data
ANDĚL · Mathematics of Chance
ANDERSON · An Introduction to Multivariate Statistical Analysis, *Third Edition*
* ANDERSON · The Statistical Analysis of Time Series
ANDERSON, AUQUIER, HAUCK, OAKES, VANDAELE, and WEISBERG · Statistical Methods for Comparative Studies
ANDERSON and LOYNES · The Teaching of Practical Statistics
ARMITAGE and DAVID (editors) · Advances in Biometry
ARNOLD, BALAKRISHNAN, and NAGARAJA · Records
* ARTHANARI and DODGE · Mathematical Programming in Statistics
* BAILEY · The Elements of Stochastic Processes with Applications to the Natural Sciences
BAJORSKI · Statistics for Imaging, Optics, and Photonics
BALAKRISHNAN and KOUTRAS · Runs and Scans with Applications
BALAKRISHNAN and NG · Precedence-Type Tests and Applications
BARNETT · Comparative Statistical Inference, *Third Edition*
BARNETT · Environmental Statistics
BARNETT and LEWIS · Outliers in Statistical Data, *Third Edition*
BARTHOLOMEW, KNOTT, and MOUSTAKI · Latent Variable Models and Factor Analysis: A Unified Approach, *Third Edition*

*Now available in a lower priced paperback edition in the Wiley Classics Library.
† Now available in a lower priced paperback edition in the Wiley-Interscience Paperback Series.

BARTOSZYNSKI and NIEWIADOMSKA-BUGAJ · Probability and Statistical Inference, *Second Edition*
BASILEVSKY · Statistical Factor Analysis and Related Methods: Theory and Applications
BATES and WATTS · Nonlinear Regression Analysis and Its Applications
BECHHOFER, SANTNER, and GOLDSMAN · Design and Analysis of Experiments for Statistical Selection, Screening, and Multiple Comparisons
BEIRLANT, GOEGEBEUR, SEGERS, TEUGELS, and DE WAAL · Statistics of Extremes: Theory and Applications
BELSLEY · Conditioning Diagnostics: Collinearity and Weak Data in Regression
† BELSLEY, KUH, and WELSCH · Regression Diagnostics: Identifying Influential Data and Sources of Collinearity
BENDAT and PIERSOL · Random Data: Analysis and Measurement Procedures, *Fourth Edition*
BERNARDO and SMITH · Bayesian Theory
BHAT and MILLER · Elements of Applied Stochastic Processes, *Third Edition*
BHATTACHARYA and WAYMIRE · Stochastic Processes with Applications
BIEMER, GROVES, LYBERG, MATHIOWETZ, and SUDMAN · Measurement Errors in Surveys
BILLINGSLEY · Convergence of Probability Measures, *Second Edition*
BILLINGSLEY · Probability and Measure, *Anniversary Edition*
BIRKES and DODGE · Alternative Methods of Regression
BISGAARD and KULAHCI · Time Series Analysis and Forecasting by Example
BISWAS, DATTA, FINE, and SEGAL · Statistical Advances in the Biomedical Sciences: Clinical Trials, Epidemiology, Survival Analysis, and Bioinformatics
BLISCHKE and MURTHY (editors) · Case Studies in Reliability and Maintenance
BLISCHKE and MURTHY · Reliability: Modeling, Prediction, and Optimization
BLOOMFIELD · Fourier Analysis of Time Series: An Introduction, *Second Edition*
BOLLEN · Structural Equations with Latent Variables
BOLLEN and CURRAN · Latent Curve Models: A Structural Equation Perspective
BOROVKOV · Ergodicity and Stability of Stochastic Processes
BOSQ and BLANKE · Inference and Prediction in Large Dimensions
BOULEAU · Numerical Methods for Stochastic Processes
* BOX · Bayesian Inference in Statistical Analysis
BOX · Improving Almost Anything, *Revised Edition*
* BOX and DRAPER · Evolutionary Operation: A Statistical Method for Process Improvement
BOX and DRAPER · Response Surfaces, Mixtures, and Ridge Analyses, *Second Edition*
BOX, HUNTER, and HUNTER · Statistics for Experimenters: Design, Innovation, and Discovery, *Second Editon*
BOX, JENKINS, and REINSEL · Time Series Analysis: Forcasting and Control, *Fourth Edition*
BOX, LUCEÑO, and PANIAGUA-QUIÑONES · Statistical Control by Monitoring and Adjustment, *Second Edition*
* BROWN and HOLLANDER · Statistics: A Biomedical Introduction
CAIROLI and DALANG · Sequential Stochastic Optimization

*Now available in a lower priced paperback edition in the Wiley Classics Library.
† Now available in a lower priced paperback edition in the Wiley-Interscience Paperback Series.

CASTILLO, HADI, BALAKRISHNAN, and SARABIA · Extreme Value and Related Models with Applications in Engineering and Science
CHAN · Time Series: Applications to Finance with R and S-Plus®, *Second Edition*
CHARALAMBIDES · Combinatorial Methods in Discrete Distributions
CHATTERJEE and HADI · Regression Analysis by Example, *Fourth Edition*
CHATTERJEE and HADI · Sensitivity Analysis in Linear Regression
CHERNICK · Bootstrap Methods: A Guide for Practitioners and Researchers, *Second Edition*
CHERNICK and FRIIS · Introductory Biostatistics for the Health Sciences
CHILES and DELFINER · Geostatistics: Modeling Spatial Uncertainty, *Second Edition*
CHOW and LIU · Design and Analysis of Clinical Trials: Concepts and Methodologies, *Second Edition*
CLARKE · Linear Models: The Theory and Application of Analysis of Variance
CLARKE and DISNEY · Probability and Random Processes: A First Course with Applications, *Second Edition*
* COCHRAN and COX · Experimental Designs, *Second Edition*
COLLINS and LANZA · Latent Class and Latent Transition Analysis: With Applications in the Social, Behavioral, and Health Sciences
CONGDON · Applied Bayesian Modelling
CONGDON · Bayesian Models for Categorical Data
CONGDON · Bayesian Statistical Modelling, *Second Edition*
CONOVER · Practical Nonparametric Statistics, *Third Edition*
COOK · Regression Graphics
COOK and WEISBERG · An Introduction to Regression Graphics
COOK and WEISBERG · Applied Regression Including Computing and Graphics
CORNELL · A Primer on Experiments with Mixtures
CORNELL · Experiments with Mixtures, Designs, Models, and the Analysis of Mixture Data, *Third Edition*
COX · A Handbook of Introductory Statistical Methods
CRESSIE · Statistics for Spatial Data, *Revised Edition*
CRESSIE and WIKLE · Statistics for Spatio-Temporal Data
CSÖRGŐ and HORVÁTH · Limit Theorems in Change Point Analysis
DAGPUNAR · Simulation and Monte Carlo: With Applications in Finance and MCMC
DANIEL · Applications of Statistics to Industrial Experimentation
DANIEL · Biostatistics: A Foundation for Analysis in the Health Sciences, *Eighth Edition*
* DANIEL · Fitting Equations to Data: Computer Analysis of Multifactor Data, *Second Edition*
DASU and JOHNSON · Exploratory Data Mining and Data Cleaning
DAVID and NAGARAJA · Order Statistics, *Third Edition*
* DEGROOT, FIENBERG, and KADANE · Statistics and the Law
DEL CASTILLO · Statistical Process Adjustment for Quality Control
DEMARIS · Regression with Social Data: Modeling Continuous and Limited Response Variables
DEMIDENKO · Mixed Models: Theory and Applications

*Now available in a lower priced paperback edition in the Wiley Classics Library.
† Now available in a lower priced paperback edition in the Wiley-Interscience Paperback Series.

DENISON, HOLMES, MALLICK and SMITH · Bayesian Methods for Nonlinear Classification and Regression
DETTE and STUDDEN · The Theory of Canonical Moments with Applications in Statistics, Probability, and Analysis
DEY and MUKERJEE · Fractional Factorial Plans
DE ROCQUIGNY · Modelling Under Risk and Uncertainty: An Introduction to Statistical, Phenomenological and Computational Models
DILLON and GOLDSTEIN · Multivariate Analysis: Methods and Applications
* DODGE and ROMIG · Sampling Inspection Tables, *Second Edition*
* DOOB · Stochastic Processes
DOWDY, WEARDEN, and CHILKO · Statistics for Research, *Third Edition*
DRAPER and SMITH · Applied Regression Analysis, *Third Edition*
DRYDEN and MARDIA · Statistical Shape Analysis
DUDEWICZ and MISHRA · Modern Mathematical Statistics
DUNN and CLARK · Basic Statistics: A Primer for the Biomedical Sciences, *Fourth Edition*
DUPUIS and ELLIS · A Weak Convergence Approach to the Theory of Large Deviations
EDLER and KITSOS · Recent Advances in Quantitative Methods in Cancer and Human Health Risk Assessment
* ELANDT-JOHNSON and JOHNSON · Survival Models and Data Analysis
ENDERS · Applied Econometric Time Series, *Third Edition*
† ETHIER and KURTZ · Markov Processes: Characterization and Convergence
EVANS, HASTINGS, and PEACOCK · Statistical Distributions, *Third Edition*
EVERITT, LANDAU, LEESE, and STAHL · Cluster Analysis, *Fifth Edition*
FEDERER and KING · Variations on Split Plot and Split Block Experiment Designs
FELLER · An Introduction to Probability Theory and Its Applications, Volume I, *Third Edition,* Revised; Volume II, *Second Edition*
FITZMAURICE, LAIRD, and WARE · Applied Longitudinal Analysis, *Second Edition*
* FLEISS · The Design and Analysis of Clinical Experiments
FLEISS · Statistical Methods for Rates and Proportions, *Third Edition*
† FLEMING and HARRINGTON · Counting Processes and Survival Analysis
FUJIKOSHI, ULYANOV, and SHIMIZU · Multivariate Statistics: High-Dimensional and Large-Sample Approximations
FULLER · Introduction to Statistical Time Series, *Second Edition*
† FULLER · Measurement Error Models
GALLANT · Nonlinear Statistical Models
GEISSER · Modes of Parametric Statistical Inference
GELMAN and MENG · Applied Bayesian Modeling and Causal Inference from Incomplete-Data Perspectives
GEWEKE · Contemporary Bayesian Econometrics and Statistics
GHOSH, MUKHOPADHYAY, and SEN · Sequential Estimation
GIESBRECHT and GUMPERTZ · Planning, Construction, and Statistical Analysis of Comparative Experiments
GIFI · Nonlinear Multivariate Analysis
GIVENS and HOETING · Computational Statistics

*Now available in a lower priced paperback edition in the Wiley Classics Library.
† Now available in a lower priced paperback edition in the Wiley-Interscience Paperback Series.

GLASSERMAN and YAO · Monotone Structure in Discrete-Event Systems
GNANADESIKAN · Methods for Statistical Data Analysis of Multivariate Observations, *Second Edition*
GOLDSTEIN · Multilevel Statistical Models, *Fourth Edition*
GOLDSTEIN and LEWIS · Assessment: Problems, Development, and Statistical Issues
GOLDSTEIN and WOOFF · Bayes Linear Statistics
GREENWOOD and NIKULIN · A Guide to Chi-Squared Testing
GROSS, SHORTLE, THOMPSON, and HARRIS · Fundamentals of Queueing Theory, *Fourth Edition*
GROSS, SHORTLE, THOMPSON, and HARRIS · Solutions Manual to Accompany Fundamentals of Queueing Theory, *Fourth Edition*
* HAHN and SHAPIRO · Statistical Models in Engineering
HAHN and MEEKER · Statistical Intervals: A Guide for Practitioners
HALD · A History of Probability and Statistics and their Applications Before 1750
† HAMPEL · Robust Statistics: The Approach Based on Influence Functions
HARTUNG, KNAPP, and SINHA · Statistical Meta-Analysis with Applications
HEIBERGER · Computation for the Analysis of Designed Experiments
HEDAYAT and SINHA · Design and Inference in Finite Population Sampling
HEDEKER and GIBBONS · Longitudinal Data Analysis
HELLER · MACSYMA for Statisticians
HERITIER, CANTONI, COPT, and VICTORIA-FESER · Robust Methods in Biostatistics
HINKELMANN and KEMPTHORNE · Design and Analysis of Experiments, Volume 1: Introduction to Experimental Design, *Second Edition*
HINKELMANN and KEMPTHORNE · Design and Analysis of Experiments, Volume 2: Advanced Experimental Design
HINKELMANN (editor) · Design and Analysis of Experiments, Volume 3: Special Designs and Applications
* HOAGLIN, MOSTELLER, and TUKEY · Fundamentals of Exploratory Analysis of Variance
* HOAGLIN, MOSTELLER, and TUKEY · Exploring Data Tables, Trends and Shapes
* HOAGLIN, MOSTELLER, and TUKEY · Understanding Robust and Exploratory Data Analysis
HOCHBERG and TAMHANE · Multiple Comparison Procedures
HOCKING · Methods and Applications of Linear Models: Regression and the Analysis of Variance, *Second Edition*
HOEL · Introduction to Mathematical Statistics, *Fifth Edition*
HOGG and KLUGMAN · Loss Distributions
HOLLANDER and WOLFE · Nonparametric Statistical Methods, *Second Edition*
HOSMER and LEMESHOW · Applied Logistic Regression, *Second Edition*
HOSMER, LEMESHOW, and MAY · Applied Survival Analysis: Regression Modeling of Time-to-Event Data, *Second Edition*
HUBER · Data Analysis: What Can Be Learned From the Past 50 Years
HUBER · Robust Statistics
† HUBER and RONCHETTI · Robust Statistics, *Second Edition*
HUBERTY · Applied Discriminant Analysis, *Second Edition*

*Now available in a lower priced paperback edition in the Wiley Classics Library.
† Now available in a lower priced paperback edition in the Wiley-Interscience Paperback Series.

HUBERTY and OLEJNIK · Applied MANOVA and Discriminant Analysis, *Second Edition*
HUITEMA · The Analysis of Covariance and Alternatives: Statistical Methods for Experiments, Quasi-Experiments, and Single-Case Studies, *Second Edition*
HUNT and KENNEDY · Financial Derivatives in Theory and Practice, *Revised Edition*
HURD and MIAMEE · Periodically Correlated Random Sequences: Spectral Theory and Practice
HUSKOVA, BERAN, and DUPAC · Collected Works of Jaroslav Hajek—with Commentary
HUZURBAZAR · Flowgraph Models for Multistate Time-to-Event Data
INSUA, RUGGERI and WIPER · Bayesian Analysis of Stochastic Process Models
JACKMAN · Bayesian Analysis for the Social Sciences
† JACKSON · A User's Guide to Principle Components
JOHN · Statistical Methods in Engineering and Quality Assurance
JOHNSON · Multivariate Statistical Simulation
JOHNSON and BALAKRISHNAN · Advances in the Theory and Practice of Statistics: A Volume in Honor of Samuel Kotz
JOHNSON, KEMP, and KOTZ · Univariate Discrete Distributions, *Third Edition*
JOHNSON and KOTZ (editors) · Leading Personalities in Statistical Sciences: From the Seventeenth Century to the Present
JOHNSON, KOTZ, and BALAKRISHNAN · Continuous Univariate Distributions, Volume 1, *Second Edition*
JOHNSON, KOTZ, and BALAKRISHNAN · Continuous Univariate Distributions, Volume 2, *Second Edition*
JOHNSON, KOTZ, and BALAKRISHNAN · Discrete Multivariate Distributions
JUDGE, GRIFFITHS, HILL, LÜTKEPOHL, and LEE · The Theory and Practice of Econometrics, *Second Edition*
JUREK and MASON · Operator-Limit Distributions in Probability Theory
KADANE · Bayesian Methods and Ethics in a Clinical Trial Design
KADANE AND SCHUM · A Probabilistic Analysis of the Sacco and Vanzetti Evidence
KALBFLEISCH and PRENTICE · The Statistical Analysis of Failure Time Data, *Second Edition*
KARIYA and KURATA · Generalized Least Squares
KASS and VOS · Geometrical Foundations of Asymptotic Inference
† KAUFMAN and ROUSSEEUW · Finding Groups in Data: An Introduction to Cluster Analysis
KEDEM and FOKIANOS · Regression Models for Time Series Analysis
KENDALL, BARDEN, CARNE, and LE · Shape and Shape Theory
KHURI · Advanced Calculus with Applications in Statistics, *Second Edition*
KHURI, MATHEW, and SINHA · Statistical Tests for Mixed Linear Models
* KISH · Statistical Design for Research
KLEIBER and KOTZ · Statistical Size Distributions in Economics and Actuarial Sciences
KLEMELÄ · Smoothing of Multivariate Data: Density Estimation and Visualization
KLUGMAN, PANJER, and WILLMOT · Loss Models: From Data to Decisions, *Third Edition*
KLUGMAN, PANJER, and WILLMOT · Solutions Manual to Accompany Loss Models: From Data to Decisions, *Third Edition*

*Now available in a lower priced paperback edition in the Wiley Classics Library.
† Now available in a lower priced paperback edition in the Wiley-Interscience Paperback Series.

KOSKI and NOBLE · Bayesian Networks: An Introduction
KOTZ, BALAKRISHNAN, and JOHNSON · Continuous Multivariate Distributions, Volume 1, *Second Edition*
KOTZ and JOHNSON (editors) · Encyclopedia of Statistical Sciences: Volumes 1 to 9 with Index
KOTZ and JOHNSON (editors) · Encyclopedia of Statistical Sciences: Supplement Volume
KOTZ, READ, and BANKS (editors) · Encyclopedia of Statistical Sciences: Update Volume 1
KOTZ, READ, and BANKS (editors) · Encyclopedia of Statistical Sciences: Update Volume 2
KOWALSKI and TU · Modern Applied U-Statistics
KRISHNAMOORTHY and MATHEW · Statistical Tolerance Regions: Theory, Applications, and Computation
KROESE, TAIMRE, and BOTEV · Handbook of Monte Carlo Methods
KROONENBERG · Applied Multiway Data Analysis
KULINSKAYA, MORGENTHALER, and STAUDTE · Meta Analysis: A Guide to Calibrating and Combining Statistical Evidence
KULKARNI and HARMAN · An Elementary Introduction to Statistical Learning Theory
KUROWICKA and COOKE · Uncertainty Analysis with High Dimensional Dependence Modelling
KVAM and VIDAKOVIC · Nonparametric Statistics with Applications to Science and Engineering
LACHIN · Biostatistical Methods: The Assessment of Relative Risks, *Second Edition*
LAD · Operational Subjective Statistical Methods: A Mathematical, Philosophical, and Historical Introduction
LAMPERTI · Probability: A Survey of the Mathematical Theory, *Second Edition*
LAWLESS · Statistical Models and Methods for Lifetime Data, *Second Edition*
LAWSON · Statistical Methods in Spatial Epidemiology, *Second Edition*
LE · Applied Categorical Data Analysis, *Second Edition*
LE · Applied Survival Analysis
LEE · Structural Equation Modeling: A Bayesian Approach
LEE and WANG · Statistical Methods for Survival Data Analysis, *Third Edition*
LePAGE and BILLARD · Exploring the Limits of Bootstrap
LESSLER and KALSBEEK · Nonsampling Errors in Surveys
LEYLAND and GOLDSTEIN (editors) · Multilevel Modelling of Health Statistics
LIAO · Statistical Group Comparison
LIN · Introductory Stochastic Analysis for Finance and Insurance
LITTLE and RUBIN · Statistical Analysis with Missing Data, *Second Edition*
LLOYD · The Statistical Analysis of Categorical Data
LOWEN and TEICH · Fractal-Based Point Processes
MAGNUS and NEUDECKER · Matrix Differential Calculus with Applications in Statistics and Econometrics, *Revised Edition*
MALLER and ZHOU · Survival Analysis with Long Term Survivors
MARCHETTE · Random Graphs for Statistical Pattern Recognition
MARDIA and JUPP · Directional Statistics

*Now available in a lower priced paperback edition in the Wiley Classics Library.
† Now available in a lower priced paperback edition in the Wiley-Interscience Paperback Series.

MARKOVICH · Nonparametric Analysis of Univariate Heavy-Tailed Data: Research and Practice
MARONNA, MARTIN and YOHAI · Robust Statistics: Theory and Methods
MASON, GUNST, and HESS · Statistical Design and Analysis of Experiments with Applications to Engineering and Science, *Second Edition*
McCULLOCH, SEARLE, and NEUHAUS · Generalized, Linear, and Mixed Models, *Second Edition*
McFADDEN · Management of Data in Clinical Trials, *Second Edition*
* McLACHLAN · Discriminant Analysis and Statistical Pattern Recognition
McLACHLAN, DO, and AMBROISE · Analyzing Microarray Gene Expression Data
McLACHLAN and KRISHNAN · The EM Algorithm and Extensions, *Second Edition*
McLACHLAN and PEEL · Finite Mixture Models
McNEIL · Epidemiological Research Methods
MEEKER and ESCOBAR · Statistical Methods for Reliability Data
MEERSCHAERT and SCHEFFLER · Limit Distributions for Sums of Independent Random Vectors: Heavy Tails in Theory and Practice
MENGERSEN, ROBERT, and TITTERINGTON · Mixtures: Estimation and Applications
MICKEY, DUNN, and CLARK · Applied Statistics: Analysis of Variance and Regression, *Third Edition*
* MILLER · Survival Analysis, *Second Edition*
MONTGOMERY, JENNINGS, and KULAHCI · Introduction to Time Series Analysis and Forecasting
MONTGOMERY, PECK, and VINING · Introduction to Linear Regression Analysis, *Fifth Edition*
MORGENTHALER and TUKEY · Configural Polysampling: A Route to Practical Robustness
MUIRHEAD · Aspects of Multivariate Statistical Theory
MULLER and STOYAN · Comparison Methods for Stochastic Models and Risks
MURTHY, XIE, and JIANG · Weibull Models
MYERS, MONTGOMERY, and ANDERSON-COOK · Response Surface Methodology: Process and Product Optimization Using Designed Experiments, *Third Edition*
MYERS, MONTGOMERY, VINING, and ROBINSON · Generalized Linear Models. With Applications in Engineering and the Sciences, *Second Edition*
NATVIG · Multistate Systems Reliability Theory With Applications
† NELSON · Accelerated Testing, Statistical Models, Test Plans, and Data Analyses
† NELSON · Applied Life Data Analysis
NEWMAN · Biostatistical Methods in Epidemiology
NG, TAIN, and TANG · Dirichlet Theory: Theory, Methods and Applications
OKABE, BOOTS, SUGIHARA, and CHIU · Spatial Tesselations: Concepts and Applications of Voronoi Diagrams, *Second Edition*
OLIVER and SMITH · Influence Diagrams, Belief Nets and Decision Analysis
PALTA · Quantitative Methods in Population Health: Extensions of Ordinary Regressions
PANJER · Operational Risk: Modeling and Analytics
PANKRATZ · Forecasting with Dynamic Regression Models
PANKRATZ · Forecasting with Univariate Box-Jenkins Models: Concepts and Cases

*Now available in a lower priced paperback edition in the Wiley Classics Library.
† Now available in a lower priced paperback edition in the Wiley-Interscience Paperback Series.

PARDOUX · Markov Processes and Applications: Algorithms, Networks, Genome and Finance
PARMIGIANI and INOUE · Decision Theory: Principles and Approaches
* PARZEN · Modern Probability Theory and Its Applications
PEÑA, TIAO, and TSAY · A Course in Time Series Analysis
PESARIN and SALMASO · Permutation Tests for Complex Data: Applications and Software
PIANTADOSI · Clinical Trials: A Methodologic Perspective, *Second Edition*
POURAHMADI · Foundations of Time Series Analysis and Prediction Theory
POWELL · Approximate Dynamic Programming: Solving the Curses of Dimensionality, *Second Edition*
POWELL and RYZHOV · Optimal Learning
PRESS · Subjective and Objective Bayesian Statistics, *Second Edition*
PRESS and TANUR · The Subjectivity of Scientists and the Bayesian Approach
PURI, VILAPLANA, and WERTZ · New Perspectives in Theoretical and Applied Statistics
† PUTERMAN · Markov Decision Processes: Discrete Stochastic Dynamic Programming
QIU · Image Processing and Jump Regression Analysis
* RAO · Linear Statistical Inference and Its Applications, *Second Edition*
RAO · Statistical Inference for Fractional Diffusion Processes
RAUSAND and HØYLAND · System Reliability Theory: Models, Statistical Methods, and Applications, *Second Edition*
RAYNER, THAS, and BEST · Smooth Tests of Goodnes of Fit: Using R, *Second Edition*
RENCHER · Linear Models in Statistics, *Second Edition*
RENCHER · Methods of Multivariate Analysis, *Second Edition*
RENCHER · Multivariate Statistical Inference with Applications
RIGDON and BASU · Statistical Methods for the Reliability of Repairable Systems
* RIPLEY · Spatial Statistics
* RIPLEY · Stochastic Simulation
ROHATGI and SALEH · An Introduction to Probability and Statistics, *Second Edition*
ROLSKI, SCHMIDLI, SCHMIDT, and TEUGELS · Stochastic Processes for Insurance and Finance
ROSENBERGER and LACHIN · Randomization in Clinical Trials: Theory and Practice
ROSSI, ALLENBY, and McCULLOCH · Bayesian Statistics and Marketing
† ROUSSEEUW and LEROY · Robust Regression and Outlier Detection
ROYSTON and SAUERBREI · Multivariate Model Building: A Pragmatic Approach to Regression Analysis Based on Fractional Polynomials for Modeling Continuous Variables
* RUBIN · Multiple Imputation for Nonresponse in Surveys
RUBINSTEIN and KROESE · Simulation and the Monte Carlo Method, *Second Edition*
RUBINSTEIN and MELAMED · Modern Simulation and Modeling
RYAN · Modern Engineering Statistics
RYAN · Modern Experimental Design
RYAN · Modern Regression Methods, *Second Edition*
RYAN · Statistical Methods for Quality Improvement, *Third Edition*
SALEH · Theory of Preliminary Test and Stein-Type Estimation with Applications

*Now available in a lower priced paperback edition in the Wiley Classics Library.
† Now available in a lower priced paperback edition in the Wiley-Interscience Paperback Series.

SALTELLI, CHAN, and SCOTT (editors) · Sensitivity Analysis
SCHERER · Batch Effects and Noise in Microarray Experiments: Sources and Solutions
* SCHEFFE · The Analysis of Variance
SCHIMEK · Smoothing and Regression: Approaches, Computation, and Application
SCHOTT · Matrix Analysis for Statistics, *Second Edition*
SCHOUTENS · Levy Processes in Finance: Pricing Financial Derivatives
SCOTT · Multivariate Density Estimation: Theory, Practice, and Visualization
* SEARLE · Linear Models
† SEARLE · Linear Models for Unbalanced Data
† SEARLE · Matrix Algebra Useful for Statistics
† SEARLE, CASELLA, and McCULLOCH · Variance Components
SEARLE and WILLETT · Matrix Algebra for Applied Economics
SEBER · A Matrix Handbook For Statisticians
† SEBER · Multivariate Observations
SEBER and LEE · Linear Regression Analysis, *Second Edition*
† SEBER and WILD · Nonlinear Regression
SENNOTT · Stochastic Dynamic Programming and the Control of Queueing Systems
* SERFLING · Approximation Theorems of Mathematical Statistics
SHAFER and VOVK · Probability and Finance: It's Only a Game!
SHERMAN · Spatial Statistics and Spatio-Temporal Data: Covariance Functions and Directional Properties
SILVAPULLE and SEN · Constrained Statistical Inference: Inequality, Order, and Shape Restrictions
SINGPURWALLA · Reliability and Risk: A Bayesian Perspective
SMALL and McLEISH · Hilbert Space Methods in Probability and Statistical Inference
SRIVASTAVA · Methods of Multivariate Statistics
STAPLETON · Linear Statistical Models, *Second Edition*
STAPLETON · Models for Probability and Statistical Inference: Theory and Applications
STAUDTE and SHEATHER · Robust Estimation and Testing
STOYAN · Counterexamples in Probability, *Second Edition*
STOYAN, KENDALL, and MECKE · Stochastic Geometry and Its Applications, *Second Edition*
STOYAN and STOYAN · Fractals, Random Shapes and Point Fields: Methods of Geometrical Statistics
STREET and BURGESS · The Construction of Optimal Stated Choice Experiments: Theory and Methods
STYAN · The Collected Papers of T. W. Anderson: 1943–1985
SUTTON, ABRAMS, JONES, SHELDON, and SONG · Methods for Meta-Analysis in Medical Research
TAKEZAWA · Introduction to Nonparametric Regression
TAMHANE · Statistical Analysis of Designed Experiments: Theory and Applications
TANAKA · Time Series Analysis: Nonstationary and Noninvertible Distribution Theory
THOMPSON · Empirical Model Building: Data, Models, and Reality, *Second Edition*
THOMPSON · Sampling, *Third Edition*
THOMPSON · Simulation: A Modeler's Approach

*Now available in a lower priced paperback edition in the Wiley Classics Library.
† Now available in a lower priced paperback edition in the Wiley-Interscience Paperback Series.

THOMPSON and SEBER · Adaptive Sampling
THOMPSON, WILLIAMS, and FINDLAY · Models for Investors in Real World Markets
TIERNEY · LISP-STAT: An Object-Oriented Environment for Statistical Computing and Dynamic Graphics
TSAY · Analysis of Financial Time Series, *Third Edition*
UPTON and FINGLETON · Spatial Data Analysis by Example, Volume II: Categorical and Directional Data
† VAN BELLE · Statistical Rules of Thumb, *Second Edition*
VAN BELLE, FISHER, HEAGERTY, and LUMLEY · Biostatistics: A Methodology for the Health Sciences, *Second Edition*
VESTRUP · The Theory of Measures and Integration
VIDAKOVIC · Statistical Modeling by Wavelets
VIERTL · Statistical Methods for Fuzzy Data
VINOD and REAGLE · Preparing for the Worst: Incorporating Downside Risk in Stock Market Investments
WALLER and GOTWAY · Applied Spatial Statistics for Public Health Data
WANG and WANG · Structural Equation Modeling: Applications Using M*plus*
WEISBERG · Applied Linear Regression, *Third Edition*
WEISBERG · Bias and Causation: Models and Judgment for Valid Comparisons
WELSH · Aspects of Statistical Inference
WESTFALL and YOUNG · Resampling-Based Multiple Testing: Examples and Methods for *p*-Value Adjustment
* WHITTAKER · Graphical Models in Applied Multivariate Statistics
WINKER · Optimization Heuristics in Economics: Applications of Threshold Accepting
WOOD WORTH · Biostatistics: A Bayesian Introduction
WOOLSON and CLARKE · Statistical Methods for the Analysis of Biomedical Data, *Second Edition*
WU and HAMADA · Experiments: Planning, Analysis, and Parameter Design Optimization, *Second Edition*
WU and ZHANG · Nonparametric Regression Methods for Longitudinal Data Analysis
YIN · Clinical Trial Design: Bayesian and Frequentist Adaptive Methods
YOUNG, VALERO-MORA, and FRIENDLY · Visual Statistics: Seeing Data with Dynamic Interactive Graphics
ZACKS · Stage-Wise Adaptive Designs
* ZELLNER · An Introduction to Bayesian Inference in Econometrics
ZELTERMAN · Discrete Distributions—Applications in the Health Sciences
ZHOU, OBUCHOWSKI, and McCLISH · Statistical Methods in Diagnostic Medicine, *Second Edition*

*Now available in a lower priced paperback edition in the Wiley Classics Library.
† Now available in a lower priced paperback edition in the Wiley-Interscience Paperback Series.